Communications
in Computer and Information Science

T0238879

Joaquim Filipe   Ana Fred (Eds.)

Agents and Artificial Intelligence

Third International Conference, ICAART 2011
Rome, Italy, January 28-30, 2011
Revised Selected Papers

 Springer

Volume Editors

Joaquim Filipe
INSTICC and IPS
Estefanilha, Setúbal, Portugal
E-mail: joaquim.filipe@estsetubal.ips.pt

Ana Fred
IST - Technical University of Lisbon
Lisbon, Portugal
E-mail: afred@lx.it.pt

ISSN 1865-0929   e-ISSN 1865-0937
ISBN 978-3-642-29965-0   e-ISBN 978-3-642-29966-7
DOI 10.1007/978-3-642-29966-7
Springer Heidelberg Dordrecht London New York

Library of Congress Control Number: 2012954489

CR Subject Classification (1998): I.2, I.2.6, H.3-5, H.2.8, C.2, F.1, J.1

Typesetting: Camera-ready by author, data conversion by Scientific Publishing Services, Chennai, India

Printed on acid-free paper

Springer is part of Springer Science+Business Media (www.springer.com)

Preface

The present book includes extended and revised versions of a set of selected papers from the Third International Conference on Agents and Artificial Intelligence (ICAART 2011), held in Rome, Italy, during January 28–30, 2011, sponsored by the Institute for Systems and Technologies of Information Control and Communication (INSTICC) and held in cooperation with the Portuguese Association for Artificial Intelligence (APPIA), the Spanish Association for Artificial Intelligence (AEPIA) and the Association for the Advancement of Artificial Intelligence (AAAI).

The purpose of the International Conference on Agents and Artificial Intelligence (ICAART) is to bring together researchers, engineers and practitioners interested in the theory and applications in these areas. The conference was organized in two simultaneous tracks: Artificial Intelligence and Agents, covering both applications and current research work within the area of agents, multi-agent systems and software platforms, distributed problem solving and distributed AI in general, including Web applications, and within the area of non-distributed AI, including the more traditional areas such as knowledge representation, planning, learning, scheduling, perception and also not so traditional areas such as reactive AI systems, evolutionary computing and other aspects of computational intelligence and intelligent systems.

ICAART 2011 received 367 paper submissions from 55 countries in all continents. Only 32 papers were published and presented as full papers, i.e., completed work (10 pages/30-minute oral presentation). In addition 81 papers reflecting work-in-progress or position papers were accepted for short presentation and another 64 contributions were accepted for poster presentation. These numbers, leading to a "full-paper" acceptance ratio of about 9% and a total oral paper presentations acceptance ratio close to 31%, show the intention of preserving a high-quality forum for the next editions of this conference.

Furthermore, ICAART 2011 included five plenary keynote lectures given by distinguished researchers: Cristiano Castelfranchi (Institute of Cognitive Sciences and Technologies), Boi Faltings (Ecole Polytechnique Federale de Lausanne), Didier Dubois (Institut de Recherche en Informatique de Toulouse), Mark Klein (MIT Center for Collective Intelligence) and Klaus Fischer (Agents and Simulated Reality, DFKI GmbH). We would like to express our appreciation to all of them and in particular to those who took the time to contribute with a paper to this book.

We thank firstly the authors, whose excellent research and development efforts are recorded here. We also thank the keynote speakers for their invaluable contribution and for taking the time to synthesize and prepare their talks. Finally, special thanks to all the members of the INSTICC team, whose collaboration was fundamental for the success of this conference.

October 2011   Joaquim Filipe
                                                                      Ana Fred

Organization

Conference Chair

Ana Fred   Technical University of Lisbon / IT, Portugal

Program Chair

Joaquim Filipe   Polytechnic Institute of Setúbal / INSTICC, Portugal

Organizing Committee

Patrícia Alves	INSTICC, Portugal
Sérgio Brissos	INSTICC, Portugal
Helder Coelhas	INSTICC, Portugal
Andreia Costa	INSTICC, Portugal
Patrícia Duarte	INSTICC, Portugal
Bruno Encarnação	INSTICC, Portugal
Frederico Fernandes	INSTICC, Portugal
Liliana Medina	INSTICC, Portugal
Raquel Pedrosa	INSTICC, Portugal
Vitor Pedrosa	INSTICC, Portugal
Daniel Pereira	INSTICC, Portugal
José Varela	INSTICC, Portugal
Pedro Varela	INSTICC, Portugal

Program Committee

Mohsen Afsharchi, Iran, Islamic Republic of
Thomas Ågotnes, Norway
Jose Aguilar, Venezuela
Natasha Alechina, UK
Frédéric Amblard, France
Plamen Angelov, UK
Costin Badica, Romania
Antonio Bahamonde, Spain
Mike Barley, New Zealand
Roman Barták, Czech Republic
Teresa M.A. Basile, Italy

Ana Lúcia C. Bazzan, Brazil
Orlando Belo, Portugal
Christoph Benzmueller, USA
Greet Vanden Berghe, Belgium
Carole Bernon, France
Daniel Berrar, Japan
Ateet Bhalla, India
Reinaldo Bianchi, Brazil
Sander Bohte, The Netherlands
Enrique Bonsón, Spain
Tibor Bosse, The Netherlands
Djamel Bouchaffra, USA

Krzysztof Patan, Poland
Juan Pavon, Spain
Wojciech Penczek, Poland
Laurent Perrussel, France
Dana Petcu, Romania
Fred Petry, USA
Eric Platon, Japan
Ramalingam Ponnusamy, India
Petrica Pop, Romania
Daowen Qiu, China
Rong Qu, UK
Franco Raimondi, UK
Luís Paulo Reis, Portugal
Alessandro Ricci, Italy
M. Birna van Riemsdijk,
  The Netherlands
Daniel Rodriguez, Spain
Juha Röning, Finland
Rosaldo Rossetti, Portugal
Fariba Sadri, UK
Manuel Filipe Santos, Portugal
Jorge Gomez Sanz, Spain
Jurek Sasiadek, Canada
Andrea Schaerf, Italy
Christoph Schommer, Luxembourg
Frank Schweitzer, Switzerland
Camilla Schwind, France
Murat Sensoy, UK
Ricardo Silveira, Brazil
Cameron Skinner, UK
Marina V. Sokolova, Spain
Adam Slowik, Poland
Safeeullah Soomro, Saudi Arabia
Armando Sousa, Portugal
Diana F. Spears, USA

Antoine Spicher, France
Caroline Sporleder, Germany
Sergiu-Dan Stan, Romania
Bruno Di Stefano, Canada
Kathleen Steinhofel, UK
Toshiharu Sugawara, Japan
Shiliang Sun, China
Pavel Surynek, Czech Republic
Ryszard Tadeusiewicz, Poland
Luke Teacy, UK
Patrícia Tedesco, Brazil
Adolfo Lozano Tello, Spain
Michael Thielscher, Australia
José Torres, Portugal
Nicolas Troquard, UK
Paola Turci, Italy
Anni-Yasmin Turhan, Germany
Paulo Urbano, Portugal
David Uthus, USA
Eloisa Vargiu, Italy
Matteo Vasirani, Spain
Laurent Vercouter, France
Jose Vidal, USA
Aurora Vizcaino, Spain
Dirk Walther, Spain
Mary-Anne Williams, Australia
Graham Winstanley, UK
Cees Witteveen, The Netherlands
T.N. Wong, China
Franz Wotawa, Austria
Bozena Wozna-Szczesniak, Poland
Seiji Yamada, Japan
Xin-She Yang, UK
Laura Zavala, USA

Auxiliary Reviewers

Ramiro Varela Arias, Spain
Bruno Beaufils, France
Yvonne Bernard, Germany
Anarosa Brandão, Brazil
Nicolas Brax, France
Sara Ceschia, Italy
Andrea Orlandini, Italy

Chi-kong Chan, China
Archie Chapman, UK
Laurence Cholvy, France
Juan José Del Coz, Spain
Richard Dobson, UK
Dariusz Doliwa, Poland
Davide Grossi, The Netherlands

Sajjad Haider, Australia
Wojciech Horzelski, Poland
Rongqing Huang, China
Björn Hurling, Germany
Shujuan Ji, China
You Ji, China
Tom Jorquera, France
Elsy Kaddoum, France
Leonidas Kapsokalyvas, UK
Lukas Klejnowski, Germany
Marek Kowal, Poland
Oscar Luaces, Spain
Pierre Marquis, France
Maxime Morge, France

Artur Niewiadomski, Poland
Rony Novianto, Australia
Andrea Orlandini, Italy
Adam Pease, USA
Domenico Redavid, Italy
Vincent Risch, France
Frans Van der Sluis, The Netherlands
Maciej Szreter, Poland
Abu Dayem Ullah, UK
Simon Williamson, UK
Michael Wittke, Germany
Zhijie Xu, China
Ioannis Zgeras, Germany

Invited Speakers

Cristiano Castelfranchi	Institute of Cognitive Sciences and Technologies (ISTC), Italy
Boi Faltings	Ecole Polytechnique Federale de Lausanne (EPFL), Switzerland
Didier Dubois	Institut de Recherche en Informatique de Toulouse (IRIT), France
Mark Klein	MIT Center for Collective Intelligence, USA
Klaus Fischer	Agents and Simulated Reality, DFKI GmbH, Germany

Table of Contents

Invited Papers

Using Incentives to Obtain Truthful Information 3
  Boi Faltings

Model Driven Design of Multiagent Systems 11
  Klaus Fischer, Stefan Warwas, and Ingo Zinnikus

Part I: Artificial Intelligence

Probabilistic Connection between Cross-Validation and Vapnik
Bounds 31
  Przemysław Klęsk

A Time-Varying Model to Simulate a Collective Decisional Problem 56
  A. Imoussaten, J. Montmain, A. Rico, and F. Rico

Setting Up Particle Swarm Optimization by Decision Tree Learning
Out of Function Features 72
  Tjorben Bogon, Georgios Poursanidis, Andreas D. Lattner, and
  Ingo J. Timm

Collision Avoidance Using Partially Controlled Markov Decision
Processes 86
  Mykel J. Kochenderfer and James P. Chryssanthacopoulos

A Restricted Model Space Approach for the Detection of Epistasis
in Quantitative Trait Loci Using Markov Chain Monte Carlo Model
Composition 101
  Edward L. Boone, Susan J. Simmons, and Karl Ricanek

Reasoning about Interest-Based Preferences 115
  Wietske Visser, Koen V. Hindriks, and Catholijn M. Jonker

Advantages of Information Granulation in Clustering Algorithms 131
  Urszula Kuzelewska

Modeling Motivations, Personality Traits and Emotional States in
Deliberative Agents Based on Automated Planning 146
  Daniel Pérez-Pinillos, Susana Fernández, and Daniel Borrajo

Symbolic State-Space Exploration and Guard Generation in Supervisory
Control Theory 161
  Zhennan Fei, Sajed Miremadi, Knut Åkesson, and Bengt Lennartson

Developing Goal-Oriented Normative Agents: The NBDI
Architecture   176
    Baldoino F. dos Santos Neto, Viviane Torres da Silva, and
    Carlos J.P. de Lucena

Evaluation of Environment Contextual Services in Multiagent
Systems   192
    Flavien Balbo, Julien Saunier, and Fabien Badeig

Data Streams Classification: A Selective Ensemble with Adaptive
Behavior   208
    Valerio Grossi and Franco Turini

On kNN Classification and Local Feature Based Similarity Functions . . .   224
    Giuseppe Amato and Fabrizio Falchi

Web Service Composition Plans in OWL-S   240
    Eva Ziaka, Dimitris Vrakas, and Nick Bassiliades

Convergence Classification and Replication Prediction for Simulation
Studies   255
    Andreas D. Lattner, Tjorben Bogon, and Ingo J. Timm

Part II: Agents

The Life of Concepts: An ABM of Conceptual Drift in Social Groups . . .   271
    Enrique Canessa, Sergio Chaigneau, and Ariel Quezada

Improving File Sharing Experience with Incentive Based Coalitions   287
    M.V. Belmonte, M. Díaz, and A. Reyna

Basics of Intersubjectivity Dynamics: Model of Synchrony Emergence
When Dialogue Partners Understand Each Other   302
    Ken Prepin and Catherine Pelachaud

Stability and Optimality in Matching Problems with Weighted
Preferences   319
    Maria Silvia Pini, Francesca Rossi, Kristen Brent Venable, and
    Toby Walsh

A Conditional Game-Theoretic Approach to Cooperative Multiagent
Systems Design   334
    Wynn Stirling

On the Design of Agent-Based Artificial Stock Markets   350
    Olivier Brandouy, Philippe Mathieu, and Iryna Veryzhenko

Defining Virtual Organizations Following a Formal Approach   365
    Sergio Esparcia and Estefanía Argente

Reinforcement Learning for Self-organizing Wake-Up Scheduling in
Wireless Sensor Networks 382
    Mihail Mihaylov, Yann-Aël Le Borgne, Karl Tuyls, and Ann Nowé

Self-Organizing Logistics Process Control: An Agent-Based Approach ... 397
    Jan Ole Berndt

Manipulation of Weighted Voting Games and the Effect of Quota 413
    Ramoni O. Lasisi and Vicki H. Allan

Distributed Consequence Finding: Partition-Based and Cooperative
Approaches 429
    Katsumi Inoue, Gauvain Bourgne, and Takayuki Okamoto

Author Index 445

Invited Papers

Using Incentives to Obtain Truthful Information

Boi Faltings

Artificial Intelligence Laboratory (LIA)
Swiss Federal Institute of Technology (EPFL), IN-Ecublens
CH-1015 Ecublens, Switzerland
boi.faltings@epfl.ch

Abstract. There are many scenarios where we would like agents to report their observations or expertise in a truthful way. Game-theoretic principles can be used to provide incentives to do so. I survey several approaches to eliciting truthful information, in particular scoring rules, peer prediction methods and opinion polls, and discuss possible applications.

1 Introduction

The internet has opened many new possibilities for gathering information from large numbers of individual agents. For example, people rate services in reputation forums, they annotate maps with location information, and they answer questions in online forums. In the future, software agents will control networks of sensors and report measurements such as air quality, radio spectrum, or traffic congestion.

An implicit assumption is that agents will make their best effort to report such information truthfully. However, when they are self-interested, this can not always be assumed. For example, in online reputation forums, leaving a rating is a time-consuming operation and most users will not do this unless they have a motive. Thus, one can often observe skewed distributions of ratings that indicate that most reviews were left by users who either loved or hated the item they rated ([1]). It is not clear whether ranking items by taking averages of such reviews is very helpful. Similar, sensors may save energy by providing inaccurate measurements or no measurements at all, or they may be manipulated to provide skewed reports that are beneficial to the interests of their owner.

To obtain better quality information, it is important to reward agents who contribute ratings and thus increase participation of agents even without ulterior motives. Such reward schemes could be useful both as incentives to human agents as well as for software agents operating sensors: rewards could finance the operation of the sensors and direct their deployment towards the most useful measurements ([2]).

Furthermore, it is possible to scale the rewards so that they specifically reward truthful reporting, and can even counter exterior incentives to report false information. These mechanisms are based on scoring rules that reward correct prediction of a future outcome once that outcome becomes known. In peer prediction methods, these rules are extended to situations where the true outcome never becomes known. Instead, they take the predictions of other agents as the ground truth to compare to. This makes truthfulness an equilibrium, i.e. the best response strategy when all other agents are also truthful. Finally, I show how to design mechanisms that achieve this independently of

J. Filipe and A. Fred (Eds.): ICAART 2011, CCIS 271, pp. 3–10, 2013.

agent beliefs and are thus easier to apply in practice, for example for encouraging truthfulness in opinion polls.

2   Truthful Reporting through Scoring Rules

In many cases, agents are asked to provide information about an outcome that will eventually become known with certainty. For example, experts may predict the weather, the
future of the economy, or the completion date of a project. When this is the case, incentives for reporting this information truthfully can be provided through *proper scoring rules* ([3]). Agents provide information in the form of a probability distribution on
different possible outcomes. Once the true outcome becomes known, they get paid a
reward that depends on how well their prediction matched the observed outcome. This
reward is computed by a scoring rule that takes the report and the true outcome as arguments. A scoring rule is called *proper* if it provides the highest expected reward exactly
when the agent reports its probability distribution truthfully.

Assume that the task is to predict which of k outcomes $o_1, .., o_k$ will actually occur,
and that an expert agent has a probability distribution $\underline{p} = (p(o_1), .., p(o_k))$ for the true
outcome. The agent reports this distribution as $\underline{q} = (q_1, .., q_k)$. We would like to provide
incentives so that it is optimal to report $\underline{q} = \underline{p}$.

This can be provided for example using the *quadratic scoring rule*:

$$pay(o_t, \underline{q}) = a + b \left(2q_t - \sum_{j=1}^{k} q_j^2 \right)$$

where o_t is the outcome that actually occured and a is a non-negative and b a positive
constant. It is straightforward to show that this scoring rule is proper in that the expected
payment:

$$E[pay](\underline{q}) = \sum_{i=1}^{k} p(o_i) pay(o_i, \underline{q})$$

$$= a + b \left[2 \sum_{j=1}^{k} p(o_j)q_j - \left(\sum_{j=1}^{k} p(o_j) \right) \left(\sum_{j=1}^{k} q_j^2 \right) \right]$$

$$= a + b(2\underline{p} \cdot \underline{q} - |\underline{q}|)$$

is maximized by maximizing $\underline{p} \cdot \underline{q}$, which is the case exactly when the vectors $\underline{p}$ and $\underline{q}$
are aligned. Thus, reporting truthfully is a dominant strategy for agents.

As an example, consider predicting whether the next day's weather will be good (g)
or bad (b) as a vector of two probabilities (p(g),p(b)). Let the scoring rule be

$$pay(o_t, \underline{q}) = 1 + \left(2q_t - \sum_{j=1}^{k} q_j^2 \right)$$

An expert's true belief could be that the weather will be good with probability 0.8, and
bad with probability 0.2. Now consider the expected payoff for reporting this distribution truthfully. If the weather turns out to be good, the expert receives a payment of

$pay(g,(0.8,0.2)) = 1 + 2 \cdot 0.8 - 0.68 = 1.92$; if it turns out to be bad, the payment is $pay(b,(0.8,0.2)) = 1 + 2 \cdot 0.2 - 0.68 = 0.72$. Thus, the expected payoff for truthfully reporting the probability distribution is:

$$0.8 pay(g,(0.8,0.2)) + 0.2 pay(b,(0.8,0.2)) = 1.68$$

Now consider a false report, for example $(0.5, 0.5)$. Now the reward in case of good and bad weather is identical and equal to $pay(g/b,(0.5,0.5)) = 1 + 2 \cdot 0.5 - 0.5 = 1.5$, and thus the expected payment is also equal to 1.5. This is significantly less than what is expected for truthful reporting.

There are other proper scoring rules, such as the logarithmic scoring rule:

$$pay(o_t, \underline{q}) = a + b \ln q_t$$

where o_t is the outcome that actually occured and a is a non-negative and b a positive constant. These may lead to lower expected payments or wider margins for truth-telling, but can have other drawbacks. For example, with the logarithmic scoring rule payments can become negative.

Proper scoring rules can also be constructed for eliciting averages and other properties of distributions. Recently, [4] have characterized the questions to which truthful answers can be elicited using scoring rules.

3   The Peer Prediction Method

Proper scoring rules can be applied whenever the ground truth that is being observed can eventually be verified. However, there are many cases where this condition is not satisfied. Consider for example ratings reported for products and services on the internet: it is not possible to independently verify whether these ratings were given truthfully. Similarly, measurements taken by sensors would often not be verifiable by other means. A similar situation exists when reporting opinions about hypothetical scenarios, such as what would happen if interest rates were raised by different degrees: since only one of these scenarios will actually be implemented, predictions about the others cannot be verified.

However, in such cases it is still possible to make truthful reporting an equilibrium strategy for agents by applying a proper scoring rule based on the prediction of another agent, called a reference report. Provided the other agent made a truthful prediction and both have the same knowledge and observing the same signals, truthful reporting is the best response. Thus, for a population of agents with the same knowledge, reporting truthfully is a Nash equilibrium. This is called the *peer prediction method* in ([5]).

As an example, consider reporting the quality of service received by a plumber. Two agents A and B both report on the quality of service they received. The key idea is that the quality of service A received will influence its expectation of the quality that B received: if A observed good service, then its belief for the probability $p(g|g)$ that B also received good service is higher than the value $p(g|b)$ if A received bad service. Assume for this example that $p(g|g) = 0.8$ and $p(g|b) = 0.4$.

Now we apply the same scoring rule mechanism we mentioned earlier, but consider B's report the ground truth. If A observed good service, its probability distribution for B's report is $(p(g|g), p(b|g)) = (0.8.0.2)$, and just like in the weather prediction example it's expected reward for the scoring rule:

$$pay(o_t, \underline{q}) = 1 + \left(2q_t - \sum_{j=1}^{k} q_j^2 \right)$$

is 1.68, provided that A's probability distribution for B's experience is indeed $(0.8.0.2)$.

If A did not experience good service, it would expect B's observation to follow a different probability distribution, in this case $(0.4, 0.6)$. If it nevertheless reports good service, the expected reward is only $0.4 \cdot 1.92 + 0.6 \cdot 0.72 = 1.2$. On the other hand, when A truthfully reports bad service, the mechanism treats this as a prediction of the probability distribution $(0.4, 0.6)$ for B's experience. The payments for truthfully reporting bad service are calculated using the probabilities $(0.4, 0.6)$ and would lead to a higher expected reward for truthful reporting of $0.4 \cdot (1 + 2 \cdot 0.4 - 0.52) + 0.6 \cdot (1 + 2 \cdot 0.6 - 0.52) = 1.52$.

Note that, contrary to the weather prediction, we are not asking A to report this probability distribution, but only whether it received good or bad service. Thus, the designer of the reward scheme needs to know how an observation influences A's beliefs about the observations of another agent B with reasonable precision in order to compute the payments. It can in part be deduced from the general expectations of the quality of service, but also involves an assumption of how the individual agents would update their beliefs in response to a positive or negative experience.

Furthermore, the original peer prediction method suffers from the weakness that truthful reporting is not the only equilibrium strategy: any strategy where agents all report the same is also a Nash equilibrium. In fact, since actual observations or predictions are likely to be noisy, the highest-paying equilibrium is always one where agents always report the same, independently of their true knowledge!

This problem can be overcome by constructing scoring rules that refer not to one, but several reference reports. [6,7] show that when at least 3 reference reports are used, truthful reporting can be made the highest-paying Nash equilibrium. Furthermore, they show that truthful reporting can be made the *only* Nash equilibrium and thus completely eliminate the problem of collusive reporting strategies.

It has recently been shown that peer prediction methods can be generalized to scenarios where agents report not on identical events, but events that are merely correlated ([8]). This makes it applicable for example to measurements in sensor networks, where different sensors measure quantities that are correlated by not equal.

4   Opinion Polls

A major weakness of the peer prediction method is that it requires all participating agents to share the same probability distribution of the reported events. If this is not the case, proper scoring rules can still be designed, but the rewards that must be paid to agents quickly become very large ([9]).

To counter this effect, it is possible to design peer prediction schemes as opinion polls that publish the current results of the poll. Agents whose probability distribution is sufficiently close to this published one will have truthful reporting as their best strategy, while agents that consider the public distribution as grossly wrong may instead be merely helpful by making reports that will drive the public poll closer to what they consider to be the true distribution.

Such a mechanism was first shown in ([10]) for aggregating opinions about a hidden signal that could be either good (g) or bad (b). At time t, the published polls shows the average fraction R_t of good reports. An agent A_i has its own probability distribution $p_i(r|s)$ that characterizes the conditional probability distribution of a reference report r given its own observation s of the signal, where the reference report is filed by another agent that observes the same signal and the same public poll. The mechanism compares the report s filed by agent A_i to a reference report r filed by another agent B, and rewards A_i if the two reports match:

- for matching a good report, the reward is $c(1 - R_t)$.
- for matching a bad report, the reward is cR_t.

where c is a positive constant to scale the average reward, for example to ensure that it compensates for the effort required to file it.

To analyze the incentives for agent A_i, we distinguish three cases:

a) A_i considers the current poll result reasonable, characterized by the fact that
   $p_i(g|b) < R_t < p_i(g|g)$.
b) A_i considers the poll result unreasonably high, characterized by the fact that $R_t \geq p_i(g|g)$, which means that no matter what A_i observes, it would always expect other agents to observe a bad signal with a higher probability than the current poll result.
c) A_i considers the poll result unreasonably low, characterized by the fact that $p_i(g|b) \geq R_t$, symmetrically on the other side.

In the case where the poll result is reasonable, the agent is best off reporting truthfully. Consider the case where it observes a good signal, then the expected rewards are:

- for reporting good (truthful):

$$p_i(g|g)c(1 - R_t) > R_t c(1 - R_t)$$

- for reporting bad (non truthful):

$$p_i(b|g)cR_t = (1 - p(g|g))cR_t < (1 - R_t)cR_t$$

Thus, the expected reward for reporting truthfully is strictly greater than the expected reward for a non-truthful report. A symmetric analysis can be made for the case of a bad observation.

As an example, consider that agents A and B both hire a plumber that according to the public reputation scheme provides good service 90% of the time, based on 10 previous reports. Suppose that A sees the plumber at work and he does a good job. Then A might consider that the current poll value is accurate or slightly too low and report good service, expecting a payment of 10/9 with a probability of higher than 0.9, so above 1 in expectation.

However, if the agent considers the poll unreasonably high, its best strategy is to report bad, independently of its own observation. While this behavior is not truthful, it can be considered *helpful* in that the agent drives the outcome of the poll closer to its own opinion. For example, suppose A observes the plumber at work and realizes that he is completely incompetent, but still by chance receives good service. Now, A might have a private probability that B will receive bad service that is much higher than the 10% that would be expected from the poll, let's say 50%. Now A would be better off reporting poor service, as its expected reward would be $1/0.1 = 10$ with probability 0.5, which is much higher than his expectation in case of truthful reporting. However, the report could still be considered helpful in that it drives the value of the opinion poll towards A's true opinion.

The advantage of this mechanism is that agents can have different and unknown prior distributions for the signal, whereas scoring rules require this distribution to be known to the mechanism designer.

5 Applications

The techniques reported here have numerous applications. The most obvious ones are forums such as reputation and review forums. Leaving such feedback is cumbersome and thus often done by agents who have ulterior motives and thus do not leave honest reports. Here is would be useful to reward raters for their effort, and it would be even better to scale these payments to encourage honest feedback.

Another range of applications is in ensuring quality of crowdsourcing. For example, consider an image labeling task as in the ESP game proposed in [11]: two people are independently asked to give keywords that describe the content of an image. They get a reward when they provide matching keywords. This game has the flaw that people will tend to use very common words, and so these have to be explicitly excluded. A more general strategy based on the opinion poll mechanism given above would be to scale the rewards according to the frequency of the matching word: a less common word would fetch a higher reward. One can imagine many other applications in crowdsourcing where rewards depend on the agreement with other worker's results.

Further applications can be found in sensor networks. The peer prediction method can be generalized to settings where agents do not measure exactly the same signal. It is sufficient that measurements are correlated in a known way ([8]). Thus, one can design a reward scheme that rewards truthful operation of a network of sensors that sense related values, for example air pollution ([2]). This could be applied in sensor networks, in particular when sensors are operated by different entities who might save cost by inaccurate measurements, or even maliciously want to manipulate measurements.

Services such as internet access, cloud computing, or wireless communications require monitoring of the quality of service. This would be most easily done by the customers themselves, but the difficulty is that they often have an incentive to misreport since they stand to gain refunds or other claims if service is deemed to be insufficient. Somewhat surprisingly, it turns out that incentive mechanisms are entirely sufficient to solve this problem, as shown in [12]. Provided that the entire user population is sufficiently large, it would take a significant coalition of users to shift the average reported quality enough to obtain a refund for poor service. However, as long as such a large coalition has not formed, a reward scheme based on peer prediction is sufficient to punish each individual user for deviating from truthful reporting, and can be realized at low cost. Thus, a lying coalition would have to be created in a coordinated fashion, and such coordinated action would be detectable by other means. This opens another wide range of applications.

6 Conclusions

The internet has enabled wide distribution of user-contributed content whose correctness cannot be verified. Much of this content is reported by agents with ulterior motives and may often not reflect the truth. I have discussed ways of providing incentives to agents to provide such content truthfully. I believe that such mechanisms are of fundamental importance for the future use of reputation forums, sensor nets and crowdsourcing applications on the internet. They also have other applications in multi-agent systems, such as service monitoring.

While work so far has shown an interesting range of mechanisms to encourage truthful reporting, many open questions remain. The biggest issue is clearly the dependence on knowledge of prior probability distributions that are not always available. The opinion poll mechanism we described is a first step but still has to be generalized to elicit more complex information than just binary signals. Also, as it stands it has little protection against collusive behavior.

Another issue is how to provide rewards. Paying monetary rewards is often not practical, and one needs to experiment with other forms of rewards, such as reputation or privileges that will be valued in similar ways as money.

Acknowledgements. I thank Radu Jurca who has worked with me on this topic for many years, and Karl Aberer for fruitful discussions on reputation and community sensing.

This work has been supported in part by Opensense project (839-401) in the Nanotera.ch program.

References

1. Hu, N., Pavlou, P.A., Zhang, J.: Can online reviews reveal a product's true quality?: empirical findings and analytical modeling of online word-of-mouth communication. In: Feigenbaum, J., Chuang, J.C.I., Pennock, D.M. (eds.) ACM Conference on Electronic Commerce, pp. 324–330. ACM (2006)
2. Aberer, K., Sathe, S., Chakraborty, D., Martinoli, A., Barrenetxea, G., Faltings, B., Thiele, L.: Opensense: Open community driven sensing of environment. In: ACM SIGSPATIAL International Workshop on GeoStreaming, IWGS (2010)
3. Savage, L.J.: Elicitation of personal probabilities and expectations. Journal of the American Statistical Association 66, 783–801 (1971)
4. Lambert, N.S., Shoham, Y.: Eliciting truthful answers to multiple-choice questions. In: Chuang, J., Fortnow, L., Pu, P. (eds.) ACM Conference on Electronic Commerce, pp. 109–118. ACM (2009)

5. Miller, N., Resnick, P., Zeckhauser, R.: Eliciting honest feedback: The peer prediction method. Management Science 51, 1359–1373 (2005)
6. Jurca, R., Faltings, B.: Collusion resistant, incentive compatible feedback payments. In: Proceedings of the ACM Conference on Electronic Commerce (EC 2007), pp. 200–209 (2007)
7. Jurca, R., Faltings, B.: Mechanisms for making crowds truthful. Journal of Artificial Intelligence Research (JAIR) 34, 209–253 (2009)
8. Witkowski, J.: Eliciting honest reputation feedback in a markov setting. In: Boutilier, C. (ed.) IJCAI, pp. 330–335 (2009)
9. Jurca, R., Faltings, B.: Robust Incentive-Compatible Feedback Payments. In: Fasli, M., Shehory, O. (eds.) TADA/AMEC 2006. LNCS (LNAI), vol. 4452, pp. 204–218. Springer, Heidelberg (2007)
10. Jurca, R., Faltings, B.: Incentives for expressing opinions in online polls. In: Proceeddings of the 2008 ACM Conference on Electronic Commerce, pp. 119–128. ACM (2008)
11. von Ahn, L., Dabbish, L.: Labeling images with a computer game. In: Proceedings of the SIGCHI Conference on Human Factors in Computing Systems, CHI 2004, pp. 319–326. ACM, New York (2004)
12. Jurca, R., Binder, W., Faltings, B.: Reliable qos monitoring based on client feedback. In: Proceedings of the 16th International World Wide Web Conference (WWW 2007), Banff, Canada, pp. 1003–1011 (2007)

Model Driven Design of Multiagent Systems

Klaus Fischer, Stefan Warwas, and Ingo Zinnikus

German Research Center for Artificial Intelligence (DFKI) GmbH
Saarbrcken, Germany
{Klaus.Fischer,Stefan.Warwas,Ingo.Zinnikus}@dfki.de

Abstract. In general software engineering modelling of software systems had a significant impact on the manner in which complex systems are designed. The Model Driven Architecture (MDA) proposed by the Object Management Group (OMG) provides a formal framework that allows to define dedicated modelling languages for different application domains. Already in the model driven design of service-oriented architectures one can identify concepts that are common in the design of such systems and what agent-based systems concerns. To directly use the MDA framework for the design of multiagent system (MAS) is therefore an obvious step. In this article we advocate the domain specific modelling language DSML4MAS for modelling MAS. However, our aim is not to just define the language, we propose a framework for DSML4MAS that allows its adaptation and dynamic development in the future. Our vision is that in the near future model repositories for model fragments that can be flexibly combined will be established and propose basic concepts that can support the development of MAS in this context. The interaction aspect is especially important in MAS design and one of the most obvious aspects where model exchange and model re-use is highly desirable. The article therefore presents the interaction aspect in more details and discusses the features that are available in the DSML4MAS.

1 Motivation

With the success of service-oriented architectures and the ever-growing connectivity of applications in the Internet, agent technologies are becoming even more attractive than they were in the past. However, many times the system design not only in agent-oriented applications is tightly bound to the execution environment. Although we are far from a state where system engineers would not care about the technologies deployed in the execution environment, it is clear that it would be highly desirable to be able to separate system design from such technologies and with this make it more sustainable regarding the evolution of software concepts. The model driven architecture proposed by the Object Management Group (OMG[1]) provides the basic ideas that can significantly improve the possibilities to maintain system designs while new technologies emerge or

---

[1] www.omg.org

J. Filipe and A. Fred (Eds.): ICAART 2011, CCIS 271, pp. 11–27, 2013.

already available technologies get adapted to new requirements. Agent technologies can contribute in this enterprise because they provide helpful abstractions for the design of complex systems.

In the following we present the ingredients of our approach. Section 2 gives an overview of the overall approach. In Section 3 we present details of the PIM4Agents metamodel that forms the core of DSML4MAS. We zoom in on the interaction aspect of PIM4Agents in Section 4 because this aspect is one of the most obvious where exchange of models and model fragment among system engineers is desirable. Section 5 presents uses cases in which we evaluate our approach and an illustrative example of the use of a concrete interaction model. Section 6 gives some pointers to related work and Section 7 draws conclusions and directions for future research.

2   Framework for Model Driven Design of Multiagent Systems

In this article we adopt a model driven approach to the design of agent-based systems. The basic ideas of the approach were developed in the EC[2] funded research projects ATHENA[3] and SHAPE[4] and are now further developed in the EC funded research project COIN[5]. The main achievement in this approach is the definition of the domain specific modelling language DSML4MAS[6] [7]. The metamodel PIM4Agents forms the core of DSML4MAS. From PIM4Agents we derive our modeling tool which is built on the Eclipse EMF/GMF technology stack [19]. However, we do not only aim at just providing the modelling language and tool support. What we want to come up with is a framework that allows to extend and refine the core metamodel by additional or more specialized concepts. For this the metamodel is separated into different parts that deal with specific concerns of the design of a multiagent system. We refer to these parts of the metamodel that form separate meaningful entities with the term *aspect*. The idea is to provide a framework that allows to specifically design and flexibly adapt the different aspects. This approach allows to extend the core of the metamodel by pluging in different realizations of the foreseen aspects. We further distinguish between the aspects into which the metamodel is separated and the different viewpoints that are supported by the modelling tool. A viewpoint in the modelling tool is defined by a diagram that displays a collection of concepts and how they relate to each other. Additionally, a tool box that allows to manipulate the concepts in the diagram is provided, e.g. add or delete new instances of a specific concept or introduce additional relations. The overall goal is to allow both a flexible definition of the aspects in the metamodel as well as a flexible definition of viewpoints in the modelling tool.

---

[2] European Commission.

[3] http://www.modelbased.net/aif/

[4] http://www.shape-project.eu/

[5] http://www.coin-ip.eu/

[6] http://sourceforge.net/projects/dsml4mas/

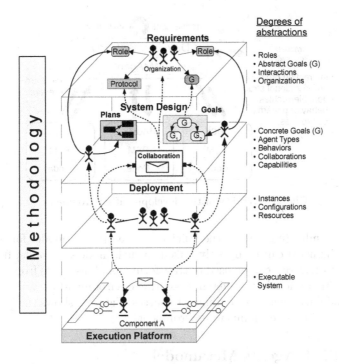

Fig. 1. Framework overview

As already mentioned, the core of our model-driven framework for developing multiagent systems is the PIM4Agents metamodel. PIM4Agents is independent of a concrete execution platform but inherently possesses different degrees of abstraction (see Figure 1). The requirements layer is the most abstract degree and covers abstract goals, roles, interactions, and organizations. The system design degree contains (i) agent types, (ii) behavior templates, (iii) concrete goals, etc. The lowest degree is the deployment layer which specifies concrete deployment configurations (e.g. agent instances and resources).

Our aim is to define for DSML4MAS a plugin framework that allows to flexibly extend or completely replace the different foreseen aspects. Additionally to the idea that parts of the metamodel can be extended by plugins with different realizations, we assume that there will be a landscape of metamodels which share a common sub-set of concepts. It is a quite safe guess that it should be possible to arrange these different metamodels in a hierarchy of specialization with a common root. At least the concept of an agent is likely to be part of any metamodel that people come up with when they want to do modelling of agent-based or multiagent systems. With a landscape of metamodels in mind, it is easy to foresee also a landscape of model repositories that hold models or model fragments according to different metamodels. In this framework collaborative modelling can be supported in the sense that system engineers can store and retrieve models or model fragments to and from model repositories.

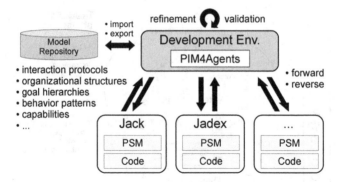

Fig. 2. Overview of the development environment

Theoretical results (e.g. about the efficiency of a specific model for a specific auction mechanism) can be directly linked to such models or model fragments. Model transformations can be used to transform model instances from one meta-model to another one or into different execution environments as well as to merge together model fragments from different sources. Figure 2 shows the basic setup for the DSML4MAS development environment.

3   The PIM4Agents Metamodel

The metamodel PIM4Agents forms the core of the domain specific modelling language DSML4MAS. The metamodel is structured into different aspects. An aspect contains a collection of concepts and definitions how these concepts relate to each other. OCL constraints are used to express semantics. There are two types of OCL constraints: (i) constraints to compute derived information, i.e. the value of a specific attribute is derived from values of other concepts and (ii) integrity constraints, e.g. the value of an integer attribute must be higher than 0 or below a given number.

PIM4Agents supports 12 aspects:

Multiagent System. Describes all basic components the MAS is composed of.
Agent. Describes single autonomous entities, the capabilities they have to solve tasks and their roles they play within the MAS.
Organization. Describes how single autonomous entities cooperate within the MAS and how complex organizational structures can be defined.
Role. Definines the requirements an agent should fulfull when it wants to engage in an organizational structure.
Interaction. Describes how the interaction between autonomous entities or organizations takes place.
Behavior. Describes how plans are composed by complex control structures and simple atomic tasks.
Information. Contains any kind of resource that is dynamically created, shared, or used by agents or organizations.

Deployment. Allows to define a MAS at instance level that can be used for startup.

Goal. Explicit representation of a goal hierarchy in form of an and/or tree representation.

Event. A stimulus the agents can react to.

Environment. Allows the agents to sense information from the outside environment and to manipulate the outside environment with actuators.

Resource. Was introduced to connect agents with a service-oriented environment.

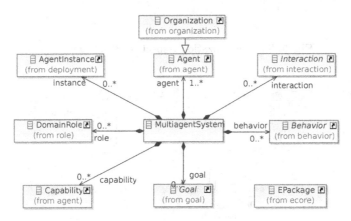

Fig. 3. The Multiagent System Aspect in PIM4Agents

Fig. 3 shows the MAS aspect which includes all major concepts a MAS is composed of. Out of these we want to briefly discuss the following: Agent, Organization, Role, Interaction, and Behavior.

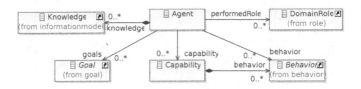

Fig. 4. The Agent Aspect of PIM4Agents

An agent (see 4) has behaviors that may be grouped to capabilities which together with the information in the information model allow the agent to achieve its goals. The agent might perform roles in an organizational structure.

An organization (see 5) is in the first place composed of roles. The agents that actually perform these roles are grouped together in concrete collaborations to be able to achieve the organization's goals. DomainRoles allow a renaming of role

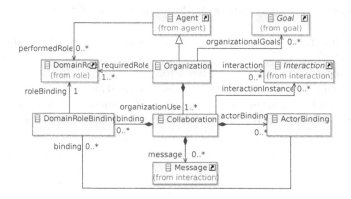

Fig. 5. The Organization Aspect of PIM4Agents

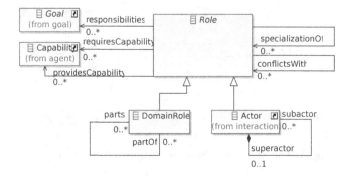

Fig. 6. The Role Aspect of PIM4Agents

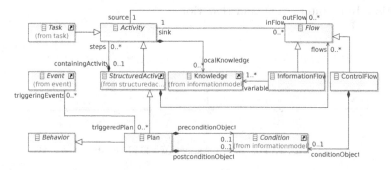

Fig. 7. The Behavior Aspect of PIM4Agents

types of the organization in collaborations. Actor bindings bind the DomainRoles to the interactions which are used in the collaboration to prescribe how an interaction has to be performed when the involved agents work towards a given goal.

Roles (see 6) are responsible for specific goals and require capabilities from the agents performing the goals that allow the agents to actually successfully achieve the goals the role is responsible for. Roles can be specialized and can be in conflict with each other, meaning that an agent that performs one role might not be allowed to perform a specific other role. DomainRoles and Actors are specialization of the more general role concept.

Figure 7 displays the Behavior aspect of the PIM4Agents metamodel. A behavior for an agent is specified by a plan. Each plan is composed of a trigger event, a pre-condition, a post-condition, and a set of activities. Activities can be complex patterns like for examples loops or simple tasks. Flows provide a sequencing among the activities. Special tasks (i.e. BeginTask and EndTask) mark for a diagram where the execution of the plan starts and where it ends. The intended semantics is that when an event arises within the agent's body that matches the trigger event, the preconditionObject is checked. If this Object returns with true, the execution of the body (i.e. the set of activities) is started at the BeginTask. Execution of the plan body ends when an EndTask is encountered. It is assumed that the postconditionObject evaluates to true when the execution of the plan body terminates.

4   Taking a Closer Look at Interactions

Regarding the design of agent interactions we take for the discussion in the article a restricted point of view. We purely concentrate on what we call contract-based interactions. This means that we assume that for the interactions which we want to take into account a predefined contract exists which the agents use when performing the interaction. The definition and analysis of interaction protocols like the Contract Net Protocol (see below) is complex and therefore it is beneficial if system engineers can flexibly adopt interaction protocols that are well understood. The approach presented in this article allows to set up repositories for model fragments (e.g. interaction protocols) and adopt them in separate designs of multiagent systems by transforming the model fragments into representations that can be directly included into the local design (e.g. capability specifications of individual agents).

To make the discussion in this article not too complicate we assume that all models are defined at design time of the system and then purely used at run time. System dynamics is purely restricted to instance level and does not include type level. However, this still allows dynamic assignment of agent instances to roles. How this role assignment is actually done is out of the scope of this article. We assume that an agent that was assigned to take a specific role on the one hand will always try its best to fulfil the obligations the role is asking for and also provides in principle all capabilities the role is asking for. At run time no explicit checks are done on whether such requirements are actually met. We assume that if such checks would be done they would be performed at type level at design time.

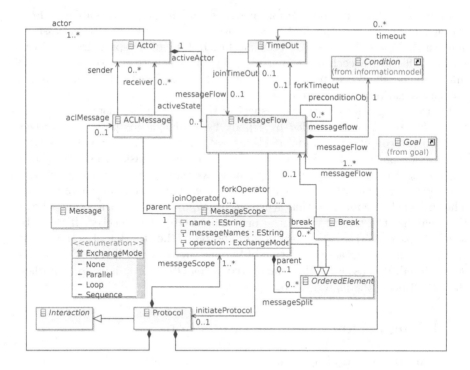

Fig. 8. The Interaction Aspect of PIM4Agents

Regarding the metamodel definitions the Interaction aspect (see Figure 8) is one of the most complex in the PIM4Agents metamodel. However, the most important concepts to understand are: Protocol, Actor, MessageFlow, and MessageScope. The protocol is actually the core concept that defines an interaction. The reason to not directly replace the concept of interaction with the concept protocol is that protocols are intuitively understood as message based interactions and we want to be open to allow in future work other forms of interaction, too.

A protocol is in the first place composed of a set of actors. For each actor a set of message flows defines the different states in which the interaction could end up. Message scopes connect two message flows of two different actors. An attribute tells whether the message flow has a fork operator (i.e. it sends a message) or a join operator which means that the message flow receives a message. The message scope refers to the concrete message that is sent from the sender (actor) to the receiver (actor). In each actor exactly one message flow is marked as initial message flow which means that the protocol execution will start for this actor in the state that is defined by this message flow. However, in the whole protocol there is only one actor for which the initial message flow has a forkOperator (i.e. the initial message flow sends a message). We call this message flow start message flow of the protocol.

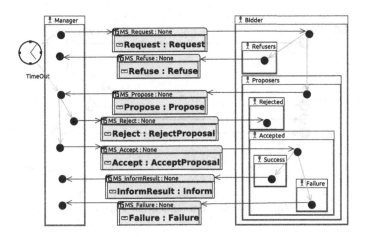

Fig. 9. The Contract Net Protocol

The core part of the contract the agents use for interaction is given with a protocol. Figure 9 shows the diagram for the Contract Net Protocol[7] [5]. In this protocol a manager agent tries to find an agent in a group of bidders that is selected to provide a specific service for the manager. We present some more details on the usefulness of this protocol in Section 5. The building blocks for protocols are (i) actors, which might be separated into sub-actors, (ii) message flows, and (ii) and messages. Actors refer to the participants of the protocols. Role bindings are used to define the requirements to agent types that actually could take the part of a specific actor in a protocol. Message flows mark specific states in the protocol execution. If a message flow has more then one exiting arc exactly one of these arcs can be chosen to continue protocol execution, which means that the outgoing arcs have an xor semantics. An additional assumption is that each message flow that receives a message spawns of an achieve goal event, where the abstract goal's name is derived from the name of the incoming message which is by definition unique, i.e. a message flow can receive at most one message. To achieve such an abstract goal might turn out to be a complex process within the agent and might very well involve the interaction with other agents which are then, behind the scenes, again organized by using contracts. Messages are defined by message types.

With these conventions the behavior that an agent needs to comply to when it engages in a specific interaction is defined in the protocol description. However, the protocol specifies the communication behavior only. All capabilities that are available from the agent's body are addressed by spawning off achieve goal events and the only direct body capability the protocol itself relies on is that the agent is able to send the messages according to the specified message types. In this

---

[7] Please do not confuse the name of the Contract Net Protocol with what we call contract based communciation. The dual use of the term *contract* is basically pure chance.

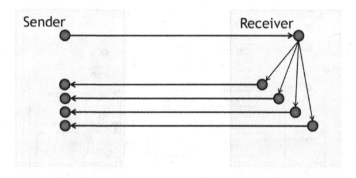

Fig. 10. Basic Communication Pattern

sense the specification of the interaction among the agents is a choreography of the capabilities of the agents that are involved in the interaction.

When it comes to executing the protocol at run time one has at least two options to choose from. In the first option the interaction protocol would be directly interpreted by a protocol interpreter that is included in the agents' bodies. In the second option the protocol is transformed into local behaviors for each of the participating agents which can be directly executed in some execution environment. The first option is more flexible but the second option is easier to implement. For this reason we use the second option. This has also the advantage that the local behavior can be produced with a model-to-model transformation at the PIM4Agents level which means that the resulting behaviors can be transformed into all different execution environment for which a transformation of PIM4Agents models is available. We therefore do not interpret the protocol model directly but transform the protocol model at design time into capabilities that provide the respective communication behavior that is required by the contract to which the protocol belongs. To achieve this a separate capability is generated from the protocol model for each of the given actors. We can identify a basic pattern which allows to already design a large number of different protocols (at least regarding those which are explicitly represented by models at design time). This basic pattern is displayed in Fig. 10. It always starts with one actor sending a message to another actor and then waiting for all answers to this message.

Behind all actors of a protocol any number of agent instances might be hiding except for the actor that contains the start message flow. Only one individual agent is allowed to play the role that is connected to this actor. Only the start message flow actually sends multiple messages to all agent instances hiding behind the actor that receives this message. All subsequent messages are exchanged in a bilateral manner. However, this means that the start message flow spans off a set of parallel interaction threads. For some protocols it is necessary to synchronize (see Fig. 11) these interaction threads. For example in the Contract Net Protocol ([5] see Fig. 9) the manager has to wait till all bidders have replied

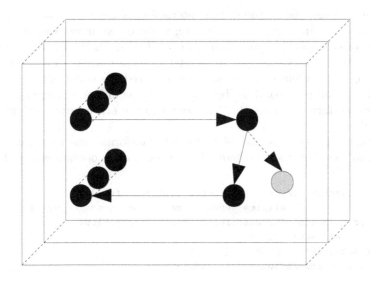

Fig. 11. Synchronization between parallel execution threads

to the call for proposals or some specified deadline has passed. Only then the protocol can proceed and the best bidder be selected. To define this kind of synchronization at the modelling level we provide the following concepts:

– For each protocol instance in each actor we maintain a context in which the state of the ongoing protocol execution can be maintained.
– In the protocol context we maintain a table which allows us to find for each message sent all replies that have been received so far.
– Message flows can be marked as synchronized. A condition specifies what needs to happen before the synchronization is successfully achieved. Such a condition can be that a specific number of replies has been received or that a specific event (e.g. a timeout) has occurred. All replies to a sent message that are received after the synchronization was successfully achieved are ignored.
– A special variable MaxMessages in the protocol context gives the number of how many messages were actually sent in the start message flow.

With this basic machinery it is already possible to design a large number of complex protocols. A limitation of the presented concepts is that it is not possible to model protocols where messages are sent out to more than one actor while the sender waits for the replies to these messages in a concurrent manner. However, because several agent instances can hide behind an actor the only limitation is that sending concurrent messages to more than one group of agents hiding behind one of the actors in the protocol diagram is not possible in one model. However, because each of the agent instances at any specific stage of the protocol execution (i.e. a specific message flow), where it receives a message, spawns off an achieve goal event to produce the message that should be sent next, in the process of achieving this event can of course again initiate the execution of an interaction protocol. So although this interaction is not visible in the original protocol diagram, any number of cascading protocol executions can result from the execution of a specific interaction protocol. To restrict sending of concurrent messages to the start message flow of an interaction protocol is therefore actually not a real restriction but enforces structure that reduces complexity.

The interaction protocol describes the interaction among the actors (i.e. the agents that perform the roles that are bound to the actors) from a centralized point of view.

To actually describe the model to model transformations from interaction protocols to local behaviors for individual agents we use operational QVT[8].

```
helper pim4agents::interaction::Actor::collectMsfs () :
                    Set(pim4agents::interaction::MessageFlow) {
    var res : Set(pim4agents::interaction::MessageFlow);
    res := self.activeState;
    self.subactor->forEach(a) {
        res := res->union(a.collectMsfs());
    };
    return res
}
```

Fig. 12. Helper of a QVT Transformation

Operational QVT offers two types of procedural concepts. The first straight forward concept are helpers. Helpers are very similar to methods of an object oriented language. The helper can be typed and in this case it basically extends the signature of the concept that is defined in the metamodel and allows computations that are useful for handling this concept. The definition of the helper also shows how the concepts in the PIM4Agents metamodel are addressed: first the metamodel is named, then the package in the metamodel, and last the concept. In the body of the helper OCL expressions can be used which allow to express complex computations in a compact manner. The helper in Figure 12 collects all message flows that are included in an actor in a given interaction protocol.

```
mapping PIM4Agents::interaction::Actor::toDomainRole () :
                            PIM4Agents::role::DomainRole {
    var msf : Set(pim4agents::interaction::MessageFlow) :=
        self.collectMsfs();
    var rmsf : Set(pim4agents::interaction::MessageFlow) :=
        msf -> select(d|d.isInitialMessageFlow or
        ((d.forkOperator <> null) and (d.MsfSuccessors(msf)->size() > 0)));
    name := 'Role' + self.name;
    providesCapability := rmsf.map toCapability(msf,rmsf);
}
```

Fig. 13. QVT Mapping Rule, Creation of a Domain Role from an Actor of a Protocol

---

[8] Query view transformation.

The second procedural concept that operational QVT offers are mappings (see Figure 13). Mappings look quite similar to helpers, however, there is an important difference. Mappings result in a link between the entity they are applied to and the entities they create. This means that a mapping can be called several times but the structures it creates are only created once and that the same structure can be mapped to different places (i.e. attributes of concepts) in the model instance that is produced as a result of the QVT transformation. In the mapping the variable msf holds all MessageFlows that are contained in the actor and rmsf is the subset of MessageFlows in the set msf that are considered relevant. Relevant MessageFlows are those where a message is sent and an answer is expected. The condition d.forkOperator <> null says that d sends a message and the helper d.MsfSuccessor(msf) returns the set of message flows that receive an answer for the message sent in d. This exactly corresponds to the situation of the actor Sender in the communication pattern shown in Figure 10. The MessageFlow sending the message would be considered relevant. The four MessageFlows in the lower part would be returned by the helper MsfSuccessors. If the communication pattern is only used once in the actors the sending MessageFlow in the actor Sender and the receiving MessageFlow in the Receiver would be marked as initial message flows where the former is called the start MessageFlow.

5   Use Cases

The presented approach for the design of multiagent systems is currently further investigated and practically used in the research projects ISReal and COIN.

The COIN project investigates collaboration and interoperability for networked enterprises. In this context we use the modelling approach presented in this article for the design of negotiation processes in enterprise systems. Negotiations occur prominently in business interactions between competing partners, but also between cooperating partners, e.g. the participants in supply chains or *virtual enterprises*. One scenario we are looking at in this work is the situation in which a production plan for collaborating partners in a supply chain has already been scheduled. In this setting there are two scenarios for which additional negotiations could be necessary while a production plan is executed: (i) for a specific step the service provider (e.g. a transportation service or a supplier of raw material) has been left open or (ii) it turns out that a pre-negotiated service provider cannot provide the agreed service. For both scenarios the Contract Net Protocol (see Fig. 9) can be used to organize the negotiation. Services the manager agent can chose from are registered and published in a *general service platform (GSP)* which provides discovery and invocation support. Services are semantically annotated using the WSMO/WSML language which facilitates ad-hoc service provisioning and execution.

In COIN, service provision can be supported at design time and/or at run time. In the first case, a process modeler can check for available services while designing the interactions and agent plans. In the second case, when a service provider drops out or e.g. cannot fulfill the required quality of service, a new service provider can be determined by retrieving a list of candidate services from the GSP and selecting the best service using the Contract Net Protocol.

The focus of ISReal is the design of agents and multiagent systems in a virtual reality settings where the agents represented by avatars that form their virtual bodies. Digital factories are one of the application areas ISReal is aiming at. The aim of the ISReal project is to develop an execution platform for semantic 3D simulations [8]. The basic idea of ISReal is to add semantic descriptions to 3D objects and specify their functionality by semantic service descriptions. Our approach is based on the semantic Web standards OWL[9], OWL-S[10], and RDFA[11]. Agents perceive the annotated facts and service descriptions and use them for reasoning and planning. The scene runs in a 3D-enabled Web browser based on XML3D[12]. We use our model-driven development environment DDE for engineering the agents that control the avatars (their virtual body) in the 3D scene. Figure 14 depicts the application of an agent interaction protocol in a 3D simulation. Agent $A1$'s target is to buy ingredients for the production of some pills on the pill filling machine shown in the background. The Contract Net Protocol is used by agent $A1$ to negotiate with the pharmacy agents $A3$, $A4$, and $A5$. In Figure 14, agent $A5$ won the auction.

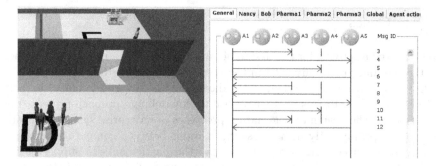

Fig. 14. Contract Net example in the ISReal scenario

6   Related Work

Communication is an important aspect of agent-based and multiagent systems and therefore has been intensively investigated. The FIPA[13] initiative was originally founded to produce specifications for software standards for heterogeneous and interacting agents and agent-based systems. One of the main achievements of

---

[9] http://www.w3.org/2004/OWL/

[10] http://www.daml.org/services/owl-s/1.2/

[11] http://www.w3.org/TR/rdfa-syntax/

[12] http://www.xml3d.org/

[13] www.fipa.org

FIPA was a standard proposal for an agent communication language (ACL). [10] gives and overview of FIPA and related activities. Although FIPA also made proposals for standards for communication protocols[14], execution of such protocols was rather neglected. In his seminal work [16] investigated formal approaches to protocol design. Declarative methods to describe protocols have the charm that the formalisms seem to be clean. However, at least in some cases it is a problem to find out whether protocols can be enacted [6]. More pragmatic approaches do not face this problems because protocol execution is directly foreseen, however, conformance to prescribed behavior is a general problem.

While the areas of general software engineering and agent-oriented software engineering lived for years without much interaction, in recent years the concepts of the two areas have grown significantly together. A good example for the common interest is the Agent Platform Special Interest Group of the Object Management Group (OMG) which has the goal to foster OMG specifications in the agent area. Other examples are the Agent Modelling Language (AML) [18] a semi-formal visual modeling language for the definition, modeling, and documentation of systems that adopt agent technologies. AML is defined as an extension of the Unified Modeling Language (UML[15]) using the most important OMG frameworks. Agent UML (AUML) [2] extended UML sequence diagrams with interaction protocols. Besides agent-based modeling approaches several methodologies were proposed (e.g. Tropos [15], Prometheus [14,17]) that provide mechanisms to support the specification, the analysis, and development of agent-based systems. Additionally proposals for metamodels for agent-based systems are on the table (e.g. Gaia [20], PASSi [4], and ADELFE [3]).

Protocol projections have been studied already early in the area of verification of communication protocols (e.g. [11]). Recently, these techniques are also used in the context of collaborations [12]. More specifically, projections are used for generating executable business processes and orchestrations from choreography descriptions (e.g. [13], [9]), restricted to languages for business process modeling.

7   Conclusions

This article presented a framework for the model driven design of MAS. The framework is built around the domain specific modelling language DSML4MAS. The core of DSML4MAS is defined by the metamodel PIM4Agents that includes platform independent concepts for the design of MAS. PIM4Agents is separated into 12 major aspects for MAS design. The general framework proposes a plugin architecture where these different aspects can be replaced or refined in a flexible manner. In the discussion of the PIM4Agents metamodel the article concentrates on the Interaction aspect because it is an obvious choice when it comes to reuse of model fragments in MAS design. The design of interaction protocols is complex and tedious and therefore reuse of well-understood protocols is highly desirable. The article presents a proposal how the Interaction aspect in DSML4MAS can

---

[14] http://www.fipa.org/repository/ips.php3
[15] www.uml.org

be realized. This approach is pragmatic and procedural with the advantage to define an operative semantics for the models in the transformation to a specific execution environment. Further research needs to be done to link the presented approach with proposals for declarative protocol specifications like for example presented in [16,6,1].

The main contribution of the work presented in this article aims at the design of agent-based systems and multiagent systems from a software engineering perspective. In this work AI aspects are not directly obvious. However, in matchmaking of queries on model repositories and in the internal reasoning of the agents AI topics are of course relevant. At least regarding reasoning work on agent technologies (e.g. the agent development tool JACK) it has been shown that modelling and reasoning can be brought together. We plan to integrate these AI aspects more deeply with our approach in future work.

Acknowledgements. The paper is based on work performed in the project COIN (EU FP7 Project 216256; www.coin-ip.eu) funded by the European Community within the IST-Programme of the 7th Framework Research Programme. The authors also thank the contribution from other partners in the COIN consortium.

References

1. Baldoni, M., Baroglio, C., Chopra, A.K., Desai, N., Patti, V., Singh, M.P.: Choice, interoperability, and conformance in interaction protocols and service choreographies. In: AAMAS (2), pp. 843–850 (2009)
2. Bauer, B., Müller, J.P., Odell, J.: Agent UML: A Formalism for Specifying Multiagent Interaction. In: Ciancarini, P., Wooldridge, M.J. (eds.) AOSE 2000. LNCS, vol. 1957, pp. 91–103. Springer, Heidelberg (2001)
3. Bernon, C., Gleizes, M.-P., Peyruqueou, S., Picard, G.: ADELFE: A Methodology for Adaptive Multi-Agent Systems Engineering. In: Petta, P., Tolksdorf, R., Zambonelli, F. (eds.) ESAW 2002. LNCS (LNAI), vol. 2577, pp. 156–169. Springer, Heidelberg (2003)
4. Chella, A., Cossentino, M., Sabatucci, L., Seidita, V.: From passi to agile passi: Tailoring a design process to meet new needs. In: Proceedings of the IEEE/WIC/ACM International Conference on Intelligent Agent Technology, IAT 2004, pp. 471–474. IEEE Computer Society, Washington, DC, USA (2004), http://dx.doi.org/10.1109/IAT.2004.59
5. Davis, R., Smith, R.G.: Negotiation as a metaphor for distributed problem solving. Artificial Intelligence 20(1) (1983)
6. Desai, N., Singh, M.P.: On the enactability of business protocols. In: Fox, D., Gomes, C.P. (eds.) AAAI, pp. 1126–1131. AAAI Press (2008)
7. Hahn, C., et al.: A platform-independent metamodel for multiagent systems. Autonomous Agents and Multi-Agent Systems 18, 239–266 (2009)
8. Kapahnke, P., Liedtke, P., Nesbigall, S., Warwas, S., Klusch, M.: ISReal: An Open Platform for Semantic-Based 3D Simulations in the 3D Internet. In: Patel-Schneider, P.F., Pan, Y., Hitzler, P., Mika, P., Zhang, L., Pan, J.Z., Horrocks, I., Glimm, B. (eds.) ISWC 2010, Part II. LNCS, vol. 6497, pp. 161–176. Springer, Heidelberg (2010)

9. Khadka, R., Sapkota, B., Ferreira Pires, L., van Sinderen, M., Jansen, S.: Model-Driven Development of Service Compositions for Enterprise Interoperability. In: van Sinderen, M., Johnson, P. (eds.) IWEI 2011. LNBIP, vol. 76, pp. 177–190. Springer, Heidelberg (2011)

10. Kone, M.T., Shimazu, A., Nakajima, T.: The state of the art in agent communication languages. Knowledge and Information Systems 2(3), 259–284 (2000), http://dx.doi.org/10.1007/PL00013712

11. Lam, S.S., Shankar, A.U.: Protocol verification via projections. IEEE Trans. Software Eng. 10(4), 325–342 (1984)

12. McNeile, A.T.: Protocol contracts with application to choreographed multiparty collaborations. Service Oriented Computing and Applications 4(2), 109–136 (2010)

13. Mendling, J., Hafner, M.: From WS-CDL choreography to BPEL process orchestration. Journal of Enterprise Information Management 21, 525–542 (2005)

14. Padgham, L., Thangarajah, J., Winikoff, M.: Tool support for agent development using the prometheus methodology. In: Proceedings of the Fifth International Conference on Quality Software, QSIC 2005, pp. 383–388. IEEE Computer Society, Washington, DC, USA (2005), http://dx.doi.org/10.1109/QSIC.2005.66

15. Penserini, L., Perini, A., Susi, A., Mylopoulos, J.: From stakeholder intentions to software agent implementations. In: Conference on Advanced Information Systems Engineering, pp. 465–479 (2006)

16. Singh, M.P.: Multiagent Systems: A Theoretical Framework for Intentions, Know-How, and Communications. LNCS, vol. 799. Springer, Heidelberg (1994), http://www.csc.ncsu.edu/faculty/mpsingh/books/MAS/

17. Thangarajah, J., Padgham, L.: Prometheus design tool. In: The 4th International Joint Conference on Autonomous Agents and Multi-Agent Systems, pp. 127–128 (2005)

18. Trencansky, I., Cervenka, R.: Agent modeling language (aml): A comprehensive approach to modelling mas. Informatica 29(4), 391–400 (2005)

19. Warwas, S., Hahn, C.: The DSML4MAS development environment. In: Proc. of the 8th Int. Conf. on Autonomous Agents and Multiagent Systems (AAMAS 2009), pp. 1379–1380. IFAAMAS (2009)

20. Zambonelli, F., Jennings, N.R., Wooldridge, M.: Developing multiagent systems: The gaia methodology. ACM Transactions on Software Engineering and Methodology 12(3), 317–370 (2003)

Part I

Artificial Intelligence

Probabilistic Connection between Cross-Validation and Vapnik Bounds

Przemysław Klęsk

Department of Methods of Artificial Intelligence and Applied Mathematics
Westpomeranian University of Technology
ul. Żolnierska 49, Szczecin, Poland
pklesk@wi.zut.edu.pl

Abstract. In the paper we analyze a connection between outcomes of the cross-validation procedure and Vapnik bounds [1,2] on generalization of learning machines. We do *not* focus on how well the measured cross-validation outcome estimates the generalization error or how far it is from the training error; instead, we want to make statements about the cross-validation result without actually measuring it. In particular we want to state probabilistically what ε-difference one can expect between the *known* Vapnik bound and the *unknown* cross-validation result for given conditions of the experiment. In the consequence, we are able to calculate the necessary size of the training sample, so that the ε is sufficiently small; and so that the optimal complexity indicated via SRM is acceptable in the sense that cross-validation, if performed, would probably indicate the same complexity. We consider a non-stratified variant of cross-validation, which is convenient for the main theorem.

Keywords. Statistical learning theory, Bounds on generalization, Cross validation, Empirical risk minimization, Structural risk minimization, Vapnik–Chervonenkis dimension.

1 Introduction and Notation

One part of the *Statistical Learning Theory* developed by Vapnik [1,2,3] is the *theory of bounds*. It provides probabilistic bounds on generalization of learning machines. The key mathematical tools applied to derive the bounds in their additive versions are Chernoff and Hoeffding inequalities[1] [2,4,5,6].

We use this theory to show a probabilistic relationship between two approaches for complexity selection: *n-fold cross-validation* (popular among practioner modelers) and *Structural Risk Minimization* proposed by Vapnik (rarely met in practice) [7,8,9,10]. We remind that SRM is $O(n)$ times faster than n-fold cross-validation (since SRM does not perform any repetitions/folds per single fixed complexity, nor testing) but less accurate,

---

[1] Chernoff inequality: $P(|\nu_I - p| \geq \varepsilon) \leq 2\exp(-2\varepsilon^2 I)$, Hoeffding inequality: $P(|X_I - EX| \geq \varepsilon) \leq 2\exp(-\frac{2\varepsilon^2 I}{B^2 - A^2})$. Meaning (respectively): observed frequencies on a sample of size I converge to the true probability as I grows large; analogically: means of a random variable (bounded by A and B) converge to the expected value. It is a *in-probability-convergence* and its rate is exponential.

J. Filipe and A. Fred (Eds.): ICAART 2011, CCIS 271, pp. 31–55, 2013.
© Springer-Verlag Berlin Heidelberg 2013

since the selection of optimal complexity is based on the guaranteed generalization risk. The bound for the guaranteed risk is expressed in terms of *Vapnik-Chervonenkis dimension*, and is a pessimistic overestimation of the *growth function*, which in turn is overestimation of the unknown *Vapnik-Chervonenkis entropy*. We formally remind these notions later in the paper. All those overestimations contribute (unfortunately) to the fact that for a fixed sample size, SRM usually underestimates the optimal complexity and chooses too simple model.

Results presented in this paper may be regarded as conceptually akin to results by Holden [11,12], where error bounds on cross-validation and so-called *sanity-check* bounds are derived. The sanity-check bound is a proof, for large class of learning algorithms, that the error of the *leave-one-out* estimate is not much worse — $O(\sqrt{h/l})$ — than the worst-case behavior of the training error estimate, where h stands for Vapnik-Chervonenkis dimension of given set of functions and l stands for the sample size. The name sanity-check refers to the fact that although we believe that under many circumstances, the leave-one-out estimate will perform better than the training error (and thus justify its computational expense) the goal of the sanity-check bound is to simply prove that it is not much worse than the training error [13].

These results were further generalized by Kearns [13,14,15] using the notion of (β_1, β_2)-*error stability*[2] rather than (β_1, β_2)-*hypothesis stability*[3] imposed on the learning algorithm.

For the sake of comparison and to set up the perspective for further reading of this paper, we highlight some differences of meaning of our results and the results mentioned above:

- we do not focus on how well the *measured* cross-validation result estimates the generalization error or how far it is from the training error in the leave-one-out case — sanity-check bounds [12,13]; instead, we want to make statements about the cross-validation result *without actually measuring it*, thus, remaining in the setting of the SRM framework.
- in particular we want to state probabilistically what ε-difference one can expect between the *known* Vapnik bound and the *unknown* cross-validation result for given conditions of the experiment,
- in the consequence, we want to be able to calculate the necessary size of the training sample, so that the ε is sufficiently small, and so that the optimal complexity indicated via SRM is acceptable in the sense that cross-validation, if performed, would probably indicate the same complexity; this statement may seem related to the notion of *sample complexity* considered e.g. by Bartlett [16,17] or Ng [18], but we do not find the sample size required for the algorithm to learn/generalize "well"

---

[2] We say that a learning algorithm has a (β_1, β_2)-error stability, if generalization errors for two models provided by this algorithm using respectively a training sample of size l and a sample with size lowered to $l-1$ are β_1-close to each other with probability at least $1 - \beta_2$. Obviously the smaller both β_1, β_2 are the more stable the algorithm.

[3] We say that a learning algorithm has a (β_1, β_2)-hypothesis stability, if the two models provided by this algorithm using respectively a training sample of size l and sample with size lowered to $l-1$ are β_1-close to each other with probability at least $1 - \beta_2$, where closeness of models is measured by some functional metrics, e.g. L_1, L_2, etc.

but rather such a sample size so that complexity selection via SRM gives similar results to complexity selection via cross-validation,

- we do not explicitly introduce the notion of error stability for the learning algorithm, but this kind of stability is implicitly derived be means of Chernoff-Hoeffding-like inequalities we write.
- we do not focus on the leave-one-out cross-validation; we consider a more general *n-fold non-stratified cross-validation* (also: more convenient for our purposes); the leave-one-out case can be read out from our results as a special case.

1.1 Notation Related to Statistical Learning Theory

We keep the notation similar to Vapnik's [2,1].

- We denote the finite set of samples as:

$$\{(\mathbf{x}_1, y_1), (\mathbf{x}_2, y_2), \ldots, (\mathbf{x}_l, y_l)\},$$

or more shortly by encapsulating pairs as

$$\{\mathbf{z}_1, \mathbf{z}_2, \ldots, \mathbf{z}_l\},$$

where $\mathbf{x}_i \in \mathbb{R}^d$ are input points, y_i are output values corresponding to them, and l is the set size. y_i differ depending on the learning task: for *classification* (pattern-recognition) $y_i \in \{1, 2, \ldots, K\}$ — finite discrete set, for *regression estimation* $y_i \in \mathbb{R}$.
- We denote the *set of approximating functions* (models) in the sense of both classification or regression estimation as:

$$\{f(\mathbf{x}, \omega)\}_{\omega \in \Omega},$$

where Ω is the domain of parameters of this set of functions, so a fixed ω can be regarded as an index of a specific function in the set.
- The *risk functional* $R: \{f(\mathbf{x}, \omega)\}_{\omega \in \Omega} \to \mathbb{R} \cup \{+\infty\}$ is defined as

$$R(\omega) = \int_{\mathbf{x} \in \mathbf{X}} \int_{y \in Y} L\Big(f(\mathbf{x}, \omega), y\Big) \underbrace{p(\mathbf{x}, y)}_{p(\mathbf{x})p(y|\mathbf{x})} dy d\mathbf{x}, \tag{1}$$

where $p(\mathbf{x})$ is the distribution density of input points, $p(y|\mathbf{x})$ is the conditional density of system/phenomenon outputs y given a fixed $\mathbf{x}$. $p(\mathbf{x}, y) = p(\mathbf{x})p(y|\mathbf{x})$ is the joint distribution density for pairs $(\mathbf{x}, y)$. In practice, $p(\mathbf{x}, y)$ is unknown but *fixed*, and hence we assume the sample $\{\mathbf{z}_1, \mathbf{z}_2, \ldots, \mathbf{z}_l\}$ to be *i.i.d.*[4] L is the so called *loss function* which measures the discrepancy between the output y and the model f. For classification, L is an indicator function:

$$L\Big(f(\mathbf{x}, \omega), y\Big) = \begin{cases} 0, & \text{for } y = f(\mathbf{x}, \omega); \\ 1, & \text{for } y \neq f(\mathbf{x}, \omega), \end{cases} \tag{2}$$

---

[4] Independent, identically distributed.

and the risk functional becomes $R(\omega) = \int_{\mathbf{x} \in \mathbf{X}} \sum_{y \in Y} L(f(\mathbf{x}, \omega), y) p(\mathbf{x}, y) \mathbf{dx}$. For regression estimation, L is usually chosen as the distance in L_2 metric:

$$L\left(f(\mathbf{x}, \omega), y\right) = \left(f(\mathbf{x}, \omega) - y\right)^2, \tag{3}$$

and the risk functional becomes $R(\omega) = \int_{\mathbf{x} \in \mathbf{X}} \int_{y \in Y} \left(f(\mathbf{x}, \omega) - y\right)^2 p(\mathbf{x}, y) dy d\mathbf{x}$.

- By ω_0 we denote the index of the best function $f(\mathbf{x}, \omega_0)$ in the set, such that:

$$R(\omega_0) = \inf_{\omega \in \Omega} R(\omega). \tag{4}$$

- Since only a finite set of samples $\{\mathbf{z}_1, \dots, \mathbf{z}_l\}$ is at disposal, we cannot count on actually finding the best function $f(\mathbf{x}, \omega_0)$. In fact, we look for its estimate with respect to the finite set of samples. We define the *empirical risk*:

$$R_{\mathrm{emp}}(\omega) = \frac{1}{l} \sum_{i=1}^{l} L(y_i, f(\mathbf{x}_i, \omega)), \tag{5}$$

and by ω_l we denote the index of the function $f(\mathbf{x}, \omega_l)$ such that:

$$R_{\mathrm{emp}}(\omega_l) = \inf_{\omega \in \Omega} R_{\mathrm{emp}}(\omega) \tag{6}$$

— *Empirical Risk Minimization* principle [1,2,19,4].

- For simplification of notation and further considerations, we introduce replacements:

$$(\mathbf{x}, y) = \mathbf{z},$$
$$L\left(f(\mathbf{x}, \omega), y\right) = Q(\mathbf{z}, \omega).$$

In other words instead of considering the set of approximating functions[5] $\{f(\mathbf{x}, \omega)\}_{\omega \in \Omega}$, we equivalently consider the *set of error functions* $\{Q(\mathbf{z}, \omega)\}_{\omega \in \Omega}$. It is a 1:1 correspondence[6]. Now, we write the true risk as:

$$R(\omega) = \int_{\mathbf{z} \in \mathbf{X} \times Y} Q(\mathbf{z}, \omega) \underbrace{p(\mathbf{z})}_{p(\mathbf{x}, y)} \mathbf{dz}$$

$$= \int_{\mathbf{Z}} Q(\mathbf{z}, \omega) dF(\mathbf{z}), \tag{7}$$

and the empirical risk as

$$R_{\mathrm{emp}}(\omega) = \frac{1}{l} \sum_{i=1}^{l} Q(\mathbf{z}_i, \omega)), \tag{8}$$

---

[5] In the sense of all learning tasks.

[6] Q is identical with L in the sense of their values. They differ only in formal posing of their domains. L works on $f(\mathbf{x}, \omega)$ and y and maps them to error values, whereas Q works directly on $\mathbf{z}$ and ω and maps them to error values.

1.2   Notation Related to Cross-Validation

In the paper, we shall consider the *non-stratified* variant of the n-fold cross-validation procedure [20]. In each single fold (iteration) we first permute the data set and then we split it at the same fixed point into two disjoint subsets — a training set and a testing set. Thus, we guarantee the randomness by permutation per each fold, and among folds we do not care to make training sets disjoint pairwise. Since permutations are independent, hence *folds are independent* as well.

Such an approach is somewhere in-between the classical n-fold cross-validation and the *bootstrapping* [21]. In the classical cross-validation, all $\binom{n}{2}$ pairs of training sets are mutually disjoint (and so are testing sets) and hence folds are dependent, whereas in the bootstrapping instead of repeatedly analyzing subsets of data set, one repeatedly analyzes the subsamples (with replacement) of the data. For more information see [22,23,24].

We introduce the following notation. I' and I'' stand for the size of training and testing sets respectively.

$$I' = \frac{n-1}{n} I,$$

$$I'' = \frac{1}{n} I.$$

Without loss of generality for theorems and proofs, let I be dividable by n, so that I' and I'' are integers.

In a single fold, let

$$\{\mathbf{z}'_1, \mathbf{z}'_2, \ldots, \mathbf{z}'_{I'}\}, \quad \{\mathbf{z}''_1, \mathbf{z}''_2, \ldots, \mathbf{z}''_{I''}\}$$

represent respectively the training set and the testing set, taken as a split of the whole permuted data set $\{\mathbf{z}_1, \mathbf{z}_2, \ldots, \mathbf{z}_I\}$. Similarly, empirical risks calculated as follows:

$$R'_{\text{emp}}(\omega) = \frac{1}{I'} \sum_{i=1}^{I'} Q(\mathbf{z}'_i, \omega), \tag{9}$$

$$R''_{\text{emp}}(\omega) = \frac{1}{I''} \sum_{i=1}^{I''} Q(\mathbf{z}''_i, \omega), \tag{10}$$

represent respectively the training error and the testing error, calculated for any function ω.

By $\omega_{I'}$ we define the function that minimizes the *empirical training risk*

$$R'_{\text{emp}}(\omega_{I'}) = \inf_{\omega \in \Omega} R'_{\text{emp}}(\omega) \tag{11}$$

when the context of discussion is constrained to single fold. When, we will need to broaden the context onto all folds, $k = 1, 2, \ldots, n$, we will write $\omega_{I',k}$ to denote the function that minimizes the empirical training risk in the k-th fold. Therefore, the final cross-validation result — an estimate of generalization error — is the mean from *empirical testing risks* R''_{emp} using functions $\omega_{I',k}$:

$$C = \frac{1}{n} \sum_{k=1}^{n} R''_{\text{emp}}(\omega_{I',k}).$$ (12)

The *independence of folds* can be formally expressed in the following way. For any two indices of folds $k \neq l$ and for any numbers A, B:

$$P(R''_{\text{emp}}(\omega_{I',k}) = A, R''_{\text{emp}}(\omega_{I',l}) = B)$$
$$= P(R''_{\text{emp}}(\omega_{I',k}) = A) \cdot P(R''_{\text{emp}}(\omega_{I',l}) = B).$$

We stress the independence once again, because later on we are going to sum up several independent probabilistic inequalities into one inequality, and we would like the result to be true with the effective probability being the product of component probabilities.

2   The Relationship for a Finite Set of Approximating Functions

2.1   Classification Learning Task

Similarly to Vapnik, let us start with the classification learning task and the simplest case of a *finite* set of N indicator functions: $\{Q(\mathbf{z}, \omega_j)\}_{\omega_j \in \Omega}$, $j = 1, 2, \ldots, N$. Not to complicate things, we will keep on writing ω_I in the sense of the optimal function minimizing the empirical risk on our finite sample of size I, instead of writing more formally e.g. $\omega_{j(I)}$[7].

Vapnik shows [1,2] that with probability at least $1 - \eta$, the following bound on the true risk is satisfied:

$$\underbrace{\int_{\mathbf{Z}} Q(\mathbf{z}, \omega_I) dF(\mathbf{z})}_{R(\omega_I)} \leq \underbrace{\frac{1}{I} \sum_{i=1}^{I} Q(\mathbf{z}_i, \omega_I)}_{R_{\text{emp}}(\omega_I)} + \sqrt{\frac{\ln N - \ln \eta}{2I}}.$$ (13)

The argument is the following:

$$P\left(\sup_{1 \leq j \leq N} R(\omega_j) - R_{\text{emp}}(\omega_j) \geq \varepsilon \right)$$
$$\leq \sum_{j=1}^{N} P\left(R(\omega_j) - R_{\text{emp}}(\omega_j) \geq \varepsilon \right) \leq N \cdot \exp(-2\varepsilon^2 I).$$

The last pass is true, since for each term in the sum Chernoff inequality is satisfied. By substituting the right-hand-side with small probability η and solving for ε, one obtains the bound:

$$R(\omega_j) - R_{\text{emp}}(\omega_j) \leq \sqrt{\frac{\ln N - \ln \eta}{2I}},$$

---

[7] In the sense that $j(I) \in \{1, \ldots, N\}$ returns the index of the minimizer given our data set of size I.

which holds true with probability at least $1 - \eta$ simultaneously *for all* functions in the set, since it holds for the worst. Hence, in particular it holds true for the function ω_l. And one gets the bound (13).

For the theorems to follow, we denote the right-hand-side in the Vapnik bound by $V = R_{\text{emp}}(\omega_l) + \sqrt{(\ln N - \ln \eta)/(2I)}$.

Theorem 1. *Let* $\{Q(\mathbf{z}, \omega_j)\}_{\omega_j \in \Omega}$, $j = 1, 2, \ldots, N$, *be a finite set of indicator functions (classification task) of size* N. *Then, for any* $\eta > 0$, *arbitrarily small, there is a small number*

$$\alpha(\eta, n) = \eta - \sum_{k=1}^{n} \binom{n}{k} (-1)^k (2\eta)^k, \tag{14}$$

and the number

$$\varepsilon(\eta, I, N, n) = \left(2\sqrt{\frac{n}{n-1}} + 1 \right) \sqrt{\frac{\ln N - \ln \eta}{2I}}$$

$$+ \left(\sqrt{n} + \sqrt{\frac{n}{n-1}} \right) \sqrt{\frac{-\ln \eta}{2I}}, \tag{15}$$

such that:

$$P\left(|V - C| \leq \varepsilon(\eta, I, N, n) \right) \geq 1 - \alpha(\eta, n). \tag{16}$$

Before we prove theorem 1, the following two remarks should be clear.

Remark 1. The value of $\alpha(\eta, n)$ is monotonous with η. I.e. the smaller η we choose, the smaller $\alpha(\eta, n)$ becomes as well. Therefore the minimum probability measure $1 - \alpha(\eta, n)$ is suitably large.

$$\lim_{\eta \to 0^+} \left(\eta - \sum_{k=1}^{n} \binom{n}{k} (-1)^k (2\eta)^k \right)$$

$$= \lim_{\eta \to 0^+} \left(\eta + 1 - \sum_{k=0}^{n} \binom{n}{k} (-1)^k (2\eta)^k \right)$$

$$= \lim_{\eta \to 0^+} \left(\eta + 1 - \underbrace{(1 - 2\eta)^n}_{\to 1} \right) = 0.$$

Remark 2. For the fixed values of η, N, n, the value of $\varepsilon(\eta, I, N, n)$ converges to zero as the sample size I grows large.

This is an important remark, because it means that both the cross-validation result C and the Vapnik bound V *converge in probability*[8] to the same value[9] as the sample size grows large. Moreover, the rate of this convergence is exponential.

Proof (**Proof of Remark 2**). Since N is fixed, we note that for $\eta \to 0^+$

$$\sqrt{\frac{\ln N - \ln \eta}{2I}} \sim \sqrt{\frac{-\ln \eta}{2I}}.$$

Therefore, for fixed η, N, n there exists a constant, say D, such that

$$\varepsilon(\eta, I, N, n) = 2\left(\sqrt{\frac{n}{n-1}} + 1\right)\sqrt{\frac{\ln N - \ln \eta}{2I}}$$

$$+ \left(\sqrt{n} + \sqrt{\frac{n}{n-1}}\right)\sqrt{\frac{-\ln \eta}{2I}} \leq D\sqrt{\frac{-\ln \eta}{2I}}.$$

Solving the inequality for η we obtain $\eta \leq \exp(-2I\varepsilon^2/D^2)$.

Having in mind the inequality (16), we now give two theorems in which the absolute value sign in $|V - C|$ is omitted. They can be viewed as the *upper* and the *lower* probabilistic bounds on C and they are derived as tighter bounds than (16). Proving these two theorems immediately implies proving the theorem 1.

Theorem 2. *With probability $1 - \alpha(\eta, n)$ or greater, the following inequality holds true:*

$$C - V \leq \left(\sqrt{\frac{n}{n-1}} - 1\right)\sqrt{\frac{\ln N - \ln \eta}{2I}}$$

$$+ \left(\sqrt{n} + \sqrt{\frac{n}{n-1}}\right)\sqrt{\frac{-\ln \eta}{2I}}. \quad (17)$$

Theorem 3. *With probability $1 - \alpha(\eta, n)$ or greater, the following inequality holds true:*

$$V - C \leq \left(2\sqrt{\frac{n}{n-1}} + 1\right)\sqrt{\frac{\ln N - \ln \eta}{2I}} + \sqrt{n}\sqrt{\frac{-\ln \eta}{2I}}. \quad (18)$$

The second result is more interesting, provided of course that the bound is positive for given constants η, I, N, n. Otherwise, we get zero or negative bound, which is trivial. The fig. 1 illustrates the sense of theorems 2 and 3.

---

[8] We say that $A(I)$ *converges in probability* to B, we write $A(I) \xrightarrow[I \to \infty]{P} B$, when for any numbers $\varepsilon > 0$, $\eta > 0$, there exists a treshold size of sample $I(\varepsilon, \eta)$, such that *for all $I \geq I(\varepsilon, \eta)$*: $P(|A(I) - B| > \varepsilon) \leq \eta$.

[9] C and V can be viewed as random variables, due to random realizations of data set $\{z_1, \ldots, z_I\}$ with joint density $p(z)$ (this affects C and V) and due to random realizations of subsets in cross-validation folds (this affects C). When the data set $\{z_1, \ldots, z_I\}$ is fixed, V is fixed too.

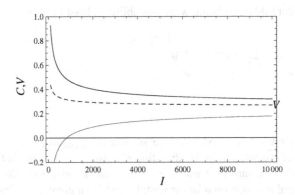

Fig. 1. Illustration of upper and lower bounds on the result of cross-validation with respect to the size of sample I. Other constants are: $\eta = 0.01 \Rightarrow 1 - \alpha(\eta) \approx 0.93$, $N = 100$, $n = 3$. With probability $1 - \alpha(\eta)$ or greater, the result C of cross-validation falls between the bounds.

Proof **(Proof of Theorem 2)**. We remind: $I' = \frac{n-1}{n}I$, $I'' = \frac{1}{n}I$.

With probability at least $1 - \eta$, the following bound on true risk holds true:

$$R(\omega_{I'}) \leq R'_{\text{emp}}(\omega_{I'}) + \sqrt{\frac{\ln N - \ln \eta}{2I'}}. \tag{19}$$

For the selected function $\omega_{I'}$, fixed from now on, Chernoff inequality is satisfied on the testing set (empirical testing risk) in either of its one-side-versions:

$$R''_{\text{emp}}(\omega_{I'}) - R(\omega_{I'}) \leq \sqrt{\frac{-\ln \eta}{2I''}}, \tag{20}$$

$$R(\omega_{I'}) - R''_{\text{emp}}(\omega_{I'}) \leq \sqrt{\frac{-\ln \eta}{2I''}}, \tag{21}$$

with probability at least $1 - \eta$ each. By joining (19) and (20) we obtain, with probability at least[10] $1 - 2\eta$ the system of inequalities:

$$R''_{\text{emp}}(\omega_{I'}) - \sqrt{\frac{-\ln \eta}{2I''}} \leq R(\omega_{I'}) \leq R'_{\text{emp}}(\omega_{I'})$$

$$+ \sqrt{\frac{\ln N - \ln \eta}{2I'}}. \tag{22}$$

---

[10] The minimum probability must be $1 - 2\eta$ rather than $(1 - \eta)^2$ (probabilistic independence case) due to correlations between inequalities. It can be also viewed as the consequence of Bernoulli's inequality.

After n independent folds we obtain, with probability at least $(1 - 2\eta)^n$:

$$\underbrace{\frac{1}{n}\sum_{k=1}^{n} R''_{\mathrm{emp}}(\omega_{I',k}) \leq \frac{1}{n}\sum_{k=1}^{n} R'_{\mathrm{emp}}(\omega_{I',k}) + \sqrt{\frac{\ln N - \ln \eta}{2I'}}}_{C}$$

$$+ \sqrt{\frac{-\ln \eta}{2I''}}. \quad (23)$$

To conclude the proof, we need to relate somehow $R'_{\mathrm{emp}}(\omega_{I',k})$ from each fold to $R_{\mathrm{emp}}(\omega_I)$. We need the relation in the direction $R'_{\mathrm{emp}}(\omega_{I',k}) \leq \cdots$, so that we can plug the right-hand-side of it into (23) and keep it true. Intuitively, one might expect that choosing an optimal function on a larger sample leads to a greater empirical risk comparing to a smaller sample, i.e. $R_{\mathrm{emp}}(\omega_I) \geq R'_{\mathrm{emp}}(\omega_{I',k})$, because it is usually easier to fit fewer data points using models of equally rich complexities. But we don't know with what probability that occurs. Contrarily, on may easily find a specific data subset for which $R_{\mathrm{emp}}(\omega_I) \leq R'_{\mathrm{emp}}(\omega_{I',k})$.

Lemma 1. *With probability* 1, *true is the following inequality:*

$$\sum_{i=1}^{I'} Q(\mathbf{z}'_i, \omega_{I'}) \leq \sum_{i=1}^{I} Q(\mathbf{z}_i, \omega_I). \quad (24)$$

On the level of sums of errors, not means, the total error for a larger sample will always surpass the total error for a smaller sample. This gives us $I' R'_{\mathrm{emp}}(\omega_{I'}) \leq I R_{\mathrm{emp}}(\omega_I)$ and further:

$$R'_{\mathrm{emp}}(\omega_{I'}) \leq \frac{n}{n-1} R_{\mathrm{emp}}(\omega_I). \quad (25)$$

Unfortunately it is of no use, because of the coefficient $\frac{n}{n-1}$. Thinking of $C - V$ in the theorem, we need a relation with coefficients 1 at both C and V.

In [2, pp. 124] we find the following helpful assertion:

Lemma 2. *With probability at least* $1 - 2\eta$:

$$\int_{\mathbf{Z}} Q(\mathbf{z}, \omega_I) dF(\mathbf{z}) - \underbrace{\inf_{1 \leq j \leq N} \int_{\mathbf{Z}} Q(\mathbf{z}, \omega_j) dF(\mathbf{z})}_{R(\omega_0)}$$

$$\leq \sqrt{\frac{\ln N - \ln \eta}{2I}} + \sqrt{\frac{-\ln \eta}{2I}} \quad (26)$$

— *the true risk for the selected function* ω_I *is not farther from the minimal possible risk for this set of functions than* $\sqrt{\frac{\ln N - \ln \eta}{2I}} + \sqrt{\frac{-\ln \eta}{2I}}$.

Proof of that statement given by Vapnik is based on two inequalities (each with probability at least $1 - \eta$), the first is (13) — we repeat it here, and the second is Chernoff inequality for the best function ω_0:

$$R(\omega_I) - R_{\text{emp}}(\omega_I) \leq \sqrt{\frac{\ln N - \ln \eta}{2I}}, \tag{27}$$

$$R_{\text{emp}}(\omega_0) - R(\omega_0) \leq \sqrt{\frac{-\ln \eta}{2I}}. \tag{28}$$

And since, by definition of ω_I, $R_{\text{emp}}(\omega_0) \geq R_{\text{emp}}(\omega_I)$, the (26) follows.

Going back to the cross-validation procedure, we notice that in each single fold the measure R_{emp} corresponds by analogy to the measure R in (26) and the measure R'_{emp} corresponds by analogy to R_{emp} therein. Obviously R is defined on an infinite and continuous space $\mathbf{Z} = \mathbf{X} \times Y$, whereas R_{emp} is defined on a discrete and finite sample $\{\mathbf{z}_1, \ldots, \mathbf{z}_I\}$, but still from the perspective of a single cross-validation fold we may view $R_{\text{emp}}(\omega_I)$ as the "target" minimal *probability* of misclassification and $R'_{\text{emp}}(\omega_{I'})$ as the observed relative *frequency* of misclassification — an estimate of that *probability*, remember that we take random subsets $\{\mathbf{z}'_1, \ldots, \mathbf{z}'_{I'}\}$ from the whole set $\{\mathbf{z}_1, \ldots, \mathbf{z}_I\}$.

We write

$$R'_{\text{emp}}(\omega_{I'}) \leq R'_{\text{emp}}(\omega_I) \leq R_{\text{emp}}(\omega_I) + \sqrt{\frac{-\ln \eta}{2I'}}. \tag{29}$$

The first inequality is true with probability 1 by definition of $\omega_{I'}$. The second is a Chernoff inequality, true with probability at least $1 - \eta$.

Now, we plug (29) into (23) and obtain with probability $1 - (-\sum_{k=1}^{n} \binom{n}{k}(-1)^k (2\eta)^k)$ $-\eta$ or greater:

$$C \leq \frac{1}{n} n \left(R_{\text{emp}}(\omega_I) + \sqrt{\frac{-\ln \eta}{2I'}} \right)$$

$$+ \sqrt{\frac{\ln N - \ln \eta}{2I'}} + \sqrt{\frac{-\ln \eta}{2I''}}$$

$$= R_{\text{emp}}(\omega_I) + \sqrt{\frac{n}{n-1}} \sqrt{\frac{\ln N - \ln \eta}{2I}}$$

$$+ \left(\sqrt{n} + \sqrt{\frac{n}{n-1}} \right) \sqrt{\frac{-\ln \eta}{2I}}$$

$$= R_{\text{emp}}(\omega_I) + \left(\sqrt{\frac{n}{n-1}} + 1 - 1 \right) \sqrt{\frac{\ln N - \ln \eta}{2I}}$$

$$+ \left(\sqrt{n} + \sqrt{\frac{n}{n-1}} \right) \sqrt{\frac{-\ln \eta}{2I}}$$

$$= V + \left(\sqrt{\frac{n}{n-1}} - 1 \right) \sqrt{\frac{\ln N - \ln \eta}{2I}}$$

$$+ \left(\sqrt{n} + \sqrt{\frac{n}{n-1}} \right) \sqrt{\frac{-\ln \eta}{2I}}.$$

This concludes the proof of theorem 2.

Proof (**Proof of Theorem 3**). The proof is analogous to the former proof, but we need to write most of the probabilistic inequalities in the different direction.

With probability at least $1 - \eta$, the following bound on true risk holds true:

$$R'_{\text{emp}}(\omega_{I'}) \leq R(\omega_{I'}) + \sqrt{\frac{\ln N - \ln \eta}{2I'}}. \tag{30}$$

By joining (30) and (21) we obtain, with probability at least $1 - 2\eta$ the system of inequalities:

$$R'_{\text{emp}}(\omega_{I'}) - \sqrt{\frac{\ln N - \ln \eta}{2I'}} \leq R(\omega_{I'}) \leq R''_{\text{emp}}(\omega_{I'})$$

$$+ \sqrt{\frac{-\ln \eta}{2I''}}. \tag{31}$$

After n independent folds we obtain, with probability at least $(1 - 2\eta)^n$:

$$\frac{1}{n} \sum_{k=1}^{n} R'_{\text{emp}}(\omega_{I',k}) - \sqrt{\frac{\ln N - \ln \eta}{2I'}} - \sqrt{\frac{-\ln \eta}{2I''}}$$

$$\leq \underbrace{\frac{1}{n} \sum_{k=1}^{n} R''_{\text{emp}}(\omega_{I',k})}_{C}. \tag{32}$$

Again as in the former proof, we need to relate $R'_{\text{emp}}(\omega_{I',k})$ from each fold to $R_{\text{emp}}(\omega_I)$, but now we need the relation to be in the direction $R'_{\text{emp}}(\omega_{I',k}) \geq \cdots$, so that we can plug the right-hand-side of it into (32) and keep it true.

We write

$$R_{\text{emp}}(\omega_I) - \sqrt{\frac{\ln N - \ln \eta}{2I'}} \leq R_{\text{emp}}(\omega'_I) - \sqrt{\frac{\ln N - \ln \eta}{2I'}}$$

$$\leq R'_{\text{emp}}(\omega'_I). \tag{33}$$

Reading it from the right-hand-side: the second is a (13)-like inequality but for discrete measures, which is true with probability at least $1 - \eta$, and the first inequality is true with probability 1 by definition of ω_I.

Now, we plug (33) into (32) and obtain with probability $1 - (-\sum_{k=1}^{n} \binom{n}{k}(-1)^k(2\eta)^k)$ $-\eta$ or greater:

$$C \geq \frac{1}{n}n\left(R_{\text{emp}}(\omega_I) - \sqrt{\frac{\ln N - \ln\eta}{2I'}}\right) - \sqrt{\frac{\ln N - \ln\eta}{2I'}}$$

$$- \sqrt{\frac{-\ln\eta}{2I''}}$$

$$= R_{\text{emp}}(\omega_I) - 2\sqrt{\frac{n}{n-1}}\sqrt{\frac{\ln N - \ln\eta}{2I}} - \sqrt{n}\sqrt{\frac{-\ln\eta}{2I}}$$

$$= R_{\text{emp}}(\omega_I) - \left(2\sqrt{\frac{n}{n-1}} - 1 + 1\right)\sqrt{\frac{\ln N - \ln\eta}{2I}}$$

$$- \sqrt{n}\sqrt{\frac{-\ln\eta}{2I}}$$

$$= V - \left(2\sqrt{\frac{n}{n-1}} + 1\right)\sqrt{\frac{\ln N - \ln\eta}{2I}} - \sqrt{n}\sqrt{\frac{-\ln\eta}{2I}}.$$

This concludes the proof of theorem 3.

Using theorems 2 and 3 we can also say what sample size I is necessary so that the the difference $C - V$ or $V - C$ is less than or equal to an imposed epsilon ε^*.

Let us denote the right-hand-sides of upper and lower bounds (17) and (18) by ε_U and ε_L respectively. Now, suppose we want to have $\varepsilon_U(\eta, I, N, n) \leq \varepsilon_U^*$. Solving it for I we get

$$I \geq \frac{1}{2\varepsilon_U^{*2}}\left(\left(\sqrt{\frac{n}{n-1}} - 1\right)\sqrt{\ln N - \ln\eta}\right.$$

$$\left. + \left(\sqrt{n} + \sqrt{\frac{n}{n-1}}\right)\sqrt{-\ln\eta}\right)^2 \quad (34)$$

Similarly, if we want to have $\varepsilon_L(\eta, I, N, n) \leq \varepsilon_L^*$.

$$I \geq \frac{1}{2\varepsilon_L^{*2}}\left(\left(2\sqrt{\frac{n}{n-1}} + 1\right)\sqrt{\ln N - \ln\eta} + \sqrt{n}\sqrt{-\ln\eta}\right)^2 \quad (35)$$

To give an example: say we have a finite set of 100 functions, $N = 100$, we perform a 5-fold cross-validation, $n = 5$, and we choose $\eta = 0.1$ and $\varepsilon_L^* = \varepsilon_U^* = 0.05$. Then it follows that we need a sample of size $I \geq 5832$ so that the cross-validation result is not worse than $V + 0.05$, whereas we need $I \geq 28314$ so that the cross-validation result is not better than $V - 0.05$. And both results are true with probability $1 - \alpha(\eta, n) \approx 0.73$ or greater.

Remark 3. For the leave-one-out cross-validation, where $n = I$, both the lower and the upper bound loosen to a constant of order $O\left(\sqrt{\frac{-\ln\eta}{2}}\right)$.

Actually, one can easily see that as we take larger samples $I \to \infty$ and we stick to the leave-one-out cross-validation $n = I$, the coefficient $\sqrt{\frac{n}{n-1}}$ standing at $\sqrt{\frac{\ln N - \ln \eta}{2I}}$ goes to 1, whereas the coefficient $\sqrt{n}$ standing at $\sqrt{\frac{-\ln \eta}{2I}}$ goes to infinity.

One might ask: for what choice of n each bound is the tightest given η, I, N? Treating for a moment n as a continuous variable, we impose the conditions:

$$\frac{\partial \varepsilon_U(\eta, I, N, n)}{\partial n} = 0, \quad \frac{\partial \varepsilon_L(\eta, I, N, n)}{\partial n} = 0,$$

and we get optimal n values:

$$n_U^* = 1 + \left(\frac{\sqrt{\ln N - \ln \eta} + \sqrt{-\ln \eta}}{\sqrt{-\ln \eta}} \right)^{\frac{2}{3}}, \tag{36}$$

$$n_L^* = 1 + \left(\frac{2\sqrt{\ln N - \ln \eta}}{\sqrt{-\ln \eta}} \right)^{\frac{2}{3}}. \tag{37}$$

Note that these values *do not depend* on the sample size I.

2.2   Regression Estimation Learning Task

Now we consider the set of *real-valued* error functions but we still stay with the simplest case when the set has a *finite* number of elements. We give theorems for the *regression estimation* learning task, analogous to the ones for the classification. We skip proofs — the only changes they would require is the assumption of the *bounded* functions, and the use of Hoeffding inequality in the place of Chernoff inequality.

Theorem 4. *Let* $\{Q(\mathbf{z}, \omega_j)\}_{\omega_j \in \Omega}$, $j = 1, 2, \ldots, N$, *be a finite set of real-valued bounded functions (regression estimation task) of size* N, $0 \le Q(\mathbf{z}, \omega_j) \le B$. *Then, for any* $\eta > 0$, *arbitrarily small, there is a small number*

$$\alpha(\eta, n) = \eta - \sum_{k=1}^{n} \binom{n}{k} (-1)^k (2\eta)^k, \tag{38}$$

and the number

$$\varepsilon(\eta, I, N, n) = \left(2\sqrt{\frac{n}{n-1}} + 1 \right) B \sqrt{\frac{\ln N - \ln \eta}{2I}}$$

$$+ \left(\sqrt{n} + \sqrt{\frac{n}{n-1}} \right) B \sqrt{\frac{-\ln \eta}{2I}}, \tag{39}$$

such that:

$$P\left(|V - C| \le \varepsilon(\eta, I, N, n) \right) \ge 1 - \alpha(\eta, n). \tag{40}$$

Theorem 5. *With probability* $1 - \alpha(\eta, n)$ *or greater, the following inequality holds true:*

$$C - V \leq \left(\sqrt{\frac{n}{n-1}} - 1 \right) B \sqrt{\frac{\ln N - \ln \eta}{2I}}$$

$$+ \left(\sqrt{n} + \sqrt{\frac{n}{n-1}} \right) B \sqrt{\frac{-\ln \eta}{2I}}. \quad (41)$$

Theorem 6. *With probability* $1 - \alpha(\eta, n)$ *or greater, the following inequality holds true:*

$$V - C \leq \left(2\sqrt{\frac{n}{n-1}} + 1 \right) B \sqrt{\frac{\ln N - \ln \eta}{2I}} + B\sqrt{n} \sqrt{\frac{-\ln \eta}{2I}}. \quad (42)$$

3   The Relationship for an Infinite Set of Approximating Functions

The simplest case with a finite number of functions in the set has been generalized by Vapnik [2,19,25] onto *infinite* sets with continuum of elements by introducing several notions of the *capacity* of the set of functions: *entropy, annealed entropy, growth function, Vapnik–Chervonenkis dimension.* We remind them in brief.

First of all, Vapnik defines $N^{\Omega}(z_1, \ldots, z_I)$ which is the number of all possible *dichotmies* that can be achieved on a fixed sample $\{z_1, \ldots, z_I\}$ using functions from $\{Q(z, \omega)\}_{\omega \in \Omega}$. Then, if we relax the sample the following notions of *capacity* can be considered:

1. expected value of $\ln N^{\Omega}$ — *Vapnik-Chervonenkis entropy:*

$$H^{\Omega}(I) = \int_{z_1 \in Z} \cdots \int_{z_I \in Z} \ln N^{\Omega}(z_1, \ldots, z_I)$$

$$\cdot p(z_1) \cdots p(z_I) dz_1 \cdots dz_I,$$

2. ln of expected value of N^{Ω} — *annealed entropy:*

$$H_{\text{ann}}^{\Omega}(I) = \ln \int_{z_1 \in Z} \cdots \int_{z_I \in Z} N^{\Omega}(z_1, \ldots, z_I)$$

$$\cdot p(z_1) \cdots p(z_I) dz_1 \cdots dz_I,$$

3. ln of supremum of N^{Ω} — *growth function*

$$G^{\Omega}(I) = \ln \sup_{z_1, \ldots, z_I} N^{\Omega}(z_1, \ldots, z_I).$$

It has been proved that:

$$G^{\Omega}(I) = \begin{cases} = \ln 2^I, & \text{dla } I \leq h; \\ \leq \ln \sum_{k=0}^{h} \binom{I}{k}, & \text{dla } I > h, \end{cases} \quad (43)$$

where h is the *Vapnik–Chervonenkis dimension*.

It has been shown [2] that

$$H^{\Omega}(I) \overset{\text{(Jensen)}}{\leq} H^{\Omega}_{\text{ann}}(I) \leq G^{\Omega}(I) \leq \ln \sum_{k=0}^{h} \binom{I}{k}$$

$$\leq \ln\left(\frac{eI}{h}\right)^{h} = h(1 + \ln\frac{I}{h}). \quad (44)$$

And the right-hand-side of (44) can be suitably inserted in the bounds to replace $\ln N$.

We mention that appropriate generalizations from the set of indicator functions (classification) onto sets of real-valued functions (regression estimation) can be found in [2] and are based on the notions of: ε-*finite net*, *set of classifiers for a fixed real-valued* f, *complete set of classifiers for* Ω.

3.1 Classification Learning Task (Infinite Set of Functions)

For shortness, we give only two theorems for bounds on $V - C$ and $C - V$, the bound on $|V - C|$ is their straightforward consequence (analogically as in previous sections).

Theorem 7. *Let* $\{Q(\mathbf{z},\omega)\}_{\omega\in\Omega}$ *be an infinite set of indicator functions with finite Vapnik–Chervonenkis dimension h. Then, with probability $1 - \alpha(\eta,n)$ or greater, the following inequality holds true:*

$$C - V \leq \left(\sqrt{\frac{n}{n-1}} - 1\right)\sqrt{\frac{h(1 + \frac{2I}{h}) - \ln\frac{\eta}{4}}{I}}$$

$$+ \left(\sqrt{n} + \sqrt{\frac{n}{n-1}}\right)\sqrt{\frac{-\ln\eta}{2I}}. \quad (45)$$

Theorem 8. *With probability $1 - \alpha(\eta,n)$ or greater, the following inequality holds true:*

$$V - C \leq \left(2\sqrt{\frac{n}{n-1}} + 1\right)\sqrt{\frac{h(1 + \frac{2I}{h}) - \ln\frac{\eta}{4}}{I}}$$

$$+ \sqrt{n}\sqrt{\frac{-\ln\eta}{2I}}. \quad (46)$$

3.2 Regression Estimation Learning Task (Infinite Set of Functions)

Again, for shortness, we give only two theorems for bounds on $V - C$ and $C - V$, the bound on $|V - C|$ is their straightforward consequence (analogically as in previous sections).

Theorem 9. *Let* $\{Q(\mathbf{z},\omega)\}_{\omega\in\Omega}$ *be an infinite set of real-valued bounded functions,* $0 \leq Q(\omega,\mathbf{z}) \leq B$, *with finite Vapnik–Chervonenkis dimension* h. *Then, with probability* $1 - \alpha(\eta,n)$ *or greater, the following inequality holds true:*

$$C - V \leq \left(\sqrt{\frac{n}{n-1}} - 1\right)B\sqrt{\frac{h(1+\frac{2I}{h}) - \ln\frac{\eta}{4}}{I}}$$

$$+ \left(\sqrt{n} + \sqrt{\frac{n}{n-1}}\right)\sqrt{\frac{-\ln\eta}{2I}}. \quad (47)$$

Theorem 10. *With probability* $1 - \alpha(\eta,n)$ *or greater, the following inequality holds true:*

$$V - C \leq \left(2\sqrt{\frac{n}{n-1}} + 1\right)B\sqrt{\frac{h(1+\frac{2I}{h}) - \ln\frac{\eta}{4}}{I}}$$

$$+ \sqrt{n}\sqrt{\frac{-\ln\eta}{2I}}. \quad (48)$$

In practice, bounds (47) and (48) can be significantly tightened by using an estimate $\widehat{B}$ in the place of the most pessimistic B. The estimate $\widehat{B}$ can be found by performing just one fold of cross-validation (instead of n folds) and bounding $\widehat{B}$ by: mean error on the testing set plus a square root implied by the Chernoff inequality:

$$\widehat{B} \leq R''_{\text{emp}}(\omega'_I) + B\sqrt{\frac{-\ln\eta_B}{2I''}}, \quad (49)$$

where η_B is an imposed small probability that (49) is not true. The reasoning behind this remark is that in practice, typical learning algorithms rarely produce functions $f(\mathbf{x},\omega_I)$, in the process of ERM, having high maximal errors. Therefore, we can insert the right-hand-side of (49) into (47) and (48) in the place of B. If this is done, then the minimal overall probability on bounds (47) and (48) should be adjusted to $1 - \alpha(\eta,n) - \eta_B$.

4 Experiments — Bounds Checks

Results of three experiments are shown in this section, for the following cases: (1) binary classification, finite set of functions, (2) binary classification, infinite set of functions, (3) regression estimation, infinite set of functions.

4.1 Set of Functions

The form of f functions, $f: [0,1]^2 \to [-1,1]$, was Gaussian-like:

$$f(\mathbf{x},\underbrace{w0,w_1,\ldots,w_K}_{\omega}) =$$

$$\max\left\{-1,\min\left\{1,w_0 + \sum_{k=1}^{K}w_k\exp\left(-\frac{\|\mathbf{x}-\mu_k\|^2}{2\sigma_k^2}\right)\right\}\right\} \quad (50)$$

where centers μ_k and widths σ_k were generated on random[11] and remained *fixed*. Therefore we have a set of functions linear in parameters $(w_0, w_1, \ldots, w_K)$. As one can see values of f where constrained by ± 1. For the classification learning task, the decision boundary was arising as the solution of $f(\mathbf{x}, w_0, w_1, \ldots, w_K) = 0$. For the regression estimation, we simply looked at the values of $f(\mathbf{x}, w_0, w_1, \ldots, w_K)$. Examples of functions from this set are shown in figures 2, 3

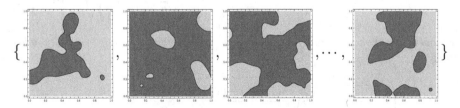

Fig. 2. Illustration of the set of functions for classification

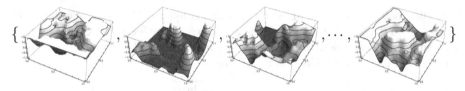

Fig. 3. Illustration of the set of functions for regression estimation

4.2   System and Data Sets

As a system $y(\mathbf{x})$ we picked on random a function from a similar class to (50) but *broader*, in the sense that the number K was greater and the range of randomness on σ_k was larger. Data sets for both classification and regression estimation were taken by sampling the system according to the joint probability density $p(\mathbf{x}, y) = p(\mathbf{x})p(y|\mathbf{x})$ where we set $p(\mathbf{x}) = 1$ — uniform distribution on the domain $[0,1]^2$ and $p(y|\mathbf{x}) = \frac{1}{\sqrt{2\pi}\sigma} \exp(-\frac{(y-y(\mathbf{x}))^2}{2\sigma^2})$ — normal noise with $\sigma = 0.1$.

4.3   Algorithm of the Learning Machine

In the case of *finite* sets of N functions, the learning machine was simply choosing the best functions as $f(\omega_l) = \arg\min_{j=1,2,\ldots,N} R_{\mathrm{emp}}(\omega_j)$ or in cross-validation folds $f(\omega_{l'}) = \arg\min_{j=1,2,\ldots,N} R'_{\mathrm{emp}}(\omega_j)$.

---

[11] Random intervals: $\mu_k \in [0,1]^2$, $\sigma_k \in [0.02, 0.1]$.

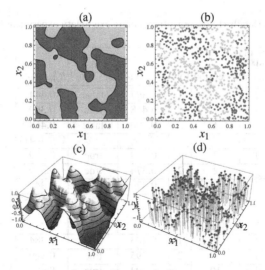

Fig. 4. System and data for classification (a, b), regression estimation (c, d)

In the case of *infinite* sets with continuum of elements, the learning machine was trained by the least-squares criterion. We remark that obviously other learning approaches can be used in this place e.g. maximum likelihood, SVM criterion [2,1,26]. If we denote the bases $\exp\left(-\frac{\|\mathbf{x}-\mu_k\|^2}{2\sigma_k^2}\right)$ by $g_k(\mathbf{x})$ and calculate the matrix of bases at data points

$$G = \begin{pmatrix} 1 & g_1(\mathbf{x}_1) & g_2(\mathbf{x}_1) & \cdots & g_K(\mathbf{x}_1) \\ 1 & g_1(\mathbf{x}_2) & g_2(\mathbf{x}_2) & \cdots & g_K(\mathbf{x}_2) \\ \vdots & \vdots & \vdots & \ddots & \vdots \\ 1 & g_1(\mathbf{x}_I) & g_2(\mathbf{x}_I) & \cdots & g_K(\mathbf{x}_I) \end{pmatrix} \tag{51}$$

we can find the optimal vector of w coefficients by the pseudo-inverse operation as follows:

$$(w_0, w_1, \ldots, w_K)^T = (G^T G)^{-1} G^T Y, \tag{52}$$

where $Y = (y_1, y_2, \ldots, y_I)^T$ is a vector of training target values.

4.4   Experiment Results and Comments

Experiments involved trying out different settings on all relevant constants such as: number of terms in approximating functions (K), number of functions (N) in the case of finite sets or VC dimension (h) in case of infinite sets, sample size (I), number of cross-validation folds (n). For each fixed setting of the constants, an experiment with repetitions was performed, during which we measured the cross-validation outcome C after each repetition. The range of these outcomes was then compared to the interval implied by the theorems we proved.

Table 1. Details (folds, repetitions) of an exemplary experiment no. 1

no. of experiment	repetition	fold	$R'_{emp}(\omega_{I'})$	is $\omega_{I'} = \omega_I$?	$R''_{emp}(\omega_{I''})$
1	1	1	0.397	false	0.444
1	1	2	0.418	true	0.369
1	1	3	0.400	false	0.468
					$C = 0.417$
1	2	1	0.359	true	0.369
1	2	2	0.374	true	0.339
1	2	3	0.370	true	0.348
					$C = 0.352$
$\vdots$	$\vdots$	$\vdots$	$\vdots$	$\vdots$	$\vdots$
1	10	1	0.403	true	0.384
1	10	2	0.395	true	0.399
1	10	3	0.394	true	0.399
					$C = 0.394$

We show the results in two tables 1 and 2. The first one gives an insight on details of a *single* exemplary experiment: results of its particular folds and repetitions. The second one shows collective results, where each row encapsulates 10 repetitions[12].

To comment on the results we first remark that before each single experiment (1-12) the whole data set was drawn once from $p(\mathbf{z})$ and remained fixed throughout repetitions. However, in the repetitions due to the non-stratified cross-validation we parted the data set (via permutations) into different training and testing subsets. That is why in the table $R_{emp}(\omega_I)$ and V are constant per experiment, whereas the cross-validation varies within some observed range. In the table 2 we also present the interval $[V - \varepsilon_L, V + \varepsilon_U]$ which is implied by the theorems.

Please note that for *all* experiments the observed range for C was contained inside $[V - \varepsilon_L, V + \varepsilon_U]$ — an empirical confirmation of theoretical results. Although the bounds are true with probability at least $1 - \alpha(\eta, n)$, in this particular experiment they held with frequency one.

In particular one can note in the table that the upper bounds $V + \varepsilon_U$ are closer to actual C outcomes, while lower bounds $V + \varepsilon_L$ are more loose — a fact we already indicated in theoretical sections. Only in the case of experiment no. 9 the lower bound we obtained was trivial. In the results one can also observe the qualitative fact that both intervals tighten with $1/\sqrt{l}$ approximately. Keep in mind that this result stops working for the 'leave-one-out' cross-validation (or a close one) and we experimented on $n = 3$ and $n = 5$.

---

[12] It was difficult to allow ourselves for more repetitions, say 100, due to large amount of results and the time-consumption of each experiment. Yet, the observed ratio 1.0 of C falling inside bounds shows that 10 repetitions was sufficient.

Table 2. Collective results — each row encapsulates 10 repetitions. Tasks: $c.$ — classification, $r.e.$ — regression estimation. We denote experiments on finite or infinite sets of functions by setting either N or h. For regression estimation we use probabilistic $\widehat{B}$ calculated as $R''_{emp}(\omega'_I) + B\sqrt{-\ln\eta_B/(2I'')}$. In all experiments $\eta = 0.2$, hence for $n = 3$ the probability that bounds are true is $1 - \alpha(\eta, n) = 0.496$ or greater and for $n = 5$ it is $1 - \alpha(\eta, n) = 0.511$ or greater.

no. of exp.	task	K	N	h	I	$R_{emp}(\omega_I)$	V	n	bounds $[V-\varepsilon_L, V+\varepsilon_U]$	observed range of C (10 repetitions)	ratio of C inside bounds
1	c.	50	10	-	10^3	0.412	0.456	3	$[0.254, 0.550]$	$[0.351, 0.445]$	1.0
2	c.	200	10	-	10^3	0.345	0.389	3	$[0.187, 0.483]$	$[0.352, 0.385]$	1.0
3	c.	200	10	-	10^4	0.369	0.383	3	$[0.319, 0.413]$	$[0.371, 0.383]$	1.0
4	c.	200	10	-	10^4	0.396	0.410	5	$[0.344, 0.442]$	$[0.386, 0.401]$	1.0
5	c.	50	100	-	10^4	0.408	0.426	3	$[0.349, 0.456]$	$[0.392, 0.418]$	1.0
6	c.	200	100	-	10^4	0.336	0.354	3	$[0.277, 0.384]$	$[0.332, 0.338]$	1.0
7	c.	50	100	-	10^5	0.401	0.407	3	$[0.383, 0.417]$	$[0.398, 0.403]$	1.0
8	c.	50	-	51	10^5	0.181	0.250	3	$[0.021, 0.267]$	$[0.181, 0.184]$	1.0
9	c.	200	-	201	10^5	0.035	0.161	3	$[-0.25, 0.185]$	$[0.035, 0.037]$	1.0
9	r.e.	50	-	51 ($\widehat{B}=0.193$)	10^4	0.172	0.209	3	$[0.078, 0.223]$	$[0.170, 0.173]$	1.0
10	r.e.	50	-	51 ($\widehat{B}=0.194$)	10^4	0.171	0.208	5	$[0.085, 0.212]$	$[0.170, 0.172]$	1.0
11	r.e.	200	-	201 ($\widehat{B}=0.020$)	10^5	0.012	0.015	3	$[0.006, 0.016]$	$[0.012, 0.013]$	1.0
12	r.e.	200	-	201 ($\widehat{B}=0.020$)	10^5	0.013	0.015	5	$[0.007, 0.016]$	$[0.012, 0.013]$	1.0

5   Experiments — SRM

In this section we show results of the *Structural Risk Minimization* approach. We consider a *structure* i.e. a sequence of nested subsets of functions: $S_1 \subset S_2 \subset \cdots \subset S_K$, where each successive $S_k = \{f(\mathbf{x}, \omega)\}_{\omega \in \Omega_k}$ is a set of functions with Vapnik-Chervonenkis dimension h_k, and we have $h_1 < h_2 < \cdots < h_K$. As the best element of the structure we choose S^* (with VC dimension h^*) for which the bound on generalization V is the smallest.

Along with observing the bound V, we observe: (1) the cross-validation result C, (2) our bounds on C, (3) the actual *true risk* R calculated as an integral according to its definition (1). We pay particular attention to how the minimum point of SRM at h^* differs from the minimum suggested by the cross-validation and the minimum of true risk (which normally in practice is unknown). We remind that obtaining the result C for each h_k is $O(n)$ times more laborious than obtaining V for each h_k. See fig. 5.

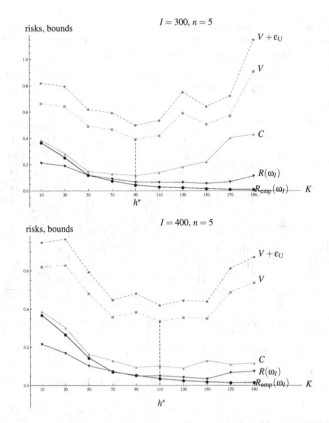

Fig. 5. SRM experiments. With $I = 300$, optimum points reached at: $h^* = 91$ (SRM), $h = 91$ (C), $h = 151$ (true risk R). With $I = 400$, optimum points reached at: $h^* = 111$ (SRM), $h = 131$ (C), $h = 151$ (true risk R).

6 Summary

In the paper we take under consideration the probabilistic relationship between two quantities: Vapnik generalization bound V and the result C of an n-fold non-stratified cross-validation. In the literature on the subject of machine learning (and SLT) typically the stated results have a different focus — namely, the relation between the *true risk* (generalization error) and either of the two quantities V, C separately. The perspective we chose was intended to:

- stay in the setting of Structural Risk Minimization approach based on Vapnik bounds,
- *not perform* the cross-validation procedure,
- be able to make probabilistic statements about closeness of SRM results to cross-validation results (if such was perfomed) for given conditions of learning experiment.

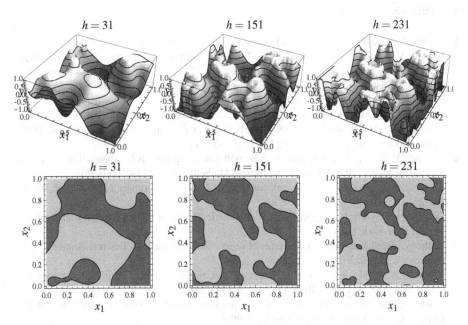

Fig. 6. Exemplary models for both regresssion estimation and classfication: under complex ($h = 31$), accurately complex — the best generalization ($h = 151$), over complex ($h = 231$).

Suitable theorems about this relationship are stated and proved. The theorems concern two learning tasks: classification and regression estimation; and also two cases as regards the capacity of the set of approximating functions: finite sets and infinite sets (but with finite Vapnik-Chervonenkis dimension).

As the sample size grows large, both C and V *converge in probability* to the same limit of true risk. The rate of convergence is exponential.

Using the theorems, one can find a threshold size of sample so that the difference $C - V$ or $V - C$ is smaller than an imposed ε. Obviously, the smaller ε for given experiment conditions, the more frequently one can expect to select the same optimal model complexity via SRM and via cross-validation (again without actually performing it).

For the special case of leave-one-out cross-validation we observe in the consequence of bounds we derived that at most a constant difference of order $O(\sqrt{-\ln\eta/2})$ between C and V can be expected.

Additionally, we showed for what number n of folds, the bounds (lower and upper) on the difference are the tightest. Interestingly, as it turns out these optimal n values *do not* depend on the sample size.

Finally, shown are experiments confirming statistical correctness of the bounds.

Acknowledgements. This work has been financed by the Polish Government, Ministry of Science and Higher Education from the sources for science within years 2010–2012. Research project no.: N N516 424938.

References

1. Vapnik, V.: The Nature of Statistical Learning Theory. Springer, New York (1995)
2. Vapnik, V.: Statistical Learning Theory: Inference from Small Samples. Wiley, New York (1995)
3. Vapnik, V.: Estimation of Dependences Based on Empirical Data. Information Science & Statistics. Springer, US (2006)
4. Cherkassky, V., Mulier, F.: Learning from data. John Wiley & Sons, Inc. (1998)
5. Hellman, M., Raviv, J.: Probability of error, equivocation and the chernoff bound. IEEE Transactions on Information Theory IT-16, 368–372 (1970)
6. Schmidt, J., Siegel, A., Srinivasan, A.: Chernoff-hoeffding bounds for applications with limited independence. SIAM Journal on Discrete Mathematics 8, 223–250 (1995)
7. Shawe-Taylor, J., et al.: A framework for structural risk minimization. In: COLT, pp. 68–76 (1996)
8. Devroye, L., Gyorfi, L., Lugosi, G.: A Probabilistic Theory of Pattern Recognition. Springer Verlag, New York, Inc. (1996)
9. Anthony, M., Shawe-Taylor, J.: A result of vapnik with applications. Discrete Applied Mathematics 47, 207–217 (1993)
10. Krzyżak, A., et al.: Application of structural risk minimization to multivariate smoothing spline regression estimates. Bernoulli 8, 475–489 (2000)
11. Holden, S.: Cross-validation and the pac learning model. Technical Report RN/96/64, Dept. of CS, University College, London (1996)
12. Holden, S.: Pac-like upper bounds for the sample complexity of leave-one-out cross-validation. In: 9th Annual ACM Workshop on Computational Learning Theory, pp. 41–50 (1996)
13. Kearns, M., Ron, D.: Algorithmic stability and sanity-check bounds for leave-one-out cross-validation. Neural Computation 11, 1427–1453 (1999)
14. Kearns, M.: A bound on the error of cross-validation, with consequences for the training-test split. In: Advances in Neural Information Processing Systems, vol. 8. MIT Press (1995)
15. Kearns, M.: An experimental and theoretical comparison of model selection methods. In: 8th Annual ACM Workshop on Computational Learning Theory, pp. 21–30 (1995)
16. Bartlett, P., Kulkarni, S., Posner, S.: Covering numbers for real-valued function classes. IEEE Transactions on Information Theory 47, 1721–1724 (1997)
17. Bartlett, P.: The sample complexity of pattern classification with neural networks: the size of weights is more important then the size of the network. IEEE Transactions on Information Theory 44 (1997)
18. Ng, A.: Feature selection, l1 vs. l2 regularization, and rotational invariance. In: 21st ACM International Conference on Machine Learning. Proceeding Series, vol. 69 (2004)
19. Vapnik, V., Chervonenkis, A.: The necessary and sufficient conditions for the consistency of the method of empirical risk minimization. Yearbook of the Academy of Sciences of the USSR on Recognition, Classification and Forecasting 2, 217–249 (1989)
20. Kohavi, R.: A study of cross-validation and boostrap for accuracy estimation and model selection. In: International Joint Conference on Artificial Intelligence, IJCAI (1995)
21. Efron, B., Tibshirani, R.: An Introduction to the Bootstrap. Chapman & Hall, London (1993)
22. Hjorth, J.: Computer Intensive Statistical Methods Validation, Model Selection, and Bootstrap. Chapman & Hall, London (1994)

23. Weiss, S., Kulikowski, C.: Computer Systems That Learn. Morgan Kaufmann (1991)
24. Fu, W., Caroll, R., Wang, S.: Estimating misclassification error with small samples via bootstrap cross-validation. Bioinformatics 21, 1979–1986 (2005)
25. Vapnik, V., Chervonenkis, A.: On the uniform convergence of relative frequencies of events to their probabilities. Dokl. Aka. Nauk. 181 (1968)
26. Korzeń, M., Klęsk, P.: Maximal Margin Estimation with Perceptron-like Algorithm. In: Rutkowski, L., Tadeusiewicz, R., Zadeh, L.A., Zurada, J.M. (eds.) ICAISC 2008. LNCS (LNAI), vol. 5097, pp. 597–608. Springer, Heidelberg (2008)

A Time-Varying Model to Simulate a Collective Decisional Problem

A. Imoussaten[1], J. Montmain[1], A. Rico[2], and F. Rico[2]

[1] LGI2P, Ecole des mines d'Alès, Site EERIE Parc Scientifique G. Bresse
30035 Nîmes cedex, France
[2] ERIC, Université Claude Bernard Lyon 1, 43 bld Du 11 novembre
69100 Villeurbanne, France
{abdelhak.imoussaten,jacky.montmain}@mines-ales.fr,
{agnes.rico,fabien.rico}@univ-lyon1.fr

Abstract. A group of agents have to decide between two alternatives. At the begining each actor have a preference and a conviction about this preference. During the debate the conviction and the preference of each agent can change under the involved social influences.

The theoretical model of this paper is presented in [7], and the influence model is the one presented in [3]. The aims is to present a new dynamical model for collective decisional problems. Moreover the presented model offers additional perspectives for purposes of controlling the debate.

Keywords. Debate, Influence, Decisional power, Choquet integral, Control, Collective decision, Social network.

1 Introduction

A group of agents is faced with a collective decision; in response, a debate has been organized to identify which alternative appears to be the most relevant following a deliberation. This study will be limited to the binary, albeit common, situation involving two options denoted $+1$ and -1. It is assumed that each agent has an inclination to choose one of the alternatives $+1$ and -1, though due to the influence of other agents this inclination may differ from the agent's actual decision [3]. In general terms, it can be considered that each time a speaker intervenes in the debate, agents may change their preference due to social influences taking place within the group. Once agents' preferences reach a point of no longer changing, then the deliberation process ends and a group decision is made. The aim of this debate is for every agent to hear the arguments of all other agents by the end of the deliberation process and then to make a final decision based on full knowledge of the facts.

This deliberation is viewed as a dynamic process with its own dynamics and where agents' beliefs and preferences evolve as arguments are exchanged. The deliberation outcome thus depends both on the order in which agents intervene in the debate to explain their opinion and on the influence a given agent may exert on a social network.

In this context, social influence is related to the statistical notion of the decisional power held by an individual within a social network, as proposed in [5] and [3].

J. Filipe and A. Fred (Eds.): ICAART 2011, CCIS 271, pp. 56–71, 2013.

One of the conclusions drawn in [3] concerns the integration of dynamic aspects into the influence model. The authors' framework is indeed a decision-making process activated after a single step of mutual influence. In reality, such mutual influence does not necessarily stop after just one step but may actually become iterative. This paper proposes a possible extension of the results presented in [3] for use in the dynamic case. The evolution of agents' beliefs throughout the debate either change or reinforce the agents' convictions relative to their initial preference. Intuitively, as well as from other standpoints, an agent's social influence depends on the relative strength of other agents' convictions. The idea for our model therefore is to define influence as a time-dependent variable.

In [7], the concepts of influence and conviction during the simulation of a debate are introduced. This article will follow up on the prior work proposed in [1]. Our main improvement here over previous efforts relates to the fact that in [7], coalitions of agents were modeled using capacities, and the change in conviction during a debate was computed with a symmetric Choquet integral, which is in fact an aggregation function typically introduced in multicriteria decision making [2]. The main drawback in [7] pertains to a lack of semantic justifications.

Reference [3] provides a formal framework to define the notion of influence, while [7] introduces the revision equations relative to agents' convictions and preferences. Moreover, [6] suggests a cybernetic interpretation to merge both of these models. The present paper is intended as a continuation of [7], with [6] also used for guidance. The main contribution of this paper is to propose the state equations of the cybernetic interpretation in order to describe the way agents' convictions may evolve over time. To achieve this goal, a capacity will be introduced to model the relative importance of agents in the debate; such a capacity is based on the decisional power of agents using the generalized Hoede-Bakker index [3,5]. Consequently, a number of simulations will be proposed to illustrate the collective decision-making process.

The paper will be organized as follows. Section 2 will briefly recall the main concepts of the models presented in [7] and [3]. From this formal framework, Section 3 will establish the state equations that serve to model the dynamic relationships between convictions and influences when a speaker-agent / listener-agent pair is isolated. Following a presentation of preference changes, Section 4 will offer a few illustrations. Lastly, Section 5 will provide the conclusion and outlook for future research.

2 Concepts and Notations

2.1 Notion of Influence in a Debate

The assumption behind our model is that an agent's influence is correlated with his capacity to alter the group decision. It addresses the concept of the "weight" of an agent's choice in a collective voting procedure. This "weight" parameter cannot be static since it needs to evolve with agent preferences, which in turn allow for the formation of certain coalitions that are more likely than others.

It is common experience that during the discussion phase, some agents will change their initial opinions. The reasons for this change, in assuming that it is not a random occurrence, may be of different types. The most natural reason is that they have been swayed by the arguments of a particular agent or group (coalition) of agents, or that they feel somewhat obliged, owing to a hierarchical, political or perhaps even more obscure reason, to follow the opinion of that particular agent or coalition. Another reason may be that they are acting in reaction against a given agent or coalition, by systematically embracing the opposite opinion. We use the generic word "influence" herein to refer to all these types of phenomena [4].

Models have been introduced into game theory in order to represent influence in social networks. The point of departure is the concept of the Hoede-Bakker index, a notion that computes the overall decisional power of an agent within a social network, which in this case is a group of n agents. This index was developed in 1982 [5]; an extended definition of decisional power was proposed in [3] and will now be summarized. The reasons behind the existence of influence phenomena, i.e. why a given individual finally changes his decision, is more a matter of the psychological sciences and lies beyond the scope of such approaches.

Let's start by considering a set of agents $\{a_1, \cdots, a_n\}$, denoted $N = \{1, \cdots, n\}$ in order to simplify notations along with a power set denoted $2^{\{a_1, \cdots, a_n\}}$. Each agent is inclined to choose either $+1$ or -1. An inclination vector, denoted i, is an n-vector consisting of $+1$ and -1. The j-th coordinate of i is thus denoted $i_{a_j} \in \{-1, +1\}$ and represents the inclination of agent a_j. Let $I = \{-1, +1\}^n$ be the set of all inclination vectors.

It can then be assumed that agents influence one another; moreover, due to influences arising in the network, the final decision of an agent may differ from his original inclination. In other words, each inclination vector $i \in I$ is transformed into a decision vector $B(i)$, where $B : I \to I, i \mapsto B(i)$ is the influence function. The coordinates of $B(i)$ are expressed by $(Bi)_{a_j}, j \in \{1, \cdots, n\}$ and $(Bi)_{a_j}$ is the decision of agent a_j. Lastly, $gd : B(I) \to \{-1, +1\}$ is a group decision function, assigned the value $+1$ if the group decision is $+1$ and the value -1 for a group decision of -1.

An influence function B may correspond to a common collective behavior. For example, in [3] a majority influence function $Maj^{[t]}$ parametrized by a real t has been introduced. More precisely, for a given $i \in I$,

$$Maj^{[t]}i = \begin{cases} 1_N \text{ if } |i^+| \geq t \\ -1_N \text{ if } |i^+| < t \end{cases}$$

where $i^+ = \{k \in N | i_k = +1\}$ and 1_N (resp. -1_N) is the vector equal to 1 (resp. -1) everywhere.

This set-up corresponds to the intuitive collective human behavior: when a majority of players have an inclination of $+1$, then all players decide $+1$. Many classifications of potential collective behavior (polarization, groupthink, mass psychology, etc.) can thus be described mathematically.

An influence function may also be defined as a simple rule. For example, the following rule may be associated with the $Guru$ function: "when a_{Guru} thinks $+1$, then all agents decide $+1$". Another example would be the opportunistic behavior, i.e.: "when most of my supervisors decide $+1$, then I decide $+1$".

It can also be anticipated that mapping $B : I \to I$ is learned from experiment. The identification of B may be perceived as a data-mining step using knowledge bases in which collective decisions have been recorded as minutes of company meetings.

Definition 1. *The Hoede-Bakker index of agent a_j is defined for a given B and a given gd by:*

$$GHB_{a_j} = \frac{1}{2^{n-1}} \sum_{\{i | i_{a_j} = +1\}} gd(B(i)).$$

The main drawback with the Hoede-Bakker index is that it blurs the actual role of the influence function, by analyzing the final decision only in terms of success and failure. The decision is successful for an agent once his inclination matches the group decision.

In [3], the authors distinguish the influence component from the group decision component and moreover propose a first modified index of decisional power, whereby the agent's decision must coincide with the group's decision to constitute a success for the agent. Lastly, these authors provided a second modified decisional power, allowing the inclination vectors to be assigned unequal probabilities.

Definition 2. *Let $p : I \rightarrow [0,1]$ be a probability distribution, with $p(i)$ being the probability of an i occurrence. The modified decisional power of agent a_j for given B, gd and p can then be expressed as:*

$$\phi_{a_j}(B, gd, p) = \sum_{\{i | (Bi)_{a_j} = +1\}} p(i).gd(B(i)) - \sum_{\{i | (Bi)_{a_j} = -1\}} p(i).gd(B(i)).$$

To conclude this summary section, for each agent a_j the probabilities of success and failure are recalled as follows:

$$SUC_{a_j}(B, gd, p) = \sum_{\{b \in I | (b)_{a_j} = gd(b)\}} p \circ B^{-1}(b)$$
$$FAIL_{a_j}(B, gd, p) = \sum_{\{b \in I | (b)_{a_j} = -gd(b)\}} p \circ B^{-1}(b).$$

Note that: $\phi_{a_j}(B, gd, p) = SUC_{a_j}(B, gd, p) - FAIL_{a_j}(B, gd, p)$.

2.2 Convictions and Preferences During a Debate

This section will present the dynamic model of the debate proposed in [7]. The influence an agent may exert on the others in the debate is modeled by a capacity over $2^{\{a_1, \cdots, a_n\}}$.

Definition 3. *A capacity v over $2^{\{a_1, \cdots, a_n\}}$ is a set function $v : 2^{\{a_1, \cdots, a_n\}} \rightarrow [0,1]$ such that $v(\emptyset) = 0$, $v(\{a_1, \cdots, a_n\}) = 1$ and $\forall A, A' \subseteq \{a_1, \cdots, a_n\}, A \subseteq A' \Rightarrow v(A) \leq v(A')$.*

The profile of an agent a_j includes his preference, importance (i.e. his capacity $v(a_j)$) and preference intensity named conviction in the following (it is to be denoted $c_{a_j} \in [0,1]$).

It is an agreed rule of the debate that agents are to speak in turns. In the proposed model set-up, the agent a_s (speaker-agent) who is currently speaking and any agent a_l (listener-agent) who is listening are formally isolated from the remainder of the group. More precisely, a capacity v_{a_l, a_s}, defined relative to the pair of agents (a_l, a_s), is introduced as follows:

$$v_{a_l,a_s}(a_l) = \frac{v(a_l)}{v(\{a_l,a_s\})}, \ v_{a_l,a_s}(a_s) = \frac{v(a_s)}{v(\{a_l,a_s\})} \text{ and } v_{a_l,a_s}(\{a_l,a_s\}) = 1.$$

The change of conviction can then be modeled using the symmetric Choquet integral, which is also called the Sipos integral. The definition of the Choquet and Sipos integrals will now be provided.

Definition 4. *Let* $c = (c_{a_1}, \ldots, c_{a_n}) \in [0,1]^n$ *be a vector of convictions,* $()$ *be a permutation on* $\{1, \ldots, n\}$ *such that* $c_{a_{(1)}} \leq \ldots \leq c_{a_{(n)}}$ *and* v *a capacity on* $2^{\{a_1,\cdots,a_n\}}$.
The Choquet integral of c *with respect to* v *is expressed as:*

$$C_v(c) = \sum_{i=1}^{n} \left[c_{a_{(i)}} - c_{a_{(i-1)}} \right] v(\{(i), \cdots, (n)\}) \qquad \text{with } c_{a_{(0)}} = 0.$$

Definition 5. *Let* $c = (c_{a_1}, \cdots, c_{a_n}) \in [-1,1]^n$ *be a vector capable of assuming negative values,* $()$ *be the permutation on* $\{1, \cdots, n\}$ *such that* $c_{a_{(1)}} \leq c_{a_{(p)}} < 0 \leq c_{a_{(p+1)}} \leq \cdots \leq c_{a_{(n)}}$ *and* v *a capacity on* $2^{\{a_1,\cdots,a_n\}}$.
The symmetric Choquet Integral of c *with respect to* v *is given by:*

$$\check{C}_v(c) = \sum_{i=1}^{p-1} [c_{a_{(i)}} - c_{a_{(i+1)}}] v(\{(1), \cdots, (i)\}) + c_{a_{(p)}} v(\{(i), \cdots, (p)\})$$

$$+ c_{a_{(p+1)}} v(\{(p+1), \cdots, (n)\}) + \sum_{i=p+2}^{n} [c_{a_{(i)}} - c_{a_{(i-1)}}] v(\{(i), \cdots, (n)\}).$$

In [7] the Sipos integral is defined on the set of agents $\{a_l, a_s\}$ and denoted $\check{C}_{v_{a_l,a_s}}$. The changes of conviction proposed can then be summarized as follows:

If agents a_l and a_s have the same preference, then one of them is more convinced, and
this situation entails two possible cases.
   – If $c_{a_s} > c_{a_l}$ then the new conviction of agent a_l becomes:
   $\check{C}_{v_{a_l,a_s}}(c_{a_s}, c_{a_l}) = c_{a_l} + (c_{a_s} - c_{a_l}) v_{a_l,a_s}(a_s)$.
   – If $c_{a_l} > c_{a_s}$ then the new conviction of agent a_l becomes:
   $\check{C}_{v_{a_l,a_s}}(c_{a_s}, c_{a_l}) = c_{a_s} + (c_{a_l} - c_{a_s}) v_{a_l,a_s}(a_l)$.
If agents a_l and a_s have different preferences, then the new conviction of agent a_l is:
   – $\check{C}_{v_{a_l,a_s}}(c_{a_s}, c_{a_l}) = -c_{a_s} v_{a_l,a_s}(a_s) + c_{a_l} v_{a_l,a_s}(a_l)$.

The main drawback to this model is its lack of semantic justification with regard to capacity v (i.e. influence is merely a normalized relative importance); in addition, the concept of conviction has not been formally defined and the revision equations are not provided in an appropriate formalism, in which time would appear explicitly (i.e. dynamic aspects).

3   Presentation of Our Dynamic Model

This section presents our dynamic model for simulating a debate outcome. To begin, let's note that within the framework of this paper, the influence function used in [3] is perceived as a disturbance function applied to the set of all possible inclination vectors.

3.1 Decisional Power and Capacities

One of the new ideas presented in this paper is the ability to design a capacity based on the decisional power included in the above model.

For any inclination vector i in I, Bi is the decision vector obtained from i whose influence is modeled by B. $gd(Bi)$ is the final decision of the group, whereby the group decision function is modeled by gd. For any $i \in I$, $gd(Bi)$ belongs to $\{-1, +1\}$, which implies that the modified decisional power for any agent a_j as denoted $\phi_{a_j}(B, gd, p)$ lies in the interval $[-1, 1]$.

Note that if the decisional power of an agent is close to -1, this means that the agent only rarely chooses the alternative to what the collective body ultimately chooses: he fails most of the time ($FAIL$). On the other hand, when his decisional power is close to 1, the agent is most often successful ($SUCC$); his decisional power therefore is strong. Hence, for any agent a_j, we can normalize $\phi_{a_j}(B, gd, p)$ in order to obtain his importance.

As an example, without any further information, the importance of agent a_j, i.e. his capacity $v(a_j)$, can be defined as follows:

Definition 6. *The importance of agent a_j for a given B, gd and p is $v_\phi(a_j) = \frac{1}{2}\phi_{a_j}(B, gd, p) + \frac{1}{2}$.*

Note that for any agent a_j, $v(a_j) \in [0, 1]$ with $v(a_j) = 0$ if and only if $\phi_{a_j}(B, gd, p) = -1$ and $v(a_j) = 1$ if and only if $\phi_{a_j}(B, gd, p) = 1$.

A capacity v_ϕ can then be generated over $2^{\{a_1, \cdots, a_n\}}$, with constraints, $\forall A, A' \subseteq \{a_1, \cdots, a_n\}$, $A \subseteq A' \Rightarrow v(A) \leq v(A')$. Without any further knowledge, it may be stated: $v_\phi(A) = \max_{a_j \in A} v(a_j)$, $\forall A \subset \{a_1, \cdots, a_n\}$ and $v_\phi(\{a_1, \cdots, a_n\}) = 1$. This last condition is necessary because it is uncertain that an agent can be found whose capacity is equal to 1.

Let's conclude this section with the following remark. The decisional power of individuals a_j on which $v_\phi : 2^{\{a_1, \cdots, a_n\}} \to [0, 1]$ is based, measures those cases where the final decision of a_j matches the group decision. An agent with considerable decisional power is expected to sway several other agents; thus, decisional power is construed as an estimation of his influence within the group, although this is not an influence index in the sense of [3].

3.2 Time-Varying Probabilities

This subsection focuses on the design of probability p as a time-varying function, to be denoted $p[k]$ at time k. Along with this time-varying probability, a time-varying extended decisional power, as presented in [3], can be computed. The following method proposes basing the probability computation on the convictions of agents with respect to the available alternatives. In this part therefore, the conviction vectors are assumed to be known. $c(k)$ (resp. $c'(k)$) denotes the conviction vector of agents w.r.t. alternative $+1$ (resp. -1) at time k:

$$c(k) = (c_{a_1}(k), \cdots, c_{a_j}(k), \cdots, c_{a_n}(k)),$$ where $c_{a_j}(k)$ is the conviction of agent a_j w.r.t alternative $+1$ at time k.

$c'(k) = (c'_{a_1}(k), \cdots, c'_{a_j}(k), \cdots, c'_{a_n}(k))$, where $c'_{a_j}(k)$ is the conviction of agent a_j w.r.t alternative -1 at time k.

Their respective computations will be provided in the next section.

The conviction of an agent concerning a given alternative is correlated with the probability that this particular agent chooses this alternative, i.e. the probability of his inclination as defined in [3].

Let $i \in I$ be an inclination vector. Each coordinate i_{a_j} is the preference of agent a_j and constitutes one of the two alternatives.

Definition 7. *Let $i \in I$ be an inclination vector. The conviction vector of i at time k is $c(i, k) = (\bar{c}_{a_1}(k), \cdots, \bar{c}_{a_n}(k))$, where for any j, $\bar{c}_{a_j}(k)$ is $c_{a_j}(k)$ if $i_{a_j} = 1$ and is $c'_{a_j}(k)$ if $i_{a_j} = -1$.*

Let $i \in I$ be an inclination vector and let's define $c^i(k) \in [0, 1]$ as an average conviction at time k for i. This value summarizes the distributions of agents' convictions in i at time k. $c^i(k)$ is an "aggregated conviction" of the group of agents for i. This aggregation should take into account the relative importance of agents and their interactions. Consequently, it seems only natural to state the following definition.

Definition 8. *Let $i \in I$ be an inclination vector and $v[k]$ be a capacity defined at time k on $2^{\{a_1, \cdots, a_n\}}$, then $c^i(k+1) = C_{v[k]}(\bar{c}_{a_1}(k), \cdots, \bar{c}_{a_n}(k))$, where $C_{v[k]}$ is the Choquet integral with respect to $v[k]$.*

The time-varying probability is built by recurrence on k. We start at time $k = 0$ and will proceed by presenting how to compute $p[k+1]$ using $p[k]$.

At time $k = 0$:

Each agent assigns a score to each alternative in the interval $[0, 1]$. For each agent, if we were to denote n_{+1} (resp. n_{-1}) as the score of $+1$ (resp. -1), then the convictions could be computed by $c_{a_j}(0) = \frac{n_{+1}}{n_{+1}+n_{-1}}$ and $c'_{a_j}(0) = \frac{n_{-1}}{n_{+1}+n_{-1}}$. We then have $c_{a_j}(0) + c'_{a_j}(0) = 1$. Initially, at time $k = 0$, if i_{a_j} is the preference of a_j then the probabilities of the agent a_j regarding his preference and the other alternative would be: $p_{a_j}(i_{a_j})[k = 0] = c_{a_j}(0)$ and $p_{a_j}(-i_{a_j})[k = 0] = 1 - c_{a_j}(0)$. We assume that before the debate starts, the inclination of each agent does not depend on the social network. The probability distribution associated with a *priori* probabilities is thus the product of the individual probabilities p_{a_j} at $k = 0$, leading to the following probability:

$$\forall i \in I, p(i)[0] = \prod_{j=1}^{n} p_{a_j}(i_{a_j})[0].$$

It is thus possible to compute the following
- the decisional power for any agent a_j at $k = 0$: $\phi_{a_j}(B, gd, p[0])$;
- the capacity $v_\phi[0]$ over $2^{\{a_1, \cdots, a_n\}}$, for $k = 0$, as proposed in Subsection 3.1: $v_\phi[0](a_j) = \frac{1}{2}\phi_{a_j}(B, gd, p[0]) + \frac{1}{2}$, and the capacity on a set A is the maximum of the capacity of agents present in the considered coalition.

How to compute $p[k+1]$, $\phi_{a_j}(B, gd, p[k+1])$, $v_\phi[k+1]$ **using** $p[k]$, $\phi_{a_j}(B, gd, p[k])$
and $v_\phi[k]$

The capacity $v_\phi[k]$ is used to compute $c^i(k+1)$, i.e. the aggregation conviction for
the inclination vector i at time $k+1$: $c^i(k+1) = C_{v_\phi[k]}(\bar{c}_{a_1}(k), \cdots, \bar{c}_{a_n}(k))$.
The time-varying probability $p[k+1]$ can then be defined as follows:

$$\forall i \in I, \quad p(i)[k+1] = \frac{c^i(k+1)}{\sum_{j \in I} c^j(k+1)}.$$

It then becomes possible to compute:

- the decisional power for any agent a_j at $k+1$: $\phi_{a_j}(B, gd, p[k+1])$;
- the capacity $v_\phi[k+1]$ over $2^{\{a_1, \cdots, a_n\}}$, at time $k+1$, as proposed in Subsection 3.1.

We have thus defined a time-varying probability. Note that the proposed method seems to be rather intuitive since it corresponds to the notion that an agent's social influence depends on the degree of assurance in the convictions of the other agents when he speaks.

3.3 Conviction State Equations

The aim of this section is to establish the state equations that serve to model the dynamic relationship between convictions and influences. Let's consider a_l to be any listener-agent and a_s a speaker-agent. Their convictions at time k for the alternative $+1$ (resp. -1) are then $c_{a_l}(k)$ and $c_{a_s}(k)$ (resp. $c'_{a_l}(k)$ and $c'_{a_s}(k)$).

Two variables are necessary to model the rhetorical quantity exchanged between the two agents a_l and a_s, namely: the difference in their conviction and their relative importance at time k, as modeled by the capacities $v_\phi[k](a_s)$ and $v_\phi[k](a_l)$.

Four rhetorical exchanges can be distinguished. These four situations are presented in the case when the agent a_l prefers alternative $+1$. Two sub-cases can then be identified for agent a_s: his preferred alternative is either the same as a_l's or the other one. Each case can be divided once again into two sub-cases: a_s's conviction is either greater or less than a_l's conviction. When agent a_l prefers alternative -1, convictions c' replace convictions c in the formula. More precisely, the equations appearing in the computation of $c_{a_l}(k+1)$ when both agents express the same preference are the same as those used to compute $c'_{a_l}(k+1)$ in the case of opposite preferences and *viceversa*. Hence, the rhetorical exchanges can be summarized by the following exchanges: synergistic exchange, revisionist exchange, and antagonistic exchange. Let's take a closer look at each of them.

Synergistic Exchange. In this case, the preference of agent a_l is reinforced by the intervention of agent a_s, who resolutely looks favorably upon the same alternative.

The conviction of agent a_l then increases, to an extent proportional to the difference between both convictions as well as to the capacity of speaker-agent a_s.

This situation, as represented in figure 1, corresponds to the case when a_l and a_s have the same preference and moreover $c_{a_s} > c_{a_l}$. The intuitive difference equation is then written:

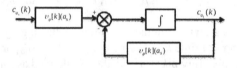

Fig. 1. Synergistic Exchange

$c_{a_l}(k+1) - c_{a_l}(k) = (c_{a_s}(k) - c_{a_l}(k))v_\phi[k](a_s)$, which is equivalent to:
$c_{a_l}(k+1) = c_{a_l}(k) + (c_{a_s}(k) - c_{a_l}(k))v_\phi[k](a_s)$.

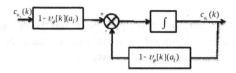

Fig. 2. Revisionist Exchange

Revisionist Exchange. In this situation, agent a_l understands the argument of agent a_s, who has the same preference but a more moderate support. Agent a_s appears to speak with restraint relative to a_l's point of view, and this exposes a_l's doubt. a_l's conviction is thus mitigated by a_s's intervention. This situation, which is depicted in figure 2, corresponds to the case when a_l and a_s have the same preference with $c_{a_l} > c_{a_s}$. The intuitive difference equation is then written as:

$c_{a_l}(k+1) - c_{a_l}(k) = (c_{a_l}(k) - c_{a_s}(k))(1 - v_\phi[k](a_l))$ which is equivalent to:
$c_{a_l}(k+1) = c_{a_s}(k) + (c_{a_l}(k) - c_{a_s}(k))v_\phi[k](a_l)$.

Agent a_l observes the indecision on the part of agent a_s who nevertheless shares his opinion: a_s contributes to a_l's doubt. The level of conviction decreases due to a_s's intervention. which is proportional on the one hand to $1 - v_\phi[k](a_l)$ (resulting from a_l's lack of assurance relative to his social position within the group) and on the other hand to the difference between both agents' convictions.

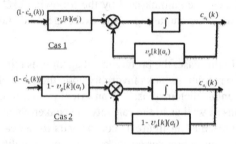

Fig. 3. Antagonistic Exchange

Antagonistic Exchange. In this situation, the two agents do not share the same preference: agent a_l nevertheless understands the advantages of a_s preference. A convincing intervention from a_s may contribute to making a_l dubious, whereas an unpersuasive intervention might on the contrary strengthen a_l's preference.

$(1 - c'_{a_s}(k))$ is a measure of a_s's hesitation and provides a_l with an estimation of the strength of a_s's opposition. Depending on the strength of this hesitation, the previous difference equations can again be used with $(1 - c'_{a_s}(k))$, yielding two situations to be distinguished (see figure 3).

An overly weak preference expressed by a_s implies weak opposition from a_l's point of view and reinforces a_l's opinion, resulting in a likely strengthening of a_l's conviction. The intuitive difference equation is then:

Case 1: $1 - c'_{a_s} \geq c_{a_l}$
$c_{a_l}(k+1) - c_{a_l}(k) = ((1 - c'_{a_s}(k)) - c_{a_l}(k))(v_\phi[k](a_s))$, which is equivalent to
$c_{a_l}(k+1) = c_{a_l}(k) + (1 - c_{a_s}(k) - c_{a_l}(k))(v_\phi[k](a_s))$.
Case 2: $1 - c'_{a_s} < c_{a_l}$.
In this case a_l's conviction weakens following a_s's intervention.
$c_{a_l}(k+1) - c_{a_l}(k) = -(c_{a_l}(k) - (1 - c'_{a_s}(k)))(1 - v_\phi[k](a_l))$, which is equivalent to $c_{a_l}(k+1) = (1 - c'_{a_s}(k)) + (c_{a_l}(k) + c'_{a_s}(k) - 1)(v_\phi[k](a_l))$.

All these various types of exchanges can be synthesized using a Sipos integral.

Proposition 1. *If agents a_s and a_l express the same preference, then:*

$$c_{a_l}(k+1) = \check{C}_{v_\phi}[k](c_{a_s}(k), c_{a_l}(k));$$

If agents a_s and a_l do not share the same preference, then:

$$c_{a_l}(k+1) = \check{C}_{v_\phi}[k](1 - c'_{a_s}(k), c_{a_l}(k)).$$

As a conclusion to this section of the paper, the decisional power ϕ provides a semantic interpretation for the capacity v in the recurrence equations presented in [7], with conviction here being related to the probability an agent will choose one alternative over the other (i.e. probability distribution over inclination vectors). The model in [7] thus becomes interpretable within a game theory framework [3]. The revision equations for conviction appear as input-output balances according to the alternatives assessment. Introducing time into the equations in [7] implies that revision equations of conviction are now seen as state equations of agents' mental perception. This new interpretation then provides a semantics for the debate model in [7]: it incorporates the notions of influence and decisional power, as proposed in [3], with a formalism close to that of dynamic models found in control theory, as suggested in [6].

4   Illustration

4.1   Preference Calculus

This section discusses how to compute preferences during the debate. Initially, each agent a_j assesses both alternatives $+1$ and -1 lying in the interval $[0, 1]$. These assessments are denoted $n_{a_j}^{+1}$ and $n_{a_j}^{-1}$, respectively. It is then possible to build initial preferences and convictions as follows: Let a in $\{-1, 1\}$

- a_j prefers alternative a with the highest score,
- a_j's conviction relative to alternative a equals $\frac{n_{a_j}^a}{n_{a_j}^a + n_{a_j}^{\bar{a}}}$.

Preference changes depend on how convictions evolve over time. For any agent a_j, it is assumed that a threshold $\epsilon_{a_j} > 0$ exists such that when the difference between two convictions lies below this threshold, then agent a_j cannot have a preference. This threshold value may be characteristic of each agent. To summarize: if $|c_{a_j} - c'_{a_j}| < \epsilon_{a_j}$, then a_j has no preference; if $|c_{a_j} - c'_{a_j}| \geq \epsilon_{a_j}$, a_j prefers the alternative with the highest conviction.

An agent without a preference cannot intervene, which is stated as one of the debate rules.

4.2 Simulations of Debates Outcome

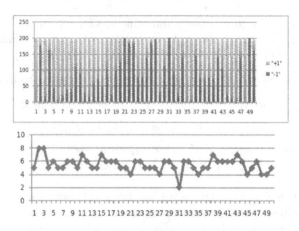

Fig. 4. Identity

In order to illustrate the principle of the above dynamic representation of a debate, the four following elementary models for influence function B have been implemented:

- B is the identity, i.e. for any inclination vector i, it can be stated that: $Bi = i$;
- B is the opposite of identity, i.e. for any inclination vector i, it is stated that: $Bi = -i$;
- B is a mass psychology effect function. More precisely, let's denote $i^\epsilon = \{k \in N | i_k = \epsilon\}$, where B satisfies the following: for each $i \in I$: $|i^\epsilon| > t$, then $i^\epsilon \subseteq (Bi)^\epsilon$, where $t \in [1, n]$ and $\epsilon = +1$ or -1;
- B is a majority influence function that models behavior of the following types: if a majority of agents have an inclination $+1$, then all agents decide $+1$; if not, all agents decide -1.

For these four cases, the group decision function gd is, a mere majority and a basic capacity is designated, as proposed in Section 1.3.

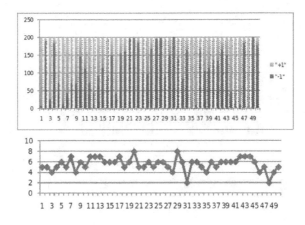

Fig. 5. Opposite of identity

Let's now consider a group of $n = 8$ agents. The initial convictions of agents relative to both alternatives are considered as variates: 50 random drawings of these 8 initial probabilities have been carried out. For each of these 50 initial conviction vectors, the order of agents' intervention in the debate can then be considered: 200 permutations are randomly selected (among the 8! possible rankings) for each initial conviction vector.

Each of the four elementary illustrations has been plotted in the following figures (i.e. one for each B function). For each of the 50 initial conviction vectors selected randomly, a bar represents the number of +1 and −1 outcomes (light gray for +1 and dark gray for −1).

Each figure is to be associated with the maximum number of rounds required to achieve the ground decision for each initial conviction vector. In the proposed simulations, this number does not exceed 8 rounds in any of the cases chosen for B.

The indifference threshold is $\epsilon = 0.01$ for any agent. Agents speak in turn according to the order generated by the 200 permutations, upon the condition that they are able to express a clear opinion, specifically: agent a_j can speak if $|c_{a_j} - c'_{a_j}| \geq \epsilon$.

For the same initial conviction vector, it can be observed that for each function B, the outcome of the debate may depend on the order the agents intervene in the debate. This type of situation can be interpreted as a weak expression of preferential contexts, whereby any perturbation is able to change the debate outcome. From this point of view, influence function B is a disturbance function for this dynamic model of a debate. As a consequence, simulations allow verifying that the order the agents intervene in the debate and their influence are both decisive variables with regard to the convergence of conviction state equations.

The social influence of an agent may thus be considered as a disturbance in the deliberation process, except if it is relevantly used by the debate manager to guide the discussion. In this latter case, social influence can be viewed as an actuator that enables controlling the debate outcome or at least accelerating its convergence. For example, when the debate outcome is practically certain (i.e. the bar is almost completely light or dark gray), then the simplest control might consist of choosing the order of agents

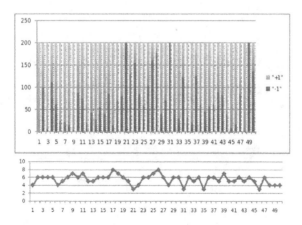

Fig. 6. Majority

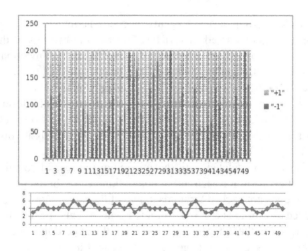

Fig. 7. Mass Psychology Effect

intervention that minimizes the maximum number of rounds. More complex controls could clearly be foreseen, yet the aim of this paper has merely been to propose a dynamic model of the debate within a framework close to control theory representations, making for a natural implementation of control techniques in the future.

4.3   Debate as a Decision-Making Process

This part of the paper will present a potential application of the dynamic model discussed herein. The aim is to apply the model like a voting system. For this example in particular, both alternatives -1 and $+1$ are not considered to be equivalent: $+1$ is the right decision, while -1 is associated with an erroneous decision. This situation could occur in classification problems when the agents are competitive classification algorithms.

The agents are expected to provide the correct answer most of the time, but they typically disagree on individual cases. One common solution is to employ a voting process in order to yield a group decision, i.e. let $d_1, d_2 \ldots d_n$ be the respective decision of the various agents, then the group decision is written as:

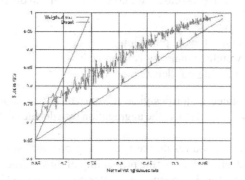

Fig. 8. Simulation-weighted vote and debate

$$
\begin{cases}
1 & \text{if } \sum_{i=1}^{n} d_i > 0 \\
-1 & \text{else}
\end{cases}
$$

For example, let the agents be 7 different classification algorithms whose success rates equal respectively: 0.6, 0.7, 0.8, 0.8, 0.6, 0.7, and 0.6. The group success rate according to a normal vote would thus be 0.86. A better aggregation process will achieve a higher success rate.

The first idea here is to use a weighted vote, i.e. let $\alpha_1, \ldots \alpha_n \in [0, 1]^n$:

$$
\begin{cases}
1 & \text{if } \sum_{i=1}^{n} \alpha_i d_i > 0 \\
-1 & \text{else}
\end{cases}
$$

One possible set of weights is the individual success rate of each agent; however, it is possible to compute the Shapley-Shubik power index [9], and our example delivers a value of $\frac{1}{7}$ for each agent. This is exactly the same value achieved in a normal vote. Since the weights do not differ considerably for a small number of agents, the sign of the weighted sum is the same as that produced during the normal vote. This finding indicates that even if some agents possess a more powerful vote, the final decision is always shared by at least 4 agents.

If we were to use our debate model as the voting process, such an outcome would not occur. The least agents also happen to be those who most readily change their point of view. More precisely, we can run the debate with identity as the B function and as success rates for convictions. It is assumed that 7 competitive classification algorithms are available and moreover that the correct solution is supposed to be alternative $+1$.

As said above, the initial probability of the 7 algorithms to choose alternative $+1$ are: 0.6, 0.7, 0.8, 0.8, 0.6, 0.7, and 0.6. The debate stops when all classification algorithms are in agreement. We will assume then that their answers are independent random variables and that 10,000 cases are studied by each agent. For each case, the agent's answer is inferred according to his probability of being correct.

Next, for each of these 10,000 cases, we compute the group decision according to 3 methods:

- the choice with a majority vote procedure,
- the choice with a weighted majority vote procedure,
- the decision derived using our debate model.

While simple and weighted majorities yield the correct answer at a rate of 86 %, our method produced a 94 % rate. Hence, the aggregation by a debate significantly increases success rate.

In order to verify this good result, we tried using different situations of the same model. For 7 agents, several values for the probability of making the right decision were randomly generated, and the 3 corresponding rates computed (results are presented in Figure 8). In this figure, both the weighted voting rate and our debate output vs. this rate are plotted. Note that the same rate for the simple vote can be obtained with very different sets of probabilities. The debate always yields a better rate, although its preferences change according to the specific probability profile. The weighted vote success rate is quite close to that of the simple vote, except for very unique probability sets where several agents (algorithms) perform much better than the others.

5 Conclusions and Outlook

The state equations derived in this paper allow simulating macroscopically the outcome of a debate according to the initial inclinations of agents and the social influences taking place within the group (whereby the influence function is a priori known). The deliberation outcome depends not only on the order in which the agents intervene in the debate to explain their opinions, but also on the influence an agent is able to exert on a social network.

The model formalism proposed in this paper is close to the one used in control theory to model the dynamic behavior of technical systems. Guiding a debate might then be seen as a control problem, whose aim could, for example, be how to reach a consensus as quickly as possible or how to reinforce one alternative over the other, etc.

A debate is thus seen as a continuous dynamic system: a state equation representation has been preferred to the multicriteria decision-making framework in [7] given that time explicitly appears in the revision of convictions. The model semantic has also been inspired from the game theory concepts proposed in [3]: influence and decisional power in a social network. In our dynamic extension, decisional power is a time-varying variable itself and may be used as the actuator signal in the debate control loop.

The system of state equations established in this paper allows stochastically simulating the outcome of a debate and effects of a control strategy on this particular issue.

One possible application of this model would obviously be to simulate a debate outcome in order to obtain certain indications regarding the final collective decision. When simulations are performed for a large number of initial agent convictions and speaker intervention rankings, the probability that outcome is ± 1 can be estimated. Hence, the dynamic influence model can be applied to either make the debate outcome more certain (this may appear to be a dishonest method when agents are actual human beings, yet remains a relevant technique when agents are artificial, such as sensors or classifiers) or modify the convergence dynamics of the debate.

References

1. Bonnevay, S., Kabachi, N., Lamure, M., Tounissoux, D.: A multiagent system to aggregate preferences. In: IEEE International Conference on Systems, Man and Cybernetics, Washington, USA, pp. 545–550 (2003)
2. Grabisch, M., Labreuche, C.: The symmetric and asymmetric choquet integrals on finite spaces for decision making. Statistical paper, vol. 43, pp. 37–52. Springer, Heudelberg (2002)
3. Grabisch, M., Rusinowska, A.: A model of influence in a social network. Business and Economics (2008)
4. Grabisch, M., Rusinowska, A.: Influence in social networks. In: COGIS 2009, Paris (2009b)
5. Hoede, C., Bakker, R.: A theory of decisional power. Journal of Mathematical Sociology 8, 309–322 (1982)
6. Imoussaten, A., Montmain, J., Rigaud, E.: Interactions in a Collaborative Decision Making Process: Disturbances or Control Variables? In: COGIS 2009, Paris (2009)
7. Rico, A., Bonnevay, S., Lamure, M., Tounissoux, D.A.: Debat modelisation with the Sipos integral. In: LFA 2004, Nantes (2004)
8. Rusinowska, A., De Swart: On some properties of the Hoede-Bakker index. Journal of Mathematical Sociology 31, 267–293 (2007)
9. Shapley, L.S.: A Value for n-Person Games. In: Kuhn, H., Tucker, A. (eds.) Contribution to the Theory of Games. Annals of Mathematics Studies Princeton, vol. II(28), pp. 303–317 (1953)

Setting Up Particle Swarm Optimization
by Decision Tree Learning Out of Function Features

Tjorben Bogon[1], Georgios Poursanidis[2],
Andreas D. Lattner[2], and Ingo J. Timm[1]

[1] Business Information System I, University of Trier, Trier, Germany
[2] Information Systems and Simulation, Goethe University Frankfurt
Frankfurt, Germany
bogon@uni-trier.de

Abstract. This work describes an approach for the computation of function features out of optimization functions to train a decision tree. This decision tree is used to identify adequate parameter settings for Particle Swarm Optimization (PSO). The function features describe different characteristics of the fitness landscape of the underlying function. We distinguish between three types of features: The first type provides a short overview of the whole search space, the second describes a more detailed view on a specific range of the search space and the remaining features test an artificial PSO behavior on the function. With these features it is possible to classify fitness functions and to identify a parameter set which leads to an equal or better optimization process compared to the standard parameter set for Particle Swarm Optimization.

Keywords. Particle swarm optimization, Machine learning, Swarm intelligence, Parameter configuration, Objective function feature computation.

1 Introduction

Metaheuristics in stochastic local search are used in numerical optimization problems in high-dimensional spaces. For varying types of mathematical functions, different optimization techniques vary w.r.t. the optimization process [16]. A characteristic of these metaheuristics is the configuration of the parameters [6]. These parameters are essential for the efficient optimization behavior of the metaheuristic but depend on the objective function, too. An efficient set of parameters influences the optimization in speed and performance. If a good parameter set is selected, an adequate solution will be found faster compared to a bad configuration of the metaheuristic. The choice of the parameters is based on the experience of the user and his knowledge about the domain or on empirical research found in literature. This parameter settings, called standard configurations, perform a not optimal but an adequate optimization behavior for most objective functions. An example for metaheuristics is the Particle Swarm Optimization (PSO). PSO is introduced by [5] and is a population-based optimization technique which is used in continuous high dimensional search spaces. PSO consists of a swarm of particles which "fly" through the search space and update their position by taking into account their own best position and depending on the topology, the best position found

J. Filipe and A. Fred (Eds.): ICAART 2011, CCIS 271, pp. 72–85, 2013.

by other particles. PSO is an example for the parameter configuration problem. If the parameters are well chosen, the whole swarm will find an adequate minimum and focus on this solution. The swarm slows down the velocity trying to get better values in the continuous search space around this found solution. This exploitation can be on a local optimum especially if the wrong parameter set is chosen and the swarm cannot escape from this local minimum. On the other hand the particles can never find the global optimum if they are too fast and never focus. This swarm behavior depends mainly on the chosen parameter and leads to solutions of different quality.

One problem in choosing the right parameters without knowledge about the objective function is to identify relevant characteristics of the function which can be used for a comparison among functions. The underlying assumption is that, e.g., a function $f_1 = x^2$ and a function $f_2 = 3x^2 + 2$ exhibit similar optimization behavior if the same parameter set for a Particle Swarm Optimization is used. In order to choose a promising parameter set, functions must be comparable with respect to certain objective function characteristics.

We describe an approach to computing features of the objective function by observing the swarm behavior. For each function we seek for a parameter set that performs better than the standard configuration and provide this set as output class for supervised learning. These data allow us to train a C4.5 decision tree [11] as classifier that computes an adequate configuration for the Particle Swarm Optimization by using function features. Experimental trials show that our decision tree classifies functions into the correct classes in many cases. This classification can be used to select promising parameter sets for which the Particle Swarm Optimization is expected to perform better in comparison to the standard configuration.

This work is structured as follows: In section 2 we describe other approaches pointing out the problem of computing good parameter sets for a metaheuristic and explain the Particle Swarm Optimization. Section 3 describes how to compute the features of a function and thereby make the functions comparable. After computing the features we describe our experimental setup and the way to build up the decision tree. Section 4 contains our experimental results for building the parameter classes to select promising parameter sets in PSO. The last section discusses our results and describes issues for future work.

2   Parameter Settings in Metaheuristics

The main difference between solving a problem with exact methods or with metaheuristics is the quality of the solution. Metaheuristics – for example, nature inspired metaheuristics [2] – have no guarantee of finding the global optimum. They focus on a point in the multidimensional search space which results to the best fitness value depending on the experience of the past optimization performance. This can be a local optimum, too. But the advantage of the metaheuristic is to find an adequate solution of a multidimensional continuous optimization problem in reasonable time [13]. This performance depends on the configuration of the metaheuristics and is an important fact of using metaheuristics. One group of metaheuristics are the population based metaheuristics. [13] defines population-based metaheuristics as nature inspired heuristics which handle more than one solution at a time. With every iteration all solutions are recomputed based on the experience of the whole population. Examples of population-based meta-heuristics are Genetic Algorithms which are an instance of Evolutionary Algortihms, Ant Colony Optimization and Particle Swarm Optimization. Different kinds of meta-heuristics exhibit varying performance on a specific kinds of problem types. They differ w.r.t. the optimization speed and the solution quality. A metaheuristic's performance is based on their configuration. Finding a good parameter set is a non-trivial task and often based on a priori knowledge about the objective function and the problem. Setting up a metaheuristic with standard parameter sets lets the optimization find a decent solution but using a parameter set which is adapted to the specific objective function might even lead to better results. In this paper we focus on PSO and try to find features character-izing the objective function in order to select an adequate parameter configuration for this metaheuristic. The optimization behavior of the particles is based on the objective function and we try identify relevant information about the function. In the following section we give a brief introduction to particle swarm optimization.

2.1 Particle Swarm Optimization

Particle Swarm Optimization (PSO) is inspired by the social behavior of flocks of birds and shoals of fish. A number of simple entities, the particles, are placed in the domain of definition of some function or problem. The fitness (the value of the objective function) of each particle is evaluated at its current location. The movement of each particle is de-termined by its own fitness and the fitness of particles in its neighborhood in the swarm. PSO was first introduced in [8]. The results of one decade of research and improve-ments to the field of PSO were recently summarized in [3], recommending standards for comparing different PSO methods. Our definition is based on [3]. We aim at contin-uous optimization problems in a search space S defined over the finite set of continuous decision variables $X_1, X_2, \ldots, X_n$. Given the set Ω of conditions to the decision vari-ables and the objective function $f : S \rightarrow \mathbb{R}$ (also called fitness function) the goal is to determine an element $s^* \in S$ that satisfies Ω and for which $f(s^*) \leq f(s)$, $\forall s \in S$ holds. $f(s^*)$ is called a global optimum.

Given a fitness function f and a search space S the standard PSO initializes a set of particles, the swarm. In a D-dimensional search space S each particle P_i consists of three D-dimensional vectors: its position $\vec{x}_i = (x_{i1}, x_{i2}, \ldots, x_{iD})$, the best position the particle visited in the past $\vec{p}_i = (p_{i1}, p_{i2}, \ldots, p_{iD})$ (particle best) and a velocity $\vec{v}_i = (v_{i1}, v_{i2}, \ldots, v_{iD})$. Usually the position is initialized uniformly distributed over S and the velocity is also uniformly distributed depending on the size of S. The move-ment of each particle takes place in discrete steps using an update function. In order to calculate the update of a particle we need a supplementary vector $\vec{g} = (g_1, g_2, \ldots, g_D)$ (global best), the best position of a particle in its neighborhood. The update function, called inertia weight, consists of two parts. The new velocity of a particle P_i is calcu-lated for each dimension $d = 1, 2, \ldots, D$:

$$v_{id}^{new} = w \cdot v_{id} + c_1 \epsilon_{1d} (p_{id} - x_{id}) + c_2 \epsilon_{2d} (g_d - x_{id}) \tag{1}$$

then the position is updated: $x_{id}^{new} = x_{id} + v_{id}^{new}$. The new velocity depends on the global best (g_d), particle best (p_{id}) and the old velocity (v_{id}) which is weighted by the inertia weight w. The parameters c_1 and c_2 provide the possibility to determine how strong a particle is attracted by the global and the particle best. The random vectors $\vec{\epsilon}_1$ and $\vec{\epsilon}_2$ are uniformly distributed over $[0, 1)^D$ and produce the random movements of the swarm.

2.2 Algorithm Configuration Problem

The general problem of configuring algorithms (algorithm configuration problem) is defined by Hutter et al. [7] as finding the best tuple θ out of all possible configurations Θ ($\theta \in \Theta$). θ represents a tuple with a concrete assignment of values for the parameter of an algorithm. Applied to metaheurisitcs the configuration of the algorithm parameters for a specific problem influences the behavior of the optimization process. Different parameter settings exhibit different performances at solving a problem. The problem to configure metaheuristics is a super ordinate problem and is analyzed for different kinds of metaheuristics. In PSO the convergence of the optimization depending on different parameter settings and different functions are analyzed by [14], [12] and [1]. But these approaches focus only on the convergence of the PSO but not on function characteristics and the relationship between the parameter configuration and the function landscape.

Different approaches to solve this algorithm configuration problem on metaheurisitcs are introduced: One approach is to find sets of adequate parameters which performs a good optimization on most different types of objective functions. This "standard parameters" are evaluated on a preset of functions to find a parameter set which leads to global good behavior of the metaheuristic. In PSO standard parameter sets are presented by [4] and [12]. Some approaches do not present a preset of parameters but change the values of the parameters during the runtime to get a better performance [10].

Another approach is introduced by Leyton-Brown et al. They try to create features which describe the underlying problem [9] and generate a model predicting the right parameters depending on the classification. They introduce several features which are grouped into nine groups. The features include, among others, problem size statistics, e.g. number of clauses and variables, and measures based on different graphical representations. This analysis is based on discrete search spaces because on continuous search spaces it is not possible to set adequate discrete values for the parameter configuration which is needed by their approach.

Our problem is to configure an algorithm working on continuous search spaces and offers infinite possibilities of parameter sets. To solve this challenge we try, similar to Leyton-Brown et al., to train a classifier with features of the fitness function landscape computed by observing swarm behavior. These features are computed and combined with the best found parameter set on the function to a training instance (see figure 1). With a trained classifier at hands we compute the features of the objective function prior to the start of the optimization process. The classifier – in our case a decision tree – classifies the function and selects the specific parameter set that is expected to perform better in the optimization process than using the standard parameters. In our first experiments, which we understand as proof of concept, we choose only a few functions which do not represent any specific types of function. We want to show that our technique is able to identify functions based on the swarm behavior provided features and thereby, select the specific parameter configuration. In order to learn the classifier

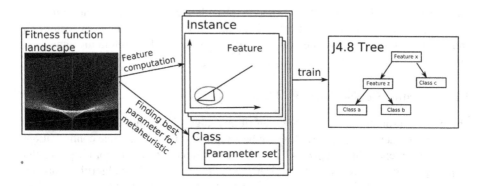

Fig. 1. Process of building a classifier

which suggests the parameter configuration, different function features are computed. These features are the basis of our training instances.

3   Computation of Function Features

Our computed features can be divided into three groups. Each group implies a distinct way of collecting information about the fitness topology of the objective function from particles. The first group *Random Probing* describes features which are calculated based on a random selection of fitness values and provides a general overview of the fitness topology. Distance-based features are calculated for the second group *Incremental Probing*. They reflect the distribution of surrounding fitness values of some pivot particles. The third group of features utilizes the dynamics of PSO to create features by using the changes of the global best fitness within a small PSO instance. The features are scale independent, i.e., that scaling the objective function by constants will not affect the feature values. By this we imply that a configuration for PSO leads to the same behavior on a function f as it shows for its scaled function $f' = \alpha f + \beta, \alpha > 0$. These three groups are based on each other which means that the pivot particle for the second group is taken from a particle of the first group to reduce the computing time. Important for all these features are the number of evaluations of the objective function. The feature computation should be only a small part of the whole optimization computation time.

3.1   Random Probing

Random Probing defines features that are calculated based on a set of $k = 100$ random particle positions which are within the initialization range of the objective function (100 particles to get a short but adequate description about the function window). Probing the objective function results in a distribution of fitness values which is used to extract three features. Trivial characteristics like mean and standard deviation cannot be used as features since they are not scale independent. That means, they will change their value if the function is scaled by constants. In order to create reliable features, the fitness values of all points are evaluated and three sets of particles (including their evaluation

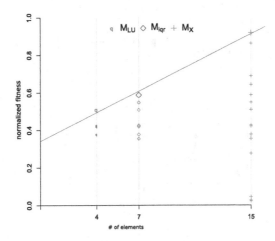

Fig. 2. Example of a random probing with the comutation of $\mu_{RP.Max}$

values) are created based upon these values. The first set is denoted M_X and contains all fitness values of the randomly selected points. The quartiles for the distribution of the fitness values are computed and the values between the upper or lower quartile are joined into the second set. This set is denoted M_{iqr}. The third set M_{LU} consists of the fitness values which are between a lower and upper boundary L and U. These boundaries are defined by $L = Q_1 + \frac{1}{2}(Q_M - Q_1)$ and $U = Q_M + \frac{1}{2}(Q_3 - Q_M)$ where Q_M denotes the median and Q_1, Q_3 denote the lower and upper quartile of M_X. For each set $M_X \supset M_{iqr} \supset M_{LU}$ the number of elements is determined.

The feature `Random Probing Min` $\mu_{RP.Min}$ is calculated based on the linear model that fits the relationship between the number of values and the minimum fitness values in each set. The straight line of the model is divided by the interquartile range of M_X. Similar to this the feature `Random Probing Max` is based on the slope of the straight line that describes the relationship of the number of elements and the maximum value of each set $M_X \supset M_{iqr} \supset M_{LU}$ (see figure 2). The slope divided by the interquartile range of M_X denoted by $\mu_{RP.Max}$ is the second feature of this group. Finally, for the feature `Random Probing Range` denoted by $\mu_{RP.Range}$ the spread, that is the difference between the maximum and the minimum value, in each set is computed. As for the other features the slope is divided by the interquartile range of M_X. All features of this group are computed based on the fitness values of the randomly selected points. For each point the objective function is evaluated once, hence, $k = 100$ evaluations are necessary for *Random Probing*.

3.2 Incremental Probing

In contrast to the features of the previous group, *Incremental Probing* is computed by the fitness values of the particle positions which are located in a defined distance to a pivotal element which we choose from the feature group above. In order to calculate the relevant fitness values, the position of a randomly selected pivot element is

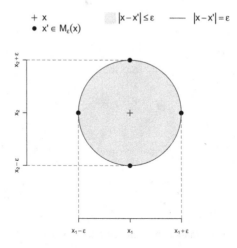

Fig. 3. Example of an incremental group in a 2 dimensional space

consecutively shifted into one dimension. The distance is given by an increment $\varepsilon > 0$ which shifts the position of the pivot element in both directions of the dimension. In each dimension i *Incremental Probing* considers two points (see figure 3 for a 2 dimensonal example). For a given pivot element $x = (x_1, \ldots x_n)$ and a given increment $\varepsilon > 0$ these positions are determined by

$$|x_i - x'_i| = \begin{cases} \varepsilon \ |i = j \\ 0 \ |i \neq j \end{cases} \tag{2}$$

where $j, i \leq n$. The increment ε is defined relatively to the domain. For instance, in a restricted n-dimensional domain $\mathbb{A} = I_1 \times \ldots \times I_n$, where the interval $I_i = [a_i, b_i]$ defines the valid subspace, the increment is applied as $\varepsilon \cdot \frac{b_i - a_i}{100}$. For each dimension the position of the pivot element is shifted into two directions. This leads to a set of $2n + 1$ points including the pivot element. The fitness value of each valid point is calculated and these fitness values are used for the extraction of objective features[1]. In this group of features, nine features are created with the use of three increments of $\varepsilon_1 = 1, \varepsilon_2 = 2$ and $\varepsilon_5 = 5$. Let n be the dimension of the domain, then $2n + 1$ evaluations are required to calculate the fitness values of the relevant points. Since three increments are used there are $(3 \times 2n) + 1$ evaluations required to calculate the features of *Incremental Probing*.

The features Incremental Min, Incremental Max and Incremental Range are computed similar to the features of *Random Probing*. For each increment the minimum, maximum and the spread of the fitness values are computed. Incremental

---

[1] In case that the point is invalid, that is it lies outside the valid domain, the evaluation of the fitness value is skipped and the fitness value of the pivot element is used instead.

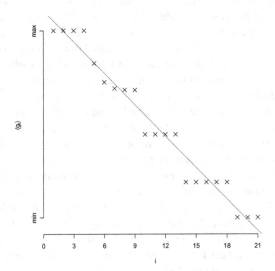

Fig. 4. Example of an incremental swarming slope where g describes the best fitness of the actual evaluation step i

`Min` describes the relationship of the minimum and the corresponding increment. There are two subtypes for this feature. $\mu_{IP.Min}$ is divided by the slope of the model's straight line by the spread of the first increment whereas the second subtype $\mu_{IP.MinQ}$ divides the slope by the interquartile range of the first increment. The features `Incremental Max` and `Incremental Range` are handled accordingly. Three additional features are created by separately looking at the fitness values of the individual increments. The fitness values of each increment is sorted in ascending order and normalized into the interval $[0, 1]$. This results in a sequence $\langle x_k \rangle = x_1, \ldots, x_k$ and we calculate a measure of linearity by

$$\mu_{IP.Fit} = \sum_{i=1}^{k} \left(x_i - \frac{i-1}{k-1} \right)^2 \tag{3}$$

where $\forall i < j : x_i < x_j$.

3.3 Incremental Swarming

The features of *Incremental Swarming* use the dynamic behavior of PSO to extract features of the objective function. Therefore, we construct a small swarm of two particles and initiate an optimization run. The particles are initialized with a defined distance to each other. We use a inertia PSO with parameter $\theta = (0.6221, 0.5902, 0.5902)$ and record the best solution found in the first $t = 20$ iterations. To get the parameter set θ we evaluated a few parameter sets empirically to find good values which lead to a fast convergence of the small swarm. The spread of the global best fitness is the difference between the first and the last fitness value. The development of the global best fitness depends on the initial positions of the particles. Consider a swarm which is initialized at a local optimum. Once a better fitness value is found, global best fitness will change. But this may not happen in the few iterations that are observed. Therefore the swarm is initialized by a pivot element chosen from a set of evaluated points. *Incremental Swarming* considers a set of $k = 100$ evaluated solutions and the position which evaluates to the worst fitness value is chosen as pivot element. This is important because if we choose a pivot element randomly, it is possible to find a local optimum and the behavior of the swarm results in no movement. The other particle is initialized in a defined distance to the pivot element. Similarly to *Incremental Probing* we use increments to define the distance between the particles. The increment values $\varepsilon_1 = 1, \varepsilon_2 = 2, \varepsilon_5 = 5$ and $\varepsilon_{10} = 10$ are used to create 20 features. For each increment the feature `Swarming Slope` describes the development of the global best fitness as a linear model that fits the relationship between the iteration and the global best fitness value (see figure 4). For the feature $\mu_{IS.Slope}$ the slope of the straight line is divided by the spread of the global best fitness. `Swarming Max Slope` describes the greatest change of the global best fitness value between two successive iterations. For normalization the value of $\mu_{IS.Max}$ is divided by the spread of the global best fitness. The other three features, which are computed for each increment, are `Swarming Delta Lin` $\mu_{IS.Lin}$, `Swarming Delta Phi` $\mu_{IS.Phi}$, and `Swarming Delta Sgm` $\mu_{IS.Sgm}$. They describe to what degree the observed development of the global best fitness value differs from sequences that represent idealistic developments. `Swarming Delta Lin` implies a measure of linearity, thus quantifies how much the observed development differs from a linear decrease of the global best fitness. Let $\langle x_t \rangle = x_0, \ldots, x_t$ denote the observed sequence of the global best fitness value. We compute this feature with equation 4.

$$\mu_{IS.Lin} = \sum_{i=0}^{k} \left(x_i - \frac{t - i + 1}{t - 1} \right)^2 \tag{4}$$

Similarly we create two additional ideal sequences and compute the features $\mu_{IS.Phi}$ and $\mu_{IS.Sgm}$ by the equations 5–6:

$$\mu_{IS.Phi} = \sum_{i=0}^{k} \left(x_i - \phi^{i-1} \right)^2 \tag{5}$$

$$\mu_{IS.Sgm} = \sum_{i=0}^{k} \left(x_i - \frac{1}{1 + \exp^{(i-1)\phi}} \right)^2 \tag{6}$$

where $\phi = \frac{2}{1+\sqrt{5}}$. The factor ϕ was selected in order to mediate between a linear and an exponential developing. The development of the global best fitness is used to calculate the features of *Incremental Swarming*. The pivot element for the initialization of the swarm is chosen from a set of k solutions and since the swarm of $m = 2$ particles is applied for $t = 20$ iterations, overall there are $k + m + mt$ evaluations of the objective function. We choose the pivot element from the set M_X which was created for the features of *Random Probing*. By this we reduce the number of additional evaluation to $m + mt = 42$.

4   Evaluation

In this section we evaluate our features and build a classifier which computes specific parameter sets for the Particle Swarm Optimization on a specific function. This optimization should have a better performance compared to the PSO on the same function with standard parameter set.

Table 1. Overview of the function pool and the initialization areas

Function	Optimum	Domain	Initialization
Ackley	$x_i = 0$	$[-32, 32]^n$	$[16, 32]^n$
Gen. Schwefel	$x_i = 420.9687$	$[-500, 500]^n$	$[-500, -250]^n$
Griewank	$x_i = 0$	$[-600, 600]^n$	$[300, 600]^n$
Rastrigin	$x_i = 0$	$[-5.12, 5.12]^n$	$[2.56, 5.12]^n$
Rosenbrock	$x_i = 1$	$[-30, 30]^n$	$[15, 30]^n$
Schwefel	$x_i = 0$	$[-100, 100]^n$	$[50, 100]^n$
Sphere	$x_i = 0$	$[-100, 100]^n$	$[50, 100]^n$

4.1   Experimental Setup

We choose 7 test functions out of the suggested test function pool from [3] and stop computing the fitness function after 300000 times. With our swarm size of 30 the number of epochs is consequently set to 10000. We define a run as a parameter set which is tested 90 times with a finite set of different seed values in order to get meaningful results. As topology of the swarm *gbest* is used. The initialization of the particle is in a defined square of the search space (see table 1). Before we start to train our classifier with the features we have to create the classes that represent specific parameter sets with a high quality of the optimization performance.

4.2   Finding the Best Parameter

In order to find the best parameter set for each function (see table 1), we start an extensive search with respect to the continuous values. We try to focus on real values with a precision of four decimal places. The standard parameter set for PSO is $\omega = (W, C_1, C_2)$ with $W = 0.72984$ and $C_1 = C_2 = 1.4962$. For the extensive examination of parameters we take into account the intervals $W \in [0, 1]$ and $C_1, C_2 \in [0, 2.5]$. We create a sequence between this interval values based on the standard value with a exponential factor of $(\frac{2}{1+\sqrt{5}})^x$ where x indicates the sequence number. We calculate 13 and 23 sequence values around the standard value and obtain a sequence of values between the intervals. Depending on the exponential factor the values close by the standard values have a lower distance to each other than the values closer to the borders of the interval. In figure 5 our configuration space of the extensive search is plotted. With all possible combinations of the single parameter values we examine $13 \times 23 \times 23 = 6877$ different parameter sets and test each of them for every function 90 times. As described above, we analyze the data of the extensive search by comparing the results of each configuration's optimization process on a function. We choose the

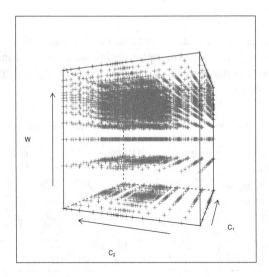

Fig. 5. Parameter sets in the configuration space

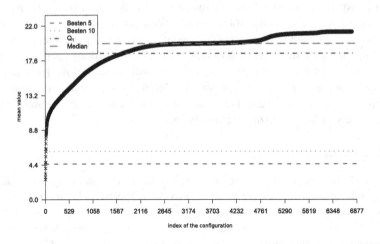

Fig. 6. Sorted Set the Mean Value of all Parameterset Results

best parameter sets for every function with respect to the best performance. The best performance is determined by the best fitness value, the mean fitness value of all 90 optimization processes and the distance to the nine other best performances within this 90 processes. This is essential because a good solution and a high distance to the other run results let this one run be an outlier. Figure 6 shows an example of a sorted sequence of the mean value of all parameter sets we tested on the "Ackley"-function. We compare the results of the specific parameter sets with the standard parameter set of [3] using our PSO implementation, and get a significantly better result (or the same if we

Table 2. Comparison of the standard parameter set against the specific best parameter set; *denotes the optimization with 9900 iterations, i.e., 297000 function evaluations

| Function | Tests | | | Reference in [3] | | Best parameter set |
	specific	specific*	standard	*gbest*	*lbest*	(W, C_1, C_2)
Ackley	**2.58**	2.62	18.34	17.6628	17.5891	(0.7893,0.3647, 2.3541)
Gen. Schwefel	**2154**	2155	3794	3508	3360	(0.7893,2.4098, 0.3647)
Griewank	**0.0135**	0.0135	0.0395	0.0308	*0.0009*	(0.6778, 2.1142, 1.3503)
Rastrigin	**6.12**	6.12	169.9	140.4876	144.8155	(0.7893,2.4098, 0.3647)
Rosenbrock	**0.851**	0.86	4.298	8.1579	12.6648	(0.7123,1.8782, 0.5902)
Schwefel	**0**	0	0	0	0.1259	more than one set
Sphere	**0**	0	0	0	0	more than one set

found the global optimum) on gbest for all functions with the selected parameter sets. Table 2 shows our results for the specific parameters for the different functions (300000 evaluations + 30000 evaluations for feature computation; denoted as "specific"), the same parameter set subtracting one percent evaluations for the feature computation (to demonstrate if we used this one percent of computation time to extract the features, i.e., a total of 300000 evaluations; "specific*") and the comparison to the standard parameter set included in our code. Additionally, the comparison to the results of the original paper of Bratton and Kennedy is included in the table.

The extensive search shows that the best specific parameter sets for the functions Gen. Schwefel and Rastrigin is comparable. The same effect is also supported by the features of both objective functions. This denotes that both functions are assigned the same class in our classifier. All the specific parameter sets are the base of our classes for each function. With the identified classes and the computed set of features for each function we can train the classifier.

4.3   Learning and Classification

As classifier we use a C4.5 decision tree. In our implementation we use WEKA's J4.8 implementation [15]. As learning input we compute 300 independent instances for each function. Each instance consists of 32 function features. The decision tree is created based upon the training data and evaluated by stratified 10-fold cross-validation (repeated 100 times). Based on the results of the extensive search we merge the classes for the objectives Gen. Schwefel and Rastrigin into one class. These functions share the same specific parameter set, i.e., the same parameter configuration performs best for both functions. Upon these six distinct classes we evaluate the model with cross validation. The mean accuracy of the 100 repetitions is 84.32% with a standard deviation of 0.29. Table 3 shows the confusion matrix of a sample classification. As it can be seen, there are 1769 of 2100 instances classified correctly (this means 15.76 percent of the instances are misclassified). The instances of the functions Ackley and Schwefel are classified correctly with an accuracy of 99.7 percent, that is only one instance of these classes is misclassified. The class for Gen. Schwefel and Rastrigin has an accuracy of 97.2 percent. The class Rosenbrock has a slightly lower accuracy, but still only 5.7 percent of its members are misclassified. The high number of incorrect instances is essentially due to the inability of the model to separate the functions Sphere and Griewank. The majority of the misclassified instances, 306 of 331, are instances of the Griewank or Sphere class that are classified as the other class.

Table 3. Confusion matrix of the cross validation

class	Ack.	Grie.	G.S./R.	Rosen.	Schwe.	Sphe.	Accuracy percent	Precision percent
Ackley	**299**	1					99.7	99.7
Griewank		**116**	1			183	38.7	48.1
Gen.Schwe./Rast.	1	1	**583**	13	2		97.2	97.2
Rosenbrock			15	**283**	1	1	94.3	95.3
Schwefel				1	**299**		99.7	99.0
Sphere		123	1			**176**	58.7	48.9

4.4 Computing Effort

Computing the features is based on the evaluated fitness value of specific positions in the search space. We restrict this calculation to 3000 which means one percent of the whole optimization process in our setting. To be comparable to the benchmark of Bratton and Kennedy we run the optimization for the specific parameter configuration for 9900 iterations leading to only 297000 fitness computations. We compare our results of the optimization with 10000 iterations to the optimization with 9900 iterations and get quite the same results as shown in table 2 (specific vs. specific*). The comparison shows minor differences in the magnitude of one percent.

5 Discussion and Future Work

In this paper we describe an approach to training a classifier which uses function features in order to select a better parameter configuration for Particle Swarm Optimization. We show how we compute the features for specific functions and describe how we get the classes of parameter sets. We include the trained classifier and evaluate the parameter configuration against a Particle Swarm Optimization with standard configuration. Our experiments demonstrate that we are able to classify different functions on basis of a few fitness evaluations and get a parameter set which leads the PSO to a significantly better optimization performance in comparison to a standard parameter set. Statistical tests (t-Tests with $\alpha = 0.05$) indicate better results for the functions where the global optimum has not been found in both settings.

The next steps are to involve all possible configurations of the PSO for example the swarm size or the neighborhood topology. These parameters are not involved in our approach because we based this work on the benchmark approach of [3]. The behavior of the swarm changes significantly if another neighborhood is chosen. To increase the size of the swarm is another task we will focus in future. Depending on the swarm size different parameter sets leads to the best optimization process. An idea is to create an abstract class of parameter sets which include different sets of predefined swarm sizes.

In order to get more information about the performance of our approach it would be interesting to allocate a fixed percentage of the whole evaluations for feature computation (e.g., 1%). In this case it would be interesting to examine the quality of the result if not all feature or features of minor quality were computed.

Another extension is to define typical mathematical function types to integrate not only one function as class but a few functions combined under a similar type of functions to get a general set of parameters. This might lead to a better generalization for the learned classifier. The problem of this task is to find a general problem class which defines typical kinds of mathematical functions.

References

1. van den Bergh, F., Engelbrecht, A.: A new locally convergent particle swarm optimiser. In: 2002 IEEE International Conference on Systems, Man and Cybernetics, vol. 3, p. 6 (October 2002)
2. Bonabeau, E., Dorigo, M., Theraulaz, G.: Swarm Intelligence: From Natural to Artificial Systems, 1st edn. Oxford University Press, USA (1999)
3. Bratton, D., Kennedy, J.: Defining a standard for particle swarm optimization. In: Swarm Intelligence Symposium, pp. 120–127 (2007)
4. Clerc, M., Kennedy, J.: The particle swarm - explosion, stability, and convergence in a multidimensional complex space. IEEE Transactions on Evolutionary Computation 6(1), 58–73 (2002)
5. Eberhart, R., Kennedy, J.: A new optimizer using part swarm theory. In: Proceedings of the Sixth International Symposium on Micro Maschine and Human Science, pp. 39–43 (1995)
6. Hoos, H., Stützle, T.: Stochastic Local Search: Foundations & Applications. Morgan Kaufmann Publishers Inc., San Francisco (2004)
7. Hutter, F., Hoos, H.H., Stutzle, T.: Automatic algorithm configuration based on local search. In: Proceedings of the Twenty-Second Conference on Artifical Intelligence (AAAI 2007), pp. 1152–1157 (2007)
8. Kennedy, J., Eberhart, R.: Particle swarm optimization. In: Proceedings of the 1995 IEEE International Conference on Neural Network, Perth, Australia, pp. 1942–1948 (1995)
9. Leyton-Brown, K., Nudelman, E., Shoham, Y.: Learning the Empirical Hardness of Optimization Problems: The Case of Combinatorial Auctions. In: Van Hentenryck, P. (ed.) CP 2002. LNCS, vol. 2470, pp. 91–100. Springer, Heidelberg (2002)
10. Pant, M., Thangaraj, R., Singh, V.P.: Particle swarm optimization using gaussian inertia weight. In: International Conference on Conference on Computational Intelligence and Multimedia Applications, vol. 1, pp. 97–102 (2007)
11. Quinlan, J.R.: C4.5: Programs for Machine Learning. Morgan Kaufmann, San Francisco (1993)
12. Shi, Y., Eberhart, R.C.: Parameter Selection in Particle Swarm Optimization. In: Porto, V.W., Waagen, D. (eds.) EP 1998. LNCS, vol. 1447, pp. 591–600. Springer, Heidelberg (1998)
13. Talbi, E.G.: Metaheuristics: From design to implementation. Wiley, Hoboken (2009), http://www.gbv.de/dms/ilmenau/toc/598135170.PDF
14. Trelea, I.C.: The particle swarm optimization algorithm: convergence analysis and parameter selection. Inf. Process. Lett. 85(6), 317–325 (2003)
15. Witten, I.H., Frank, E.: Data Mining: Practical machine learning tools and techniques, 2nd edn. Morgan Kaufmann, San Francisco (2005)
16. Wolpert, D., Macready, W.: No free lunch theorems for optimization. IEEE Transactions on Evolutionary Computation 1(1), 67–82 (1997)

Collision Avoidance Using Partially Controlled Markov Decision Processes*

Mykel J. Kochenderfer and James P. Chryssanthacopoulos

Lincoln Laboratory, Massachusetts Institute of Technology
244 Wood Street, MA 02420, Lexington, U.S.A
{mykelk,chryssanthacopoulos}@ll.mit.edu

Abstract. Optimal collision avoidance in stochastic environments requires accounting for the likelihood and costs of future sequences of outcomes in response to different sequences of actions. Prior work has investigated formulating the problem as a Markov decision process, discretizing the state space, and solving for the optimal strategy using dynamic programming. Experiments have shown that such an approach can be very effective, but scaling to higher-dimensional problems can be challenging due to the exponential growth of the discrete state space. This paper presents an approach that can greatly reduce the complexity of computing the optimal strategy in problems where only some of the dimensions of the problem are controllable. The approach is applied to aircraft collision avoidance where the system must recommend maneuvers to an imperfect pilot.

Keywords: Markov decision processes, Dynamic programming, Collision avoidance.

1 Introduction

Manually constructing a robust collision avoidance system, whether it be for an autonomous or human-controlled vehicle, is challenging because the future effects of the system cannot be known exactly. Due to their safety-critical nature, collision avoidance systems must maintain a high degree of reliability while minimizing unnecessary path deviation. Recent work has investigated formulating the problem of collision avoidance as a Markov decision process (MDP) and solving for the optimal strategy using dynamic programming (DP) [14,15,25]. One limitation of this approach is that the computation and memory requirements grow exponentially with the dimensionality of the state space. Hence, these studies focused on MDP formulations that capture only a subset of the relevant state variables at the expense of impaired performance.

This paper presents a new approach for significantly reducing the computation and memory requirements for partially controlled Markov decision processes. The approach involves decomposing the problem into two separate subproblems, one controlled and one uncontrolled, that can be solved independently offline using dynamic programming.

---

* This work is sponsored by the Federal Aviation Administration under Air Force Contract #FA8721-05-C-0002. Opinions, interpretations, conclusions, and recommendations are those of the authors and are not necessarily endorsed by the United States Government.

J. Filipe and A. Fred (Eds.): ICAART 2011, CCIS 271, pp. 86–100, 2013.

During execution, the results from offline computation are combined to determine the approximately optimal action.

The approach is demonstrated on an airborne collision avoidance system that recommends vertical maneuvers to a pilot. Although the pilot may maneuver horizontally, it is assumed that the system does not influence the horizontal motion. The problem is naturally represented using seven state variables, which is impractical to solve with a reasonable level of discretization. By carefully decomposing the problem into two lower-dimensional problems, a solution can be obtained quickly and stored in primary memory. The optimized system is compared with the Traffic Alert and Collision Avoidance System (TCAS), currently mandated worldwide on all large transport aircraft [22].

The next section summarizes related work on collision avoidance. Section 3 reviews Markov decision processes. Section 4 describes the solution method and outlines the required assumptions. Section 5 applies the method to airborne collision avoidance. Section 6 evaluates the method in simulation with TCAS as a baseline. Section 7 concludes and outlines further work.

2   Related Work

A common technique for collision avoidance in autonomous and semi-autonomous vehicles is to define conflict zones for each obstacle and then use a deterministic model, such as linear extrapolation, to predict whether a conflict will occur [4,10,6]. If conflict is anticipated, the system selects the maneuver that provides minimal path deviation while preventing conflict. Such an approach requires little computation and can prevent collision much of the time, but it lacks robustness because the deterministic model ignores the stochastic nature of the environment. Although one may mitigate collision risk to some extent by artificially enlarging the conflict zones to accommodate uncertainty in the future behavior of the vehicles, this approach frequently results in unnecessary path deviation. The TCAS collision avoidance logic adopts an approach along these lines but incorporates many hand-crafted, heuristic rules to enhance robustness.

Several other approaches to collision avoidance can be found in the literature that do not use a probabilistic model of vehicle behavior, including potential field methods [13,11,12] and rapidly expanding random trees [19,18,23]. However, avoiding collision with a high degree of reliability while keeping the rate of path deviation low requires the use of a probabilistic model that accounts for future state uncertainty. Several methods have been suggested that involve using a probabilistic model to estimate the probability of conflict and to choose the maneuver that keeps the probability of conflict below some set threshold [26,5,15]. One limitation of these threshold-based approaches is that they do not model the effects of delaying the avoidance maneuver. In many cases, it can be beneficial to observe how the encounter develops before committing to a particular maneuver. The dynamic programming approach pursued in this work, in contrast, takes into account every possible future sequence of actions taken by the collision avoidance system and their outcomes when making a decision.

3   Markov Decision Processes

An MDP is defined by a transition function T and cost function C. The probability of transitioning from state s to state s' after executing action a is given by $T(s, a, s')$. The immediate cost when executing a from s is given by $C(s, a)$. In this paper, the state space S and action space A are finite [21,3].

A policy is a function π that maps states to actions. The expected sum of immediate costs when following π for K steps starting from state s is denoted $J_K^\pi(s)$, often called the cost-to-go function. The solution to an MDP with a horizon of K is a policy π_K^* that minimizes the cost to go from every state.

One way to compute π_K^* is to first compute J_K^*, the cost-to-go function for the optimal policy, using a dynamic programming algorithm known as value iteration. The function $J_0^*(s)$ is set to zero for all states s. If the state space includes terminal states with immediate cost $C(s)$, then $J_0^*(s) = C(s)$ for those terminal states. The function $J_k^*(s)$ is computed from J_{k-1}^* as follows:

$$J_k^*(s) = \min_a \left[C(s, a) + \sum_{s'} T(s, a, s') J_{k-1}^*(s') \right]. \tag{1}$$

The iteration continues until horizon K.

The expected cost to go when executing a from s and then continuing with an optimal policy for $K - 1$ steps is given by

$$J_K^*(s, a) = C(s, a) + \sum_{s'} T(s, a, s') J_{K-1}^*(s'). \tag{2}$$

An optimal policy may be obtained directly from $J_K^*(s, a)$:

$$\pi_K^*(s) = \arg \min_a J_K^*(s, a). \tag{3}$$

If the state space contains continuous variables, which is common for collision avoidance problems, the state space can be discretized using a multi-dimensional grid or simplex scheme [9]. The transition function $T(\mathbf{x}, a, \mathbf{x}')$ in continuous space can be translated into a discrete transition function $T(s, a, s')$ using a variety of sampling and interpolation methods [15]. Once the state space, transition function, and cost function have been discretized, $J^*(s, a)$ may be computed for each discrete state s and action a. For a continuous state $\mathbf{x}$ and action a, $J^*(\mathbf{x}, a)$ may be approximated using, for example, multilinear interpolation. The best action to execute from continuous state $\mathbf{x}$ is simply $\arg \min_a J^*(\mathbf{x}, a)$.

Discretizing the full state space can result in a large number of discrete states, exponential in the number of dimensions, which makes computing J^* infeasible for many problems. This "curse of dimensionality" [1] has led to a variety of different approximation methods [20].

4   Partial Control

This paper explores a new solution technique for partially controlled MDPs that is applicable to certain collision avoidance problems. It may be applied to interception-seeking or goal-oriented problems as well by incorporating negative costs. So long as the problem satisfies a set of assumptions, this solution method will provide a finite-horizon solution. The approach involves independently solving a controlled subproblem and an uncontrolled subproblem and combining the results online to identify the approximately optimal action.

4.1 Assumptions

It is assumed that the state is represented by a set of variables, some controlled and some uncontrolled. The state space of the controlled variables is denoted S_c, and the state space of the uncontrolled variables is denoted S_u. The state of the controlled variables at time t is denoted $s_c(t)$, and the state of the uncontrolled variables at time t is denoted $s_u(t)$. The solution technique may be applied when the following three assumptions hold:

1. The state $s_u(t+1)$ depends only upon $s_u(t)$. The probability of transitioning from s_u to s'_u is given by $T(s_u, s'_u)$.
2. The episode terminates when $s_u \in G \subset S_u$ with immediate cost $C(s_c)$.
3. In nonterminal states, the immediate cost $c(t+1)$ depends only upon $s_c(t)$ and $a(t)$. If the controlled state is s_c and action a is executed, the immediate cost is denoted $C(s_c, a)$.

4.2 Controlled Subproblem

Solving the controlled subproblem involves computing the optimal policy for the controlled variables under the assumption that the time until s_u enters G, denoted τ, is known. In an airborne collision avoidance context, τ may be the number of steps until another aircraft comes within 500 ft horizontally. Of course, τ cannot be determined exactly from $s_u(t)$ because it depends upon an event that occurs in the future, but this will be addressed by the uncontrolled subproblem (Section 4.3).

The cost to go from s_c given τ is denoted $J_\tau(s_c)$. The series $J_0, \ldots, J_K$ is computed recursively, starting with $J_0(s_c) = C(s_c)$ and iterating as follows:

$$J_k(s_c) = \min_a \left[C(s_c, a) + \sum_{s'_c} T(s_c, a, s'_c) J_{k-1}(s'_c) \right]. \tag{4}$$

The expected cost to go from s_c when executing a for one step and then following the optimal policy is given by

$$J_k(s_c, a) = C(s_c, a) + \sum_{s'_c} T(s_c, a, s'_c) J_{k-1}(s'_c). \tag{5}$$

The K-step expected cost to go when $\tau > K$ is denoted $J_{\bar{K}}$. It is computed by initializing $J_0(s_c) = 0$ for all states and iterating equation 4 to horizon K. The series $J_0, \ldots, J_K, J_{\bar{K}}$ is saved in a table in memory, requiring $O(K|A||S_c|)$ entries.

4.3 Uncontrolled Subproblem

Solving the uncontrolled subproblem involves using the probabilistic model of the uncontrolled dynamics to infer a distribution over τ for each uncontrolled state s_u. This distribution is referred to as the entry time distribution because it represents the distribution over the time for s_u to enter G. The probability that s_u enters G in τ steps is denoted $D_\tau(s_u)$ and may be computed using dynamic programming. The probability that $\tau = 0$ is given by

$$D_0(s_u) = \begin{cases} 1 & \text{if } s_u \in G, \\ 0 & \text{otherwise.} \end{cases} \tag{6}$$

The probability that $\tau = k$, for $k > 0$, is computed from D_{k-1} as follows:

$$D_k(s_u) = \begin{cases} 0 & \text{if } s_u \in G, \\ \sum_{s'_u} T(s_u, s'_u) D_{k-1}(s'_u) & \text{otherwise.} \end{cases} \tag{7}$$

Depending on s_u, there may be some probability that s_u does not enter G within K steps. This probability is denoted $D_{\bar{K}}(s_u)$ and may be computed from $D_0, \ldots, D_K$:

$$D_{\bar{K}}(s_u) = 1 - \sum_{k=0}^{K} D_k(s_u). \tag{8}$$

The sequence $D_0, \ldots, D_K, D_{\bar{K}}$ is stored in a table with $O(K|S_u|)$ entries. Multilinear interpolation of the distributions may be used to determine $D_\tau(\mathbf{x}_u)$ at an arbitrary continuous state $\mathbf{x}_u$.

4.4 Online Solution

After $J_0, \ldots, J_K, J_{\bar{K}}$ and $D_0, \ldots, D_K, D_{\bar{K}}$ have been computed offline, they are used together online to determine the approximately optimal action to execute from the current state. For any discrete state s in the original state space, $J_K^*(s, a)$ may be computed as follows:

$$J_K^*(s, a) = D_{\bar{K}}(s_u) J_{\bar{K}}(s_c, a) + \sum_{k=0}^{K} D_k(s_u) J_k(s_c, a), \tag{9}$$

where s_u is the discrete uncontrolled state and s_c is the discrete controlled state associated with s. Combining the controlled and uncontrolled solutions online in this way requires time linear in the size of the horizon. Multilinear interpolation can be used to estimate $J_K^*(\mathbf{x}, a)$ for an arbitrary state $\mathbf{x}$, and from this the optimal action may be obtained.

The memory required to store $J_K^*(s, a)$ is $O(|A||S_c||S_u|)$. However, the method in this section allows the solution to be represented using $O(K|A||S_c| + K|S_u|)$ storage, which can be a tremendous savings when $|S_c|$ and $|S_u|$ are large. For the collision avoidance problem discussed in the next section, this method allows the cost table to be stored in 500 MB instead of over 1 TB. The offline computational savings are even more significant.

An alternative to using dynamic programming for computing the entry time distribution offline is to use Monte Carlo to estimate the entry time distribution online. A Monte Carlo approach does not require the uncontrolled variables to be discretized and does not require $D_0, \ldots, D_K, D_{\bar{K}}$ to be stored in memory. However, using Monte Carlo increases the amount of online computation. For problems where the conflict region is small, the number of samples required to produce an adequate estimate of the distribution may be large, though importance sampling can help improve this estimate with fewer samples [8].

5 Airborne Collision Avoidance System

This section demonstrates the approach from the previous section on an MDP representing an airborne collision avoidance problem. In this problem, the collision avoidance system issues resolution advisories to pilots who then adjust their vertical rate to avoid coming within 500 ft horizontally and 100 ft vertically of an intruding aircraft. This section considers a simplified version of the collision avoidance problem in which one aircraft equipped with a collision avoidance system, called the own aircraft, encounters only one other unequipped aircraft, called the intruder aircraft. The remainder of the section outlines the assumptions and decomposes the problem into controlled and uncontrolled subproblems.

5.1 Assumptions

In this problem, s_c represents the state of the vertical motion variables, and s_u represents the state of the horizontal motion variables. This problem defines coming within 500 ft horizontally and 100 ft vertically of an intruder as a conflict.

The first assumption in Section 4.1 requires that $s_u(t+1)$ depend only upon $s_u(t)$. In this collision avoidance problem, it is assumed that pilots randomly maneuver horizontally, and that the advisories issued by the collision avoidance system do not influence the horizontal motion.

The second assumption requires the episode to terminate when s_u enters G. In this problem, G is the set of states where there is a horizontal conflict, defined to be when an intruder comes within 500 ft horizontally. The immediate cost when this occurs is given by $C(s_c)$, which is one when the intruder is within 100 ft vertically and zero otherwise. In simulation, the episode does not terminate when s_u enters G, since entering G does not necessarily imply that there has been a conflict (e.g., the two aircraft may have safely missed each other by 1000 ft vertically). However, it is generally sufficient to plan up to the moment where s_u enters G because adequate separation at that moment usually indicates that the encounter is resolved.

The third assumption requires that for states where $s_u \notin G$ the immediate cost function depends on the controlled state variables and the action. As outlined in Section 5.2, the nonterminal cost function only depends on the advisory state and the advisory being issued.

5.2   Controlled Subproblem

The controlled subproblem, formulated as an MDP, is defined by the available actions, the dynamics, and the cost function. The dynamics are determined by the pilot response model and aircraft dynamic model. The cost function takes into account both safety and operational considerations. In addition to describing these components of the MDP, this section discusses the resulting optimal policy.

Resolution Advisories. The airborne collision avoidance system may choose to issue one of two different initial advisories: climb at least 1500 ft/min or descend at least 1500 ft/min. Following the initial advisory, the system may choose to either terminate, reverse, or strengthen the advisory. An advisory that has been reversed requires a vertical rate of 1500 ft/min in the opposite direction of the original advisory. An advisory that has been strengthened requires a vertical rate of 2500 ft/min in the direction of the original advisory. After an advisory has been strengthened, it can then be weakened to reduce the required vertical rate to 1500 ft/min in the direction of the original advisory.

Dynamic Model. The state is represented using four variables:

- h: altitude of the intruder relative to the own aircraft,
- $\dot{h}_0$: vertical rate of the own aircraft,
- $\dot{h}_1$: vertical rate of the intruder aircraft, and
- s_{RA}: the state of the resolution advisory.

The discrete variable s_{RA} contains the necessary information to model the pilot response, which includes the active advisory and the time to execution by the pilot. Five seconds are required for the pilot to begin responding to an initial advisory. The pilot then applies a 1/4 g acceleration to comply with the advisory. Subsequent advisories are followed with a 1/3 g acceleration after a three second delay. When an advisory is not active, the pilot applies an acceleration selected at every step from a zero-mean Gaussian with 3 ft/s^2 standard deviation. At each step, the intruder pilot independently applies a random acceleration from a zero-mean Gaussian with 3 ft/s^2 standard deviation.

The continuous state variables are discretized according to the scheme in Table 1. The discrete state transition probabilities were computed using sigma-point sampling and multilinear interpolation [15]. This discretization scheme produces a discrete model with 213 thousand discrete states.

Table 1. Controlled Variable Discretization

Variable	Grid Edges
h	$-1000, -900, \ldots, 1000$ ft
$\dot{h}_0$	$-2500, -2250, \ldots, 2500$ ft/min
$\dot{h}_1$	$-2500, -2250, \ldots, 2500$ ft/min

Cost Function. An effective collision avoidance system must satisfy competing objectives, including maximizing safety and minimizing the rate of unnecessary alerts. These objectives are encoded in the cost function. In addition to incurring a cost for conflict, it is desirable to incur a cost for other events such as alerting or changing an advisory, as shown in Table 2. A small negative cost is awarded at every step the system is not alerting to provide an incentive to discontinue alerting after resolution of the encounter.

Table 2. Event Costs

Conflict	Alert	Strengthening	Reversal	Clear of Conflict
1	0.001	0.009	0.01	$-1 \cdot 10^{-4}$

Optimal Policy. The optimal cost-to-go tables $J_0, \ldots, J_K, J_{\bar{K}}$ were computed offline in less than two minutes on a single 3 GHz Intel Xeon core using a horizon of $K = 39$ steps. Storing only the values for the valid state-action pairs requires 263 MB using a 64-bit floating point representation. Figure 1 shows a plot of the optimal policy through a slice of the state space where the own aircraft is initially climbing at 1500 ft/min, the intruder is level, and no alert has been issued. The blue region indicates where the logic will issue a descend advisory, and the green region indicates where the logic will issue a climb advisory. The optimal policy will sometimes issue a climb even when the intruder is above. This occurs when the aircraft are closely separated in altitude and little time remains until potential conflict. Because the own aircraft is already climbing, there is insufficient time to accelerate downward to avoid conflict. Climbing above the intruder is more effective. Another notable feature of the plot is that no advisory is issued when $\tau \leq 5$ s. Because an advisory has no effect until five seconds after it is issued, alerting less than five seconds prior to conflict is ineffective.

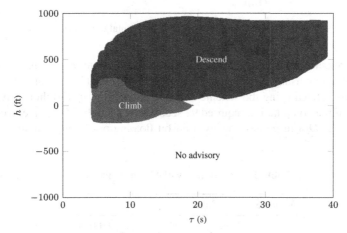

Fig. 1. Optimal action plot for $\dot{h}_0 = 1500$ ft/min, $\dot{h}_1 = 0$ ft/min, $s_{\text{RA}} =$ "no advisory"

5.3   Uncontrolled Subproblem

The uncontrolled subproblem involves estimating the distribution over τ (i.e., the time until the aircraft are separated less than 500 ft horizontally) given the current state. This section describes the horizontal dynamics and three methods for estimating the entry time distribution.

Dynamic Model. The aircraft move in the horizontal plane in response to independent random accelerations generated from a zero-mean Gaussian with a standard deviation of $3\,\text{ft/s}^2$. The motion can be described by a three-dimensional model, instead of the typical four-dimensional (relative positions and velocities) model, due to rotational symmetry in the dynamics. The three state variables are as follows:

- r: horizontal range to the intruder,
- r_v: relative horizontal speed, and
- θ_v: difference in the direction of the relative horizontal velocity and the bearing of the intruder.

These variables are illustrated in Figure 2.

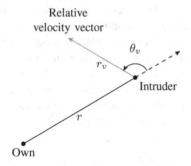

Fig. 2. Three-variable model of horizontal dynamics

Dynamic Programming Entry Time Distribution. The entry time distribution can be estimated offline using dynamic programming as discussed in Section 4.3. The state space was discretized using the scheme in Table 3, resulting in 730 thousand discrete states. The offline computation required 92 seconds on a single 3 GHz Intel Xeon core. Storing $D_0, \ldots, D_{39}$ in memory using a 64-bit floating point representation requires 222 MB.

Table 3. Uncontrolled Variable Discretization

Variable	Grid Edges
r	$0, 50, \ldots, 1000, 1500, \ldots, 40000\,\text{ft}$
r_v	$0, 10, \ldots, 1000\,\text{ft/s}$
θ_v	$-180°, -175°, \ldots, 180°$

Monte Carlo Entry Time Distribution. Monte Carlo estimation can be used online to estimate the entry time distribution as explained at the end of Section 4.3. The experiments in this paper use 100 Monte Carlo samples to estimate τ.

Simple Entry Time Distribution. A point estimate of τ can be obtained online as follows. The range rate is given by

$$\dot{r} = r_v \cos(\theta_v). \tag{10}$$

If the aircraft are converging in range, then τ can be approximated by $r/|\dot{r}|$. Otherwise, τ is set beyond the horizon.

6   Results

This section evaluates the performance of the collision avoidance system using simulated encounters and compares it against the current version of TCAS, Version 7.1.

6.1   Encounter Initialization

Encounters are initialized in the horizontal plane by randomly and independently generating the initial ground speeds of both aircraft, s_0 and s_1, from a uniform distribution between 100 and 500 ft/s. The horizontal range between the aircraft is initialized to $r = t_{\text{target}}(s_0 + s_1) + u_r$, where u_r is a zero-mean Gaussian with 500 ft standard deviation. The parameter t_{target}, nominally set to 40 s, controls how long until the aircraft come into conflict.

The bearing of the intruder aircraft with respect to the own aircraft, χ, is sampled from a zero-mean Gaussian distribution with a standard deviation of $2°$. The heading of the intruder with respect to the heading of the own aircraft, β, is sampled from a Gaussian distribution with a mean of $180°$ and a standard deviation of $2°$. When $\beta = 180°$, the intruder is heading directly toward the own aircraft.

The initial vertical rates $\dot{h}_0$ and $\dot{h}_1$ are drawn independently from a uniform distribution spanning -1000 and 1000 ft/min. The initial altitude of the own aircraft, h_0, is set to 43,000 ft. The initial altitude of the intruder is $h_0 + t_{\text{target}}(\dot{h}_0 - \dot{h}_1) + u_h$, where u_h is a zero-mean Gaussian with 25 ft standard deviation.

6.2   Example Encounter

Figure 3 shows an example encounter comparing the behavior of the system using the DP entry time distribution against the TCAS logic. Figure 4 shows the entry time distribution computed using the three methods of Section 5.3 at the first alerting point ($t = 17$ s) of the DP logic in the example encounter.

Seventeen seconds into the encounter, the DP logic issues a descend to pass below the intruder. The expected cost to go for issuing a descend advisory is approximately 0.00928, lower than the expected cost to go for issuing a climb advisory (0.0113) or for not issuing an advisory (0.00972). The DP entry time distribution at this time has a conditional mean $E[\tau \mid \tau < 40\,\text{s}]$ of approximately 12.01 s, and a considerable portion of the probability mass ($\sim 40\%$) is assigned to $\tau \geq 40$ s. The Monte Carlo entry time distribution, in comparison, has less support but a comparable conditional mean of 17.12 s. Only 15% of the probability mass is concentrated on $\tau \geq 40$ s. The point estimate of τ using the simple method is 21.65 s.

After the descend advisory is issued, the intruder begins to increase its descent, causing the DP logic to reverse the descend to a climb 20 seconds into the encounter. The pilot begins the climb maneuver three seconds later. Once the aircraft are safely separated, the DP logic discontinues the advisory at $t = 31$ s. The minimum horizontal separation is 342 ft, at which time the vertical separation is 595 ft. No conflict occurs.

TCAS initially issues a climb advisory four seconds into the encounter because it anticipates, using straight-line projection, that by climbing it can safely pass above the intruder. Nine seconds later, when the own aircraft is executing its climb advisory, TCAS reverses the climb to a descend because it projects that maintaining the climb will not provide the required separation. TCAS strengthens the advisory three seconds later, but fails to resolve the conflict. The aircraft miss each other by 342 ft horizontally and 44 ft vertically. Although the TCAS logic alerts earlier and more often, the DP logic still outperforms it in this example encounter.

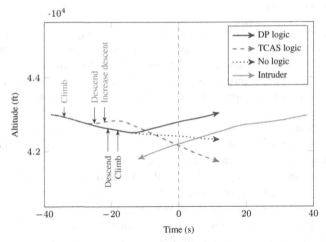

Fig. 3. Example encounter comparing the system with the DP entry time distribution against TCAS

6.3 Performance Evaluation

Table 4 summarizes the results of simulating the DP logic and the TCAS logic on one million encounters generated by the model of Section 5. The table summarizes the number of conflicts, alerts, strengthenings, and reversals.

As the table shows, the DP logic can provide a much lower conflict rate while significantly reducing the alert rate. The Monte Carlo entry time distribution results in a greater number of conflicts, but it alerts less frequently than the other methods. Increasing the number of samples used generally improves performance but increases online

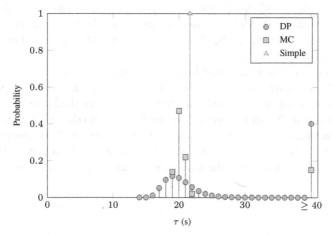

Fig. 4. Entry time distribution computed using dynamic programming (DP), Monte Carlo (MC), and the simple (Simple) methods at $t = 17$ s

computation time. The DP logic using the simple point estimate of τ resolves all but one conflict while rarely reversing or strengthening the advisory, but alerts more frequently than Monte Carlo.

Table 4. Performance Evaluation

	DP Logic (DP Entry)	DP Logic (MC Entry)
Conflicts	2	11
Alerts	540,113	400,457
Strengthenings	39,549	37,975
Reversals	1242	747

	DP Logic (Simple Entry)	TCAS Logic
Conflicts	1	101
Alerts	939,745	994,640
Strengthenings	26,485	45,969
Reversals	129	193,582

6.4 Safety Curve

The results of Section 6.3 considered the performance of the system optimized using the fixed event costs of Table 2. Figure 5 shows the safety curves for the DP logic and TCAS when different parameters are varied.

The DP logic safety curves were produced by varying the cost of alerting from zero to one while keeping the other event costs fixed. Separate curves were produced for the three methods for estimating the entry time distribution. The upper-right region of the plot corresponds to costs of alerting near zero and the lower-left region corresponds to costs near one.

The safety curve for TCAS was generated by varying the sensitivity level of TCAS. The sensitivity level of TCAS is a system parameter of the logic that increases with altitude. At higher sensitivity levels, TCAS will generally alert earlier and more aggressively to prevent conflict.

The safety curves show that the DP logic can exceed or meet the level of safety provided by TCAS while alerting far less frequently. The safety curves can aid in choosing an appropriate value for the cost of alerting that satisfies a required safety threshold.

Figure 5 also reveals that the DP and Monte Carlo methods for estimating τ offer similar performance and that they both outperform the simple method, especially when the cost of alerting is high and the logic can only alert sparingly to prevent conflict. In the upper-right region of the plot, the three methods are nearly indistinguishable.

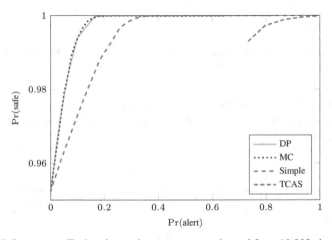

Fig. 5. Safety curves. Each point on the curves was estimated from 10,000 simulations.

7   Conclusions and Further Work

This paper presented a method for solving large MDPs that satisfy certain assumptions by decomposing the problem into controlled and uncontrolled subproblems that can be solved independently offline and recombined online. The method was applied to airborne collision avoidance and was compared against TCAS, a system that was under development for several decades and has a proven safety record.

The experiments demonstrate that the collision avoidance logic that results from solving the MDP using the method presented in this paper reduces the risk of collision by a factor of 50 while issuing fewer alerts than TCAS in the simulated encounters. The system reverses less than 1% of the time that TCAS reverses, and the system strengthens less frequently as well. It should be emphasized that further simulation studies using more realistic encounter models are required to quantify the expected performance of the DP logic [16].

Real collision avoidance systems have imperfect sensors, which results in state uncertainty. TCAS currently relies on radar beacon surveillance, which results in somewhat significant uncertainty in the intruder bearing. When state uncertainty is significant, the uncertainty must be taken into account when choosing actions. With a sensor model, the problem may be transformed into a partially observable Markov decision process (POMDP) and solved approximately using various methods [7,24,17].

Another area of further research involves introducing coordination between aircraft. If both aircraft have a collision avoidance system on board, then safety can be enhanced by coordinating their maneuvers. If either the sensor measurements are perfect or the communication between aircraft is perfect and unlimited, then the problem can be modeled as a larger MDP. Otherwise, the problem turns into a Decentralized POMDP (Dec-POMDP), which are, in general, impractical to solve exactly [2]. Further research will investigate the performance of MDP-derived policies and strategies for leveraging the structure of the problem to reduce the complexity of finding an acceptable solution.

Acknowledgements. This work is the result of research sponsored by the TCAS Program Office at the Federal Aviation Administration. The authors appreciate the support provided by the TCAS Program Manager, Neal Suchy. This work has benefited from discussions with Leslie Kaelbling and Tomas Lozano-Perez from the MIT Computer Science and Artificial Intelligence Laboratory.

References

1. Bellman, R.E.: Adaptive control processes: A guided tour. Princeton University Press (1961)
2. Bernstein, D.S., Zilberstein, S., Immerman, N.: The complexity of decentralized control of Markov decision processes. In: Conference on Uncertainty in Artificial Intelligence, pp. 32–37. Morgan Kaufmann (2000)
3. Bertsekas, D.P.: Dynamic Programming and Optimal Control, 3rd edn., vol. 1. Athena Scientific, Belmont (2005)
4. Bilimoria, K.D.: A geometric optimization approach to aircraft conflict resolution. In: AIAA Guidance, Navigation, and Control Conference and Exhibit, Denver, Colo. (2000)
5. Carpenter, B.D., Kuchar, J.K.: Probability-based collision alerting logic for closely-spaced parallel approach. In: AIAA 35th Aerospace Sciences Meeting, Reno, NV (January 1997)
6. Chamlou, R.: Future airborne collision avoidance—design principles, analysis plan and algorithm development. In: Digital Avionics Systems Conference (2009)
7. Chryssanthacopoulos, J.P., Kochenderfer, M.J.: Accounting for state uncertainty in collision avoidance. Journal of Guidance, Control, and Dynamics 34(4), 951–960 (2011)
8. Chryssanthacopoulos, J.P., Kochenderfer, M.J., Williams, R.E.: Improved Monte Carlo sampling for conflict probability estimation. In: AIAA Non-Deterministic Approaches Conference, Orlando, Florida (2010)
9. Davies, S.: Multidimensional triangulation and interpolation for reinforcement learning. In: Mozer, M.C., Jordan, M.I., Petsche, T. (eds.) Advances in Neural Information Processing Systems, vol. 9, pp. 1005–1011. MIT Press, Cambridge (1997)
10. Dowek, G., Geser, A., Muñoz, C.: Tactical conflict detection and resolution in a 3-D airspace. In: 4th USA/Europe Air Traffic Management R&D Seminar, Santa Fe, New Mexico (2001)
11. Duong, V.N., Zeghal, K.: Conflict resolution advisory for autonomous airborne separation in low-density airspace. In: IEEE Conference on Decision and Control, December 10-12, vol. 3, pp. 2429–2434 (1997)

12. Eby, M.S., Kelly, W.E.: Free flight separation assurance using distributed algorithms. In: IEEE Aerospace Conference, March 6-13, vol. 2, pp. 429–441 (1999)
13. Khatib, O., Maitre, J.F.L.: Dynamic control of manipulators operating in a complex environment. In: Symposium on Theory and Practice of Robots and Manipulators, pp. 267–282. Elsevier, Udine (1978)
14. Kochenderfer, M.J., Chryssanthacopoulos, J.P.: A decision-theoretic approach to developing robust collision avoidance logic. In: IEEE International Conference on Intelligent Transportation Systems, Madeira Island, Portugal (2010)
15. Kochenderfer, M.J., Chryssanthacopoulos, J.P., Kaelbling, L.P., Lozano-Perez, T.: Model-based optimization of airborne collision avoidance logic. Project Report ATC-360, Massachusetts Institute of Technology, Lincoln Laboratory (2010)
16. Kochenderfer, M.J., Edwards, M.W.M., Espindle, L.P., Kuchar, J.K., Griffith, J.D.: Airspace encounter models for estimating collision risk. Journal of Guidance, Control, and Dynamics 33(2), 487–499 (2010)
17. Kurniawati, H., Hsu, D., Lee, W.: SARSOP: Efficient point-based POMDP planning by approximating optimally reachable belief spaces. In: Robotics: Science and Systems (2008)
18. Kuwata, Y., Fiore, G.A., Teo, J., Frazzoli, E., How, J.P.: Motion planning for urban driving using RRT. In: IEEE/RSJ International Conference on Intelligent Robots and Systems, Sepember 22-26, pp. 1681–1686 (2008)
19. LaValle, S.M.: Rapidly-exploring random trees: A new tool for path planning. Tech. Rep. 98-11, Computer Science Department, Iowa State University (October 1998)
20. Powell, W.B.: Approximate Dynamic Programming: Solving the Curses of Dimensionality. Wiley, Hoboken (2007)
21. Puterman, M.L.: Markov Decision Processes: Discrete Stochastic Dynamic Programming. Wiley series in probability and mathematical statistics. Wiley, New York (1994)
22. RTCA: Minimum operational performance standards for Traffic Alert and Collision Avoidance System II (TCAS II), DO-185b. RTCA, Inc., Washington, D.C. (June 2008)
23. Saunders, J., Beard, R., Byrne, J.: Vision-based reactive multiple obstacle avoidance for micro air vehicles. In: American Control Conference, June 10-12, pp. 5253–5258 (2009)
24. Smith, T., Simmons, R.G.: Point-based POMDP algorithms: Improved analysis and implementation. In: Uncertainty in Artificial Intelligence (2005)
25. Temizer, S., Kochenderfer, M.J., Kaelbling, L.P., Lozano-Pérez, T., Kuchar, J.K.: Collision avoidance for unmanned aircraft using Markov decision processes. In: AIAA Guidance, Navigation, and Control Conference, Toronto, Canada (2010)
26. Yang, L.C., Kuchar, J.K.: Prototype conflict alerting system for free flight. Journal of Guidance, Control, and Dynamics 20(4), 768–773 (1997)

A Restricted Model Space Approach for the Detection of Epistasis in Quantitative Trait Loci Using Markov Chain Monte Carlo Model Composition

Edward L. Boone[1,*], Susan J. Simmons[2], and Karl Ricanek[2]

[1] Virginia Commonwealth University, Richmond, Virginia, U.S.A.
[2] University of North Carolina Wilmington, Wilmington, North Carolina, U.S.A.
elboone@vcu.edu, {simmonssj,ricanekk}@uncw.edu
http://faceaginggroup.com/

Abstract. Epistasis or the interaction between loci on a genome that controls a quantitative trait is of great interest to geneticists. This work presents a powerful Bayesian method utilizing Markov chain Monte Carlo model composition approach using restricted spaces is developed for identifying epistatic effects in Recombinant Inbred Lines (RIL) in plant studies. This method produces both posterior activation probabilities and posterior conditional activation probabilities. The method is verified through a simulation study and applied to an *Arabidopsis thaliana* data set with cotyledon as the quantitative trait.

Keywords. Quantitative trait loci, Epistasis, Bayesian statistics, Markov chain Monte Carlo model composition.

1 Introduction

Quantitative Trait Loci (QTL) analysis determines which region(s) on a genome explains or controls a quantitative trait. However, in many instances an iteraction between regions or loci may provide a better explanation for a trait than regions having a strictly additive influence. This interaction between loci on a genome is known as *epistasis*. To study QTL in plant species, organisms generated by recombinant inbreeding are often used. Recombinant Inbred Lines (RIL) are plants that have been repeatedly mated with siblings and themselves in order to create an inbred line whose genetic structure is a combination of the original parent lines. These RILs provide a mechanism to reduce environmental and individual effects. Furthermore, by utilizing RILs, the alleles at each loci are homozygous and help simplify the search for QTL. For a complete review of RILs see Broman [1].

Several methods have been developed to detect and evaluate epistatic effects for continuous traits. Multiple Interval Mapping (MIM) proposed by [10] is based on fitting a multiple regression model that has both main effect terms as well as interactions and employs a non-Bayesian search method. Carlborg et al. [5] use a genetic algorithm to search for the loci and epistatic effects. Hensen et al. [9] propose a theoretical framework for higher order interactions. Kao et al. [11] use the framework of [6] to partition

---

* Corresponding author.

J. Filipe and A. Fred (Eds.): ICAART 2011, CCIS 271, pp. 101–114, 2013.
© Springer-Verlag Berlin Heidelberg 2013

the variance for known main and epistatic effects in order to understand the contribution of each with no search method. Zeng et al. [18] and [12] use [6] partition the variance when epistatic effects with multiple alleles are present however no search method is presented in this work. Hanlon and Lorenz [8] use a optimization approach to find combinations of epistatic effects that best represent the trait of interest based on squared error distance.

To avoid the issue of model selection, Broman and Speed [2] use Markov Chain Monte Carlo Model Composition (MC^3) to search for the main effects (additive models) that contribute to the trait. This procedure is a variant of reversible jump Markov chain Monte Carlo by [7]. Boone et al. [4] extend this to restricted model spaces to allow for situations where a genome contains more loci than plant lines. Yi et al. in [14], [15], [16], [17] use the MC^3 framework with various restrictions on the model space to search for main and epistatic effects. However, [14], [15], [16], [17] and the R/qtlbim software of [13] do not require that the main effect terms corresponding to the epistatic effects be present in the model. Furthermore, [2], [14], [15], [16], [17] and [13] employ information criteria such as AIC or BIC as the basis for the MC^3 search. Boone et al. [3] show that while BIC is an asymptotically correct approximation for posterior model probabilities, in the low to moderate sample size cases, BIC performs poorly.

This work uses activation probabilities, defined in Section 2.2 for each of the main and epistatic effects to determine the marginal posterior probability of each effect regardless of which model is chosen. Figure 1 shows an example heatmap of the activation probabilities that may occur when epistasis is present. Activation probabilities along the diagonal correspond to the main effects of the loci. The off diagonal activation probabilities correspond to epistatic effects. Notice that by looking along the diagonal the main effects appear to be at locus 12, locus 26 and locus 35 as the $(12, 12)$, $(26, 26)$ and $(35, 35)$ regions have high probability. Furthermore one can look at the off diagonal regions and see that loci 12 and 26 appear to have an epistatic effect denoted by high probability in the $(12, 26)$ region. However, loci 12 and 35 and loci 26 and 35 do not appear to have an epistatic effect due to low probability in the regions common to $(12, 35)$ and $(26, 35)$ on the heatmap.

Section 2 defines the model, basic search strategy, activation probabilities and conditional activation probabilities. Section 2.3 explains the neighborhood definition and search strategy under restricted model spaces. Section 3 gives a simulation study showing the efficacy of the method for detecting both main effects and two-way interaction effects. Section 4 considers the *Arabidopsis Thaliana* as an example. The dataset for this model organism has 158 lines of RIL and 38 markers (loci) and cotelydon opening angle is the quantitative trait of interest.

2   Bayesian Model Search

2.1   Model Definition

Let y_i be the quantitative trait value for the ith observation. For each of the p loci $l_1, l_2, ..., l_p$ the parentage of the allele is recorded as A if the allele came from parent A and B if the allele came from parent B. However, in some instances the allele is not

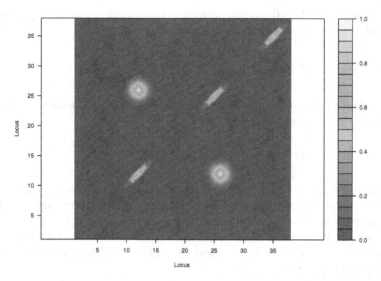

Fig. 1. Simulated heatmap of activation probabilities for main effect and epistatic effects. Activation probabilities along the diagonal correspond to main effects and off diagonal correspond to epistatic effects.

determined which needed to be reflected in the analysis. For the ith observation and locus l_j this information can be coded into X_{ij} as:

$$X_{ij} = \begin{cases} 1, & \text{allele } l_j \text{ is from parent A} \\ -1, & \text{allele } l_j \text{ is from parent B} \\ 0, & \text{allele } l_j \text{ is undetermined} \end{cases} \tag{1}$$

Here the X_{ij} correspond to the main effects. For the epistatic effects (two-way interaction) this produces the interaction between loci l_j and l_k as:

$$X_{ij}X_{ik} = \begin{cases} 1, & \text{alleles } l_j \text{ and } l_k \text{ from same parent} \\ -1, & \text{alleles } l_j \text{ and } l_k \text{ from different parents} \\ 0, & \text{allele } l_j \text{ or } l_k \text{ is undetermined} \end{cases} \tag{2}$$

Using a traditional first order model with a two-way multiplicative interaction terms the model is defined as:

$$y_i = \mu + \sum_{j=1}^{p} \beta_j X_{ij} I_{P_c}(l_j) + \sum_{k<j} \beta_{jk} X_{ij} X_{ik} I_{P_c}(l_j) I_{P_c}(l_k) + \epsilon_i. \tag{3}$$

where $\epsilon_i \sim N(0, \sigma_c^2)$, P_c is the set of loci l_j in model M_c, and I_{P_c} is an indicator function that takes the value 1 if $l_j \in P_c$ and 0 otherwise. Here β_j corresponds to the main effect of locus l_j and β_{jk} is the epistatic effect between loci l_j and l_k.

2.2 Bayesian Model Averaging

In a model space $\mathcal{M}$ with $|\mathcal{M}|$ models, the posterior probability of model M_c given the data $\mathbf{D}$ can be computed via Bayes' Theorem:

$$P(M_c|\mathbf{D}) = \frac{P(M_r)P(\mathbf{D}|M_c)}{\sum_{t=1}^{|\mathcal{M}|} P(M_t)P(\mathbf{D}|M_t)}. \tag{4}$$

The marginal probability of the data $\mathbf{D}$ given model M_c, $P(\mathbf{D}|M_c)$ is involved in computing (4) and can be calculated using:

$$P(\mathbf{D}|M_c) = \int P(\theta_c|M_c)P(\mathbf{D}|\theta_c, M_c)d\theta_c, \tag{5}$$

where θ_c is the parameter vector corresponding to model M_c. Evaluating the integral in (5) can be complicated. Approximations such as the Laplace approximation and the approximation based on Schwarz Bayesian Information Criterion (BIC) could be employed. However, in the linear model case, as in equation (3), where the coefficient vector for model M_c, $\beta_c \sim N(\mu_c, V_c)$ and $\sigma_c^2 \sim Inv - \chi^2(\nu, \lambda)$ prior is used, an analytic expression for (5) is:

$$\begin{aligned}
P(\mathbf{D}|\mu_c, V_c, \nu, X_c, M_c) = & \frac{\Gamma\left(\frac{\nu+n}{2}\right)(\nu\lambda)^{\frac{\nu}{2}}}{\pi^{\frac{n}{2}}\Gamma\left(\frac{\nu}{2}\right)|I + X_cV_cX_c'|^{1/2}} \\
& \times [\lambda\nu + (Y - X_c\mu_c)' \\
& \times (I + X_cV_cX_c')^{-1} \\
& \times (Y - X_c\mu_c)]^{-\frac{\nu+n}{2}},
\end{aligned} \tag{6}$$

where μ_c and V_c are the mean vector and variance-covariance matrix, respectively, and ν and λ are the degrees of freedom, and scale parameter, respectively. This work will employ (6) for computing (5) versus any information criterion based on approximations.

In cases where the model space is sufficiently large, calculating (5) for each model is computationally infeasible. A stochastic search through the model space can be performed using a metropolis-hastings approach. This can be accomplished by constructing neighborhoods around the current model M_c. Typically, the neighborhoods $nbd(M_c)$ consist of all models with one additional term than model M_c and all models with one less term than model M_c. For a candidate model $M_t \in nbd(M_c)$ the probability, α, of acceptance of model M_t is given by.

$$\alpha = \min\left\{1, \frac{P(M_t)P(\mathbf{D}|M_t)}{P(M_c)P(\mathbf{D}|M_c)}\frac{q(M_t|M_c)}{q(M_c|M_t)}\right\}, \tag{7}$$

where $q(M_t|M_c)$ is the probability that the candidate model is M_t is selected for consideration given the current state is model M_c. Note the neighborhood structure mentioned above is not appropriate when the main effect terms are required to be in the model whenever an epistatic term is in the model. [14], [15], [16], [17] and [13] allow the neighborhood to be all models with one main effect term more or less than M_c and all models with one epistatic effect more or less than M_c. These previously proposed models do not require a main effect to be present in the model when its interaction term is included. This may induce alaising, and therefore the proposed methodology herein ensures main effects are in the model prior to any interaction effects. A further restriction on previous methods is that candidate models can only include loci that are near to loci in the current model. Our methodology does not require this restrriction and allows more flexibility.

Once the posterior model probabilities have been computed activation probabilities can be used to assess the impact of loci X_j and can be computed via:

$$P(\beta_j \neq 0|\mathbf{D}) = \sum_{c=1}^{|\mathcal{M}|} P(\beta_j \neq 0|\mathbf{D}, M_c)P(M_c|\mathbf{D}). \tag{8}$$

Activation probabilities are different from the traditional p-value in that large values indicate significance versus small values. In addition, activation probabilities do not depend on a specific model as do p-values. The activation probabilities can be calculated via MC^3 as defined in section 2.3.

Activation probabilities will have a problem detecting two-way interactions when the main effect terms are required to be in the model in order for the two-way interaction term to be present. This induces the following inequalities:

$$P(\beta_{jk}|\mathbf{D}) \leq P(\beta_j|\mathbf{D}) \tag{9}$$
$$P(\beta_{jk}|\mathbf{D}) \leq P(\beta_k|\mathbf{D}).$$

Hence, using the standard activation probabilities for two-way interaction effects will produce probabilities that are damped. In order to amplify the activation probabilities of the two-way interaction effects one can use conditional activation probabilities. Conditional activation probabilities can also be obtained by:

$$P(\beta_{jk} \neq 0|\beta_j \neq 0, \beta_k \neq 0, \mathbf{D}) \tag{10}$$
$$= \frac{P(\beta_{jk} \neq 0, \beta_j \neq 0, \beta_k \neq 0|\mathbf{D})}{P(\beta_j \neq 0, \beta_k \neq 0|\mathbf{D})},$$

provided that $P(\beta_j \neq 0, \beta_k \neq 0|\mathbf{D}) > 0$. In practice one should only consider conditional activation probabilities when both $P(\beta_j|\mathbf{D})$ and $P(\beta_k|\mathbf{D})$ are considerably large. In cases where $P(\beta_j|\mathbf{D})$ or $P(\beta_k|\mathbf{D})$ are small then unreasonably large inflations to the conditional activation probabilities will occur and hence the result in incorrect inferences.

2.3 Restricted Model Space

A simple approach to defining the neighborhoods of a model M_c is to include all models that add an additional term or drop an existing term. However, this violates a model that require both main effect terms need to be present in the model in order for the corresponding two-way interaction to be added. Furthermore, the model need not contain all interaction terms possible. Notice this creates a large model space. For the first order models with p predictors the size of the model space is 2^p. However with the addition of interaction terms, the size grows considerably more. In a dataset with 30 loci, a full model with all first order terms and two-way interaction terms will have 465 terms. This can be prohibitively large for most datasets and algorithms. If the model space is restricted to $r < p$ predictors and the corresponding epistasis terms, then any model considered will not have nearly as many terms. If r is chosen wisely, then the researcher can ensure that each model under consideration has sufficient degrees of freedom for parameter estimation.

Furthermore, cases where linear dependencies exist among the predictors estimation can be complicated. One approach to address this issue is to assign $P(M_c) = 0$ to all models where linear dependencies exist among the predictors. Hence removing all multicollinear models from consideration. Any time there are multicollinear terms an index will need to be created in order to keep track of any *aliased* terms. This aliasing can cause problems when there is a large effect size for the aliased terms.

The use of restricted model spaces allows for the assessment of all candidate variables, however it restricts the number of candidate variables that may be simultaneously considered in a single model. [14], [15], [16], [17] and [13] use two restrictions one for the number of main effect terms and one for the number of epistatic terms allowed in the model simultaneously. They also give a simple guideline to determine the size of each restriciton. They suggest to choose the restriction $r = m + 2\sqrt{m}$ where m is the a priori expected number of main effects. Similarly the same formula can be employed where m is the expected number of epistatic effect. While this is an easily determined guideline, in practice and is shown, anecdotally, in Section 4.1 that the restriction size does not seem to have a great impact on the resulting inferences from the proposed method. However, one should note that if the restriction is set very small the stochastic search will have a difficult time moving around the model space and hence the algorithm will take a long time to converge.

To search through the restricted model space, MC^3 can be employed using equation (7). Note that $q(M_t|M_c)$ must be determined to move through the sample space. Let $nbd(M_c)$ be all models with one main effect term more, one valid interaction term more, one main effect term less and one interaction term less than model M_l. Denote adding a main effect term as AMT, adding an interaction effect term as AIT, dropping a main effect term as DMT and dropping an interaction effct term as DIT. The probability of each of these actions depends on the attributes of the current model M_c. Let γ_c and ϕ_c be the number of main effect terms and number of interaction terms in M_c, respectively. In order to ensure that all models in $nbd(M_c)$ are equally likely, the probability of each action, AMT, AIT, DMT and DIT need to be determined. Let $\Omega = \{AMT, AIT, DMT, DIT\}$ be an action space. Once these probabilities have been calculated, the following procedure allows for each of the models in $nbd(M_c)$ to be candidate models. First determine, $P(AMT)$, $P(AIT)$, $P(DMT)$ and $P(DIT)$, and choose an action with the corresponding probability. Then select with equal probability a model that is in $nbd(M_c)$ and corresponds to the action chosen. This procedure ensures that all models in $nbd(M_c)$ have equal probability. Having all models in $nbd(M_c)$ equally likely will be necessary in computing $q(M_c|M_t)$.

For $\gamma_c = 0$, only a main effect term may be added since no interaction terms are in the model. Hence the probability distribution for Ω is:

$$P(AMT) = 1, P(DMT) = 0,$$
$$P(AIT) = 0, P(DIT) = 0. \tag{11}$$

For $\gamma_c = 1$, one of the $p - 1$ main effect terms not in the model may be added or the one main effect term in the model may be dropped and no interaction terms are allowed in this model. Hence the probability distribution for Ω is:

$$P(AMT) = \frac{p-1}{p}, P(DMT) = \frac{1}{p},$$
$$P(AIT) = 0, P(DIT) = 0. \tag{12}$$

For $2 \le \gamma_c < r$, no restrictions are involved. Hence, all actions in Ω are allowed. Hence, the probability distribution for Ω is:

$$P(AMT) = \frac{p-\gamma_c}{p+\binom{\gamma_c}{2}}, \quad P(AIT) = \frac{\binom{\gamma_c}{2}-\phi_c}{p+\binom{\gamma_c}{2}},$$
$$P(DMT) = \frac{\gamma_c}{p+\binom{\gamma_c}{2}}, \quad P(DIT) = \frac{\phi_c}{p+\binom{\gamma_c}{2}}. \tag{13}$$

For $\gamma_c = r$, due to the restriction that no more than r main effect terms may be in a model at a single time, no main effect terms may be added. However, main effect terms may be dropped and interaction terms may be added or dropped. Hence, the probability distribution for Ω is:

$$P(AMT) = 0, P(AIT) = \frac{\binom{r}{2} - \phi_r}{\binom{r}{2} + k},$$
$$P(DMT) = \frac{r}{\binom{r}{2}+r}, P(DIT) = \frac{\phi_c}{\binom{r}{2}+r}. \tag{14}$$

Since each model in $nbd(M_c)$ is equally likely to be sampled, $q(M_t|M_c)$ can easily be calculated. For example, let M_t and M_c be such that $\gamma_t = \gamma_c + 1$ where $\gamma_t < r$ and $\gamma_c > 2$. Then this corresponds to the action AMT and the probability of candidate model M_t given that the current model is M_c is one out of the number of models in $nbd(M_c)$, specifically, $q(M_t|M_c) = \left(p + \binom{\gamma_c}{2}\right)^{-1}$ and similarly $q(M_c|M_t) = \left(p + \binom{\gamma_t}{2}\right)^{-1}$. Hence the ratio of the probability of candidate models for this case is:

$$\frac{q(M_t|M_c)}{q(M_c|M_t)} = \frac{p + \left(\dfrac{\gamma_t}{2}\right)}{p + \left(\dfrac{\gamma_c}{2}\right)}. \tag{15}$$

3   Simulation Study

To validate this approach, loci information from *Arabidopsis thaliana* Bay-0 × Shah-dara was used. Figure 2 illustrates the genetic map of the *Arabidopsis thaliana* Bay-0 × Shahdara,which has five chromosomes and a total of 38 markers. For this simulation study, 158 lines were used.

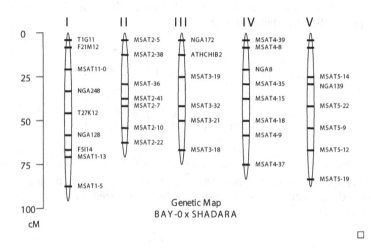

Fig. 2. Genetic map of the *Arabidopsis Thaliana* Bay-0 by Shahdara

Using the loci matrix from the *Arabidopsis thaliana* dataset two loci X_A and X_B were randomly selected from the possible loci and the following model was used to generate the data:

$$y_i = \delta X_{Ai} + \delta X_{Bi} + \delta X_{Ai} X_{Bi} + \epsilon_i, \tag{16}$$

where δ is the effect size, $\epsilon_i \sim N(0, 1)$. Each dataset contained a sample size of 158 observations. Effect sizes of $0, 1/2, 1, 3/2, 2, 5/2, 3, 7/2, 4, 9/2$ and 5 were generated. Each of these effect sizes was repeated 10 times.

Using the data set and the method proposed the following probabilties were cal-culated: $P(X_A|D)$, $P(X_B|D)$, $P(X_{AB}|D)$ and $P(X_{AB}|X_A, X_B, D)$. These were calculated for 110 simulated data sets. Using the following prior distributions $\beta_j \sim N(0, 200)$ and $\sigma^2 \sim \chi^2(1)$ for the model parameters and $P(M_i)$ is uniform over the all models subject to the restriction of $r = 10$. For each simulated data set 5 chains of 16,000 samples were taken from the posterior distribution of the models, with the first

1,000 samples discarded as burn-in samples. The activation probabilities were calculated using the remaining 75,000 samples.

Figure 3 shows boxplots for the main effect activation probabilities versus the effect size from the simulated datasets. Notice that for effect sizes of 0 and 1/2 the activation probabilities are low indicating that not much evidence exists for the main effect at that locus. However, for effect sizes at and above 1 the activation probabilites are quite high, typically above 0.8. It should be noted that activation probabilities are not associated with the idea of a p-value and hence cannot be interpreted as such. Furthermore, the choice of cutoff values for activation probabilities and what is deemed statistically significant, in the Type I and Type II error sense, has not been studied. However, we should notice that the activation probabilities for effect sizes at and above 1 are much larger than those when the effect size is 0. Hence, one could feel confident that the locus is important for influencing the observed trait.

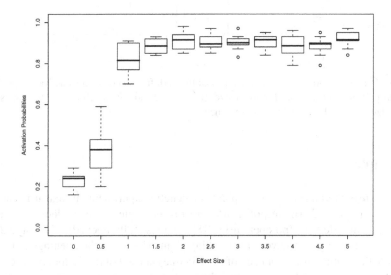

Fig. 3. Boxplots of main effect activation probabilities for effect sizes 0, 1/2, 1, 3/2, 2, 5/2, 3, 7/2, 4, 9/2 and 5 using simulated data sets

Figure 4 shows boxplots for activation probabilities of the epistatic effects and the conditional activation probabilities for epistatic effects versus the effect sizes of 0, 1/2, 1, 2 and 4. Notice that in both plots that both the activation probabilities and the conditional activation probabilities are low for effect sizes 0 and 1/2 indicating that the epistatic effect of the two loci have no minimal effect on the observed trait. However, notice that for effect sizes larger than 1 the conditional activation probabilities are considerably higher than the standard activation probabilities. Again there has been no studies of cutoff values for activation probabilities nor conditional activation probabilities. Looking at both the activation probabilities and conditional activation probabilities with reference to effect size 0 one could feel confident that the two loci work in combination to influence the observed trait.

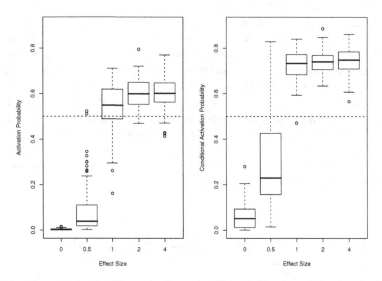

Fig. 4. Boxplots of epistatic effect activation probabilities (left) and conditional activation probabilities for epistatic effects (right) for effect sizes 0, 1/2, 1, 2 and 4 using 100 simulated data sets per effect size. Dashed line at $1/2$ is for reference.

4   Example

The *Arabidopsis thaliana* is a model plant for genetic experiments in that it is easily genetically manipulated. The response of interest is the angle of the cotyledon opening that ranges from 0 degrees (no opening) to 180 degrees (fully opened). The cotyledon is the is the first embryonic leaves on a seedling plant. The wider the opening angle the more viable the mature plant.For each of the 158 lines at each of the 38 loci (markers) a value of 1 or -1 corresponding to whether the marker at that location came from parent A or parent B, respectively (unknown loci information is coded with a 0). With this data and using an unrestricted model space the largest model would have 741 terms. Hence, many models can not be fit. By restricting the model space to $r = 10$ the largest model would have 55 terms and thus, all models have enough observations to be estimated.

  To determine if any aliasing between the main effects and interaction terms occured the data was screened. This screening showed that no interaction terms are aliased with any main effect term. Hence, conclusions about the main effect terms will not be confounded with any epistatic effects. An additional screen of the data was performed to determine if there is aliasing between any interaction effects. Aliased interaction effects were noted for consideration during posterior inferences.

  For this example, the exact marginal posterior probability of the data given model M_c was computed using equation (6) and the proposed MC^3 method with restricted model space was utilized. For each model under consideration the prior distributions for the model parameters were defined as: $\beta_j \sim N(0, 200)$ for all j and $\beta_{jk} \sim N(0, 200)$ for

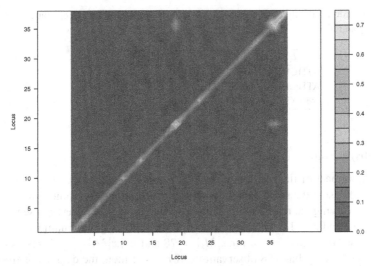

Fig. 5. Heatmap of activation probabilities for main effect and epistatic effects for the *Arabidopsis thaliana* Bay-0 × Shahdara at the 38 loci. Activation probabilities along the diagonal correspond to main effects and off diagonal correspond to epistatic effects.

all interaction terms jk in model i; for σ^2, $\lambda = 1$ and $\nu = 1$ are used. Note that when $\nu = 1$ and $lambda = 1$ the $Inv - \chi^2_{\nu,\lambda}$ has infinite mean and variance. Hence, should be relatively uninformative. For each model M_c where multicollinearity does not occur, the prior probability $P(M_c)$ is chosen uniformly across this space. Thus, a priori, no model is preferred over another.

Using the restriction $r = 10$, 25 chains of 11,000 were run with a burn in of 1,000 samples using overdispersed starting models resulting in 250,000 samples. The number of visits to model M_c was recorded and the probability of model given the data $P(M_c|D)$ was estimated as the number of visits to model M_c divided by the length of the chain. The probabilities appeared to converge after 15 chains, indicating convergence. Using these probabilities the activation probabilities $P(\beta_j \neq 0|\mathbf{D})$ and $P(\beta_{jk} \neq 0|\mathbf{D})$ are computed for each main and epistatic effect, respectively. Figure 5 shows a heat map of the epistatic probabilities. Notice that the highest activation probability locus on the heat map is at locus 18 (ATHCHIB2) and no epistatic, off diagonal, effects have high activation probability.

The activation probabilities for the 3 highest loci, ATHCHIB2, MSAT5-9 and MSAT5-22 are as follows: $P(ATHCHIB2|D) = 0.741$, $P(MSAT5 - 9|D) = 0.481$ and $P(MSAT5 - 22|D) = 0.445$. This suggests that the following epistatic effects should be considered: $ATHCHIB2 \times MSAT5 - 9$, $ATHCHIB2 \times MSAT5 - 22$ and $MSAT5 - 9 \times MSAT5 - 22$. The marginal epistatic activation probabilities and the conditional epistatic activation probabilities of these interactions are shown in Table 1. Previous studies have shown ATHCHIB2 to be a locus associated with cotelydon opening [4]. Hence the results agree with biological expectations. In order to more accurately locate the locus associated with cotelydon opening a dense map of genes near ATHCHIB2 should be undertaken.

Table 1. Activation probabilities and conditional activation probabilities of epistatic effects between locus l_j and locus l_k

l_i	l_j	$P(l_{ij}|\mathbf{D})$	$P(l_{ij}|l_i, l_j, \mathbf{D})$
ATHCHIB2	MSAT5-9	0.070	0.243
ATHCHIB2	MSAT5-22	0.061	0.191
MSAT5-9	MSAT5-22	0.062	0.135

4.1 Sensitivity Study

To assess the impact of the restriction on the resulting inferences a simulation study was conducted. Using the same data set and prior distributions, different analyzes were performed by varying the restriction paramter r. Four choices of r were chosen: $r = 8$, $r = 10$, $r = 12$ and $r = 14$. This corresponds to the following maximum model sizes, $max|M_i| = 36$, $max|M_i| = 55$, $max|M_i| = 78$ and $max|M_i| = 105$, respectively. Since the data set only has 158 observations, if $r > 14$ then, the degrees of freedom associated with the maximum model would be less than 50.

Table 2 shows the results of the sensitivity study. The marginal posterior probability for locus i, l_i and $P(l_i|D)$ is given for the five loci with the highest probability. Notice that regardless of the restriction setting, locus $ATHCHIB2$ is deemed most probable. This agrees with previous studies using this data set. In addition, the second and third most probable loci agree across restriction settings. However, these loci have low probability. One other trait to notice is that no epistatic terms appear in the highest marginal loci. The fact that no epistatic effects appear significant is not a result of the restriction set.

Table 2. Activation probabilities of locus i, l_i for restriction parameter $r = 8, 10, 12$ and 14

$r = 8$		$r = 10$	
$max|M_i| = 36$		$max|M_i| = 55$	
l_i	$P(l_i|D)$	l_i	$P(l_i|D)$
ATHCHIB2	0.699	ATHCHIB2	0.751
MSAT5-9	0.415	MSAT5-9	0.481
MSAT5-22	0.402	MSAT5-22	0.445
MSAT5-12	0.291	MSAT5-12	0.331
MSAT1-5	0.271	MSAT2-36	0.327
$r = 12$		$r = 14$	
$max|M_i| = 78$		$max|M_i| = 105$	
l_i	$P(l_i|D)$	l_i	$P(l_i|D)$
ATHCHIB2	0.735	ATHCHIB2	0.733
MSAT5-9	0.509	MSAT5-9	0.516
MSAT5-22	0.473	MSAT5-22	0.511
MSAT2-36	0.380	MSAT2-36	0.441
MSAT1-5	0.373	MSAT1-5	0.434

5   Discussion

The proposed method for detecting epistasis has the ability to determine which main effects as well as which two-way interaction effects are present in a dataset as evidenced by the simulation study. The method was applied to the *Arabidopsis thaliana* data and no epistatic effects were found with respect to cotyledon opening angle. However, the known locus for controlling cotyledon opening was detected, ATHCHIB2. The search method was employed in a situation where the number of parameters in the full model far exceeded the number of observations, but by placing restrictions on the parameter space each model under consideration had sufficient degrees of freedom for estimating parameters.

A study of epistatic models which do not require the first order terms to be present should be considered as well. This may allow for better detection of epistatic effects as the model search does not need to first add a main effect in order to later include the epistatic term. In this case the model space would be reduced by 2^p models. However, if all other interaction terms are equally likely to be added to the model, the Metropolis-Hastings step may have low acceptance probability and convergence of the MC^3 algorithm may be slow. In addition, any loci that have effects that are not in interaction with other loci may not be detected. Hence, reducing the utility of the method.

Caution should be used when using restricted model spaces. The method works best when it is believed that only a few loci control the trait of interest. In cases where it is believed that a large number of loci control the trait of interest, especially when this exceeds the restriction on the model space, then the search method maybe come very ineffective at assessing both the main effect as well as the epstatic effects. Since the models at the restriction boundary will have high posterior model probabilites it may be difficult to move through regions of lower probability towards even more probable models. In most cases in genetics it is believed that only a few loci control the trait of interest.

References

1. Broman, K.W.: The Genomes of Recombinant Inbred Lines. Genetics 169, 1133–1146 (2005)
2. Broman, K.W., Speed, T.P.: A model selection approach for the identification of quantitative trait loci in experimental crosses. J.R. Statist. Soc. B 64, 641–656 (2002)
3. Boone, E.L., Ye, K., Smith, E.P.: Assessment of two approximation methods for computing posterior model probabilities. Computational Statistics & Data Analysis 48, 221–234 (2005)
4. Boone, E.L., Simmons, S.J., Ye, K., Stapleton, A.E.: Analyzing quantitative trait loci for the Arabidopsis thaliana using Markov chain monte carlo model composition with restricted and unrestricted model spaces. Statistical Methodology 3, 69–78 (2006)
5. Carlborg, O., Andersson, L., Kinghorn, B.: The Use of a Genetic Algorithm for Simultaneous Mapping of Multiple Interacting Quantitative Trait Loci. Genetics 155, 2003–2010 (2000)
6. Cockerham, C.: An extension of the concept of partitioning hereditary variance for the analysis of covariances among relatives when epistasis is present. Genetics 39, 859–882 (1954)
7. Green, P.J.: Reversible jump Markov chain Monte Carlo computation and Bayesian model determination. Biometrika 82, 711–732 (1995)

8. Hanlon, P., Lorenz, A.: A computational method to detect epistatic effects contributing to a quantitative trait. J. Thoer. Biol. 235, 350–364 (2005)

9. Hansen, T.F., Wagner, G.P.: Modeling genetic architecture: a multilinear theory of gene interaction. Theor. Popul. Biol. 59, 61–86 (2001)

10. Kao, C.H., Zeng, Z.B., Teasdale, R.D.: Multiple Interval Mapping for Quantitative Trait Loci. Genetics 152, 1203–1216 (1999)

11. Kao, C.H., Zeng, Z.B.: Modeling Epistasis of Quantitative Trait Loci Using Cockerham's Model. Genetics 160, 1243–1261 (2002)

12. Wang, T., Zeng, Z.-B.: Models and partition of varieance for quantitative trait loci with epistasis and linkage disequilibrium. BMC Genetics 7, 9 (2006)

13. Yandell, B.S., Mehta, T., Samprit, B., Shriner, D., Venkataraman, R., Moon, J.Y., Neeley, W.W., Wu, H., von Smith, R., Yi, N.: R/qtlbim: QTL with Bayesian Interval Mapping in experimental crosses. Bioinformatics 23, 641–643 (2007)

14. Yi, N., Xu, S., Allison, D.B.: Bayesian Model Choice and Search Strategies for Mapping Interacting Quantitative Trait Loci. Genetics 165, 867–883 (2003)

15. Yi, N., Yandell, B.S., Churchill, G.A., Allison, D.B., Eisen, E.J., Pomp, D.: Bayesian model selection for genome-wide epistatic quantitative trait loci analysis. Genetics 170, 1333–1344 (2005)

16. Yi, N., Samprit, B., Pomp, D., Yandell, B.S.: Bayesian Mapping of Genomewide Interacting Quantitative Trait Loci for Ordinal Traits. Genetics 176, 1855–1864 (2007)

17. Yi, N., Shriner, D., Samprit, B., Mehta, T., Pomp, D., Yandell, B.S.: An efficient Bayesian model selection approach for interacting quantitative trait loci models with many effects. Genetics 176, 1865–1877 (2007)

18. Zeng, Z.-B., Wang, T., Zou, W.: Modeling quantitative trait loci and interpretation of models. Genetics 169, 1711–1725 (2005)

Reasoning about Interest-Based Preferences

Wietske Visser, Koen V. Hindriks, and Catholijn M. Jonker

Man-Machine Interaction Group, Delft University of Technology
Delft, The Netherlands
{wietske.visser,k.v.hindriks,c.m.jonker}@tudelft.nl

Abstract. In decision making, negotiation, and other kinds of practical reasoning, it is necessary to model preferences over possible outcomes. Such preferences usually depend on multiple criteria. We argue that the criteria by which outcomes are evaluated should be the satisfaction of a person's underlying interests: the more an outcome satisfies his interests, the more preferred it is. Underlying interests can explain and eliminate conditional preferences. Also, modelling interests will create a better model of human preferences, and can lead to better, more creative deals in negotiation. We present an argumentation framework for reasoning about interest-based preferences. We take a qualitative approach and provide the means to derive both ceteris paribus and lexicographic preferences.

1 Introduction

We present an approach to qualitative, multi-criteria preferences that takes underlying interests explicitly into account. Reasoning about interest-based preferences is relevant in decision making, negotiation, and other types of practical reasoning. Since our long-term goal is the development of a negotiation support system, the motivations and examples in this paper are mainly taken from the context of negotiation, but the main ideas apply equally well in other contexts.

The goal of a negotiation support system is to help a human negotiator reach a better deal in negotiation. The quality of a deal is determined for a large part by the user's personal preferences. A deal generally consists of multiple issues. For example, when applying for a new job, some issues are the position, the salary, and the possibility to work part-time. For a complete deal, negotiators have to agree on the value for every issue. The satisfaction of a negotiator with a possible outcome depends on his preferences.

Since the number of possible outcomes is typically very large (exponential in the number of issues), it is not feasible to have the user express his preferences over all possible outcomes directly. It is common to compute or derive preferences over possible outcomes from preferences over the possible values of issues and a weighing or importance ordering of the issues. One of the best-known approaches is multi-criteria utility theory [1], a quantitative approach where preferences are expressed by numeric utilities. Since such quantities are hard for humans to provide, qualitative approaches have been proposed too, e.g. [2]. Our approach is also of a qualitative nature.

In this paper we argue that issues alone are not enough to derive outcome preferences. Instead, we will focus on modelling underlying interests and their relation to issues.

J. Filipe and A. Fred (Eds.): ICAART 2011, CCIS 271, pp. 115–130, 2013.

There are several reasons for taking interests into account. First, underlying interests can explain and eliminate conditional preferences. Consider the following example. If it rains, I prefer to take my umbrella, but if it doesn't, I prefer not to take it. This is a conditional preference; my preference over taking my umbrella depends on the circumstance of rain. Underlying interests can explain such conditional preferences: I prefer to take my umbrella when it rains because I do not want to get wet, and I prefer not to take it when it's dry because I don't want to carry things unnecessarily. If we take such interests as criteria on which to base preference, we can eliminate conditional preferences entirely. We will get back to this in more detail later. Second, interest-based negotiation is said to lead to better outcomes than position-based negotiation [3,4]. By understanding one's own and the other party's reasons behind a position and discussing these interests, people are more likely to find more creative options in a negotiation and by that reach a mutually acceptable agreement more easily. A well-known example is that of the two sisters negotiating about the division of an orange. They both want the orange, and end up splitting it in half. Had they known each other's underlying interests, they would have reached a better deal: one sister only needed the peel to make a cake and would gladly have let the other sister have all of the flesh for her juice. Third, thinking about underlying interests is a very natural, human thing to do. Interests are what really matters to people, they are what drive them in their decisions and opinions. Taking underlying interests explicitly into account will result in a better model of human preferences. Such a model is also suited for explanation of the reasoning and advice of a support system.

This last point brings us to the motivation for using argumentation to reason about interest-based preferences. Reasoning by means of arguments is a very human type of reasoning. People often base their decisions on (mental) lists of arguments in favour of and against certain decisions. Therefore argumentation is suitable for explanation of a system's reasoning to a human user. Another advantage of argumentation is that it is a kind of defeasible reasoning. It is able to reason with incomplete, uncertain and contradictory information. Finally, argumentation can be used to (try to) persuade the opponent during negotiation (but this is outside the scope of this paper).

The paper is organised as follows. In Section 2 we introduce and discuss the most important concepts that we will use throughout the paper. Then, in Section 3, we give an overview of existing approaches to preferences and underlying interests. We give some more details about qualitative multi-criteria preferences in Section 4. In Section 5 we motivate the explicit modelling of underlying interests, illustrated with examples. Our own approach is presented in Section 6. Finally, Section 7 concludes the paper.

2   Concepts

Before we go on, we will clarify some important concepts that we will use. In negotiation, *issues* are the matters which are under negotiation. An issue is a concrete, negotiable aspect such as monthly salary or number of holidays. Every issue has a set or range of possible values. The value of an issue in a given instance can be objectively determined (e.g. €2400, 30 days). Issues and their possible values typically depend on the domain. Besides the issues under negotiation, there may be other properties of a deal that influence preferences. For example, the location of the company that you are applying to work for can be very important, because it determines the duration of your daily commute, but it is hardly negotiable. Still, such properties are important in negotiation. If, for example, you already got an offer from another company near your home, you will only consider offers that are better taking the location into account.

A *possible outcome* or possible deal has a specified value for every issue. All bids made during a negotiation are possible outcomes. For example, a possible outcome could be a job contract for the position of programmer, with a salary of €3000 gross per month, with 25 holidays, for the duration of one year with the possibility of extension. Any other assignment to the issues would constitute a different outcome. It is the user's preferences over such possible outcomes that we are interested in.

With *criteria* we mean the features on which a preference between outcomes is based. It is common to base preferences directly on the negotiated issues; in that case the issues are the criteria. In this paper we argue that not issues, but underlying interests should be used as criteria.

Many terms are used for what we consider to be *underlying interests*, such as fundamental objectives, values, concerns, goals and desires. In our view, an interest can be any kind of motivation that leads to a preference. Essentially, a preference depends on how well your interests are met in the outcomes to be compared. The degree to which interests are met is influenced by the issues, but there is not necessarily a one-to-one relation between issues and interests. For example, an applicant with childcare responsibilities will have the interest that the children are taken care of after school. This interest can be met by various different issues, for example part-time work, the possibility to work from home, a salary that will cover childcare expenses, etc. One issue may also contribute to multiple interests. Many issues that deal with money do so, because the interests different people have for using the money will be diverse.

3 Related Work

Existing literature about preferences is abundant and very diverse. In this section we briefly discuss the approaches that are most closely related to our interests.

Interest-based negotiation is discussed in [4]. However, this approach has a particular view on negotiation as an allocation of indivisible and non-sharable resources. The resources are needed to carry out plans to reach certain goals. Even though the goals can be seen as underlying interests, it is hard to model e.g. negotiation about a job contract as an allocation of resources. Salary might be an allocation of money, but other issues, like position or start date, cannot be translated as easily into resources.

Argumentation about preferences has been studied extensively in the context of *decision making* [5,6,7,8]. The aim of decision making is to choose an action to perform. The quality of an action is determined by how well its consequences satisfy certain criteria. For example, [5] present an approach in which arguments of various strengths in favour of and against a decision are compared. However, it is a two-step process in which argumentation is used only for epistemic reasoning. In our approach, we combine reasoning about preferences and knowledge in a single argumentation framework.

Within the context of argumentation, an approach that is related to underlying inter-ests is *value-based argumentation* [9,10]. Values are used in the sense of 'fundamental social or personal goods that are desirable in themselves' [10], and are used as the basis for persuasive argument in practical reasoning. In value-based argumentation, ar-guments are associated with values that they promote. Values are ordered according to importance to a particular audience. An argument only defeats another argument if it attacks it and the value promoted by the attacked argument is not more important than the value promoted by the attacker. We will illustrate this with a little example. Con-sider two job offers a and b. a offers a higher salary, but b offers a better position. We can construct two mutually attacking preference arguments, A: 'I prefer job offer a over job offer b because it has a higher salary', and B: 'I prefer job offer b over job offer a because it has a better position'. In Dung-style argumentation frameworks [11], there is no way to choose between two mutually attacking arguments (unless one is defended and the other is not). In value-based argumentation, we could say that preferring a over b promotes the value of wealth (w), and preferring b over a promotes the value of status (s), and e.g. wealth is considered more important than status. In this case A defeats B, but not the other way around.

In this framework, every argument is associated with only one value, while in many cases there are multiple values or interests at stake. [12] define so-called *value-specification argumentation frameworks*, in which arguments can support multiple values, and preference statements about values can be given. However, the preference between arguments is not derived from the preference between the values promoted by the arguments. Besides, there is no guarantee that a value-specification argumentation framework is consistent, i.e., some sets of preference statements do not correspond to a preference ordering on arguments.

In value-based argumentation, we cannot argue about what values are promoted by the arguments or the ordering of values; this mapping and ordering are supposed to be given. But these might well be the conclusion of reasoning, and might be defeasi-ble. Therefore, it would be natural to include this information at the object level. [13] describe some argument schemes regarding the influence of certain perspectives on val-ues. However, for the aggregation of multiple values, they assume a given order on sets of values, whereas we want to derive such an order from an order on individual values.

4   Qualitative Multi-criteria Preferences

Regardless of whether we take issues or interests as criteria, we need to be able to model multiple criteria. In any realistic setting, preferences are determined by multiple criteria and the interplay between them. Therefore we shortly introduce two well-known approaches to multi-criteria preferences which we will use in our framework.

One approach is *ceteris paribus* ('all else being equal') comparison. One outcome is preferred to another ceteris paribus, if it is better on some criteria and the same on all other criteria. This approach has been widely used since [14]. Also [15] derive prefer-ences from sets of goals in a ceteris paribus way. In [16], ceteris paribus comparison is combined with conditional preferences in a graphical preference language called CP-nets. The preference order resulting from ceteris paribus comparison is not complete; an

Table 1. Satisfaction of issues and interests

	a. Issues				**b.** Interests		
	high salary	high position	full-time		wealth	status	family time
a	✓	✓	✓	a	✓	✓	✗
b	✓	✓	✗	b	✓	✓	✓
c	✓	✗	✓	c	✓	✓	✗
d	✓	✗	✗	d	✓	✓	✓
e	✗	✓	✓	e	✗	✓	✗
f	✗	✓	✗	f	✗	✓	✓
g	✗	✗	✓	g	✗	✗	✗
h	✗	✗	✗	h	✗	✗	✓

outcome satisfying criterion G but not H cannot be compared to an outcome satisfying H but not G.

Another well-known approach is the *lexicographic* preference ordering (see e.g. [2], where it is denoted #). Here, preferences over outcomes are based on a set of relevant criteria, which are ranked according to their importance. The importance ranking of criteria is defined by a total preorder $\succeq$, which yields a stratification of the set of criteria into importance levels. Each importance level consists of criteria that are equally important. The lexicographic preference ordering first considers the highest importance level. If some outcome satisfies more criteria on that level than another, then the first is preferred over the second. If two outcomes satisfy the same number of criteria on this level, the next importance level is considered, and so on. Two outcomes are equally preferred if they satisfy the same number of criteria on every level.

We use a slightly more abstract definition of preference that covers both ceteris paribus and lexicographic preferences. Let C be a set of binary criteria, ordered according to importance by a preorder $\succeq$. If $P \succeq Q$ and not $Q \succeq P$, we say that P is strictly more important than Q and write $P \succ Q$. If $P \succeq Q$ and $Q \succeq P$, we say that P is equally important as Q and write $P \approx Q$. C can be divided into equivalence classes induced by $\approx$, which we call importance levels. An importance level L is said to be more important than L' iff the criteria in L are more important than the criteria in L'. Let $\mathcal{O}$ be a set of outcomes, and *sat* a function that maps outcomes $a \in \mathcal{O}$ to sets of criteria $C_a \in 2^C$. If $P \in sat(a)$, we say that a satisfies P.

Definition 1. (Preference). An outcome a is *strictly preferred* to another outcome b if it satisfies more criteria on some importance level L, and for any importance level L' on which b satisfies more criteria than a, there is a more important level on which a satisfies more criteria than b. An outcome a is *equally preferred* as another outcome b if both satisfy the same number of criteria on every importance level.

The least specific importance order possible is the identity relation, in which case the importance levels are all singletons and no importance level is more important than any other. In this case, the preference definition is equivalent to ceteris paribus preference (if a is preferred to b ceteris paribus, there are no criteria that b satisfies but a does not). If the importance order is a total preorder, the definition is equivalent to lexicographic

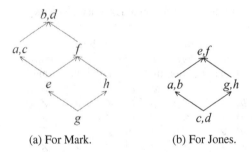

(a) For Mark.   (b) For Jones.

Fig. 1. Ceteris paribus preference orderings (arrows point towards more preferred outcomes)

preference. In general, the more information about the relative importance of interests is known, the more preferences can be derived. We note that lexicographic preferences subsume ceteris paribus preferences in the sense that if one outcome is preferred to another ceteris paribus, it is also preferred lexicographically, regardless of the importance ordering on criteria.

5 Modelling Interests

We will illustrate the ideas presented in this paper by means of an example. Mark has applied for a job at a company called Jones. After the first interview, they are ready to discuss the terms of employment. There are three issues on the table: the salary, the position, and whether the job is full-time or part-time. All possible outcomes are listed in Table 1a. After some thought, Mark has determined that the interests that are at stake for him are wealth, status, and time with his family. A high position will give status. A high salary will provide both wealth and status. A part-time job will give him time to spend with his family. Table 1b shows which interests each of the outcomes satisfies.

All information is encoded in a knowledge base, which consists of three parts.

• *Facts* about the properties of the outcomes to be compared. When comparing offers in negotiation, these may be the values for each issue, or any other relevant properties. Facts are supposed to be objectively determined.

• A set of *interests* of a negotiator. Underlying interests are personal and subjective, although they can sometimes be assumed by default. Interests may vary according to importance. If no importance ordering is given, the ceteris paribus principle can be used to derive preferences. The more information about the relative importance of interests is known, the more preferences can be derived. If there is a total preorder of interests according to importance, a complete preference ordering over possible outcomes can be derived using the lexicographic principle.

• *Rules* relating issues and other outcome properties to interests. These rules can be very subjective, e.g. some people consider themselves very wealthy if they earn €3000 gross salary per month, while for others this may be a pittance. Even so, there can still be default rules that apply in general, e.g. that a high salary promotes wealth for

Table 2. Outcomes in the evening dress example

	a. Issues				**b.** Interests good combi-nation
	jacket	pants	shirt		
i	b	b	w	i	✗
j	b	b	r	j	✓
k	b	w	w	k	✓
l	b	w	r	l	✗
m	w	b	w	m	✓
n	w	b	r	n	✗
o	w	w	w	o	✗
p	w	w	r	p	✓

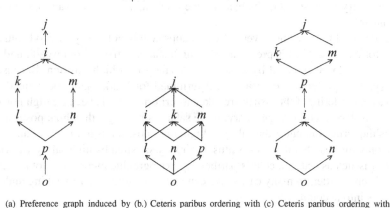

(a) Preference graph induced by CP-net.

(b.) Ceteris paribus ordering with interests.

(c) Ceteris paribus ordering with interests, good combination most important.

Fig. 2. Preference orderings (arrows point towards more preferred outcomes)

the employee. The relation between issues and interests does not have to be one-to-one. There may be multiple issues that can satisfy an interest, some issues may satisfy multiple interests at once, or a combination of issues may be needed to fulfill an interest. As is common in defeasible reasoning, there may be exceptions to rules. For example, one might say that a high position ensures status in general, but this effect is cancelled out if the job is badly paid.

With the inference scheme of defeasible modus ponens (see scheme 1 in Table 4), arguments can be constructed that derive statements about what interests are satisfied by possible outcomes, based on their issue values and the rules relating issues to interests. The conclusions from these arguments are summarized in Table 1b. If we compare the possible outcomes ceteris paribus, we can construct a partial preference order for Mark, with b and d being the most preferred options, and g the least preferred (see Figure 1a). This preference order is not complete. To determine Mark's preference between a and c on the one hand and f on the other hand, we need to know whether wealth or family time is more important to him. If wealth is more important, Mark will prefer a or c.

If family time is more important, he will prefer f. Similarly, to determine a preference between e and h, we need to know whether status or family time is more important.

The company Jones has two major interests: it needs a manager and it has to cut back on expenses. These interests relate directly (one-to-one) to high position and low salary. The ceteris paribus preference ordering for Jones is displayed in Figure 1b.

The Added Value of Interests. It may seem that using interests next to issues just introduces an extra layer in reasoning. From the issues and the relations between issues and interests, we derive the interests that are met by outcomes, and from that we derive preferences. Would it not be easier to derive the preferences directly from the issues? We could just state that Jones has the interests of high position and low salary, optionally with an ordering between them, and we would be able to derive Jones' preferences from that. This is because in this case there is a one-to-one relation between interests and issues: every interest is met by exactly one issue, and every (relevant) issue meets exactly one interest.

There are good reasons, however, why this approach is not always a good solution. Consider for example Mark's preferences. A high salary satisfies both wealth and status, and status can be satisfied by either a high salary or a high position. Because of this, the (partial) preference ordering we determined for Mark cannot be defined as a ceteris paribus ordering if the issues are taken as criteria. This is because high position as criterion is dependent on high salary: if the salary is not high, then high position is a distinguishing criterion, but if the salary is high, high position is not relevant anymore, since the only interest that it serves, status, is already satisfied by high salary. So with a fixed set of issues as criteria, ceteris paribus or lexicographic models cannot represent every preference order. In many cases, this can be solved intuitively by taking underlying interests into account.

There are other approaches to deal with this matter. Instead of assuming independence of the criteria, one can also model conditional preferences, where criteria may be dependent on other criteria. A well-known approach to represent conditional preferences is CP-nets [16], which is short for conditional ceteris paribus preference networks. A CP-net is a graph where the nodes are variables (comparable to our notion of issues). Every node is annotated with a conditional preference table, which lists a user's preferences over the possible values of that variable. If such preferences are conditional (dependent on other variables), each condition has a separate entry in the table, and the variables that influence the preference are parent nodes of this variable in the graph. In [16], an example of conditional preference is given regarding an evening dress. A man unconditionally prefers black to white as a colour for both the jacket and the pants. His preference between a white and a red shirt is conditioned on the combination of jacket and pants. If they have the same colour, he prefers a red shirt (for a white shirt will make his outfit too colourless). If they are of different colours, he prefers a white shirt (because a red shirt will make his outfit too flashy). The complete assignments (outcomes in our terminology) are listed in Table 2a. The preference graph induced by the CP-net for this example is displayed in Figure 2a.

We propose to replace the variables the preferences over which are conditional with underlying interests – the reason for the dependency. In the evening dress example, the underlying interest is that the colours of jacket, pants and shirt make a good

Table 3. The knowledge base for the example

$highsal(c)$	$I_M(wealth)$	$highsal(x) \Rightarrow wealth(x)$
$\neg highpos(c)$	$I_M(status)$	$highsal(x) \Rightarrow status(x)$
$full\text{-}time(c)$	$I_M(family)$	$highpos(x) \Rightarrow status(x)$
$\neg highsal(f)$		$\neg full\text{-}time(x) \Rightarrow family(x)$
$highpos(f)$	$I_J(manager)$	$highpos(x) \Rightarrow manager(x)$
$\neg full\text{-}time(f)$	$I_J(cutback)$	$\neg highsal(x) \Rightarrow cutback(x)$

combination, which in this case is defined by being neither too colourless nor too flashy. The satisfaction of this interest by the different outcomes is listed in Table 2b. The variables jacket and pants are unconditional, so they can remain as criteria. If we take jacket, pants, and good combination as criteria, we can construct the preference graph in Figure 2b, using the ceteris paribus principle. The difference with the preferences induced by the CP-net is that in the CP-net case, outcome i is more preferred than k and m, and p is less preferred than l and n, while in the interest-based case they are incomparable. This is due to the fact that in CP-nets, conditional preferences are implicitly considered less important than the preferences on the variables they depend on ([16], p. 145). In fact, if we would specify that both jacket and pants are more important than a good combination, our preference ordering would be the same as in Figure 2a. But the interest approach is more flexible; it is possible to specify any (partial) importance ordering on interests. For example, we could also state that a good combination is more important than either the jacket or the pants, which results in the preference ordering in Figure 2c. In our opinion, there is no a priori reason to attach more importance to unconditional variables as is done in the CP-net approach.

6 Argumentation Framework

In this section, we present an argumentation framework (AF) for reasoning about qualitative, interest-based preferences. An abstract AF in the sense of Dung [11] is a pair $\langle \mathcal{A}, \rightarrow \rangle$ where $\mathcal{A}$ is a set of arguments and $\rightarrow$ is a defeat relation (informally, a counterargument relation) among those arguments. To define which arguments are justified, we use Dung's [11] preferred semantics.

Definition 2. (Preferred Semantics) . A *preferred extension* of an AF $\langle \mathcal{A}, \rightarrow \rangle$ is a maximal (w.r.t. $\subseteq$) set $S \subseteq \mathcal{A}$ such that: $\forall A, B \in S : A \nrightarrow B$ and $\forall A \in S$: if $B \rightarrow A$ then $\exists C \in S : C \rightarrow B$. An argument is credulously (sceptically) *justified* w.r.t. preferred semantics if it is in some (all) preferred extension(s).

Informally, a preferred extension is a coherent point of view that can be defended against all its attackers. In case of contradictory information, there will be multiple preferred extensions, each advocating one point of view. The contradictory conclusions will be credulously, but not sceptically justified.

We instantiate an abstract AF by specifying the structure of arguments and the defeat relation.

Table 4. Inference schemes

$$1 \quad \frac{L_1,\ldots,L_k,\sim L_l,\ldots,\sim L_m \Rightarrow L_n \quad L_1 \quad \ldots \quad L_k \quad \sim L_l \quad \ldots \quad \sim L_m}{L_n} \; DMP$$

$$2 \quad \overline{\sim L} \; asm(\sim L)$$

$$3 \quad \frac{L}{asm(\sim L) \text{ is inapplicable}} \; asm(\sim L)uc$$

$$4 \quad \overline{sat(a,[P]_\alpha,0)} \; count(a,[P]_\alpha,\varnothing)$$

$$5 \quad \frac{P_1(a) \quad \ldots \quad P_n(a) \quad P_1 \approx_\alpha \ldots \approx_\alpha P_n \quad I_\alpha(P_1) \quad \ldots \quad I_\alpha(P_n)}{sat(a,[P_1]_\alpha,n)} \; count(a,[P_1]_\alpha,\{P_1,\ldots,P_n\})$$

$$6 \quad \frac{P_1(a) \quad \ldots \quad P_n(a) \quad P_1 \approx_\alpha \ldots \approx_\alpha P_n \quad I_\alpha(P_1) \quad \ldots \quad I_\alpha(P_n)}{count(a,[P_1]_\alpha,S \subset \{P_1,\ldots,P_n\}) \text{ is inapplicable}} \; count(a,[P_1]_\alpha,S)uc$$

$$7 \quad \frac{sat(a,[P]_\alpha,n) \quad sat(b,[P']_\alpha,m) \quad P \approx_\alpha P' \quad n > m}{pref_\alpha(a,b)} \; prefinf(a,b,[P]_\alpha)$$

$$8 \quad \frac{sat(a,[Q]_\alpha,n) \quad sat(b,[Q']_\alpha,m) \quad Q \approx_\alpha Q' \succ_\alpha P \quad n < m}{prefinf(a,b,[P]_\alpha) \text{ is inapplicable}} \; prefinf(a,b,[P]_\alpha)uc$$

$$9 \quad \frac{sat(a,[P]_\alpha,n) \quad sat(b,[P']_\alpha,m) \quad P \approx_\alpha P' \quad n = m}{eqpref_\alpha(a,b)} \; eqprefinf(a,b,[P]_\alpha)$$

$$10 \quad \frac{sat(a,[Q]_\alpha,n) \quad sat(b,[Q']_\alpha,m) \quad Q \approx_\alpha Q' \quad n \neq m}{eqprefinf(a,b,[P]_\alpha) \text{ is inapplicable}} \; eqprefinf(a,b,[P]_\alpha)uc$$

Arguments. Arguments are built from formulas of a logical language, that are chained together using inference steps. Every inference step consists of premises and a conclusion. Inferences can be chained by using the conclusion of one inference step as a premise in the following step. Thus a tree of chained inferences is created, which we use as the formal definition of an argument (cf. e.g. [17]).

Definition 3. (Argument). An *argument* is a tree, where the nodes are inferences, and an inference can be connected to a parent node if its conclusion is a premise of that node. Leaf nodes only have a conclusion (a formula from the knowledge base), and no premises. A subtree of an argument is also called a *subargument*. `inf` returns the last inference of an argument (the root node), and `conc` returns the conclusion of an argument, which is the same as the conclusion of the last inference.

Definition 4. (Language). Let $\mathcal{P}$ be a set of predicate names with typical elements P,Q; $\mathcal{O}$ a set of outcome names with typical elements a,b; α an audience; and n a non-negative integer. The *input language* $\mathcal{L}_{KB}$ and full *language* $\mathcal{L}$ are defined as follows.

Table 5. Example arguments

```
A  highsal(c)   highsal(x) ⇒ wealth(x)
   ──────────────────────────────
        wealth(c)                  I_M(wealth)
   ─────────────────────────────────────────────────────────────────────────
        sat(c,[wealth]_M,1)          sat(f,[wealth]_M,0)   wealth ≈_M wealth   1>0
B  ¬full-time(f)   ¬full-time(x) ⇒ family(x)            pref_M(c,f)          ───── α
   ──────────────────────────────
        family(f)                  I_M(family)
   ─────────────────────────────────────────────────────────────────────────
        sat(f,[family]_M,1)          sat(c,[family]_M,0)   family ≈_M family   1>0
C  highsal(c)   highsal(x) ⇒ wealth(x)                   pref_M(f,c)          ───── β
   ──────────────────────────────
        wealth(c)                  I_M(wealth)
   ─────────────────────────────────────────────────────────────────────────
        sat(c,[wealth]_M,1)          sat(f,[wealth]_M,0)   wealth ≻_M family   1≠0
D  ¬full-time(f)   ¬full-time(x) ⇒ family(x)        β is inapplicable
   ──────────────────────────────
        family(f)                  I_M(family)
   ─────────────────────────────────────────────────────────────────────────
        sat(f,[family]_M,1)          sat(c,[family]_M,0)   family ≻_M wealth   1≠0
                            α is inapplicable
```

$$\varphi \in \mathcal{L}_{KB} ::= L \mid I_\alpha(P) \mid P \succ_\alpha Q \mid P \approx_\alpha Q \mid$$
$$L_1,\ldots,L_k, \sim L_l,\ldots,\sim L_m \Rightarrow L_n$$

where $L_i = P(a)$ or $\neg P(a)$.

$$\psi \in \mathcal{L} ::= \varphi \in \mathcal{L}_{KB} \mid \sim L \mid sat(a,[P]_\alpha,n) \mid$$
$$pref_\alpha(a,b) \mid eqpref_\alpha(a,b)$$

We make a distinction between an input and full language. A knowledge base, which is the input for an argumentation framework, is specified in the input language. The input language allows us to express facts about the criteria that outcomes (do not) satisfy, statements about interests of an audience and their importance ordering, and defeasible rules. The knowledge base for the job contract example (the facts restricted to outcomes c and f) is displayed in Table 3. Other formulas of the language that are not part of the input language, e.g. expressing a preference between two outcomes, can be derived from a knowledge base using inference steps that build up an argument (such formulas are not allowed in a knowledge base because they might contradict derived statements).

Inferences. Table 4 shows the inference schemes that are used. The first inference scheme is called defeasible modus ponens. It allows to infer conclusions from defeasible rules. The next two inference rules define the meaning of the weak negation $\sim$. According to inference rule 2, a formula $\sim \varphi$ can always be inferred, but such an argument will be defeated by an undercutter built with inference rule 3 if φ is the case. Inference schemes 4 and 5 are used to count the number of interests of equal importance (according to audience α) as some interest P_1 that outcome a satisfies. This type of inference is inspired by accrual [18], which combines multiple arguments with the same conclusion into one accrued argument for the same conclusion. Although our application is different, we use a similar mechanism. Inference scheme 4 can be used when an outcome satisfies no interests. It is possible to construct an argument that does not count all interests that are satisfied, a so-called non-maximal count. But we want all interests to be counted, otherwise we would conclude incorrect preferences. To ensure that only maximal counts are used, we provide an inference scheme to construct arguments that undercut non-maximal counts (inference scheme 6). An argument of this type says that any count which is not maximal is not applicable. Inference scheme 7 says that an outcome a is preferred over an outcome b if the number of interests of a certain importance level that a satisfies is higher than the number of interests on that same level that b satisfies. Inference scheme 8 undercuts scheme 7 if there is a more important level than that of P on which a and b do not satisfy the same number of interests. Finally, inference schemes 9 and 10 do the same as 7 and 8, but for equal preference.

Defeat. The most common type of defeat is rebuttal. An argument rebuts another argument if its conclusion contradicts conclusion of the other argument. Conclusions contradict each other if one is the negation of the other, or if they are preference or importance statements that are incompatible (e.g. $pref_\alpha(a,b)$ and $pref_\alpha(b,a)$, or $pref_\alpha(a,b)$ and $eqpref_\alpha(a,b)$). Defeat by rebuttal is mutual. Another type of defeat is undercut. An undercutter is an argument for the inapplicability of an inference used in another argument. Undercut works only one way. Defeat is defined recursively, which means that rebuttal can attack an argument on all its premises and (intermediate) conclusions, and undercut can attack it on all its inferences.

Definition 5. (Defeat) An argument A *defeats* an argument B ($A \rightarrow B$) if $conc(A)$ and $conc(B)$ are contradictory (*rebuttal*), or $conc(A) = $ 'inf(B) is inapplicable' (*undercut*), or A defeats a subargument of B.

Let us return to the example. With the information from the knowledge base, the arguments A and B in Table 5 can be formed. A advocates a preference for c, based on the interest wealth. B advocates a preference for f, based on the interest family. Without an ordering on these interests, no decision between these arguments can be made. But if *wealth* $\succ_M$ *family* is known, argument C can be made, which undercuts B. Similarly, with *family* $\succ_M$ *wealth*, argument D can be made, which undercuts A.

Validity. If some conditions in the input knowledge base (KB) hold, it can be shown that the proposed argumentation framework models ceteris paribus and lexicographic preference. In the following, we consider a single audience and leave out the subscript α.

Condition 1. Let C be a set of interests to be used as criteria, with importance order $\succeq$.
(1) For all P, '$I(P)$' is in KB iff $P \in C$.
(2) For all $P \in C$, a, '$P(a)$' is a conclusion of a sceptically justified argument iff a satisfies P.
(3) The relative importance among interests is
   (a) a total preorder,
   (b) the identity relation,
and for all $P, Q \in C$, '$P \succ Q$' is in KB iff $P \succ Q$, and '$P \approx Q$' is in KB iff $P \approx Q$.

Theorem 1. (i) If conditions 1.1, 1.2 and 1.3a hold, then $pref(a,b)$ (resp. $eqpref(a,b)$)) is a sceptically justified conclusion of the argumentation framework iff a is strictly (resp. equally) preferred over b according to the lexicographic preference ordering.
(ii) If conditions 1.1, 1.2 and 1.3b hold, then $pref(a,b)$ (resp. $eqpref(a,b)$)) is a sceptically justified conclusion of the argumentation framework iff a is strictly (resp. equally) preferred over b according to the ceteris paribus preference ordering.

Proof. We prove the theorem for strict preference. The same line of argument can be followed for equal preference.
(i) $\Leftarrow$: Suppose a is strictly lexicographically preferred over b. This means that there is an importance level on which a satisfies more interests (say, $P_1, \ldots, P_n$) than b (say, $P'_1, \ldots, P'_m$, $n > m$), and on all more important levels, a and b satisfy an equal number of interests. In this case, we can construct the following arguments, where the first two arguments are subarguments of the third (note that these arguments can also be built if m is equal to 0, by using the empty set count).

$$\frac{P_1(a) \quad \ldots \quad P_n(a) \quad I(P_1) \quad \ldots \quad I(P_n) \quad P_1 \approx \ldots \approx P_n}{sat(a, [P_1], n)}$$

$$\frac{P'_1(b) \quad \ldots \quad P'_m(b) \quad I(P'_1) \quad \ldots \quad I(P'_m) \quad P'_1 \approx \ldots \approx P'_m}{sat(b, [P'_1], m)}$$

$$\frac{sat(a, [P_1], n) \quad sat(b, [P'_1], m) \quad P_1 \approx P'_1 \quad n > m}{pref(a,b)}$$

We will now try to defeat this argument. Premises of the type $P(a)$ are justified by condition 1.2. Premises of the type $I(P)$ and $P_1 \approx P_2$ cannot be defeated (conditions 1.1 and 1.3a). There are three inferences we can try to undercut (the last inference of the argument and the last inferences of two subarguments). For the first count, this can only be done if there is another P_j such that $I(P_j)$ and $P_j \approx P$ and $P_j \notin \{P_1, \ldots, P_n\}$ and $P_j(a)$ is the case. However, $P_1 \ldots P_n$ encompass all interests that a satisfies on this level, so count undercut is not possible. The same argument holds for the other count. At this point it is useful to note that these two counts are the only ones that are undefeated. Any lesser count will be undercut by the count undercutter that takes all of $P_1 \ldots P_n$ (resp. $P'_1 \ldots P'_m$) into account. Such an undercutter has no defeaters, so any non-maximal count is not justified. The undercutter of $prefinf(a, b, [P_1])$ is based on two counts. We have seen that any non-maximal count will be undercut. If the maximal counts are used, we have $n = m$ for undercutter arguments that use $Q \succ P$, since we have that on all more important levels than $[P_1]$, a and b satisfy an equal number of interests. So the undercutter inference rule cannot be applied since $n \neq m$ is not true. For that reason, a rebutting argument with conclusion $pref(b,a)$ will not be justified. This means that for every possible type of defeat, either the defeat is inapplicable or the defeater is itself defeated by undefeated arguments. This means that the argument is sceptically justified.
$\Rightarrow$: Suppose that a is not strictly lexicographically preferred over b. This means that for all importance levels $[P]$, either a does not satisfy more interests than b on that

level, or there exists a more important level where b satisfies more interests than a. This means that any argument with conclusion $pref(a,b)$ (which has to be of the form above) is either undercut by $count(b,[P],S)uc$ because it uses a non-maximal count, or by $prefinf(a,b,[P])uc$ because there is a more important level where a preference for b over a can be derived. This means that any such argument will not be sceptically justified.

(ii) $\Leftarrow$: Suppose a is strictly ceteris paribus preferred over b. This means that there is (at least) one interest, let us say P, that a satisfies and b does not, and there are no interests that b satisfies and a does not. In this case, we can construct the following argument.

$$\frac{\dfrac{P(a) \quad I(P)}{sat(a,[P],1)} \quad \dfrac{}{sat(b,[P],0)} \quad P \approx P \quad 1 > 0}{pref(a,b)}$$

Premise $P(a)$ is justified by condition 1.2. Premise $I(P)$ cannot be defeated (condition 1.1). Note that, since there is no importance ordering specified, counts can only include 0 or 1 interest(s). So the first count cannot be undercut, because there are no other interests that are equally important as P (condition 1.3b). The second count cannot be undercut because b does not satisfy P. Since there are no interests that b satisfies but a does not, the last inference can only be undercut by an undercutter that uses a non-maximal count and so will be undercut itself.

$\Rightarrow$: Suppose a is not strictly ceteris paribus preferred over b. This means that either there is no interest that a satisfies but b does not, or there is some interest that b satisfies and a does not. In the first case, the only arguments that derive a preference for a over b have to use non-maximal counts and hence are undercut. In the second case, any argument that derives a preference for a over b is rebut by the following argument,

$$\frac{\dfrac{Q(b) \quad I(Q)}{sat(b,[Q],1)} \quad \dfrac{}{sat(a,[Q],0)} \quad Q \approx Q \quad 1 > 0}{pref(b,a)}$$

and is not sceptically justified. $\square$

7   Conclusions

In this paper we have made a case for explicitly modelling underlying interests when reasoning about preferences in the context of practical reasoning. We have presented an argumentation framework for reasoning about qualitative interest-based preferences that models ceteris paribus and lexicographic preference.

In the current framework, we have only considered Boolean issues and interests. While this suffices to illustrate the main points discussed in this paper, multi-valued scales would be more realistic. Such an approach would open the way to modelling different degrees of (dis)satisfaction of an interest. For example, [5] take into account the level of satisfaction of goals on a bipolar scale. In the Boolean case, the lexicographic preference ordering is based on counting the number of interests that are satisfied by outcomes. This is no longer possible if multi-valued scales are used. In that case, we could count interests that are satisfied to a certain degree (like e.g. [5]), or compare outcomes in a pairwise fashion and count the number of interests that one outcome satisfies to a higher degree than another (like e.g. [7,13]).

Currently, we suppose that the interests and importance ordering among them are given in a knowledge base. We can make our framework more flexible by allowing such statements to be derived in a way that is similar to the derivation of statements about the satisfaction of interests.

We would also like to look into the interplay between different issues promoting or demoting the same interest. For example, a high salary and a high position both lead to status, but together they may lead to even more status. Or a low salary may promote cutback, but providing a lease car will demote it. Do these effects cancel each other out? The principles that play a role here are related to the questions posed in the context of accrual of arguments [18].

Since our long-term goal is the development of an automated negotiation support system, we plan to look into negotiation strategies that are based on qualitative, interest-based preferences as described here, as opposed to utility-based approaches currently in use. For the same reason, we plan to implement the argumentation framework for reasoning about interest-based preferences that we have presented here. Another interesting question in this context is how interest-based preferences can be elicited from a human user.

Acknowledgements. We thank Henry Prakken for useful comments on earlier drafts of this paper. This research is supported by the Dutch Technology Foundation STW, applied science division of NWO and the Technology Program of the Ministry of Economic Affairs. It is part of the Pocket Negotiator project with grant number VICI-project 08075.

References

1. Keeney, R.L., Raiffa, H.: Decisions with multiple objectives: preferences and value trade-offs. Cambridge University Press (1993)
2. Brewka, G.: A rank based description language for qualitative preferences. In: 16th European Conference on Artificial Intelligence (ECAI 2004), pp. 303–307 (2004)
3. Keeney, R.L.: Value-Focused Thinking: A Path to Creative Decisionmaking. Harvard University Press (1992)
4. Rahwan, I., Pasquier, P., Sonenberg, L., Dignum, F.: On the benefits of exploiting underlying goals in argument-based negotiation. In: 22nd Conference on Artificial Intelligence (AAAI 2007), pp. 116–121 (2007)
5. Amgoud, L., Bonnefon, J.F., Prade, H.: An Argumentation-Based Approach to Multiple Criteria Decision. In: Godo, L. (ed.) ECSQARU 2005. LNCS (LNAI), vol. 3571, pp. 269–280. Springer, Heidelberg (2005)
6. Amgoud, L., Prade, H.: Using arguments for making and explaining decisions. Artificial Intelligence 173(3-4), 413–436 (2009)
7. Ouerdane, W., Maudet, N., Tsoukiàs, A.: Argument schemes and critical questions for decision aiding process. In: Besnard, P., Doutre, S., Hunter, A. (eds.) Computational Models of Argument (COMMA 2008). Frontiers in Artificial Intelligence and Applications, pp. 285–296. IOS Press (2008)

8. Ouerdane, W., Maudet, N., Tsoukiàs, A.: Argumentation theory and decision aiding. In: Ehrgott, M., Figueira, J.R., Greco, S. (eds.) New Trends in Multiple Criteria Decision Analysis. Springer, Heidelberg (2010)

9. Bench-Capon, T.J.M.: Persuasion in practical argument using value based argumentation frameworks. Journal of Logic and Computation 13(3), 429–448 (2003)

10. Bench-Capon, T., Atkinson, K.: Abstract argumentation and values. In: Rahwan, I., Simari, G.R. (eds.) Argumentation in Artificial Intelligence, pp. 45–64. Springer, Heidelberg (2009)

11. Dung, P.M.: On the acceptability of arguments and its fundamental role in nonmonotonic reasoning, logic programming and n-person games. Artificial Intelligence 77, 321–357 (1995)

12. Kaci, S., van der Torre, L.: Preference-based argumentation: Arguments supporting multiple values. International Journal of Approximate Reasoning 48(3), 730–751 (2008)

13. Van der Weide, T., Dignum, F., Meyer, J.J., Prakken, H., Vreeswijk, G.: Practical Reasoning using Values: Giving Meaning to Values. In: McBurney, P., Rahwan, I., Parsons, S., Maudet, N. (eds.) ArgMAS 2009. LNCS, vol. 6057, pp. 79–93. Springer, Heidelberg (2010)

14. Von Wright, G.H.: The Logic of Preference: An Essay. Edinburgh University Press (1963)

15. Wellman, M.P., Doyle, J.: Preferential semantics for goals. In: 9th National Conference on Artificial Intelligence (AAAI 1991), pp. 698–703 (1991)

16. Boutilier, C., Brafman, R.I., Domshlak, C., Hoos, H.H., Poole, D.: CP-nets: A tool for representing and reasoning with conditional ceteris paribus preference statements. Journal of Artificial Intelligence Research 21, 135–191 (2004)

17. Vreeswijk, G.A.W.: Abstract argumentation systems. Artificial Intelligence 90(1-2), 225–279 (1997)

18. Prakken, H.: A study of accrual of arguments, with applications to evidential reasoning. In: 10th International Conference on Artificial Intelligence and Law (ICAIL 2005), pp. 85–94 (2005)

Advantages of Information Granulation in Clustering Algorithms

Urszula Kużelewska

Faculty of Computer Science, Bialystok University of Technology
Wiejska 45a, 15-521 Bialystok, Poland
u.kuzelewska@pb.edu.pl
http://www.wi.pb.edu.pl

Abstract. Clustering is a part of data mining domain. Its task is to identify groups consisting of similar data objects according to defined similarity criterion. One of the most common problems in this field is the time complexity of algorithms. Reducing the time of processing is particularly important due to constantly growing size of present databases. Granular computing (GrC) techniques create and/or process data portions, called granules, identified with regard to similar description, functionality or behavior. An interesting characteristic of granular computation is the ability to create multi-perspective view of data depending on the resolution level required. Data granules identified on different levels of resolution form a hierarchical structure expressing relations between the objects of data. Granular computing includes methods from various areas with the aim of supporting human in better understanding of analyzed problems and generated results.

The proposed solution of clustering is based on processing granulated data in the form of hyperboxes. The results are compared with the clustering of point-type data with regard to complexity, quality and interpretability.

Keywords: Knowledge discovery, Data mining, Information granulation, Granular computing, Clustering, Hyperboxes.

1   Introduction

Cluster analysis is organizing a collection of patterns (usually represented as a vector of measurements, or a point in a multi-dimensional space) into clusters based on their similarity [5]. The points within one cluster are more similar to one another than to any other points from the remaining clusters. The term "similar" can be different for various clustering algorithms and the type of data used, but usually means a reverse of a distance between the points, Euclidean for continuous attributes. Partitioning methods have had wide applications, among others, in pattern recognition, image processing, statistical data analysis and knowledge discovery.

There are many challenges met by clustering methods such as: differences in cluster size or density, arbitrary shapes of clusters, presence of noise or outliers and detecting data of no clusters present [4]. Another issue when discussing clustering algorithms is time complexity. This is particularly important when dealing with large databases.

J. Filipe and A. Fred (Eds.): ICAART 2011, CCIS 271, pp. 131–145, 2013.

Granular computing is a new multidisciplinary theory rapidly developing in recent years. The most common definitions of GrC [10], [14] include an assumption of computing with information granules, that is collections of objects, which exhibit similarity in terms of their properties or functional appearance. Although the term is new, the ideas and concepts of GrC have been used in many fields under different names: information hiding in programming, granularity in artificial intelligence, divide and conquer in theoretical computer science, interval computing, cluster analysis, fuzzy and rough set theories, neutrosophic computing, quotient space theory, belief functions, machine learning, databases, and many others. According to the more universal definition, granular computing may be considered a label of a new field of multi-disciplinary study dealing with theories, methodologies, techniques and tools which make use of granules in the process of problem solving [2].

Distinguishable aspect of GrC is a multi-perspective standpoint on data. Multi-perspective means diverse levels of resolution depending on saliency features or grade of details of a studied problem. Data granules, which are identified on different levels of resolution form a hierarchical structure expressing relations between data objects. This structure can be used to facilitate investigation and helps to understand complex systems. Understanding of analyzed problem and attained results is the main aspect of human-oriented systems. In addition, there are also definitions of granular computing focused on systems supporting human beings [2]. According to definitions mentioned above, such methodology allows to ignore irrelevant details and concentrate on the essential features of the systems to make them more understandable.

There have been many attempts to solve problems with data granulation. To give a few examples: knowledge exploration in spatio-temporal databases [8], intelligent fault detection system [7], image segmentation [13], data mining [11]. In [1] the approach to data granulation based on approximating data by multi-dimensional hyperboxes is presented. The hyperboxes represent data granules formed from the data points focusing on maximizing density of information present in the data. It benefits from the improvement of computational performance, among others. The algorithm is described in the following sections.

This article examines an approach to data clustering based on processing granules of data in the form of hyperboxes. This solution is characterized by reduced time in contrary to processing point-type data. Experiments have been performed on several multi-dimensional data sets containing different numbers of clusters. They have been examined both the time of data clustering and the quality of results measured by quality indices. The article also discusses the way of creating hierarchical structure of data containing levels of point-type object clusters as well as groups of hyperboxes.

This paper is organized as follows: next section, Section 2, describes the method of hyperboxes creation, Section 3 contains description of clustering methods: traditional - partitioning (Section 3.1) and hierarchical (Section 3.2) and one of recently proposed - SOSIG (Section 3.3). The following part, Section 4, describes indices for assessment clustering results. Section 5 reports on collected data sets and executed experiments. The last section concludes the article.

2   Data Granulation

The method of granulation is based on maximization of information density from point-type data. There are hyperboxes created, which cover areas densely populated by data objects. The hyperboxes (referred as I) are multi-dimensional structures described by a pair of values a and b for every dimension. The point a_i and b_i represent minimal and maximal value of the granule in i-th dimension respectively, thus, width of i-th dimensional edge equals $|b_i - a_i|$.

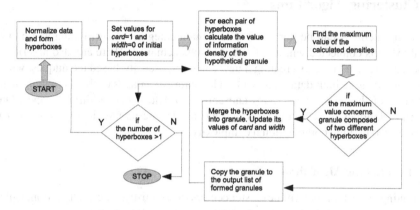

Fig. 1. Algorithm of hyperboxes construction

The main steps of the algorithm are presented in Figure 1. Information density can be expressed by Equation 1:

$$\sigma = \frac{card(I)}{\phi(width(I))},\tag{1}$$

where $card(I)$ denotes the number of data points belonging to hyperbox I and $\phi(width(I))$ is a function of hyperboxes width described by Equation 2. Belonging to a hyperbox means, that the values of point attributes are between or equal the minimal and maximal values of the hyperbox attributes. For that reason there is a necessity to recalculate cardinality in every case of forming a new larger granule from a combination of two granules. Maximization of σ is a problem of balancing the possible shortest dimensions against the greatest cardinality of formed granule I.

In case of multi-dimensional granules as a function of hyperboxes width the function from Equation 2 is applied:

$$\phi(u) = \exp(K \cdot \max_i(u_i) - \min_j(u_j)), i,j = 1,\dots,k\tag{2}$$

where k represents a number of dimensions, $u = (u_1, u_2, \dots, u_k)$ and $u_i = width([a_i, b_i])$ for $i,j = 1,\dots,k$. The points a_i and b_i denote minimal and maximal value in i-th dimension respectively. Constant K originally equals 2, however, in the experiments different values of a given as parameter K have been used used.

This algorithm assumes processing of both hyperboxes and point-type data. To make it possible, new data are characterized by $2 \cdot n$ values in comparison to original data. The first n attributes describe minimal, whereas the following n maximal values for every dimension. To assure topological "compatibility" point-type data and hyperboxes dimensionality of the data is doubled initially. Computational complexity of this algorithm is $O(N^3)$. However, in every step of the method, the size of data is decreased by 1, which in practice reduces the general complexity significantly.

3   Clustering Algorithms

Clustering methods can be divided into two main categories: partitioning and hierarchical [5]. It is connected with the form of results achieved. There are also algorithms proposed in recent 20 years creating a new category - density-based techniques, where groups are areas of high data density [4]. The main advantages of them are: automatic evaluation of the number of groups and ability to detect their arbitrary shapes. One of such techniques is SOSIG [11], which has all the merits mentioned before, but is characterized by high computational complexity.

3.1   Partitioning Algorithms

Partitioning algorithms determine cluster centers to optimize a criterion function, which is the most common the sum-squared-error function (see Equation 3).

$$E = \sum_{i=1}^{nc} \sum_{x \in C_i} |x - c_i|^2,$$ (3)

where c_i denotes the center of cluster C_i. Partitioning methods are based on iterative process relocating objects among the groups to minimize the value of the criterion function. The most popular partitioning method is k-means [5], dividing data into required k groups.

All partitioning methods suffer from some difficulties. Among other things, the number of clusters is required a priori. The form of criterion function limits the shapes of detected groups to spherical shapes. Additionally, there are problems when clusters are varied in size. However, time complexity of partitioning algorithms is very low - $O(n)$, which makes them the most often used methods in exploration of large databases.

3.2   Hierarchical Algorithms

Hierarchical clustering algorithms [4] create a kind of cluster tree, called dendrogram, showing relationship among the objects. Grouping into the desired number of clusters is achieved by cutting the dendrogram at an appropriate level. On account of the method of distance calculation between clusters they can be divided mainly into single-link (hsl) and complete-link (hcl) approaches. In hsl method dissimilarity between groups is calculated as distance between two nearest points from the groups. On the contrary, in hcl approach the dissimilarity is measured as the longest distance between points from different groups. Although computationally expensive, this is a very popular technique, overcoming limitations such as differences in size or shape of clusters. In this technique time complexity is about $O(n^2)$.

3.3 Self-Organizing System for Information Granulation

SOSIG (Self-Organizing System for Information Granulation) algorithm is a system designed for detecting granules (groups) present in data. It takes a parameter $rg \in [0, 1]$ denoting required level of details (resolution) of generated result. Originally it has been designed to deal with point-type data, but in order to reduce its high time complexity it has been adapted to hyperboxes processing.

Using partitioning of both hyperboxes and data objects, it is possible to create a three-level structure, which can be helpful to understand relations between data objects. The main level of the hierarchy is defined by clusters of hyperboxes and the third lowest level consists of point-type data clusters. The middle level is composed of granules containing these third level clusters, which are entirely located in one of the hyperboxes.

SOSIG creates a network structure of connected objects (according to a chosen approach: points or hyperboxes) forming clusters. The organization of the system, including the objects as well as the connections, is constructed on the basis of relationships between input data without any external supervision. The structure elements are representatives of input data, that is, an individual object from the structure stands for one or more object from the input set. As a result, the number of representatives is much smaller than input data without losing information.

To have convenient and compact notation, let us assume input data are defined as an information system $IS = (U, A)$ [9], where $U = \{x_1, \ldots, x_n\}$ is a set of objects and $A = \{a_1, \ldots, a_k\}$ is a set of attributes. The result generated by SOSIG is also described by an information system $IS' = (Y, A \cup \{a_{gr}\})$, where the last attribute a_{gr} : $Y \rightarrow \{1, \ldots, nc\}$ denotes the label of generated cluster and $card(Y) \leq card(U)$ and $\forall x \in U \exists y \in Y (\delta(x, y) < NR)$. Parameter NR in general defines the area of objects interactions and is re-calculated in every iteration according to current connections in the network. The steps of the main (learning) part of SOSIG are shown in Algorithm 1. For a detailed description of the method see [11] and [6].

A measure of usefulness of objects is the similarity level expressed by Equation 4, which defines a degree of closeness between the examined representative point y and the most similar point x from the training data. Only training points from the neighborhood of y (defined by NR) are considered.

$$s_l(y) = NR - \min(\{\delta(y, x) : x \in U\}) \tag{4}$$

To calculate the distance between hyperboxes measure expressed by Equation 5 was used:

$$d(I_A, I_B) = (\|a_B - a_A\| + \|b_B - b_A\|)/2 \tag{5}$$

where $\|a_B - a_A\|$ and $\|a_B - a_A\|$ denote the sum of subtractions of minimal (a) and maximal (b) values of granules I_A and I_B respectively in every dimension. The equation has been introduced in [1].

Algorithm 1. Main steps of SOSIG algorithm

Data:
- $IS = (U, A)$ - an information system, where $U = \{x_1, \ldots, x_n\}$ is a set of objects and $A = \{a_1, \ldots, a_k\}$ is a set of attributes,
- $\{\delta_a : a \in A\}$ - a set of distance function of form $\delta_a : V_a \times V_a \rightarrow [0, \infty)$, where V_a is a set of values for attribute $a \in A$ and a global distance function $\delta : U \times U \rightarrow [0, \infty)$ defined by $\delta(x, y) = fusion(\delta_{a_1}(a_1(x), a_1(y)), \ldots, \delta_{a_k}(a_k(x), a_k(y)))$
- $size_{net} \in \{0, 1, \ldots, card(U)\}$ - initial size of the network, $rg \in [0, 1]$ - resolution of granulation,

Result: $IS' = (Y, A \cup \{a_{gr}\})$ - an information system, where the last attribute $a_{gr} : Y \rightarrow \{1, \ldots, nc\}$ denotes label of generated granule and $card(Y) \leq card(U)$ and $\forall x \in U \exists y \in Y \delta(x, y) < NR$

begin

  $[NR_{init}, Y] \longleftarrow$ initialize$(U, A, size_{net})$;
  for $y_i, y_j \in Y, i \neq j$ **do** /*form clusters*/
    if $\delta(y_i, y_j) < NR_{init}$ **then** connect(y_i, y_j);

  $NR \longleftarrow NR_{init}$;
  while $\neg$stopIterations(Y) **do**
    for $y \in Y$ **do**
      $\Delta(y) = (\delta(y, x))_{x \in U}$; /*calculate distances between input data*/;
      $s_l(y) = NR - \min \Delta(y)$;/*similarity level of the object from the network*/;

    delete(U, A, Y); /*remove redundant network objects*/;
    for $y_i, y_j \in Y, i \neq j$ **do** /* reconnect objects*/
      if $\delta(y_i, y_j) < NR$ **then** connect(y_i, y_j);
      $a_{gr}(y_i) \longleftarrow 0; a_{gr}(y_j) \longleftarrow 0$;

    $grLabel \longleftarrow 1$;
    for $y_i \in Y$ **do** /*label objects*/
      if $a_{gr} = 0$ **then** $a_{gr}(y_i) \longleftarrow grLabel$;
      for $y_j \in Y, j \neq i$ **do**
        if connected(y_i, y_j) **then** $a_{gr}(y_j) \longleftarrow grLabel$;
      $grLabel \longleftarrow grLabel + 1$;

    for $y_i \in Y$ **do**
      /*calculate the nearest neighbor for every objects*/;
      $\delta_{NN}(y_i) = \min(\{\delta(y_i, y_j) : y_j \in Y \& j \neq i\})$;

    $NR \longleftarrow rg \cdot \frac{\sum_{y \in Y} \delta_{NN}(y)}{card(Y)}$;/*new value of NR*/;
    if $\neg$stopIterations(Y)/*test the stopping condition */ **then**
      joinNotRepresented(U, Y, NR, Δ);
      adjust(Y, U, A, NR);

To control the size of the network there is a removal step, where useless objects are removed. It affects redundant objects from the network representatives. The term redundant applies to the points with the same input object (from U) in their neighborhood. The best points stay in the network and also the ones which are not redundant for other input data. The remaining objects are re-connected and labeled. A granule is determined by the edges between the objects in the structure. The components of the same granule (group) have equal labels.

The last step is to apply a procedure of adjusting of all network objects, where values of attributes of some objects are slightly modified (depending on a similarity level of objects and the type of attributes). This procedure allows to adjust network objects in the attained solution to the examined problem.

It must be emphasized, that algorithm SOSIG does not require a number of clusters to be given. On contrary to partitioning and hierarchical methods groups are identified automatically, which eliminates the inconvenient step of assessing and selecting the best result from a set of potential clusterings.

4 Clustering Validation

Together with specification of elementary granules it is necessary to define measures of granule quality [12]. The aim of clustering techniques is detecting of granules, which are possibly the most compact and separable. To evaluate compactness and separability of discovered clusters there are proposed statistics, so-called internal validity indices. Validity indices are designed to estimate the quality of obtained partitioning. Assessment of the most optimal result requires calculation of validity indices for different values of algorithm parameters, which usually is a number of clusters. The most commonly used indices are Dunn and Dunn-like statistics and Davies-Bouldin (DB) index [3]. Their advantage is indicating no trends with respect to the number of clusters. Therefore, the minimum (DB) or maximum (Dunn) value indicates the most optimal partition. The Dunn's value for specified number of granules nc is defined by Equation 6. Let U be a set of objects and let C_i be a cluster, where $i = 1, \ldots, nc$.

$$D_{nc} = \min_{i=1,\ldots,nc} \left\{ \min_{j=i+1,\ldots,nc} \left(\frac{d(C_i, C_j)}{\max_{k=1,\ldots,nc} diam(C_k)} \right) \right\} \qquad (6)$$

where $d(C_i, C_j)$ is the dissimilarity function between two clusters C_i and C_j defined as

$$d(C_i, C_j) = \min_{x \in C_i, y \in C_j} d(x, y) \qquad (7)$$

and $diam(C)$ is a diameter of a cluster defined as follows:

$$diam(C) = \max_{x,y \in C} d(x, y) \qquad (8)$$

Following the above definition the index value is large for compact clusters situated significantly far from one another. DB index is expressed by Equation 9. It is defined for the number of clusters, which equals nc.

$$DB_{nc} = \frac{1}{nc} \sum_{i=1}^{nc} \left(\max_{j=1,\ldots,nc, j \neq i} R_{ij} \right) \qquad (9)$$

where

$$R_{ij} = \frac{stdev\,(C_i) + stdev\,(C_j)}{d(C_i, C_j)} \tag{10}$$

where $stdev\,(C_i)$ $(stdev\,(C_j))$ denotes standard deviation of a cluster C_i $(C_j$, respectively). The standard deviation of a cluster i is given by Equation 11.

$$stdev\,(C_i) = \frac{1}{|C_i|} \sqrt{\sum_{x \in C_i} (d\,(x, \overline{x}))^2} \tag{11}$$

where $\overline{x}$ is a centroid of the cluster and $d\,(x, \overline{x})$ is an Euclidean distance between the point x and the centroid $\overline{x}$.

DB index measures the average similarity between each cluster and its most similar one, thus it is desirable to minimize this value.

When there is appropriate partitioning available, external validity measures can be used. External indices take into account a membership of points belonging to the generated (C) and compared (P) structure [3]. One example is $Rand$ statistic (R), which has values between 0 and 1. High values of this index indicate great similarity between C and P.

Let U be a set of objects $U = \{x_1, \ldots, x_n\}$ and original C and compared partitioning P are composed of r clusters - $C = \{c_1, \ldots, c_r\}$ and $P = \{p_1, \ldots, p_r\}$. $Rand$ index is defined by Equation 12:

$$R = \frac{a + b}{a + b + c + d} \tag{12}$$

where a, b, c and d are defined as follows:

- a is the number of pairs of elements in U which are in the same set in C and in the same set in P,
- b - the number of pairs of elements in U which are in different sets in P and in different sets in C,
- c - the number of pairs of elements in U which are in the same set in P and in different sets in C,
- d is the number of pairs of elements in U which are in different sets in P and in the same set in C.

5   Experiments

The experiments focus on comparing results of detecting groups in two approaches: when data are points and hyperboxes. There are the following algorithms used: k-means, hcl, hsl and SOSIG. For methods, which require a number of groups as a parameter there are given original values from Table 1. In every case the following are compared: time of clustering (Table 3) and values of validity indices (Tables 4 and 5). In case of SOSIG algorithm, because of its ability to detect a number of clusters, the numbers of detected groups are also examined (Table 2). The interpretability of clusterings created on the basis of SOSIG results has also been taken into consideration (Tables 6 and 7).

5.1 Description of Datasets

There are several data sets in the experiments, shown in Table 1. There are synthetically generated ($norm2D2gr$, $sph2D6gr$, $sph10D4gr$) and real data ($irises$). The sets are various with regard to the number of objects, dimensionality and the existed number of groups. Column $number\ of\ groups$ contains the number of clusters present in the data according to the subjective human perception based on the separation and compactness of the groups. However, the $irises$ data set contains real data delivered with a priori class attribute. For this reason the value of group number for this data is related to the number from the decision attribute.

Table 1. Data sets used in the experiments

data set	number of dimensions	number of points	number of hyperboxes	number of groups
norm2D2gr	2	200	51	2
sph2D6gr	2	300	70	6
irises	4	150	94	3
sph10D4gr	10	200	13	4

5.2 Results of Experiments

Algorithm SOSIG detects a number of clusters automatically. The number of groups identified this way in described above data sets is presented in Table 2. When the result consists of groups of highly variable sizes, only a number of main groups is presented there. Partitioning of $irises$ set contains two levels (low and high resolution), which is visible in all the following tables. In the result there are 2 clusters when granulation is performed on low resolution level, whereas in high resolution level one large cluster is split in two smaller ones and additionally, there are 5 significantly smaller groups. The results considering both levels of granulation are shown in the same cell of the tables where the first value corresponds to low and the second to high level of resolution. Clustering of $irises$ hyperboxes is composed of only one level with 4 main and 6 additional smaller groups. In clustering results of the remaining data sets the number of groups corresponds to each other for both types of processed data.

Table 2. Results of clustering of point-type and granulated data with respect to the number of identified groups

data set	number of groups	
	point-type data	granulated data
norm2D2gr	2	2
sph2D6gr	6	6
irises	2, 3	4
sph10D4gr	4	4

The results presented in Table 3 consider the run time (in seconds) of the algorithms examined on a one-off basis. This is the average time of 50 runs of the methods calculated for clustering original data as well as hyperboxes. The last column of the table contains the quotient of the values. It can be seen, that the processing of granulated data is significantly (up to about 40 times in case of SOSIG and 14 times in case of the remaining algorithms) faster than processing original point-type objects. The most acceleration is visible when the number of objects in data is great and considerably predominate the number of attributes.

Table 3. Average time (in seconds) of clustering hyperboxes and point-type data

data set	algorithm	point-type data t_{pd}	granulated data t_{gd}	t_{pd}/t_{gd}
norm2D2gr	SOSIG	0.360	0.040	9
	k-means	0.062	0.047	1.32
	hcl	0.110	0.032	3.44
	hsl	0.125	0.031	4.03
sph2D6gr	SOSIG	0.930	0.080	11.63
	k-means	0.187	0.094	2.0
	hcl	0.266	0.047	5.66
	hsl	0.250	0.032	7.81
irises	SOSIG	0.870, 0.800	0.790	1.01
	k-means	0.141	0.125	1.13
	hcl	0.078	0.046	1.70
	hsl	0.094	0.047	2.0
sph10D4gr	SOSIG	0.270	0.010	38.57
	k-means	0.156	0.047	3.32
	hcl	0.141	0.016	8.81
	hsl	0.219	0.015	14.6

Comparing the results of clustering algorithms one can notice the most increased speed for hierarchical algorithms and SOSIG. As it has been mentioned, hierarchical algorithms arouse scientists' interest due to their better clustering ability in comparison to less complex partitioning methods. However, their time complexity is greater. The same applies to SOSIG. Processing granulated data in advance can be a way of enabling them to cluster large size databases in reasonable time.

Obviously, the total time of clustering is influenced by the time of data preprocessing, particularly when the algorithm of data preparation is complex. However, in the experiments described in this paper this time is not taken into consideration for two reasons. First of all, the number of objects in preparing a set is decreasing by one in every iteration, which practically reduces the time complexity of pre-processing procedure. In addition, in case of algorithms, which take a number of groups as an input parameter, data should be clustered at least several times to evaluate the number of clusters present in this data. In this case single preparation of data has significantly less importance in comparison to multiple data clustering.

To compare results of clustering regarding the most compact and separable partitioning two internal indices: DB and $Dunn's$ have been chosen. In addition, external measure R has been selected since there were a priori group labels available. This index is convenient to assess clusters quality as well as differences between partitioning results as its range is in interval [0,1]. In general, validity indices are not universal. However, this is the most popular tool for assessing clustering results [3]. Simultaneous comparison of several of them can give a quite objective result. The evaluation of grouping results is shown in Tables 4 and 5.

Table 4. Results of clustering of $norm2D2gr$ and $sph2D6gr$ set in form of point-type and hyperboxes

		$norm2D2gr$ set		$sph2D6gr$ set	
algorithm	index	point-type data	granulated data	point-type data	granulated data
SOSIG	R	0.96	0.98	0.99	0.99
	DB	0.06	0.08	0.03	0.01
	Dunn's	0.16	0.74	0.51	1.38
k-means	R	0.98	1.0	0.76	0.99
	DB	0.07	0.09	0.05	0.01
	Dunn's	0.26	0.50	0.06	1.33
hcl	R	0.98	1.0	0.87	0.99
	DB	0.07	0.09	0.03	0.01
	Dunn's	0.26	0.50	0.51	1.33
hsl	R	0.50	1.0	0.50	1.0
	DB	0.07	0.09	0.03	0.01
	Dunn's	0.26	0.50	0.51	1.33

When studying values of the indices it can be noticed, that in the most cases data granulation did not influence negatively the condition of clustering. Clusterings of $norm2D2gr$, $sph2D6gr$ and $sph10D4gr$ in form of hyperboxes performed by all of the algorithms are characterized by comparable or better values of the internal indices. In case of R index there can be also noticed increase of quality (up to 50%) for hyperbox results. For $irises$ set the values of the internal indices are better for point-type clustering. However, for this type of input data R index is smaller for hcl and k-means algorithms.

Table 6 contains detailed description of groups detected in clustering of $irises$ hyperbox data. The final result is composed of 10 clusters. However, due to considerable differences in their size the result focuses on the main 3 granules. The $apriori$ decision attribute is composed of 3 classes: Iris-setosa (I-S), Iris-versicolor (I-Ve) and Iris-virginica (I-Vi). The set is described by 4 attributes: sepal-length (SL), sepal-width (SW), petal-length (PL) and petal-width (PW). The granule gr_1 contains 13 smaller granules (hyperboxes) and all of them belong to class Iris-setosa. The other granule (gr_3) has comparable size (15 objects) and contains only objects from Iris-versicolor class. The largest granule gr_2 consists of 36 hyperboxes. It is not homogenous with respect of class attribute due to 31% of the objects come from Iris-versicolor class and 69% from Iris-virginica.

Attention has to be focused on the attributes resulted from doubling of dimensions. These features are related to minimal and maximal values of the original attributes. As

Table 5. Results of clustering of *irises* and *sph*10*D*4*gr* set in form of point-type and hyperboxes

algorithm	index	irises set		sph10D4gr set	
		point-type data	granulated data	point-type data	granulated data
SOSIG	R	0.91	0.72	1.0	1.0
	DB	0.14, 0.12	0.2	0.01	0.0001
	Dunn's	0.39, 0.19	0.25	7.83	9.29
k-means	R	0.79	0.82	1.0	1.0
	DB	0.18	0.29	0.01	0.0001
	Dunn's	0.06	0.09	7.83	9.29
hcl	R	0.77	0.81	1.0	1.0
	DB	0.18	0.28	0.01	0.0001
	Dunn's	0.11	0.15	7.83	9.29
hsl	R	0.78	0.70	1.0	1.0
	DB	0.13	0.21	0.01	0.0001
	Dunn's	0.20	0.22	7.83	9.29

Table 6. Main level of *irises* data hierarchy composed of clustering result of hyperboxes set. Table contains 3 main granules.

granule id/ granule size	class distribution	attributes	minimal values	maximal values	$diff_{Avg}$
gr₁/13	100% I-S	SL	4.4-5.4	4.8-5.5	0.25
		SW	3.0-3.7	3.1-3.9	0.15
		PL	1.0-1.5	1.5-1.9	0.29
		PW	0.1-0.4	0.1-0.5	0.12
gr₂/36	31% I-Ve	SL	5.6-7.1	5.6-7.1	0.04
	69% I-Vi	SW	2.5-3.4	2.5-3.4	0.03
		PL	4.3-6.0	4.4-6.0	0.07
		PW	1.4-2.5	1.4-2.5	0.02
gr₃/15	100% I-Ve	SL	5.2-6.1	5.2-6.2	0.11
		SL	5.2-6.1	5.2-6.2	0.11
		SW	2.3-2.9	2.3-3.0	0.08
		PL	3.5-4.7	3.6-4.7	0.17
		PW	1.0-1.4	1.1-0.5	0.08

a consequence there appears an additional feature - the difference between the maximal and minimal value of particular variables. Average differences are presented in Table 6 in column $diff_{Avg}$. Granule gr_1 is characterized by the widest range of all attributes, granule gr_2 contains flowers with the smallest size of petals and sepals. Finally, granule gr_3 is composed of irises with narrow and long petals and sepals.

Table 7 presents granules from the second level of data relationship hierarchy. The granules are hyperboxes identified in the first phase of the granulation. In the table the greatest 3 hyperboxes (denoted as gr_{ij}) from every granule of the main level were selected. The second-level granules from the top-level granules gr_1 and gr_3 have larger

Table 7. Second level of *irises* data hierarchy composed of hyperboxes (only selected objects are presented)

main granule	granule id	size	class distribution	minimal values of attributes		$diff_{Avg}$
gr_1	gr_{11}	15	100% I-S	SL	5.0	0.5
				SW	3.4	0.3
				PL	1.3	0.4
				PW	0.2	0.2
	gr_{12}	15	100% I-S	SL	4.6	0.5
				SW	3.3	0.3
				PL	1.0	0.7
				PW	0.2	0.3
	gr_{13}	9	100% I-S	SL	4.8	0.2
				SW	3.0	0.2
				PL	1.2	0.4
				PW	0.1	0.2
gr_2	gr_{21}	5	100% I-Ve	SL	6.4	0.3
				SW	2.9	0.2
				PL	4.3	0.4
				PW	1.3	0.2
	gr_{22}	4	100% I-Vi	SL	6.4	0.1
				SW	3.0	0.2
				PL	5.1	0.4
				PW	1.8	0.2
	gr_{23}	4	100% I-Vi	SL	5.9	0.3
				SW	2.8	0.2
				PL	4.8	0.3
				PW	1.8	0.0
gr_3	gr_{31}	14	100% I-Ve	SL	5.6	0.5
				SW	2.7	0.3
				PL	3.9	0.8
				PW	1.2	0.3
	gr_{32}	8	100% I-Ve	SL	5.7	0.5
				SW	2.6	0.3
				PL	3.5	0.8
				PW	1.0	0.3
	gr_{33}	6	100% I-Vi	SL	5.4	0.3
				SW	2.8	0.2
				PL	4.1	0.4
				PW	1.3	0.2

size and the range of their attributes values is greater in contrast to the granules belonging to gr_2. It shows that granules gr_1 and gr_3 are more compact and have greater regions of even information density. It can be noticed that the hyperboxes are homogenous with regard to the class attribute.

6   Conclusions

The article presents an approach to time complexity reduction in the process of clustering data. The idea is based on preparation of point-type input data to multidimensional granules in the form of hyperboxes. Formation of the granules maximizes information density transferred by the hyperboxes. The experiments showed the advantage of the presented approach: significant time reduction of granulated data clustering in comparison to point-type partitioning. It is particularly visible when data contain large number of objects. Additionally, the quality of clustering result has not deteriorated when coping with granulated data, on the contrary - in most of the cases the quality has increased. This is connected with the generalization ability of the presented method.

In case of SOSIG algorithm, clustering process can be performed on different resolution of data. Clustering of hyperboxes has been executed without changing the resolution. A three-level structure of data has been constructed by joining original point (third down level) in hyperboxes (second level), whereas the top level contains dividing of hyperboxes into clusters. Partitioning at the top level of hyperboxes granulation (clustering) is composed of the same number of groups as partitioning point-type data. The quality of created clusters is also comparable due to the similar values of quality indices are similar.

The process of hyperbox creation is a type of aggregation operation, therefore the major benefit of the presented method is shortening the time of cluster creation in comparison to the processing point-type data. It is particularly effective when data contain large number of objects. Hyperboxes also determine additional level of relationship existing within data. Finally, the description of granules is more comprehensible since the hyperboxes contain minimal and maximal values of attributes.

Acknowledgements. The experiments have been performed on the computer cluster at Faculty of Computer Science, Bialystok University of Technology.

This work was supported by Grant No. S/WI/5/08.

References

1. Bargiela, A., Pedrycz, W.: Classification and Clustering of Granular Data. In: IFSA World Congress and 20th NAFIPS International Conference, vol. 3, pp. 1696–1701 (2001)
2. Bargiela, A., Pedrycz, W.: Granular Computing: an Introduction. Kluwer Academic Publishers, Boston (2002)
3. Halkidi, M., Batistakis, Y.: On clustering validation techniques. Journal of Intelligent Information Systems 17(2/3), 107–145 (2001)
4. Han, J., Kamber, M.: Data Mining: Concepts and Techniques. Morgan Kaufmann (2000)
5. Jain, A.K., Murty, M.N., Flynn, P.J.: Data clustering: a review. ACM Computing Surveys 31(3), 264–323 (1999)
6. Kużelewska, U.: Clustering with granular information processing. In: Proceedings of ICAART Conference, vol. 1, pp. 89–97. SciTePress (2011)
7. Li, F., Xie, J., Xie, K.: Granular computing theory in the application of fault diagnosis. In: Proceeding of Control and Decision Conference CCDC, pp. 595–597 (2008)

8. Peters, J.F., Skowron, A., Stepaniuk, J.: Information Granules in Spatial Reasoning. In: Terano, T., Chen, A.L.P. (eds.) PAKDD 2000. LNCS (LNAI), vol. 1805, pp. 380–383. Springer, Heidelberg (2000)

9. Pawlak, Z.: Rough Sets. Theoretical Aspects of Reasoning about Data. Kluwer Academic Publishers (1991)

10. Stepaniuk, J.: Rough Granular Computing in Knowledge Discovery and Data Mining. Springer, Heidelberg (2008)

11. Stepaniuk, J., Kużelewska, U.: Information Granulation: A Medical Case Study. In: Peters, J.F., Skowron, A., Rybiński, H. (eds.) Transactions on Rough Sets IX. LNCS, vol. 5390, pp. 96–113. Springer, Heidelberg (2008)

12. Skowron, A., Stepaniuk, J.: Modeling of High Quality Granules. In: Kryszkiewicz, M., Peters, J.F., Rybiński, H., Skowron, A. (eds.) RSEISP 2007. LNCS (LNAI), vol. 4585, pp. 300–309. Springer, Heidelberg (2007)

13. Vachkov, G.: Similarity analysis of images based on information granulation and fuzzy decision. In: International IEEE Conference on Intelligent Systems, pp. 1014–1021 (2008)

14. Zadeh, L.A.: A new direction in AI: Toward a computational theory of perceptions. AI Magazine 22(1), 73–84 (2001)

Modeling Motivations, Personality Traits and Emotional States in Deliberative Agents Based on Automated Planning

Daniel Pérez-Pinillos, Susana Fernández, and Daniel Borrajo

Departamento de Informática, Universidad Carlos III de Madrid,
Avda. de la Universidad, 30, Leganés (Madrid), Spain
{daniel.perez,susana.fernandez,daniel.borrajo}@uc3m.es
http://www.plg.inf.uc3m.es

Abstract. There is a wide variety of applications that require modeling the behaviour of virtual agents. Some of these applications aim at human interaction, such as virtual assistants, and others aim at simulation of human behavior, such as games or robotics. Most of these applications require not only some level of intelligent behavior, but also a display of realistic human behavior. This has led to the definition and use of models that integrate features like emotions, personality traits, preferences and motivations. Most of this work has been carried out in the context of reactive architectures. Thus, the reasoning on the emotional state of agents is only performed for the very next future, generating behavior that is myopic for middle or long term goals. In this paper, we propose instead a deliberative model based on automated planning that integrates all these features for long term reasoning.

Keywords: Planning, Agent, Emotion, Personality, Motivation, Preference, Decision making.

1 Introduction

In many domains, the behaviour of any agent can be seen as a sequential decision-making process, i.e. the cognitive process results in the selection of a course of actions to fulfill some goals. The decision making process is a continuous process integrated with the interaction with the environment where individual decisions must be examined in the context of a set of needs and preferences that the agent has. Recent theories state that human decision-making is also influenced by marker signals that arise in bioregulatory processes, including those that express themselves in emotions and feelings [11]. Probably, this is one of the reasons why the work on reasoning about emotions is becoming increasingly relevant, specially in contexts such as assistive technology, user interfaces, or virtual agents [4,14].

In spite of the wide variety of points of view that have been used to study emotions, it seems there is some agreement to consider emotion as an inborn and subjective reaction to the environment, with an adaptive function, and accompanied of several organic, physiological and endocrine changes [17]. Another point of agreement is that emotions

J. Filipe and A. Fred (Eds.): ICAART 2011, CCIS 271, pp. 146–160, 2013.

are an outstanding factor in humans, because they modify and adapt their usual behavior. In the development of systems that interact with persons, as human behavior simulators, emotions can not be ignored, because, on one hand, they may help on this interaction and, on the other hand, they constitute a decisive part of human reasoning and behavior. This is specially true when reasoning about sequential decision-making, as in medium-long term planning, where the sequence of decisions can be influenced by the emotional state of agents.

Emotions are also very related to characteristics of human personality. In contemporary psychology, there are five factors or dimensions of personality, called the Big Five factors [19], which have been scientifically defined to describe human personality at the highest level of organization. The Big Five traits are also referred to as a purely descriptive model of personality called the Five Factor Model [10,24]. The Big Five factors are: openness to experience, conscientiousness, extraversion, agreeableness and neuroticism (opposite to emotional stability). Each of these factors has a more specific set of features among which there is a correlation.

In the present work, a model of long term reasoning based on emotions and factors of personality has been designed. It follows some ideas introduced in [1] using concepts that already appeared in other works, like motivations and the use of drives to represent basic needs [7,8]. The main novelty of our model is the use of automated planning for providing long term deliberation on effects of actions taking into account not only the agents goals, but also the impact of those actions in the emotional state of the agent.

We have defined a planning domain model that constitutes the reasoning core of an agent in a virtual and multi-agent world [15]. It is a game oriented towards the use of Artificial Intelligence controlled Bots, and it was designed as a test environment of several Artificial Intelligence techniques. The game borrows the idea from the popular video game THE SIMS. Each agent controls a character that has autonomy, with its own drives, goals, and strategies for satisfying those goals. In this implementation, we introduce the concept of how an agent prefers some actions and objects depending on its preferences, its personality traits and its emotional state, and the influence of those actions on long term achievement of goals. Thus, agents solve problems improving the quality of the solution, achieving better emotional states.

The remainder of the paper describes the model design, the description of the domain that implements the model, the empirical results that validate the model, the related work and the conclusions derived from the work, together with future research lines.

2   Model Design

Our aim in this work is to include emotions and human personality traits in a deliberative system, that uses automated planning in order to obtain more realistic and complex behavior of agents. These behaviors are necessary to implement a wide variety of applications such as agents that help users to change their way of life, systems related with marketing and advertising, educational programs, systems that play video games or automatically generate text. The goal is to show that the use of emotional features, with the establishment of preferences about certain actions and objects in its environment, improves the performance of a deliberative agent by generating better plans.

In the virtual world, an agent tries to cater for its needs, its motivations, through specific actions and interacting with different objects. Five basic needs have been identified for the agent, which are easily identifiable in human beings: hunger, thirst, tiredness, boredom and dirtiness. Along with the first three, widely used in many systems, we have added dirtiness and boredom, which are more domain-specific to add a wider variety of actions and get richer behaviors. These basic needs increase over time, so their values increase as time goes by. Thus, the agent always needs to carry out actions to maintain its basic needs values within reasonable limits.

To cater for each of these basic needs, the agent must perform actions. For example, it can drink to satisfy its thirst or sleep to recover from fatigue. There are different actions to cater for the same need, and the agent prefers some actions over others. Thus, the agent may choose to read a book or play a game to reduce boredom. Besides, the effects of those actions can be different depending on its emotional state. It will receive more benefit from applying more active actions when its emotional state is more aroused and more passive or relaxed actions when it is calm.

To carry out each of these actions, the agent needs to use objects of specific types. Thus, it will need food to eat, a ball to play or a book to read. There are different objects of each type in its environment and the agent has preferences over them. When an agent executes an action with an object, its emotional state is modified depending on the agent personality, and preferences and activations for this object.

We have chosen to implement a model widely-accepted in psychology that represents the emotional state of an agent as a two-dimensional space of two qualities: valence and arousal [13]. Valence ranges from highly positive to highly negative, whereas arousal ranges from calming or soothing to exciting or agitating. The first one is a measure of the pleasantness or hedonic value, and the second one represents the bodily activation. Other models use a set of independent emotions, which requires defining a group of basic emotions. However, not all combinations of values for these emotions are a valid emotional state (e.g. the combination of maximum values in the emotions of joy and anger is not a realistic emotional state). In general, the valence and arousal model can be shown to be equivalent to the explicit representation of the usual set of emotions of other computational cognitive simulations, though it requires a simpler representation and reasoning. For instance, an emotion such as happiness can be represented as high valence and high arousal. Both models are recognized and defended by experts in psychology, but we prefer the second alternative because it makes processing easier and prevent invalid states. In our model, the valence and the arousal are modified by the execution of actions, so both values are modified when an agent executes an action with an object, depending on the agent preference and activation for this object, the personality traits and the emotional state. Our goal is that the agent generates plans to satisfy its needs and to achieve the most positive value of valence.

3   Domain Description

In order to use domain-independent planning techniques, we have to define a domain model described in the standard language PDDL [16]. This domain should contain all the actions that the agent can perform in order to achieve the goals. Automated planning can be described as a search for a solution on a problem space where, the states are represented using a set of predicates, functions and types, and the actions are described with a set of preconditions and effects that model the state transitions. An action is applicable only if all its preconditions hold in the current state and executing the action changes the current state by adding and deleting the action effects. A problem is specified as an initial state (true literals in the starting state) and a set of goals. Also, an optimization metric (as in our case valence, arousal and/or total time) can be defined. Our domain has been designed based on the previous concepts of drive, emotion, preference, activation and personality traits to represent each agent of the virtual world. Now, we will define the different concepts composing the model, in automated planning terms.

3.1 Drives

As already said, we use five drives: hunger, thirst, tiredness, dirtiness and boredom. Drives are represented in the domain through functions. The ideal value for all drives is established at zero. So, when a drive has a value of zero, its need is totally satisfied. Any other value means the intensity of the need and the distance to the ideal value. The value of each drive is increased as time goes by to represent the need rise. To reduce it, the agent has to carry out some action. For instance, the agent must eat to reduce the drive hunger. Given that the drives increase with time, every time an action is executed, one or more drives will be decreased, but the rest will be increased. Thus, the planning task becomes hard if we want all drives to be fulfilled (below a given threshold).

3.2 Objects

Objects describe the different elements of the virtual world. Objects may be of two kinds: resources (or physical objects) and rooms. Resources represent objects needed to carry out the actions to cater for needs; for instance, food, balls, books, etc. Rooms describe physical spaces, where the agents may move and where resources are placed. Both kinds of objects are represented as planning types and several instances of them will be present in each problem. Also resources may be of two kinds: fungible resources and non-fungible resources.

3.3 Personality Traits

Personality traits describe the agents personality and are based on the Big Five factors model (openness to experience, conscientiousness, extraversion, agreeableness and neuroticism). Openness to experience involves active imagination, aesthetic sensitivity, preference for variety and intellectual curiosity. Openness is modeled as a higher preference for new experiences, i.e., an agent with high openness (open-minded) tends to use and prefer new objects to known objects, while an agent with low openness will tend to prefer known objects to new objects. Neuroticism represents the degree of emotional stability of the agent. The bigger the neuroticism is, the smaller the emotional stability is. So, neuroticism is implemented as the variation factor of the emotional state. Thus, the emotional state of a neurotic agent will vary more suddenly than a stable one when actions are applied, as described later.

Conscientiousness includes elements such as self-discipline, carefulness, thoroughness, organization, deliberation and need for recognition. We implement conscientiousness as a factor in the decrements of the drives due to action executions, representing how meticulous the agent is in carrying out the action. Thus, an agent with a high value of conscientiousness gets a bigger effect when applying actions (a bigger decrease of the involved drive). But, similarly, the other drives will also increase proportionately to the conscientiousness value as time passes. The conscientiousness value also influences the duration of the actions performed by agents. For instance, the actions performed by a meticulous agent take more time than the ones performed by a careless agent.

The last two factors, extraversion and agreeableness, are related to social interaction. Thus, they will be used in future versions of the system that include multiple agents and interactions among them. Personality traits are represented in the domain through functions.

3.4   Emotional State

The agents emotional state is determined by two components: valence and arousal. Valence represents whether the emotional state of the individual is positive or negative and to which degree. Arousal represents the bodily activation or agitation. We represent them in the domain as PDDL functions. Since we want to obtain plans that maximize the valence, we have to define the planning problems metric accordingly. Even if PDDL allows generic functions to be defined as metrics, most current planners can only deal with metrics that are defined over minimizing an increasingly monotonous function (no action can have an effect that decreases its value), since metrics are considered in PDDL as costs and each action has an associated cost.

In our model, objects used in the actions can cause valence both to increase (when the agent likes the object) or decrease (when it does not like it). Therefore, it is not possible to use the valence directly as the problem metric. Instead, we define an increasingly monotonous function, v-valence, that the planner tries to minimize. Each action increases v-valence, with positives values between 0 and 10 depending on the preference for the object used, in the following amount:

$$\Delta v = \left(\frac{n}{n_{max}}\right) \times \left(p_{max} - \frac{(p_a + p_o)}{2}\right)$$

where v is the value of v-valence, n the agent neuroticism, n_{max} the maximum possible value for neuroticism, p_{max} the maximum possible value for a preference, p_a the agent preference for the executed action and p_o the agent preference for the used object. In case the object is new to the agent, p_o=-1 and we replace p_o for the value of the agent openness. Thus, this value can be used as a metric alone or combined with others such as the duration of the plan.

3.5 Preferences

Preferences describe the agent personal likes for each physical object of its environment. They are represented as PDDL functions of the form:

```
(= (preference apple) 5)
```

These values are not modified during the planning process and they are between zero, for the detested objects, and ten, for the favourite ones. Preferences can also describe the agent personal likes for each action. They are represented as PDDL functions of the form:

```
(= (read-preference) 5)
```

Again, these values are not modified during the planning process and they are between zero, for the detested actions, and ten, for the favourite ones. Preferences affect the direction and degree of changes on the value of the valence, produced by the effects of actions.

3.6 Activations

Activations describe the effect over the agent arousal for each physical object of its environment. They are represented as PDDL functions of the form:

```
(= (activation apple) 5)
```

These values are not modified during the planning process and they are between zero, for the objects that relax, and ten, for the objects that agitate. Activations can also describe the effect over the agent arousal for each action. They are represented as PDDL functions of the form:

```
(= (read-activation) 5)
```

Again, these values are not modified during the planning process and they are between zero, for the actions that relax, and ten, for the actions that agitate.

3.7 Actions

Actions defined in the domain describe activities that the agent may carry out. Each action has a simulated duration (time spent in the virtual world). This duration is determined from a standard time that takes to execute the corresponding action and the agent's value of conscientiousness. There are five types of actions:

- Actions to cater for its needs: Each one of these actions needs one object of a specific type to decrease in one unit its corresponding drive value. In this group of actions, we have defined: eat, drink, sleep, bath, shower, play, read, watch and listen. Some of these actions require that the agent has taken the object used, like eat, drink or read. Others, however, only require that the object is located in the same room of the agent, like bath or sleep. In addition, some actions such as eat and drink decrease the available amount of the object used.

In Figures 1 and 2, we show two examples of this action type. In each of these actions, the agent needs to have the appropriate objects in order to carry out the respective action; e.g. it needs food to eat and a readable object (a book or newspaper) to read. We can see that the related drive decreases a quantity, depending on the agent conscientiousness. On the other hand, the changes in the agent emotional state (valence and arousal) depend on the agent preferences and activations over the action to perform and the used object, and its personality traits. Thus, we have an integrated model of these concepts, that can affect how actions are combined in order to solve the agents problems.

```
(:action READ
    :parameters    (?reading-object - reading-object)
    :precondition (and (in ?room)(taken ?reading-object)(not (time-goes-by)))
    :effect
    (and
        (time-goes-by)
        (assign (action-time) (* (conscientiousness) (read-duration)))
        (decrease (boredom) (conscientiousness))
        (when (and (< (boredom) 0))
            (and (assign (boredom) 0)))
        (when (and (< (preference ?reading-object) 0))
            (and
                (increase (valence)
                    (* (/ (neuroticism) (max-neuroticism))
                       (- (/ (+ (preference ?reading-object) (read-preference)) (max-preference)) 1)))
                (increase (v-valence)
                    (* (/ (neuroticism) (max-neuroticism))
                       (- (max-preference) (/ (+ (preference ?reading-object) (read-preference)) 2))))))
        (when (and (> (preference ?reading-object) 0))
            (and
                (increase (valence)
                    (* (/ (neuroticism) (max-neuroticism))
                       (- (/ (+ (openness) (read-preference)) (max-preference)) 1)))
                (increase (v-valence)
                    (* (/ (neuroticism) (max-neuroticism))
                       (- (max-preference) (/ (+ (openness) (read-preference)) 2))))))
        (increase (arousal)
            (* (/ (neuroticism) (max-neuroticism))
               (- (/ (+ (activation ?reading-object) (read-activation)) (max-activation)) 1)))
        (increase (v-arousal)
            (* (/ (neuroticism) (max-neuroticism))
               (- (max-activation) (/ (+ (activation ?reading-object) (read-activation)) 2))))))
```

Fig. 1. Example of action (READ) to cater for the boredom need

- TAKE and LEAVE actions: the agent uses them to take and leave objects required to perform some actions, like eat or drink.
- BUY action: the agent uses it to purchase new resources. Agents must be in a shop and the resource must be available to be bought.
- GO action: allows the agents to move as Figure 3 shows.
- TIME-GOES-BY action: It is a fictitious action (Figure 4) that represents the influence of the course of time over the value of the drives. Its execution produces an increase on all drives, so that it simulates the passing of time. The increment depends on the last action duration (action-time function added in action effects). We also force the planner to be executed after every other action application (through the time-goes-by predicate).

All actions (except for TIME-GOES-BY) modify (in their effects) the emotional state that depend on the agent preferences, activations and personality traits. Along with the metric of the problem, this allows us to model the agents behaviour. So, there are no

```
(:action EAT
    :parameters    (?food - food ?room - room)
    :precondition (and (in ?room)(taken ?food)(> (quantity ?food) 0)(not (time-goes-by)))
    :effect
        (and
            (time-goes-by)
            (assign (action-time) (* (conscientiousness) (go-duration))))
            (decrease (hunger) (conscientiousness))
            (decrease (quantity ?food) (conscientiousness))
            (when (and (< (hunger) 0))
                (and (assign (hunger) 0)))
            (when (and (< (preference ?food) 0))
                (and
                    (increase (valence)
                        (* (/ (neuroticism) (max-neuroticism))
                            (- (/ (+ (preference ?food) (eat-preference)) (max-preference)) 1)))
                    (increase (v-valence)
                        (* (/ (neuroticism) (max-neuroticism))
                            (- (max-preference) (/ (+ (preference ?food) (eat-preference)) 2))))))
            (when (and (> (preference ?food) 0))
                (and
                    (increase (valence)
                        (* (/ (neuroticism) (max-neuroticism))
                            (- (/ (+ (openness) (eat-preference)) (max-preference)) 1)))
                    (increase (v-valence)
                        (* (/ (neuroticism) (max-neuroticism))
                            (- (max-preference) (/ (+ (openness) (eat-preference)) 2))))))
            (increase (arousal)
                (* (/ (neuroticism) (max-neuroticism))
                    (- (/ (+ (activation ?food) (eat-activation)) (max-activation)) 1)))
            (increase (v-arousal)
                (* (/ (neuroticism) (max-neuroticism))
                    (- (max-activation) (/ (+ (activation ?food) (eat-activation)) 2))))))
```

Fig. 2. Example of an action (EAT) to cater for a need (hunger)

hard constraints on our model. All agents can perform all actions, but they prefer (soft constraints) the ones that better suit their preferences, personality and current emotional state.

3.8   Goals

The agent motivation is to satisfy its basic needs, so goals consist of a set of drives values that the agent has to achieve. As an example, goals may consist of the achievement of need values (and emotional variables) that are under a given threshold. They could be very easily combined with other kinds of standard planning goals, creating other kinds of domains. For instance, we could define strategy games where agents should accomplish some tasks, taking into account also their needs.

4   Experiments

We report here the results obtained with the proposed model comparing its performance to a reactive model. In the case of the deliberative model, we have used an A* search technique with the well-known domain-independent heuristic of FF [22]. This heuristic is not admissible, but even if it does not ensure optimality, it is good enough for our current experimentation. In the case of the reactive model, we have used a function to choose the best action at each step (to cover the drive with the higher value, i.e. the worse drive). These search techniques have been implemented in an FF-like planner, SAYPHI [12].

```
(:action GO
    :parameters   (?place-from - place ?place-to - place)
    :precondition (and (in ?place-from)(not (time-goes-by)))
    :effect
      (and
        (time-goes-by)
        (assign (action-time) (* (conscientiousness) (go-duration)))
        (increase (valence)   (* (/ (neuroticism) (max-neuroticism))
                              (- (/ (* (go-preference) 2) (max-preference)) 1)))
        (increase (arousal)   (* (/ (neuroticism) (max-neuroticism))
                              (- (/ (* (go-activation) 2) (max-activation)) 1)))
        (increase (v-valence) (* (/ (neuroticism) (max-neuroticism))
                              (- (max-preference) (go-preference))))
        (increase (v-arousal) (* (/ (neuroticism) (max-neuroticism))
                              (- (max-activation) (go-activation))))
        (not (in ?place-from))
        (in ?place-to)))
```

Fig. 3. GO action

```
(:action TIME-GOES-BY
    :parameters ()
    :precondition (and (time-goes-by))
    :effect (and
                (increase (boredom) (* 0.1 (action-time)))
                (assign (boredom) (min (max-drive) (boredom)))
                (increase (dirtiness) (* 0.1 (action-time)))
                (assign (dirtiness) (min (max-drive) (dirtiness)))
                (increase (hunger) (* 0.1 (action-time)))
                (assign (hunger) (min (max-drive) (hunger)))
                (increase (thirst) (* 0.1 (action-time)))
                (assign (thirst) (min (max-drive) (thirst)))
                (increase (tiredness) (* 0.1 (action-time)))
                (assign (tiredness) (min (max-drive) (tiredness)))
                (increase (total-time) (action-time))
                (assign (action-time) 0)
                (not (time-goes-by))))
```

Fig. 4. TIME-GOES-BY action

4.1   Experimental Setup

In the first experiment, we have defined several kinds of problems for this domain. In each problem, we have established a specific initial need in one of the drives, which are called dominant drives. Each of these dominant drives will have a initial value higher than the rest of drives. Also, we have defined a problem where all five drives are dominant drives. The goal is to fulfill all the agent needs, so we have defined it as having a value below a threshold for all drives. Furthermore, for each action, the agent has three objects to choose from, with varying degrees of preference: preferred, indifferent and hated, and a new object (the agents do not have an "a priori" preference for this object) for testing openness. In this experiment, all actions have the same standard duration.

The experiments were performed with four different personality models: (1) a standard personality (average values in all traits), (2) a neurotic personality (high value of neuroticism and average values for the rest), (3) an open-minded personality (high value of openness and average values for the rest) and (4) a meticulous personality (high value of conscientiousness and average values for the rest).

In the second experiment, we have also established a dominant drive in each problem and the agent has three objects of each type. In this case, the targets are again getting a value below a threshold. But, we added another goal: to achieve a valence value above a threshold, so that we can consider that the agent is in a good state after the execution of the plan. We have established different standard durations for actions: instantaneous (as `take` and `leave` objects), short duration (as `drink`, `go` and `shower`), medium duration (as `eat`, `bath`, `play`, `watch`, `read`, `listen` and `buy`) and long duration (`sleep`). The metric was minimizing the total duration of the plan. The experiments were performed with two different personalities according to the conscientiousness value: careless and meticulous, because conscientiousness weights the standard duration of the actions.

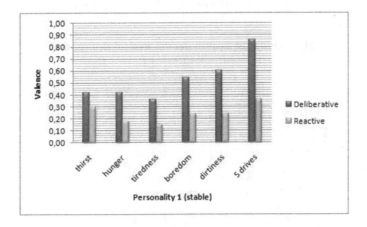

Fig. 5. Quality of the plans for the stable agent

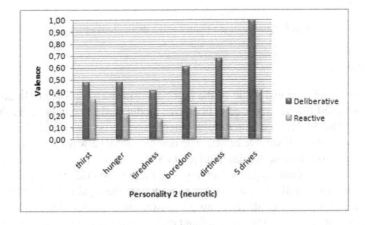

Fig. 6. Quality of the plans for the neurotic agent

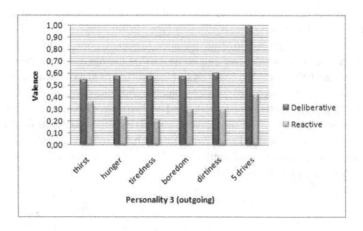

Fig. 7. Quality of the plans for the open-minded agent

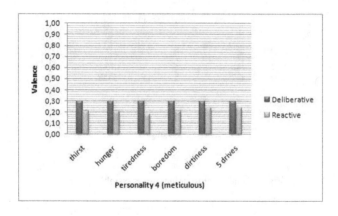

Fig. 8. Quality of the plans for the meticulous agent

4.2 Results

Figures 5 to 8 show the end value of the (valence) for each problem. In all cases, the value obtained by the proposed deliberative model is significantly better than the reactive one. This is due to a better employment of the buy action and the reduction on go actions of the deliberative model. The reactive model always tries to satisfy the need associated to the most dominant drive at each time. So, for instance, if reducing the current dominant drive requires drinking, and there is no drink in the current agent room, then the agent will move to another room where the drinking action can be accomplished. However, the deliberative model reasons on a medium-long term, so if the need in another drive, not being the dominant one, can be satisfied in the current room, the plan will prefer to reduce it now, even if the dominant drive increases a bit. Most previous work on emotional agents would mimic the reactive model, while our model is able to take into account future recompenses in an integrated way with other agents

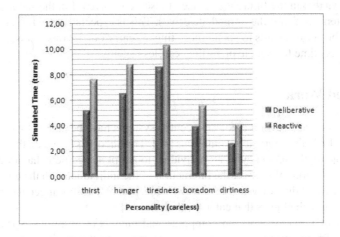

Fig. 9. Simulated time of the plans for the careless agent

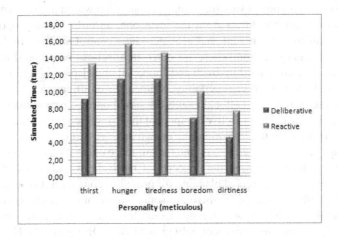

Fig. 10. Simulated time of the plans for the meticulous agent

goals. We also see that if the personality tends to be more neurotic, then the deliberative model is even better than the reactive one, since actions effects are increased, and drives increase more acutely. When the agent is open minded, it is more likely to choose using new objects to satisfy its curiosity, so that the resulting valence values are higher than in the other cases. When it has a meticulous personality, we can see that actions decrease drives values faster. Therefore, the agent needs less actions for its needs, and we can see that the resulting values are lower.

Figures 9 and 10 show the end value of the (total-time) for each problem. In all cases, the value obtained by the proposed deliberative model is significantly lower than the reactive one. This is because, once again, the deliberative agent quickly reaches the goals to meet its other needs. For instance, it saves repeated travels to buy items in the store by buying several items in the store when it goes there. So, it buys extra items, when it knows that in the future it will need those items, even if in the state, it does not need those items to fulfill the immediate drives. This is especially true for a meticulous personality, because it tends to spend more time to carry out actions. Thus, meticulous agents take less time to execute plans than careless agents.

5 Related Work

During the last years, several emotion-oriented systems have been developed, that normally follow Frijda's theory about emotions [18]. This theory is based on the hypothesis that emotions are the tendency of an individual to adopt a specific behavior according to its needs. Emotions also cover the interaction of the individual with the environment. For instance, individuals try to move away objects that put in danger their survival, while they approach objects that cater for their needs [6].

Examples of previous work on computational models of emotions is the work of Cañamero [7,8] that proposes a homeostatic approach to the motivations model. She creates a self-regulatory system, very close to natural homeostasis, that connects each motivation to a physiological variable, which is controlled within a given range. When the value of that variable differs from the ideal one, an error signal proportional to the deviation, called drive, is sent, and activates some control mechanism that adjusts the value in the right direction. There are other architectures based on drives, as the Dorner's PSI architecture used by Bach and Vuine [3] and also by Lim [23], that offer a set of drives of different type, as certainty, competence or affiliation.

Most of these works on emotional agents are based on reactive behaviors. When a drive is detected, it triggers a reactive component that tries to compensate its deviation, taking into account only the following one or two actions. Thus, there is no inference being done on medium-long term goals and the influence of emotions on how to achieve those goals. Our model borrows the concepts of motivations and drives to represent basic needs, but it uses automated planning for providing long term deliberation.

Regarding deliberative models, there are some works on emotions based on planning, but mainly oriented to storytelling. Examples are emergent narrative in FEARNOT! [2] and the interactive storytelling of Madame Bovary on the Holodeck [9]. The work of Gratch and coauthors [20,21] shows a relevant application of emotional models to different research areas in artificial intelligence and autonomous agents design, endowing them with an ability to think and engage in socio-emotional interactions with human users. Other models, as Rizzo's works [25], combine the use of emotions and personality to assign preferences to the goals of a planning domain model, but the changes in the emotional state happen in another module. Thus, they are not really used in the reasoning process. A similar integration of a deliberative and a reactive model is the one in [5] where the emotions reasoning is performed again by the reactive component. Opposite to all these approaches, there are no hard constraints on our model. All our agents can perform all actions, but they prefer (soft constraints) the ones that better suit their preferences, personality and current emotional state.

6   Conclusions and Future Work

This work proposes a model of long term reasoning integrating emotions, drives, preferences and personality traits in autonomous agents, based on AI planning. The emotional state is modeled as two functions: valence and arousal. This two-dimensional model has been chosen because it is simpler and offers the same representation capabilities as the rest of emotional models. Anyhow, it is not difficult now to integrate any other emotional model. Thus, actions produce variations in the valence depending on the agent personality and agent preferences. The goal is to generate plans that maximize the valence, while satisfying the agent needs or drives. Given that current planners only deal with monotonous functions as metric functions, we converted the non-monotonous valence into a monotonous one, `v-valence`. The results of the experiments show that the quality of the solutions (measured as the value of the `valence`) improves when the deliberative model is used compared to the reactive one. Thus, the increase in the quality of the solutions implies a more realistic behavior of the agent.

The proposed model is the first step in the development of a richer and more complex architecture. In the next future, we would like to include new actions in the domain, especially those related to the processes of social interaction, by including some component that reasons about multi-agent interaction and collaboration. Another future work is to model the idea of well-being, which will focus the agent to keep all its needs below a certain level along time. The physiological well-being of the agent will influence its emotional state altering the value of valence. This idea is very related to the idea of continuous planning to control the behaviour of virtual agents [1].

References

1. Avradinis, N., Aylett, R.S., Panayiotopoulos, T.: Using Motivation-Driven Continuous Planning to Control the Behaviour of Virtual Agents. In: Balet, O., Subsol, G., Torguet, P. (eds.) ICVS 2003. LNCS, vol. 2897, pp. 159–162. Springer, Heidelberg (2003)
2. Aylett, R.S., Louchart, S., Dias, J., Paiva, A., Vala, M.: Fearnot!: an experiment in emergent narrative, pp. 305–316 (2005)
3. Bach, J., Vuine, R.: The AEP Toolkit for Agent Design and Simulation. In: Schillo, M., Klusch, M., Müller, J., Tianfield, H. (eds.) MATES 2003. LNCS (LNAI), vol. 2831, pp. 38–49. Springer, Heidelberg (2003)
4. Bates, J.: The role of emotion in believable agents. Communications of the ACM 37, 122–125 (1994)
5. Blythe, J., Reilly, W.S.: Integrating reactive and deliberative planning for agents. Technical report (1993)
6. Breazeal, C.: Biological Inspired Intelligent Robots. SPIE Press (2003)
7. Cañamero, D.: Modeling motivations and emotions as a basis for intelligent behavior. In: First International Symposium on Autonomous Agents (Agents 1997), pp. 148–155. The ACM Press, New York (1997)
8. Cañamero, D.: Designing Emotions for Activity Selection in Autonomous Agents. MIT Press (2003)
9. Cavazza, M., Lugrin, J., Pizzi, D., Charles, F.: Madame bovary on the holodeck: Immersive interactive storytelling. In: Proceedings of the ACM Multimedia 2007. The ACM Press, Augsburg (2007)

10. Costa, P., McCrae, R.: Revised NEO Personality Inventory (NEO-PI-R) and NEO Five-Factor Inventory (NEO-FFI) manual. Psychological Assessment Resources, Odessa (1992)
11. Damasio, A.R.: Descartes' Error. Putnam Press, New York (1994)
12. de la Rosa, T., García Olaya, A., Borrajo, D.: Using Cases Utility for Heuristic Planning Improvement. In: Weber, R.O., Richter, M.M. (eds.) ICCBR 2007. LNCS (LNAI), vol. 4626, pp. 137–148. Springer, Heidelberg (2007)
13. Duffy, E.: An explanation of emotional phenomena without the use of the concept of emotion. Journal of General Psychology 25, 283–293 (1941)
14. Fellous, J.-M., Arbib, M.A.: Who needs emotions?: the brain meets the robot, vol. Series. Oxford University Press, Oxford (2005)
15. Asensio, J., Jiménez, M., Fernández, S., Borrajo, D.: A Social and Emotional Model for Obtaining Believable Emergent Behaviors. In: Dochev, D., Pistore, M., Traverso, P. (eds.) AIMSA 2008. LNCS (LNAI), vol. 5253, pp. 395–399. Springer, Heidelberg (2008)
16. Fox, M., Long, D.: Pddl2.1: An extension to pddl for expressing temporal planning domains. Journal of Artificial Intelligence Research 20, 61–124 (2003)
17. Frijda, N.H.: The laws of emotion. American Psychologist 43, 349–358 (1988)
18. Frijda, N.H.: Emotions in robots. In: Roitblat, H.L., Meyer, J.-A. (eds.) Comparative Approaches to Cognitive Science, pp. 501–516. The MIT Press, Cambridge (1995)
19. Goldberg, L.: The structure of phenotypic personality traits. American Psychologist. 48, 26–34 (1993)
20. Gratch, J.: Why you should buy an emotional planner. In: Proceedings of the Agents 1999 Workshop on Emotion-based Agent Architectures (EBAA 1999) and ISI Research Report, pp. 99–465 (1999)
21. Gratch, J., Rickel, J., André, E., Badler, N., Cassell, J., Petajan, E.: Creating interactive virtual humans: Some assembly required. IEEE Intelligent Systems 17, 54–63 (2002)
22. Hoffmann, J.: FF: The fast-forward planning system. AI magazine 22, 57–62 (2001)
23. Lim, M.Y., Aylett, R.S., Jones, C.M.: Emergent Affective and Personality Model. In: Panayiotopoulos, T., Gratch, J., Aylett, R.S., Ballin, D., Olivier, P., Rist, T. (eds.) IVA 2005. LNCS (LNAI), vol. 3661, pp. 371–380. Springer, Heidelberg (2005)
24. McCrae, O., John, R.R.: An introduction to the five-factor model and its applications. Journal of Personality 2, 175–215 (1992)
25. Rizzo, P., Veloso, M., Miceli, M., Cesta, A.: Personality-driven social behaviors in believable agents. In: Proceedings of the AAAI Fall Symposium on Socially Intelligent Agents, pp. 109–114. AAAI Press (1997)

Symbolic State-Space Exploration and Guard Generation in Supervisory Control Theory

Zhennan Fei, Sajed Miremadi, Knut Åkesson, and Bengt Lennartson

Automation Research Group, Department of Signals and Systems
Chalmers University of Technology, Göteborg, Sweden
{zhennan,miremads,knut,bengt.lennartson}@chalmers.se

Abstract. Supervisory Control Theory (SCT) is a model-based framework for automatically synthesizing a supervisor that minimally restricts the behavior of a plant such that given specifications is fulfilled. The main obstacle which prevents SCT from having a major industrial breakthrough is that the supervisory synthesis, consisting of a series of reachability tasks, suffers from the state-space explosion problem. To alleviate this problem, a well-known strategy is to represent and explore the state-space symbolically by using Binary Decision Diagrams. Based on this principle, an alternative symbolic state-space traversal approach, depending on the disjunctive partitioning technique, is presented in this paper. In addition, the approach is adapted to the prior work, the guard generation procedure, to extract compact propositional formulae from a symbolically represented supervisor. These propositional formulae, referred to as guards, are then attached to the original model, resulting in a modular and comprehensible representation of the supervisor.

Keywords: Supervisory control theory, State-space exploration, Binary decision diagrams, Partitioning techniques, Propositional formulae.

1 Introduction

The analysis of reactive systems has been paid much attention by researchers and scientists in the computer science community. One of the classic methods to analyze reactive systems is utilizing formal verification techniques, such as model checking, to verify whether the considered system fulfills specifications. Nevertheless, from the control engineering point of view, instead of verifying the correctness of the system, a controller which guarantees that the system always behaves according to specifications is preferred. Supervisory Control Theory (SCT) [1,2] provides such a control-theoretic framework to design a device, called the *supervisor*, for reactive systems, referred to as Discrete Event Systems (DESs). Given the model of a DES to be controlled, the *plant*, and the intended behavior, the *specification*, the supervisor can be automatically synthesized, guaranteeing that the closed-loop system fulfills given specifications. SCT has been applied for various applications in different areas such as automated manufacturing lines and embedded systems [3,4,5].

Generally, a supervisor is a function that, given a set of events, restricts the plant to execute desired events according to the specification. A typical issue is how to realize

J. Filipe and A. Fred (Eds.): ICAART 2011, CCIS 271, pp. 161–175, 2013.

such a control function efficiently and represent it appropriately. Since the synthesis task involves a series of reachability computations, as DESs becoming more complicated, the traditional explicit state-space traversal algorithm may be intractable due to *the state-space explosion problem.* By using binary decision diagrams (BDD) [6,7], the supervisor can be represented and computed symbolically such that the state-space explosion problem is alleviated to some extent. However, the symbolic computation is not a silver bullet. Transforming from the traditional explicit state-space traversal algorithm into a BDD-based computation scheme does not guarantee that the algorithm will become remarkably efficient. Thus numerous researches have been performed to improve the efficiency of symbolic computations. In this paper, we mainly focus on partitioning techniques, which decompose the state-space into a set of structural components and utilize these partitioned components to realize efficient reachability computations.

With BDD-based traversal algorithms, some larger DESs could be solved without causing the state-space explosion. Meanwhile, another problem is arising from the BDD representation of the resultant supervisor. Since the original models have been reformulated and encoded, it is cumbersome for the users to relate each state with the corresponding BDD variables. Therefore, it is more convenient and natural to represent the supervisor in a form similar to the models. In [8], a promising approach is presented, where a set of minimal and tractable logic expressions, referred to as guards, are extracted from the supervisor and attached to the original models of the closed-loop system. However, this approach computes the supervisor symbolically based on the conjunctive partitioning technique. This might lead to the state-space explosion, due to the large number of intermediate BDD nodes.

The main contribution of this paper is adapting a symbolic supervisory synthesis approach to the guard generation procedure, to make it applicable for industrially interesting applications. The approach automatically synthesizes a supervisor by taking the advantage of the disjunctive partitioning technique. The monolithic state-space is then split into a set of simpler components and the reachability search is performed structurally with a set of heuristic decisions. Moreover, the guard generation procedure is tailored to use the partitioned structure to extract the simplified guards and attach them to the original models. Finally, a comparison of algorithm efficiency between two partitioning techniques is made by applying them to a set of benchmark examples.

The paper is organized as follows: For the readers who might be unfamiliar with Supervisory Control Theory, Section 2 gives an informal and brief explanation. Section 3 provides some preliminaries that are used throughout the paper. The symbolic supervisory synthesis and the guard generation procedure will be discussed in detail in Section 4 and 5. In Section 6, we apply what we have discussed and implemented to several real case studies. Finally, we end up with some conclusions in Section 7.

2   Motivating Example

For readers who might be unfamiliar with SCT, the following simple example gives a brief overview and states what the exact problem this paper is about to solve.

Example 1. Consider a resource booking problem where two industrial robots need to book two resources in opposite order to carry out their tasks. To avoid collisions, a constraint requires that two robots are not allowed to occupy two zones simultaneously.

Figure 1 shows one way to model the system as the state machines, or *deterministic finite automata*. Figure 1a and 1b depict the robot (plant) models and Fig. 1c and 1d depict the resource (specification) models. The states having an incoming arrow from outside denote the beginning of the task, while the states having double circles, called *marked states*, denote the accomplishment of the task. The event $use^A_{R_1}$ means that Robot A uses Resource 1. The other events can be interpreted similarly. The goal of the SCT is to automatically synthesize a *minimally restrictive* supervisor from these modular models. Traditionally, to do this, the algorithm starts with the composition (formally described in Section 3.1) of all the automata as the initial candidate supervisor S_0 (Fig. 1e). Then the undesirable states will be removed iteratively. Generally, undesirable states can either be blocking or uncontrollable. A state is blocking when no marked state can be reached, while uncontrollable states are defined in Section 3.1. In Fig. 1e, we have one blocking state $\langle q^A_2, q^B_2, q^C_2, q^D_2 \rangle$, which depicts the situation where Robot A has booked Resource 1 and is trying to book Resource 2, while Robot B has booked Resource 2 and is trying to book Resource 1. In such case, none of the robots can do other movements, which is a blocking situation. After removing the blocking state together with the associated transitions, a non-blocking supervisor is produced.

It can be observed that for such a simple example, the composed automaton contains 9 states. With a DES getting more complicated, the composed automaton will become significantly larger. To alleviate this problem, a well known strategy is to represent the state space symbolically by using *Binary Decision Diagrams* (BDD). In [8], based on this principle, an alternative approach is presented, where guards are generated to prevent the controlled system to reach undesirable states. The advantage of this approach is that it never constructs the composed automaton, which means that an incomprehensible BDD representation of the supervisor is avoided. Instead, the approach characterizes a supervisor by a set of minimal guards that are attached to the original models to represent the supervisor behavior. Figure 2 shows the application of the guard generation to the example, where the variables v_A, v_B, v_C, v_D are introduced to hold the current states of the corresponding automata.

The intention of this paper is to improve the guard generation procedure by introducing an alternative symbolic approach. This approach, which is based on the disjunctive partitioning technique, partitions the transition function into a set of simple but structural components. These components, having the disjunctive connection relation between each other, therefore can be used to search the state-space without constructing a total transition function for the composed automaton. Besides, to keep the intermediate number of BDD nodes as small as possible, the approach includes a set of selection heuristics to search the state-space in a structural way.

3   Preliminaries

This section provides some preliminaries which are used throughout the rest of the paper.

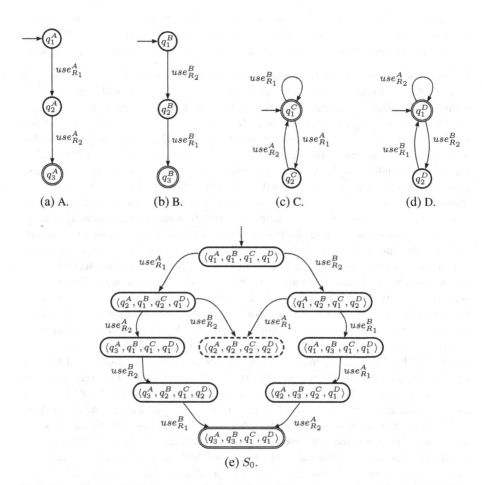

Fig. 1. Example 1. 1a-1b) Robot automata A and B, 1c-1d) resource automata C and D, and (1e) a supervisor candidate S_0.

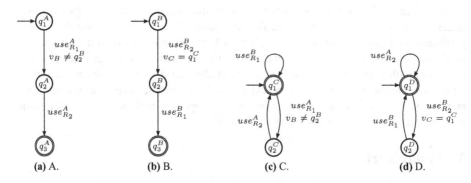

Fig. 2. Guards representing the behavior of the supervisor for Example 1

3.1 Supervisory Control Theory

Generally, a DES can either be described by textual expressions, such as regular expressions or graphically by for instance Petri nets or automata. In this paper, we focus on deterministic finite automata.

Definition 1. *A deterministic finite automaton (DFA), is a five-tuple:*

$$(Q, \Sigma, \delta, q_{init}, Q_m)$$

where:

- Q *is a finite set of states;*
- Σ *is a non-empty finite set of events;*
- $\delta: Q \times \Sigma \to Q$ *is a partial transition function which expresses the state transitions;*
- $q_{init} \in Q$ *is the initial state;*
- $Q_m \subseteq Q$ *is a set of marked or accepting states.*

The composition of two or more automata is realized by the *full synchronous composition* [9].

Definition 2. *Let $A^i = (Q^i, \Sigma^i, \delta^i, q^i_{init}, Q^i_m), i = 1, 2$ be two DFAs. The full synchronous composition of A^1 and A^2 is*

$$A^1 \parallel A^2 = (Q^{1\parallel2}, \Sigma^1 \cup \Sigma^2, \delta^{1\parallel2}, q^{1\parallel2}_{init}, Q^1_m \times Q^2_m)$$

where:

- $Q^{1\parallel2} \subseteq Q^1 \times Q^2$;
- $q^{1\parallel2}_{init} = \langle q^1_{init}, q^2_{init} \rangle$;
- $\delta^{1\parallel2}(\langle q^1, q^2 \rangle, \sigma) = \begin{cases} \delta^1(q^1, \sigma) \times \delta^2(q^2, \sigma) & \text{if } \sigma \in \Sigma^1 \cap \Sigma^2 \\ \delta^1(q^1, \sigma) \times \{q^2\} & \text{if } \sigma \in \Sigma^1 \backslash \Sigma^2 \\ \{q^1\} \times \delta^2(q^2, \sigma) & \text{if } \sigma \in \Sigma^2 \backslash \Sigma^1 \\ \text{undefined} & \text{otherwise.} \end{cases}$

As described in Section 1, the goal of SCT is to automatically synthesize a minimally restrictive supervisor S, which guarantees the behavior of the plant P always fulfills the given specification Sp. Here if the plant is given as a number of sub-plants $P_1, \ldots, P_n$, the plant can be obtained by performing the full synchronous composition operation on these sub-plants. Thus $P = P_1 \parallel \ldots \parallel P_n$. Similarly, $Sp = Sp_1 \parallel \ldots \parallel Sp_m$.

In SCT, events in the alphabet Σ can either be controllable or uncontrollable. Hence, Σ can be divided into two disjoint subsets, the controllable event set Σ_c and the uncontrollable event set Σ_u. The supervisor is only allowed to restrict controllable events from occurring in the plant.

Additionally, given a plant P and a specification Sp, two properties [1,2] that the supervisor ought to have are:

- *Controllability*: Let Σ_u be the set of uncontrollable event set. The supervisor S is never allowed to disable any uncontrollable event that might be generated by the plant P.
- *Non-blocking*: This is a progress property enforced by the supervisor S, which guarantees that at least one marked state is always reachable in the closed-loop system, $S \parallel P$.

3.2 Binary Decision Diagrams (BDD)

Binary decision diagrams (BDD), used for representing Boolean functions, can be extended to symbolically represent states, events and transitions of automata. In contrast to explicit representations, which might be computationally expensive in terms of time and memory, BDDs often generate compact and operation-efficient representations.

A binary decision diagram is a directed acyclic graph (DAG) consisting of two kinds of nodes: *decision nodes* and *terminal nodes*. Given a set of Boolean variables V, a BDD is a Boolean function $f: 2^V \rightarrow \{0, 1\}$ which can be recursively expressed using Shannon's decomposition [10]. Besides, a variable v_1 has a lower (higher) *order* than variable v_2 if v_1 is closer (further) to the root and is denoted by $v_1 \prec v_2$. The variable ordering will impact the number of BDD nodes. However, finding an optimal variable ordering of a BDD is a NP-complete problem [11]. In this paper, a simple but powerful heuristic based on Aloul's Force algorithm [12] is used to compute a suitable static variable ordering.

Symbolic Representation of Automata. The BDD data structure can be extended to also represent models such as automata. The key point is to make use of *characteristic functions*.

Given a finite state set U as universe, for every $S \subseteq U$, the characteristic function can be defined as follows:

$$\chi_S(\alpha) = \begin{cases} 1 & \alpha \in S \\ 0 & \alpha \notin S \end{cases} \tag{1}$$

Set operations can be equivalently carried on corresponding characteristic functions. For example, $S_1 \cup S_2$, $(S_1, S_2 \subseteq U)$ can be mapped equivalently to $\chi_{S_1} \vee \chi_{S_2}$, since $S_1 \cup S_2 = \{\alpha \in U \mid \alpha \in S_1 \vee \alpha \in S_2\}$.

The elements of a finite set can be expressed as a Boolean vector. So a set with n elements, requires a Boolean vector of length $\lceil \log_2 n \rceil$. Just like the case of coding the states in a set, binary encoding of the transition function δ follows the same rule but with the difference that the transition function distinguishes between source-states and target states. Hence, we need two Boolean vectors with different sets of Boolean variables to express the domain of source-states and target-states respectively.

4 BDD-Based Partitioning Computation

The *safe-state* algorithm, an efficient supervisor synthesis algorithm, formally defined in [13], is used in this paper. The algorithm creates the supervisor by first building the candidate $S_0 = P \parallel Sp$, then removing states from Q^{S_0} until the remaining safe states are both non-blocking and controllable.

As Algorithm 1 shows, given a set of forbidden states Q_x, the algorithm computes the set of safe states Q^S by iteratively removing the blocking states (RestrictedBackward in line 5) and the uncontrollable states (UncontrollableBackward in line 6). Note that after the termination of the algorithm, not all of the safe states are reachable from the initial state. Therefore, a forward reachability search is needed to exclude the safe states which are not reachable. The safe-state algorithm is discussed in more detail in [13].

Algorithm 1. The safe-state synthesis algorithm.

```
1: input : Q_x, Q^{S_0}
2: let X_0 := Q_x, k := 0;
3: repeat
4:    k := k + 1;
5:    Q' := RestrictedBackward(Q_m, X_{k-1});
6:    Q'' := UncontrollableBackward(Q^{S_0}\Q');
7:    X_k := X_{k-1} ∪ (Q'');
8: until X_k = X_{k-1}
9: return Q^{S_0}\X_k
```

4.1 Efficient State Space Search

Not surprisingly, the backward and forward reachability searches turn out to be the bottle-neck of the algorithm presented above. The problem with the intuitive reachability is that for a large and complicated modular DES, the BDD representation of the total transition function δ^{S_0} is often too large to be constructed. The natural way to tackle the complexity of the transfer function is to split it into a set of less complex partial functions with a connection between them. Such methods are based on conjunctive and disjunctive partitioning techniques.

Conjunctive Representation. Conjunctive partitioning, introduced in [14,15], is an approach to represent synchronous digital circuits where all transitions happen simultaneously. In the context of DES, the conjunctive partitioning of the full synchronous composition can be achieved by adding self-loops to the automata for events that are not included in their original alphabets. This leads to a situation where all automata have equal alphabet. Therefore, the conjunctive transition function $\hat{\delta}^i$ for the automaton A^i and the total transition function can be defined as follows:

$$\hat{\delta}^i(q^i, \sigma) = \begin{cases} \delta^i(q^i, \sigma) & \text{if } \delta^i(q^i, \sigma) \text{ is defined} \\ q^i & \text{if } \sigma \notin \Sigma^i \\ \text{undefined otherwise} . \end{cases} \quad (2)$$

$$\delta = \bigwedge_{1 \leq i \leq n} \hat{\delta}^i . \quad (3)$$

By making use of (2) and (3), we can search the state-space without constructing the total transition function. Algorithm 2 applies this technique for the forward reachability search. Assuming that the automaton set $A = \{A_1, \ldots, A_n\}$ and the state $q = \langle q^1, q^2, \ldots, q^n \rangle$, the algorithm explores the target state $\acute{q}$ by performing each conjunctive transition function $\hat{\delta}^i$ with arguments (the local state q^i and the event $\sigma \in \Sigma$) to get each local target state $\acute{q}^i$.

Disjunctive Representation. The conjunctive partitioning of the transition relation works well for formal verification of synchronous digital circuits. However, because of the asynchronous feature of the full synchronous composition, the intermediate states (Q_{k-1}) can still cause the explosion problem when performing the reachability search,

---

Algorithm 2. Conjunctive forward reachability algorithm.

1: **input** : $Q_{init}, \{\hat{\delta}^1, \ldots, \hat{\delta}^n\}, \Sigma$
2: **let** $Q_0 := Q_{init}, k := 0$;
3: **repeat**
4:     $k := k + 1$;
5:     $Q_k := Q_{k-1} \cup \{\acute{q} \mid \exists q \in Q_{k-1}, \exists \sigma \in \Sigma, \forall i \in \{1, \ldots, n\}$ such that $\hat{\delta}^i(q^i, \sigma) = \acute{q}^i\}$;
6: **until** $Q_k = Q_{k-1}$
7: **return** Q_k

---

which prevents the conjunctive partitioning technique from being applied to large systems. The disjunctive partitioning, explained subsequently, on the other hand, is then shown to be an appropriate partitioning technique for SCT.

Assuming $A = \{A_1, \ldots, A_n\}$ and $q = \langle q^1, \ldots, q^n \rangle$, the disjunctive transition function $\check{\delta}^i$ of A^i, is defined based on the event $\sigma \in \Sigma^i$ and the dependency set $D(A^i)$:

$$D(A^i) = \{A^j \in A \mid \exists A^i \in A \text{ where } \Sigma^i \cap \Sigma^j \neq \emptyset\} . \tag{4}$$

$$\check{\delta}^i(q, \sigma) = \left(\bigwedge_{A^j \in D(A^i)} \zeta^{i,j}(q^j, \sigma) \right) \wedge \left(\bigwedge_{A^k \notin D(A^i)} q^k \overset{\sigma}{\leftrightarrow} q^k \right) . \tag{5}$$

$$\zeta^{i,j}(q^j, \sigma) = \begin{cases} \delta^j(q^j, \sigma) & \text{if } \sigma \in \Sigma^i \cap \Sigma^j \\ q^j & \text{otherwise} . \end{cases} \tag{6}$$

Additionally, the total transition function is defined as:

$$\delta = \bigvee_{1 \leq i \leq n} \check{\delta}^i . \tag{7}$$

The construction of the dependency set for each automaton can be obtained through calculating which automaton shares any event with it. Taking Example 1 as an example, for the automaton A, since it shares the events $use_{R_1}^A$, $use_{R_2}^A$ with the automaton C and the event $use_{R_2}^A$ with the automaton D, $D(A)$ can be constructed as follows:

$$D(A) = \{A, C, D\}.$$

Besides, the total transition function defined for the state $\langle q_1^A, q_1^B, q_1^C, q_1^D \rangle$ and the event $use_{R_1}^A$ can be obtained by computing $\check{\delta}^A$ and $\check{\delta}^C$, since $use_{R_1}^A$ only belongs to Σ^A and Σ^C. By using (5) and (6), it can be inferred that

$$\delta(\langle q_1^A, q_1^B, q_1^C, q_1^D \rangle, use_{R_1}^A) = \check{\delta}^A(\langle q_1^A, q_1^B, q_1^C, q_1^D \rangle, use_{R_1}^A)$$

$$= \check{\delta}^C(\langle q_1^A, q_1^B, q_1^C, q_1^D \rangle, use_{R_1}^A) = \langle q_2^A, q_1^B, q_2^C, q_1^D \rangle.$$

Notice that the disjunctive transition function represented in BDDs, is shown explicitly here to easily understand.

4.2   Workset Based Strategies

In Section 4.1, we suggested the use of partitioning techniques to deal with the large number of intermediate BDD nodes. However, using partitioning techniques alone is not enough to yield efficient BDD-based reachability algorithms. In [16], it has been shown that random structural reachability search yields poor compression of intermediate BDD nodes. In order to improve these algorithms to substantially reduce the number of intermediate BDD nodes, it is vital to search the state space in a structural and efficient way. Here we introduce a simple algorithm, Algorithm 3, which is formally defined in [13]. The workset algorithm maintains a set of active disjunctive transition functions W_k. These active transition functions are selected one at a time for the local reachability search. If there is any new state found for the currently selected transition relation, then all of its *dependent transition functions* (8) will be added in W_k. Notice that in Algorithm 3, "$\cdot$" can be any event, since we don't care about the specific events as long as it is defined in $\breve{\delta}^i$.

$$E(\breve{\delta}^i) = \{\breve{\delta}^j \mid A^j \in D(A^i)\backslash\{A^i\}\} \ . \tag{8}$$

Algorithm 3. Workset forward reachbility algorithm.

1: **input** : $Q_{init}, \{\breve{\delta}^1, \ldots, \breve{\delta}^n\}$
2: **let** $W_0 := \{\breve{\delta}^1, \ldots, \breve{\delta}^n\}, Q_0 := Q_{init}, k := 0;$
3: **repeat**
4:    $\mathbb{H}$: Pick and remove a transition $\breve{\delta}^i \in W_k;$
5:    $k := k + 1;$
6:    $Q_k := Q_{k-1} \cup \{\acute{q} \mid \exists q \in Q_{k-1}, \breve{\delta}^i(q, \cdot) = \acute{q}\};$
7:    **if** $Q_k \neq Q_{k-1}$ **then**
8:       $W_k := W_{k-1} \cup E(\breve{\delta}^i);$
9:    **end if**
10: **until** $W = \emptyset$
11: **return** Q_k

Selection Heuristics. In Algorithm 3, $\mathbb{H}$ denotes the heuristics of selecting the next disjunctive transition function for the reachability search such that the number of intermediate BDD nodes is computed as small as possible. How a disjunctive transition function $\breve{\delta}^i$ is chosen among those in the working set has great influence on the performance of the algorithm. Here we suggest a series of simple heuristics that have been implemented and seem to work well for real-world problems. In Section 6, those heuristics will be applied to a benchmark example to compare how they influence the performance of the workset algorithm.

To find a good heuristic, a two-stage selection rule was implemented, as Fig. 3 shows. Using this method, a complex selection procedure can be described as a combination of two selection rules. In the current implementation, the first stage H_1 selects a subset $W' \subset W$ to be sent to H_2 using one of the following rules:

- MaxF: Choose the automata with the largest dependency set cardinality.
- MinF: The opposite of above.

In case W' is not a singleton, the second stage H_2 is used to choose a single disjunctive transition function $\check{\delta}^i$ among W'. In the experiment, the following shown heuristics can significantly reduce the number of intermediate BDD nodes for some relatively large problems.

- Reinforcement learning (R) [17]: Choose the best transition relation based on the previous activity record.
- Reinforcement learning + Tabu (RT) [18]: Same as the reinforcement learning with the difference that using tabu search for the selection policy.

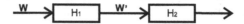

Fig. 3. The two stage selection heuristics for the workset algorithm

5   Supervisor as Guards

As mentioned in Section 1, given a supervisor represented as a BDD, it is cumbersome for the users to relate each state to the corresponding BDD variables. Therefore, it is more convenient and natural to represent the supervisor in a form similar to the original models. In this section, the guard generation procedure, originating from [8], is discussed and combined with the BDD-based disjunctive partitioning approach in Section 4.1.

The guard generation procedure, being dependent on three kinds of state sets, extracts a set of compact guards indicating under which conditions the event can be executed without violating the specifications. These guards are then attached to the original model to represent the supervisor.

5.1   Computation of the Basic State Sets

Concerning the states that are retained or removed after the synthesis process, the states that enable an arbitrary event σ can be divided into three *basic state sets*: forbidden state set, allowed state set and don't care state set.

The forbidden state set, denoted by Q_f^σ, is the set of states in the supervisor where the execution of σ is defined for S_0, but not for the supervisor. The allowed state set, denoted by Q_a^σ, is the set of states in the supervisor where the execution of σ is defined for the supervisor. In other word, for each event σ, Q_a^σ represents the set of states where event σ must be allowed to be executed in order to end up in states belonging to the supervisor.

In order to obtain compact and simplified guards, inspired from the Boolean minimization techniques, another set of states, denoted by Q_{dc}^σ, which describes a situation where executing σ will not impact the result of the synthesis, is utilized to minimize the guards.

Algorithms 4 and 5 presented below show how to compute the forbidden states Q_f^σ and the allowed states Q_a^σ by making use of the disjunctive transition functions. Note that Q^S and Q_x denote the resultant supervisor states and all the forbidden states yielded from Algorithm 1. The don't care state set, Q_{dc}^σ can be defined as the complement of the union of Q_a^σ and Q_f^σ. The proof can be found in [8].

5.2   Guard Generation

Based on the basic state sets, guards can be extracted. For every automaton in the DES, a new variable v is introduced to hold the current state of the automaton. For each event σ, the following propositional function, $G^\sigma : Q^{A^1} \times Q^{A^2} \times \ldots \times Q^{A^n} \to \mathbb{B}$ is defined as:

$$G^\sigma \langle v_{A^1}, v_{A^2}, \ldots v_{A^n} \rangle = \begin{cases} \text{true} & \langle v_{A^1}, v_{A^2}, \ldots v_{A^n} \rangle \in Q_a^\sigma \\ \text{false} & \langle v_{A^1}, v_{A^2}, \ldots v_{A^n} \rangle \in Q_f^\sigma \\ \text{don't care otherwise} . \end{cases} \tag{9}$$

where $\mathbb{B}$ is the set of Boolean values and v_{A^i} represents the current state of automaton A^i. In particular, σ is allowed to be executed from the state $\langle v_{A^1}, v_{A^2}, \ldots v_{A^n} \rangle$ if the guard is true.

By applying minimization methods of Boolean functions (utilizing the don't care state set) and certain heuristics, the generated guards can be simplified. The procedure is discussed in details in [8].

Algorithm 4. Computation of Q_f^σ.

1: **input** : $\sigma, Q_x, Q^S, \{\check{\delta}^1, \ldots, \check{\delta}^n\}$
2: **let** $Q_f^\sigma := \emptyset$;
3: **for all** A^i if $\sigma \in \Sigma^i$ **do**
4:     $Q_f^\sigma := Q_f^\sigma \cup \{q \mid \exists \acute{q} \in Q_x, \check{\delta}^i(q, \sigma) = \acute{q}\}$;
5: **end for**
6: **let** $Q_f^\sigma := Q_f^\sigma \cap Q^S$;
7: **return** Q_f^σ

Algorithm 5. Computation of Q_a^σ.

1: **input** : $\sigma, Q^S, \{\check{\delta}^1, \ldots, \check{\delta}^n\}$
2: **let** $Q_a^\sigma := \emptyset$;
3: **for all** A^i if $\sigma \in \Sigma^i$ **do**
4:     $Q_a^\sigma := Q_a^\sigma \cup \{q \mid \exists \acute{q} \in Q^S, \check{\delta}^i(q, \sigma) = \acute{q}\}$;
5: **end for**
6: **let** $Q_a^\sigma := Q_a^\sigma \cap Q^S$;
7: **return** Q_a^σ

---

6 Case Studies

What we have discussed in the previous sections has been implemented and integrated in the supervisory control tool Supremica [19] which uses JavaBDD [20] as BDD package. In this section, the implemented program will be applied to a set of relatively complicated examples[1].

6.1 Benchmark Examples

A set of benchmark examples is briefly described as follows.

Automated Guided Vehicles. An AGV system, described in [21], is a simple manufacturing system where five automated guided vehicles transport material between stations. As the routes of the vehicles cross each other, single-access zones are introduced to avoid collisions.

Parallel Manufacturing Example. The Parallel Manufacturing Example, introduced in [22], consists of three manufacturing units running in parallel. The system is modeled in three layers in a hierarchical interface-based manner.

The Transfer Line. The Transfer Line $TL(n, m)$, introduced as a tutorial example in [23], defines a very simple factory consisting of a series of identical cells. Each cell contains two machines and two buffers, one between the machines and one before a testing unit which decides whether the work piece should be sent back to the first machine for further processing, or if it should be passed to the next cell. The capacity of each buffer is m, which is usually chosen to be either 1 or 3.

The Extended Cat and Mouse. An extended cat and mouse problem [8], which is more complicated than the transfer line model, generalizes the classic one presented in [1]. The extended version makes it possible to generate problem instances of arbitrary size, where n and k denote the number of levels and cats respectively.

6.2 Approach Evaluation

In this section, we evaluate the approach from two aspects. First, a comparison between two partitioning techniques is made by analyzing the statistical data from Fig. 1. In addition, the extended cat and mouse example with multiple instances is utilized to investigate how the choice of heuristics in the workset algorithm influences the time efficiency.

Conjunctive vs. Disjunctive. Figure 1 shows the result of applying two partitioning techniques for the examples explained above. The supervisors synthesized for these examples are both non-blocking and controllable and the safe states are reachable. It is observed that both of the partitioning based algorithms can handle the AGV and the Parallel Manufacturing example, for which the number of reachable states is up to 10^7.

---

[1] The experiment was carried out on a standard Laptop (Core 2 Duo processor, 2.4 GHz, 2GB RAM) running Ubuntu 10.04.

Table 1. Non-blocking and controllability synthesis

Model	Reachable States	Supervisor states	Conjunctive Synthesis		Workset Algorithm	
			BDD Peak	Computation Time (s)	BDD Peak	Computation Time (s)
AGV	22929408	1148928	9890	6.50	2850	0.87
Parallel Man	5702550	5702550	12363	2.47	2334	1.57
Transfer line (1,3)	64	28	17	0.05	13	0.10
Transfer line (5,3)	1.07×10^9	8.49×10^4	2352	1.69	299	0.59
Transfer line (10,3)	1.15×10^{18}	6.13×10^{13}	31022	48.36	1257	3.89
Transfer line (15,3)	1.23×10^{27}	4.42×10^{20}	–	–	3032	12.80
Cat and mouse (1,1)	20	6	43	0.02	31	0.05
Cat and mouse (1,5)	605	579	2343	0.08	273	0.09
Cat and mouse (5,1)	1056	76	848	0.30	305	0.30
Cat and mouse (5,5)	6.91×10^9	3.15×10^9	–	–	15964	20.86

* - denotes memory out.

However, with DESs getting larger and more complicated, the conjunctive partitioning technique is not capable of synthesizing non-blocking and controllable supervisors any more. The disjunctive partitioning, on the other hand, could successfully explore the state space within acceptable time. In addition, the column "BDD Peak", the maximal number of BDD nodes during the reachability computation shows that the disjunctive partitioning together with heuristic decisions can effectively reduce the number of intermediate BDD nodes.

Heuristics. Table 2 shows the computation time for synthesizing non-blocking supervisors of the extended cat and mouse with different instances. Different combinations of heuristics, presented in Section 4.2, are chosen to test the performance of the workset algorithm. Empirically, for the models with relatively large dependency sets, the heuristic pair (MaxF,RT) seems to be a good choice, although it hasn't been formally proved. Observing the results from Table 2, the workset algorithm can handle problem instances with either a large number of levels n or cats k rather well. However, with both numbers increasing, the computation time increases rapidly no matter which heuristic pair is chosen.

Table 2. Computation time for non-blocking supervisors with different heuristics

Cat and mouse (n, k)	Computation Time (s)			
	Workset(MaxF,R)	Workset(MaxF,RT)	Workset(MinF,R)	Workset(MinF,RT)
$(1, 1)$	0.04	0.06	0.05	0.05
$(1, 5)$	0.30	0.27	0.33	0.36
$(5, 1)$	0.08	0.08	0.09	0.08
$(5, 5)$	3.15	2.90	3.85	3.42
$(1, 10)$	0.67	0.66	0.75	0.73
$(7, 7)$	21.4	17.6	25.5	22.9
$(10, 1)$	0.23	0.20	0.24	0.23
$(10, 7)$	100.3	88.5	136.4	138.0

7 Conclusions

In this paper, we improved and extended our previous work, the guard generation procedure to make it applicable for industrially interesting applications. More specifically, the content of the paper can be summarized as follows:

- Introduce the partitioning techniques to split the BDD representation of $\delta^{Sp\|P}$ into a set of smaller but structural components.
- To alleviate the problem that the intermediate number of BDD nodes might still be huge during the reachability exploration, we introduce the workset algorithm together with a set of simple heuristics to search the state-space in a structured and efficient way.
- The guard generation procedure is tailored to make use of the partitioned transition functions and the synthesized supervisor to compute the basic state sets for an event.
- The presented approach is applied to a set of benchmark examples to be evaluated.

It is concluded that the disjunctive partitioning, with appropriate heuristics, is suitable for solving large modular supervisory control problems. There are several directions towards which we could extend our approach. For instance, additional heuristics could be applied to the workset algorithm, to further decrease the number of intermediate BDD nodes. Moreover, it is possible to combine with more sophisticated synthesis techniques, such as compositional techniques, to substantially improve the algorithm efficiency.

References

1. Ramadge, P.J.G., Wonham, W.M.: The Control of Discrete Event Systems. Proceedings of the IEEE 77, 81–98 (1989)
2. Cassandras, C.G., Lafortune, S.: Introduction to Discrete Event Systems, 2nd edn. Springer, Heidelberg (2008)
3. Balemi, S., Hoffmann, G.J., Gyugyi, P., Wong-Toi, H., Franklin, G.F.: Supervisory Control of a Rapid Thermal Multiprocessor. IEEE Transactions on Automatic Control 38, 1040–1059 (1993)
4. Feng, L., Wonham, W.M., Thiagarajan, P.S.: Designing Communicating Transaction Processes by Supervisory Control Theory. Formal Methods in System Design 30, 117–141 (2007)
5. Shoaei, M.R., Lennartson, B., Miremadi, S.: Automatic Generation of Controllers for Collision-Free Flexible Manufacturing Systems. In: 6th IEEE Conference on Automation Science and Engineering, pp. 368–373 (2010)
6. Akers, S.B.: Binary Decision Diagrams. IEEE Transactions on Computers 27, 509–516 (1978)
7. Bryant, R.E.: Symbolic Manipulation with Ordered Binary Decision Diagrams. ACM Computing Surveys 24, 293–318 (1992)
8. Miremadi, S., Akesson, K., Lennartson, B.: Extraction and Representation of a Supervisor Using Guards in Extended Finite Automata. In: 9th International Workshop on Discrete Event Systems, pp. 193–199 (2008)
9. Hoare, C.A.R.: Communicating Sequential Processes. Communications of the ACM 21, 666–677 (1985)

10. Shannon, C.E., Weaver, W.: The Mathematical Theory of Communication. University of Illinois Press (1949)
11. Bollig, B., Wegener, I.: Improving the Variable Ordering of OBDDs Is NP-Complete. IEEE Transactions on Computers 45, 993–1002 (1996)
12. Aloul, F.A., Markov, I.L., Sakallah, K.A.: Force: A Fast and Easy-To-Implement Variable-Ordering Heuristic. In: 13th ACM Great Lakes symposium on VLSI, pp. 116–119 (2003)
13. Vahidi, A., Fabian, M., Lennartson, B.: Efficient Supervisory Synthesis of Large Systems. Control Engineering Practice 14, 1157–1167 (2006)
14. Burch, J.R., Clarke, E.M., Long, D.E.: Symbolic Model Checking with Partitioned Transition Relations. In: International Conference on Very Large Scale Integration, vol. A-1, pp. 49–58 (1991)
15. Burch, J.R., Clarke, E.M., Long, D.E., Mcmillan, K.L., Dill, D.L.: Symbolic Model Checking for Sequential Circuit Verification. IEEE Transactions on Computer-Aided Design of Integrated Circuits and Systems 13, 401–424 (1994)
16. Byröd, M., Lennartson, B., Vahidi, A., Åkesson, K.: Efficient Reachability analysis on Modular Discrete-Event Systems using Binary Decision Diagrams. In: 8th International Workshop on Discrete Event Systems, pp. 288–293 (2006)
17. Kaelbling, L.P., Littman, M.L., Moore, A.W.: Reinforcement Learning: A Survey. Journal of Artificial Intelligence Research 4, 237–285 (1996)
18. Glover, F., Laguna, M.: Tabu Search. Journal of the Operational Research Society 5 (1997)
19. Åkesson, K., Fabian, M., Flordal, H., Malik, R.: Supremica – An Integrated Environment for Verification, Synthesis and Simulation of Discrete Event Systems. In: 8th International Workshop on Discrete Event Systems, pp. 384–385 (2006)
20. JavaBDD, http://javabdd.sourceforge.net
21. Holloway, L.E., Krogh, B.H.: Synthesis of Feedback Control Logic for a Class of Controlled Petri Nets. IEEE Transactions on Automatic Control 35, 514–523 (1990)
22. Leduc, R.J.: Hierarchical Interface-Based Supervisory Control. In: 40th IEEE Conference on Decision and Control, pp. 4116–4121 (2002)
23. Murray Wonham, W.: Notes on Control of Discrete Event Systems. University of Toronto (1999)

Developing Goal-Oriented Normative Agents: The NBDI Architecture

Baldoino F. dos Santos Neto[1], Viviane Torres da Silva[2], and Carlos J.P. de Lucena[1,*]

[1] Computer Science Department, PUC-Rio, Rio de Janeiro, Brazil
[2] Computer Science Department, Fluminense Federal University (UFF)
Rio de Janeiro, Brazil
{bneto,lucena}@inf.puc-rio.br,
viviane.silva@ic.uff.br

Abstract. In open multi-agent systems norms are mechanisms used to restrict the behaviour of agents by defining what they are obligated, permitted or prohibited to do and by stating stimulus to their fulfillment such as rewards and discouraging their violation by pointing out punishments. In this paper we propose the NBDI architecture to develop goal-oriented normative agents whose priority is the accomplishment of their own desires while evaluate the pros and cons associated with the fulfillment or violation of the norms. The BDI architecture is extended by including norms related functions to check the incoming perceptions (including norms), select the norms they intend to fulfill based on the benefits they provide to the achievement of the agent's desires and intentions, and decide to cope or not with the norms while dropping, retaining or adopting new intentions. The applicability of our approach is demonstrated through an non-combatant evacuation scenario implemented by using the Normative Jason platform.

Keywords: Norms and BDI agents.

1 Introduction

Normative regulation is a mechanism that aims to cope with the heterogeneity, autonomy and diversity of interests among the different members of an open multi-agent system establishing a set of norms that ensures a desirable social order [5].

Such norms regulate the behaviour of the agents by indicating that they are obligated to accomplish something in the world (obligations) [6], permitted to act in a particular way (permissions) and prohibited from acting in a particular way (prohibitions) [6]. Moreover, norms may define rewards to their fulfillment and may state punishments in order to discourage their violation[6].

In this paper we consider that agents are goal-oriented entities that have the main purpose of achieving their desires while trying to fulfill the system norms. In this context, the paper presents an abstract architecture to build agents able to deal with the norms

___
* The present work has been partially funded by the Spanish project "Agreement Technologies" (CONSOLIDER CSD2007-0022,INGENIO 2010) and by the Brazilian research councils CNPq under grant 303531/2009-6 and FAPERJ under grant E-26/110.959/2009.

J. Filipe and A. Fred (Eds.): ICAART 2011, CCIS 271, pp. 176–191, 2013.
© Springer-Verlag Berlin Heidelberg 2013

of a society in an autonomous way. The NBDI (Norm-Belief-Desire-Intention) architecture extends the BDI (Belief-Desire-Intention) architecture [8] by including norms related functions to support normative reasoning. The agents built according to the proposed architecture have: *(i)* a review function of norms and beliefs used to check the incoming perceptions (including norms), *(ii)* a norm selection function to select the norms they intend to fulfill based on the benefits they provide to the achievement of the agent's desires and intentions, and to identify and solve conflicts among the selected norms, and *(iii)* a norm filter where the agents decide to cope or not with the norms while dropping, retaining or adopting new intentions.

We demonstrate the applicability of the NBDI architecture through a non-combatant evacuation scenario where the tasks related to review, select and filter norms are implemented by using the Normative Jason platform [7] that already provides support to the implementation of BDI agents and a set of normative functions able to check if the agent should adopt or not a norm, evaluate the pros and cons associated with the fulfillment or violation of the norm, check and solve conflicts among norms, and choose desires and plans according to their decisions of fulfilling or not a norm.

The paper is structured as follows. In Section 2 we outline the background about norms that is necessary to follow the paper. In Section 3 we present the non-combatant evacuation scenario where norms are defined to regulate the behaviour of rescue agents. In 4 the proposed NBDI normative agent architecture is explained and exemplified by using the proposed scenario. Section 5 demonstrates the applicability of the NBDI architecture. Section 6 summarizes relevant related work and, finally, Section 7 concludes and presents some future work.

2 Norms

In this work, we adopt the representation for norms described in [7], as shown below:

norm (Addressee, Activation, Expiration, Rewards,Punishments, DeonticConcept, State)

where *Addressee* is the agent or role responsible for fulfilling the norm, *Activation* is the activation condition for the norm to become active, *Expiration* is the expiration condition for the norm to become inactive, *Rewards* are the rewards to be given to the agent for fulfilling a norm, *Punishments* are the punishments to be given to the agent for violating a norm, *DeonticConcept* indicates if the norm states an obligation or a prohibition, and *State* describes a set of states being regulated.

3 Scenario: Rescue Operation

The applicability of the architecture proposed in this paper is demonstrated by using the simplified non-combatant evacuation scenario. In such scenario agents have the goals to plan the evacuation of members of a Non-Governmental Organisation (NGO) that are in hazardous location and, to do so, they can use different resources that help to evacuate the members, such as: *(i)* helicopters, *(ii)* troops and *(iii)* land-based helicopters. Considering that such resources are limited, we have a *Commander Agent* that is responsible

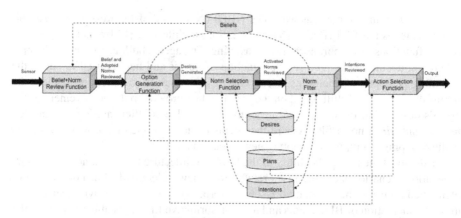

Fig. 1. NBDI Architecture

to control the use of the resources regulating the behaviour of the agents according to the norms 1, 2 and 3:

Norm 1

Addressee. *Rescue Entity.*

Activation. NGO workers are stranded in a hazardous location.

Expiration. NGO workers are stranded in a safe location.

DeonticConcept. Obligation.

State. To evacuate NGO workers.

Rewards. The *Commander Agent* gives more troops to *Rescue Entity.*

Rewards. The *Commander Agent* gives land-based helicopters to *Rescue Entity.*

Punishments. (obligation) *Rescue Entity* is obligated to return to the *Commander Agent* part of their troops.

Norm 2

Addressee. *Rescue Entity.*

Activation. The weather is bad.

Expiration. The weather is good.

DeonticConcept. Prohibition.

State. To evacuate NGO workers.

Punishments. (obligation) *Rescue Entity* is obligated to return to the *Commander Agent* part of their helicopters or land-based helicopters.

Norm 3

Addressee. *Rescue Entity.*

Activation. The weather is bad.

Expiration. The weather is good.

DeonticConcept. Prohibition.

State. To use helicopters.

Rewards. The *Commander Agent* gives more troops or land-based helicopters to *Rescue Entity.*

Punishments. (obligation) *Rescue Entity* is obligated to return to the *Commander Agent* part of their troops.

4   NBDI Architecture

The NBDI (Norm-Belief-Desire-Intention) architecture extends the BDI architecture to help agents on reasoning about the system norms. Norm is considered as a primary concept that influences the agent while reasoning about its beliefs, desires and intentions. The extensions we have made are represented in the NBDI architecture by the following components, as illustrated in Figure 4: *Belief+Norm Review Function*, *Norm Selection Function*, *Norm Filter* and *Plans* base[1] (that stores the plans of the agent).

In a nutshell, the NBDI architecture (Figure 4) works as follows. The agent perceives information about the world by using its sensors. The sensed information is the input of the *Belief+Norm Review Function*, an extension of *Belief Review Function* [8] defined in the BDI architecture that is responsible for reviewing the *Beliefs* base taking into account the current perception and ones already stored in the base.

In this work, we consider that norms are also stored in the *Beliefs* base, so, besides performing the original functionality of the *Belief Review Function*, the *Belief+Norm Review Function* is also responsible for: *(i)* in case the current perception is a norm, reviewing the sets of adopted norms by comparing the information loaded in the new norm with the norms and beliefs already stored in the base; and *(ii)* updating the sets of adopted and activated norms, considering that some may become active and others inactive due to the incoming perceptions.

Next, the *Option Generation Function* updates the agent's desires, and also their priorities. Such adaptation must consider both agent's current beliefs and intentions, and must be opportunistic, i.e., it should recognize when environmental circumstances change advantageously to offer the agent new ways of achieving intentions, or the possibility of achieving intentions that were otherwise unachievable [8]. Note that this function works exactly as the original function described in the BDI architecture. This function does not consider the norms stored in the *Beliefs* base while updating the agent desires because the agent must be able to generate new desires or adapt the existing ones without the influence of the norms. Our architecture considers that the agent is an *autonomous goal-oriented entity that fulfils the system norms if it decides to do so*.

After reviewing the beliefs, desires, activated and adopted norms, the *Norm Selection Function* is executed in order to *(i)* evaluate the activated norms in order to select the ones that the agent has the intention to fulfil; and *(ii)* identify and solve the conflicts among these norms.

Next, the *Norm Filter*, an extension of *Filter* [8] defined in the BDI architecture, selects the desires that will become intentions taking into account the norms the agent wants to fulfil. The plans that will achieve the intentions are also selected by following the norms the agent wants to fulfil.

Finally, the *Action Selection Function* is responsible for performing the actions specified by the intention. The next subsections detail the components added to the original BDI architecture, the one that was extended and a set of algorithms that demonstrate how such components can be implemented.

---

[1] Plans are composed by actions and states that the agent has the desire to achieve.

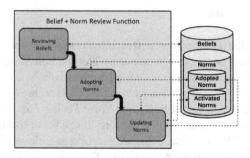

Fig. 2. Belief+Norm Review Function

Algorithm 1. Adopting Norms

Require: *Beliefs* base N: norms stored in the beliefs base
Require: agent: informations about the agent, such as: name and role
Require: *Beliefs* base NA: adopted norms stored in the beliefs base
Require: NN: new norms
 1: **for all** newNorm in NN **do**
 2:     x = true
 3:     **for all** n in N **do**
 4:         **if** (n == newNorm) **then**
 5:             x = false
 6:         **end if**
 7:     **end for**
 8:     **if** x ∩ $((agent.Name$ == $newNorm.Addresse)$ ∪ $((agent.Role$ == $newNorm.Addresse))$ **then**
 9:         NA.**add(newNorm)**
10:     **end if**
11: **end for**

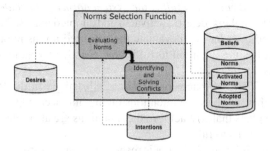

Fig. 3. Norm Selection Function

4.1 Belief+Norm Review Function

Besides performing the original functionality of the *Belief Review Function*, which is the revision of the beliefs (represented by *Reviewing Beliefs* task), the *Belief+Norm Review Function* (Figure 2), helps the agent on recognizing its responsibilities towards

Algorithm 2. Updating Norms

Require: *Beliefs* base NAD: adopted norms stored in the beliefs base
Require: *Beliefs* base NAC: activated norms stored in the beliefs base
Require: P: new perceptions
```
 1: for all n in (NAD ∪ NAC) do
 2:    for all p in P do
 3:       if (n.Activation.unify(p)) then
 4:          NAD.remove(n)
 5:          NAC.add(n)
 6:       else
 7:          if (n.Expiration.unify(p)) then
 8:             NAC.remove(n)
 9:             NAD.add(n)
10:          end if
11:       end if
12:    end for
13: end for
```

---

other agents by adopting new norms that specify such responsibilities (represented by *Adopting Norms* task) and updates the sets of activated and adopted norms (represented by *Updating Norms* task).

Adopting Norms (AN). This task recognizes from the set of receiving perceptions the ones that describe norms. After recognizing the norms, such function reviews the set of adopted norms applying the following verifications: (i) it checks if the new norm unifies with one of the norms already adopted, i.e., if the incoming norm already exists in the agent *Belief Base* (*Algorithm 1* from line 2 to 7), and (ii) it verifies if the agent is the addressee of the norm, i.e., if the field *Addressee* of the new norm unifies with the agent role or agent name, also stored as a belief in the *Belief Base* (*Algorithm 1* line 8). Finally, such function updates the set of adopted norms in the *Belief Base* if the new norm does not already exist and the agent is the addressee of the norm (*Algorithm 1* line 9).

With the aim to exemplify the use of this task, let's consider the scenario presented in Section 3 where two groups of agents are leaded by *Agent A* and *Agent B* playing the role *Rescue Entity*. When these entities receive information about the three system norms, the **AN** task is executed checking if the norms are not stored yet in the agent's belief base and comparing the addressee information with the role being played by the agents.

Updating Norms (UN). UN task updates the set of activated and adopted norms checking if the fields *Activation* and *Expiration* of the norm unifies with the beliefs of the agent. If the activation conditions unify with the beliefs, the adopted norm is activated (*Algorithm 2* from line 3 to 6). If the expiration conditions unify with the beliefs, the norm is deactivated and stored as an adopted norm (*Algorithm 2* from line 7 to 10).

Following the example above, if the weather of the area operated by one of the two rescue entities is bad, both norms 2 and 3 are activated, since the activation condition of both norms is "The weather is bad". If the norms are activated, the rescue entity must not rescue NGO members and must not use helicopters. Both norms are deactivated when the expiration condition unifies with the information about a good weather stored in the agent's belief base.

4.2   Norm Selection Function

The main goal of the *Norm Selection Function* (Figure 3) is to select the norms that the agent has the intention to fulfil. In order to do this, such function performs two tasks: 1) *Evaluating Norms* and 2) *Identifying and Solving Conflicts*. The first task helps the agent on selecting, from the set of activated norms, the norms that it has the intention to fulfil and the ones it has the intention to violate. The function evaluates the benefits of fulfilling or violating the norms, i.e., it checks how close the agent gets of achieving its desires if it decides to fulfil or if it decides to violate the norms. The function groups the activated norms in two sub-sets: norms to be fulfilled and norms to be violated. Finally, the second task of this function identifies and solves the conflicts among the norms that the agent has the intention to fulfil and among the ones that the agent has the intention to violate.

Evaluating the Norms (EN). In order to evaluate the benefits of the fulfilment or violation of a norm according to the agent's desires and intentions, the steps below should be followed: (**Step 1**) In case of obligations, it checks if the state described in the norm is equal to one of the states that the agent has desire (or intention) to achieve. In affirmative cases, the contribution is positive and the function $g(n.DeonticConcept, n.State)$ returns a value indicating the level of norm's contribution that is calculated according to the priority of the desire that is similar to the state described by norm. The function receives as parameters $n.DeonticConcept$ representing the deontic concept type, i. e., obligation or prohibition, and $n.State$ representing the state that is been regulated. In any other case, the contribution is zero since it does not disturb the achievement of the agent's desires or intentions. Such step is represented in *Algorithm* 3 from line 2 to 8. (**Step 2**) In case of prohibitions, it checks if the state described in the norm is equal to one of the states that the agent has desire (or intention) to achieve. In affirmative cases, the contribution is negative since it disturbs the achievement of the agent's desires or intentions and the function $g(n.DeonticConcept, n.State)$ calculates the absolute value of the contribution. In any other case, the prohibition will contribute neutrally. Such step is represented in *Algorithm* 3 between lines 9 and 15. (**Step 3**) After analyzing the state being regulated, this step considers the influence that the rewards have to the achievement of the agent's desires. We consider that rewards can never influence the agent negatively but always positively or neutrally since they give permissions to achieve a set of states. Such step is represented in *Algorithm* 3 by line 16. Function $r(n.Rewards)$ verifies the desires (or intentions) that are equal to the rewards and returns a value indicating the contribution that is the sum of the priorities of the agent's desires benefited by the rewards. (**Step 4**) Finally, the punishments are evaluated in order to check if they will influence the achievement of the agent's desires and intentions negatively or positively. **(Step 4.1)**In case the punishment states a prohibition and the state being prohibited is one of the agent desires or intentions, the punishment will influence negatively since it will disturb the agent of achieving one of its desires. If it is not the case, the punishment will not influence. Such step is represented in *Algorithm* 4 from line 2 to 8. Function *g(n.punishments.DeonticConcept)* returns a absolute value indicating the contribution of the fulfilment of the prohibition to the achievement of the agent's desires and intentions. **(Step 4.2)** In case the punishment states an obligation and the state being obliged is one of the agent desires or intentions, the punishment will influence positively (or neutrally) since such state will already be achieved by the agent. If it is not the case, the punishment will not influence. Such step is represented in *Algorithm* 4 from line 9 to 15. Function *g(n.punishments.DeonticConcept)* returns a value indicating the contribution of the fulfilment of the obligation to the achievement of the agent's desires and intentions.

Algorithm 3. Evaluating the fulfilment

Require: *Desires* base D
Require: *Intentions* base I
Require: *Beliefs* base N: norms stored in the beliefs base
 1: $x = 0$
 2: **if** n.DeonticConcept == Obligation **then**
 3:     **for all** d in $(D \cup I)$ **do**
 4:         **if** n.State == d **then**
 5:             $x = x + g$ (n.DeonticConcept, n.State)
 6:         **end if**
 7:     **end for**
 8: **end if**
 9: **if** n.DeonticConcept == Prohibition **then**
10:     **for all** d in $(D \cup I)$ **do**
11:         **if** n.State == d **then**
12:             $x = x - g$ (n.DeonticConcept, n.State)
13:         **end if**
14:     **end for**
15: **end if**
16: $x = x + r$ (n.Rewards)
17: **return** x

Note that it is necessary to individually check the contribution of the fulfilment and violation of each norm to the achievement of the agent desires (or intentions). Such step is represented in *Algorithm* 5 in the lines 2 and 3.

After checking how each norm can contribute to the achievement of the agent's desires and intentions, the function helps the agent on deciding which are the norms that it should fulfil, i.e., the norms whose contribution coming from its fulfilment is greater than the contribution coming from its violation. Steps 1, 2 and 3 return the contribution of the norm if the agent chooses to fulfil it (*Algorithm* 3 line 17) and 4.1 and 4.2 return

---

Algorithm 4. Evaluating the violation

Require: Desires base D
Require: Intentions base I
Require: *Beliefs* base N: norms stored in the beliefs base
1: x = 0
2: **if** n.punishment == Prohibition **then**
3:      **for all** d in $(D \cup I)$ **do**
4:          **if** n.punishment.state == d **then**
5:              x = x - g (n.punishments.DeonticConcept, n.punishments.State)
6:          **end if**
7:      **end for**
8: **end if**
9: **if** n.punishment == Obligation **then**
10:      **for all** d in $(D \cup I)$ **do**
11:          **if** n.punishment.state == d **then**
12:              x = x + g (n.punishments.DeonticConcept, n.punishments.State)
13:          **end if**
14:      **end for**
15: **end if**
16: **return** x

Algorithm 5. Reasoning about norms (Main)

Require: *fulfilSet* NF: norms stored in the fulfil set
Require: *vilateSet* NV: norms stored in the violate set
Require: Norms base N
1: **for all** Norm n in N **do**
2:      fulfil = Execute Algorithm 3 using n
3:      violate = Execute Algorithm 4 using n
4:      **if** fulfil >= violate **then**
5:          NF.**add(n)**
6:      **else**
7:          NV.**add(n)**
8:      **end if**
9: **end for**

---

its contribution if the agent chooses to violate the norm (*Algorithm* 4 line 16). Therefore, *Algorithm* 3 should be used to evaluate the contribution of the fulfilment of the norm to the agents' desires/intentions and *Algorithm* 4 should be used to calculate the contribution of the violation of the norm to the agents' desires/intentions.

If the contribution for fulfilling the norm is greater than or equal to the contribution for violating the norm, the norm is selected to be fulfilled and added to the sub-set *Fulfill* of the activated norms. Such step is represented in *Algorithm* 5 from line 4 to 6. Otherwise, it is selected to be violated and added to the sub-set *Violate* of the activated norms. Such step is represented in *Algorithm* 5 from line 6 to 8.

In order to exemplify the applicability of the EN task, let's consider the rescue operation scenario. The evaluation of the benefits of fulfilling and violating the three norms

Algorithm 6. Detecting Conflicts

Require: *fulfilSet* NF: norms stored in the fulfil set
Require: *vilateSet* NV: norms stored in the violate set
 1: **for all** Norm n1 in NF **do**
 2:     **for all** Norm n2 in NF **do**
 3:         **if** n1.State == n2.State and (n1.DeonticConcept == Obligation and n2.DeonticConcept == Prohibition) or (n2.DeonticConcept == Obligation and n1.DeonticConcept == Prohibition) **then**
 4:             Execute Algorithm 7 using n1 and n2
 5:         **end if**
 6:     **end for**
 7: **end for**
 8: **for all** Norm n1 in NV **do**
 9:     **for all** Norm n2 in NV **do**
10:         **if** n1.State == n2.State and (n1.DeonticConcept == Obligation and n2.DeonticConcept == Prohibition) or (n2.DeonticConcept == Obligation and n1.DeonticConcept == Prohibition) **then**
11:             Execute Algorithm 7 using n1 and n2
12:         **end if**
13:     **end for**
14: **end for**

Algorithm 7. Solving Conflicts

Require: *fulfilSet* NF: norms stored in the fulfil set
Require: *vilateSet* NV: norms stored in the violate set
 1: fulfiln1 = Execute Algorithm 3 using n1
 2: violaten2 = Execute Algorithm 4 using n2
 3: fulfiln2 = Execute Algorithm 3 using n2
 4: violaten1 = Execute Algorithm 4 using n1
 5: **if** fulfiln1 + violaten2 >= fulfiln2 + violaten1 **then**
 6:     NF.**remove(n2)**
 7:     NV.**remove(n1)**
 8: **else**
 9:     NF.**remove(n1)**
10:     NV.**remove(n2)**
11: **end if**

are shown in Tables 1, 3 and 2 that indicates the contribution of each norm element to the achievement of the agent goals. We consider that any norm element generates the same contribution that is 1.

After analysing the contribution of the norms shown in Tables 1, 3 and 2, Norm 1 is included in the set of norms to be fulfilled since the contribution for fulfilling it is equal to "+3" and greater than the contribution for violating it that is equal to "-1". Norm 2 is also included in the fulfil set since the contribution for fulfilling it is equal to "-1" and greater than the contribution for violating it that is equal to "-2". And, finally, Norm 3 is included in the fulfil set since the contribution for fulfilling it is equal to "+1" and

Table 1. Evaluating norm 1

Norm	Contribution		
	positive	neutral	negative
Obligation	1	0	0
Reward	1	0	0
Reward	1	0	0
Punishments			
Obligation	0	0	1

Table 2. Evaluating norm 3

Norm	Contribution		
	positive	neutral	negative
Prohibition	0	0	1
Reward	1	0	0
Reward	1	0	0
Punishments			
Obligation	0	0	1

Table 3. Evaluating norm 2

Norm	Contribution		
	positive	neutral	negative
Prohibition	0	0	1
Punishments			
Obligation	0	0	1
Obligation	0	0	1

greater than the contribution for violating it that is equal to "-1". It indicates that the agent has the intention to fulfil the three norms.

Detecting and Solving Conflicts (DSC). If two different norms (one being an obligation and the other one a prohibition) specify the same state, it is important to check their status, i.e., to check if they are in the set of norms that will be violated or fulfilled since they may be in conflict. If the agent intends to fulfil the obligation but does not intend to fulfil the prohibition, these norms are not in conflict. The same can be said if the agent intends to fulfil the prohibition and to violate the obligation. On the other hand, if the agent intends to fulfil both norms or to violate both norms, they are in conflict and it must be solved. Such step is represented in *Algorithm* 6.

For instance, in case of conflicts between two norms that the agent intends to fulfil or violate, the one with highest contribution to the achievement of the agent's desires (and intentions) can be selected. If the contributions have equal values we can choose anyone. That is, if the contribution coming from the fulfilment of the first norm (Steps 1, 2 and 3) plus the contribution coming from the violation of the second norm (Steps 4.1 and 4.2) is greater to or equal than the contribution coming from the fulfilment of the second norm plus the contribution coming from the violation of the first norm, the first norm is selected to be fulfilled and the second one to be violated. It is represented in *Algorithm* 7 from line 5 to 8. In case the opposite happens, the second norm is selected to be fulfilled and the first to be violated, as described in *Algorithm* 7 from line 8 to 11.

Considering the norms evaluated in the *EN* function, a conflict between Norm 1 and 2 is detected and should be solved. The conflict is solved by selecting Norm 1 to be fulfilled and Norm 2 to be violated since the contribution coming from the fulfilment of the first norm (+3) plus the contribution coming from the violation of the second norm

(-2) is greater than the contribution coming from the fulfilment of the second norm (-1) plus the contribution coming from the violation of the first norm (-1).

4.3   Norm Filter

The *Norm Filter* (Figure 4) is executed in order to drop any intention that does not bring benefits to the agent, retains intentions that are still expected to have a positive overall benefit and adopt new intentions, either to achieve existing intentions, or exploit new opportunities. In order to accomplish theses tasks and besides performing the original functionality of the *Filter function*, such function performs two additional steps.

Selecting Desires (SD). In this step the filter selects the desires that will become intentions taking into account the norms the agent wants to fulfil. The desires are selected according to their priorities and the norms may increase or decrease such priorities. If the agent has a desire to achieve a state and there is a norm that obliges the agent of achieving such state, the desire priority is increased according to the importance of the norm (represented in *Algorithm* 8 from line 3 to 5). If the agent has a desire to achieve a state and there is a norm that prohibits the agent of achieving such state, the desire priority is decreased according to the importance of the norm (represented in *Algorithm* 8 from line 5 to 9). If there is not any norm related to the desires, its priority is not modified. Finally, the function *getDesireHighestPriority()* (*Algorithm* 8 line 12) returns the desire with highest priority.

By applying this function to our example, the goal "Evacuating the members of a NGO to a safe location" is selected because such goal has highest priority since it receives a positive influence of Norm 1.

Selecting Plans (SP). After selecting the desires with highest priorities, i.e., after generating the agent intentions, the agent needs to select the plans that will achieve such intentions. Like in the selection of desires, the selection of plans will also be influenced by the norms.

While selecting a plan it is important to make sure that such plan will achieve the state described in the obligation norm and that will not achieve an state being prohibited. Therefore, the plans that achieve a given intention are ordered according to their priorities. If the state described by an obligation norm is equal to one of the states included in the plan, the norm increases the priority of such plan (represented in *Algorithm* 9 from line 4 to 6). Otherwise, if the state described by an prohibition norm is equal to one of the states included in the plan, the norm decreases the priority of such plan (represented in *Algorithm* 9 from line 6 to 10).

Let's consider that the desires of the agents in our example with highest priority is "Evacuating the members of a NGO to a safe location", that the agent has the intention to fulfil Norm 3, and that the priority of plans that uses helicopters has decrease. When SP step is executed it selects the plan with highest priority that tries to rescue the NGO workers and that will not use helicopters to do so.

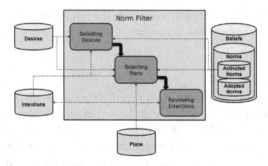

Fig. 4. Norm Filter

Algorithm 8. Selecting Desires

Require: *fulfilSet* NF: norms stored in the fulfil set
Require: Desires base D
Require: Intentions base I
 1: **for all** Norm n in *NF* **do**
 2:   **for all** d in $(D \cup I)$ **do**
 3:     **if** $(n.State == d) \cap (n.DeonticConcept == Obligation)$ **then**
 4:       d.annotatePriority(+1)
 5:     **else**
 6:       **if** $(n.State == d) \cap (n.DeonticConcept == Prohibition)$ **then**
 7:         d.annotatePriority(-1)
 8:       **end if**
 9:     **end if**
10:   **end for**
11: **end for**
12: **return** getDesireHighestPriority()

---

5 Implementing the NBDI Architecture

The NBDI architecture proposed in section 4 was implemented by translating the functions related to review, select and filter norms proposed in such architecture to the Normative Jason platform, described in [7]. The platform provides a set of normative functions, as follows: (i) the Norm Review function helps the agent on recognizing its responsibilities towards other agents by incorporating the norms that specify such responsibilities. Such function implements the *Algorithm* 1 described in the **(AN)** step; (ii) the main task of the Updating Norm function is to update the set of activated and adopted norms. It implements the algorithm *Algorithm* 2 described in the **(UN)** step; (iii) the Evaluating Norm function helps the agent on selecting the norms that it has the intention to fulfil and the ones it has the intention to violate by executing the *Algorithm* 5 and, consequently, *Algorithms* 3 and 4 described in the **(EN)** step; (iv) the Detecting and Solving Conflicts function checks and solves the conflicts among the norms. It executes *Algorithms* 6 and 7 described in the **(ISC)** step; (v) the Selecting Desires function selects the desires that will become intentions taking into account their priorities and executes *Algorithm* 8 described in the **(SD)** step; and, finally, (vi) the Selecting Plans function chooses a single applicable plan from the set of options based on their priorities and executes *Algorithm* 9 described in the **(SP)** step. The applicability of the NBDI architecture and its implementation was demonstrated by the developing of the non-combatant evacuation scenario, presented in section 3.

Algorithm 9. Selecting Plan

Require: *fulfilSet* NF: norms stored in the fulfil set
Require: P: all plans that achieve the selected desire
 1: **for all** Norm n in *NF* **do**
 2:     **for all** p in P **do**
 3:         **for all** state in p **do**
 4:             **if** $(n.State == state) \cap (n.DeonticConcept == Obligation)$ **then**
 5:                 p.annotatePriority(+1)
 6:             **else**
 7:                 **if** $(n.State == state) \cap (n.DeonticConcept == Prohibition)$ **then**
 8:                     p.annotatePriority(-1)
 9:                 **end if**
10:             **end if**
11:         **end for**
12:     **end for**
13: **end for**
14: **return** getPlanHighestPriority()

---

6   Related Work

Our work was influenced by the architecture proposed in [2]. Such architecture to build normative agents also contemplates functions to deal with the adoption of norms and the influence of norms on the selection of desires and plans. However, our work presents details about the verifications that must be satisfied in order to agents adopt norms, the evaluations that must be made to select the norms the agents intend to fulfil and violate and a guidelines to help agents on selecting plans according to the norms they want to fulfil and violate. The BOID (Belief-Obligation-Intention-Desire) architecture proposed in [1] is an extension of the BDI architecture that considers the influence of beliefs, obligations, intentions and desires on the generation of the agent desires. The BOID architecture applies the notion of agent types to help on the generation of the desires. Thus, their approach could have been used in the **(SD)** function being proposed in our paper since this function is the one responsible to select the desires. Instead of basing the selection of desires on the agent type, we have used the norm contribution and the priority of the desires (and intentions) to provide a quantifiable solution to the selection of the agent desire. The approach described in [4] proposes an architecture to build norm-driven agents whose main purpose is the fulfilment of norms and not the achievement of their goals. In contrast, our agents are desire-driven that take into account the norms but are not driven by them. In [6] the authors provide a technique to extend BDI agent languages by enabling them to enact behaviour modification at runtime in response to newly accepted norms, i.e., it consists of creating new plans to comply with obligations and suppressing the execution of existing plans that violate prohibitions. However, they have not considered the desires and plans priorities. In our work we consider that obligations and prohibitions may increase or decrease the priority of a desire or a plan, and that the selection of desires and plans are based on their priorities. The agents built according to the architecture presented in [5] are able to evaluate the effects of norms on their desires helping then on deciding to comply or not with the norms. This architecture is based on the BDI architecture whose properties have been expanded to include normative reasoning. In the *Norm Review* process being proposed, we extend some of the verifications defined in [5], such as: our architecture checks *(i)* if the norm was not adopted already and *(ii)* if the agent is the addressee of the norm. Besides, in the *Norm selection* process, although the approach proposed in [5] evaluates the positive and negative effects of norms on the agent desires, it does not consider the influence of rewards in such evaluation. The authors in [3] present concepts, and their relations, that are used for modelling autonomous agents in an environment that is governed by some (social) norms. Although such approach considers that the selection of desires and plans should be based on their priorities and that such priorities can be influenced by norms, it does not present a complete strategy with a set of verification in the norm review process, and strategies to evaluate, identify and solve conflicts between norms such as our work does.

7   Conclusions

This paper proposes an extension to the BDI architecture called NBDI to build goal-oriented agents able to: (i) check if the agent should adopt or not a norm, (ii) evaluate the pros and cons associated with the fulfilment or violation of the norm, (iii) check and solve conflicts among norms, and (iv) choose desires and plans according to their decisions of fulfilling or not a norm.

By implementing the algorithms from 1 to 9 and using the Normative Jason platform, the applicability of NBDI architecture could be verified in the example presented in Section 3. Such agents are responsible to plan the evacuation of people that are in hazardous location, check the incoming perceptions (including norms), select the norms they intend to fulfil based on the benefits they provide to the achievement of the agent's desires and intentions, identify and solve conflicts among the selected norms, and decide to cope or not with the norms while dropping, retaining or adopting new intentions. We are investigating the need for extenting the AgentSpeak language with new predicates that better represent the norms.

References

1. Beavers, G., Hexmoor, H.: Obligations in a bdi agent architecture. In: IC-AI (2002)
2. Castelfranchi, C., Dignum, F., Jonker, C., Treur, J.: Deliberative normative agents: Principles and architecture. In: Proc. of the 6th Int. Workshop on Agent Theories, Architectures, and Languages (1999)
3. Dignum, F.: Autonomous agents and social norms. In: Proc. of the Workshop on Norms, Obligations and Conventions, pp. 56–71 (1996)

4. Kollingbaum, M.: Norm-Governed Practical Reasoning Agents. PhD thesis, University of Aberdeen (2005)
5. Lopez-Lopez, F., Marquez, A.: An architecture for autonomous normative agents. In: Proc. of the 5th Int. Conf. in Computer Science (2004)
6. Meneguzzi, F., Luck, M.: Norm-based behaviour modification in bdi agents. In: Proc. of 8th Int. Conf. on Autonomous Agents and Multiagent Systems (2009)
7. Neto, B.F.S., da Silva, V.T., de Lucena, C.J.P.: Using jason to develop normative agents. In: Proc. of the XX Brazilian Symposium on Artificial Intelligence (2010)
8. Weiss, G.: Multiagent Systems: A Modern Approach to Distributed Modern Approach to Artificial Intelligence. Massachusetts Institute of Technology (1999)

Evaluation of Environment Contextual Services in Multiagent Systems

Flavien Balbo[1,2], Julien Saunier[3], and Fabien Badeig[1,2]

[1] Université Paris-Dauphine, LAMSADE, Place du Maréchal de Lattre de Tassigny
Paris, France
[2] Université Paris Est, IFSTTAR, GRETTIA
2 Rue de la Butte Verte, 93160 Noisy le Grand, France
[3] Université Paris Est, IFSTTAR, LEPSiS, 58 Bld Lefèbvre, 75015 Paris, France
{balbo,badeig}@lamsade.dauphine.fr, saunier@ifsttar.fr

Abstract. The environment is a powerful first-order abstraction in Multi-Agent Systems (MAS), as well as a critical building block. The agents interact in their environment and the effects of their actions are observed and evaluated through this environment. The local complexity of the agents depends on its management of the interaction and action processes. If the environment carries out a part of this processes, the complexity of the agents is reduced. This delegating process implies a centralization of a part of the MAS computations inside the environment and therefore a flexible way to exchange information and to coordinate the agents.

In this paper, we present the modeling of an environment which supports both communication services and simulation services: multi-party communications (communication) and contextual activation (simulation). We evaluate the cost of these environment services and compare it to the execution of the same tasks inside the agents. The evaluation and comparison are done theoretically and empirically for communication and simulation. We also investigate the clustering of the agents in several environments.

Keywords: Environment, Evaluation, Communication, Activation.

1 Introduction

In [16], the authors give the following definition of a multi-agent system (MAS) environment: *"The environment is a first-class abstraction that provides the surrounding conditions for agents to exist and that mediates both the interaction among agents and the access to resources"*. The environment is a critical building block in the multi-agent systems that encapsulates its own responsibilities. The agents interact in their environment and the effects of their actions are observed and evaluated through this environment. Thus, the environment provides observability and accessibility services.

The relation between the environment and the agents is traditionally based on a perception - decision - action cycle [15] which is repeated by the agents during their life time. The three phases of the cycle are executed in each agent. The perception being the agent ability to observe its environment, its result is the computation of what can be

J. Filipe and A. Fred (Eds.): ICAART 2011, CCIS 271, pp. 192–207, 2013.

called *contexts*. A context is defined as a set of information that can be used to characterize a situation. This set of information includes percepts and messages that are obtained by the agent and its own state. Thanks to the context computation, the agent decides to execute the suitable action. The relation between the environment and the agents based on this perception - decision - action cycle does not exploit the potential of the environment that remains a static entity. This relation can be enhanced by delegating the perception process to the environment [1,8,13,17]. The objective is to support advanced features that are difficult to obtain without the environment.

We have proposed an environment modeling and an environment framework that are based on the delegation of the perception process to the environment. Our proposition supports two advanced features: 1) Multi-Party Communications (MPC) [3] takes into account dyadic interaction (one to one), group interaction (one to many) and overhearing (many to one/many) within the same interaction process; 2) Contextual activation [9] applied to the simulation process consists in activating the agents directly according to their context. In our proposition, despite the delegation process to the environment, the agents keep their autonomy because the decision process and the action process remain in the agents. The perception phase of the cycle is modified to take into account the relation with the environment. That means that the environment computes the contexts for the agents, and that the agents can modify this computation process to suit their needs: the agents decide which contexts are important for them and act by modifying in consequence the computation process that is executed by the environment. In the MPC case, the environment mediates the communication: the context is the information about the agent receiver, the message and any other environment information which is relevant to this communication. if the context is validated, the environment addresses the messages to the related agents. In the simulation case, the environment is the scheduler that manages the agent activation: the context is the subset of information accessible to the agents. The environment activates an agent in a specific context and according to this context, the agent performs the suitable action.

To adapt the perception process to each agent, the environment needs 'tools' to compute their contexts. In our proposition, these tools are *filters* that reify the relation between an agent (described in the environment) and its contexts. A filter contains the constraints on the context related to an action of an agent. This action is the reception of a message for communication filters and a specific agent action to perform after validation for activation filters. The agents modify their relation to the environment thanks to the addition and removal of their filters in the environment. This process is dynamic and the environment activates only its current filters. These choices (addition/removal) belong to the decision process that is managed by each agent.

The externalization of the computation of the perception process improves qualitatively the design of a MAS. The dynamic filter management by the agents gives more flexibility to the MAS design. More details on qualitative improvement are given in [13] for communication and in [1] for simulation. The counterpart is a centralization of a part of the MAS processes inside the environment. In this paper, we evaluate the cost of this centralization for computing agent contexts, in comparison to the execution of equivalent solutions for communication and simulation.

Section 2 motivates the delegating process of tasks to the environment for communication and simulation and introduces an illustrative example with different examples issued of a crisis management application. Section 3 presents the environment model. Section 4 provides a theoretical evaluation. In section 5, we provide an empirical evaluation, and we conclude in section 6.

2   Motivations

2.1   Communication

Recent research on multi-party communications [3,13] shows how multi-agent communications can take advantage of the complexity of the human communication process. The main issue in supporting MPC is to take into account dyadic interaction (one to one), group interaction (one to many) and overhearing (many to one/many) within the same interaction process. The sender does not know all the agents that might be interested in its messages. For example, an agent can listen to messages without the agreement/knowledge of the sender through overhearing. For a recipient, the usefulness of a message may depend on the context of the sender, the context of the message, and the context of the recipient itself.

These challenges are related to the way the recipients are chosen. MPC requires knowledge of the needs of both the sender and the recipients. From the sender viewpoint (direct interaction), it is a connection problem: which agents are related to my message? The problem is to map the senders needs (information, capabilities, resources, ...) to the address of related agents. From the recipient viewpoint (indirect interaction), it is a data extraction problem: which messages are related to me? The problem is to map the recipients needs to the content of the messages. For each message, these problems have to be simultaneously solved by the communication infrastructure. The environment is able to solve these problems by mediating the communication in order to find all the receivers of each message. Nevertheless, in the cognitive agents community, few works explicitly present the environment as an interaction support. For direct interaction, the environment is often associated to an infrastructure that supports point to point communication. For indirect interaction, cognitive agents use specific services that are based on the management of a shared collection of data (*e.g.* [11]) that may be understood as a part of the environment. There is therefore a separation between the solutions to realize direct and indirect interaction although the environment provides a suitable framework to unify them [12].

If the computation of the context does not take into account ambient conditions, it can classically be done inside the agents. This solution implies that the agents receive all messages and filter them. An evaluation of the environment support consists in the cost comparison of the context filtering either in each agent after the reception or in the environment during the transmission process.

2.2   Agent-Based Simulation

In a simulation, the scheduling policy defines the activation order of the agents. Once activated, the agents behave according to their context. How the agents are activated and what information is available to compute the agent context depend on the agent-based simulation (ABS) framework.

In most ABS frameworks, at each simulation step, a scheduler activates sequentially all the agents and each activated agent computes its local context to choose one action to execute. The classical ABS frameworks are designed to support this activation process. For example, in the platforms CORMAS [2] and MASON [10], the scheduler activates a standard method for each agent. This method is specialized by the designer to adapt the agent behavior. In the Logo-based multi-agent platforms such as *TurtleKit* [7] or the STARLOGO system[1], an agent has an automaton that determines the next action to perform.

The choice of these platforms to be agent-oriented, with a light environment support, implies that the computation of the context is repetitive because it is computed in each agent at each time cycle during the simulation execution. This computation is done in each agent even if the agent context does not change between two time cycles and/or if several agents share the same context during a time cycle. Here, the evaluation consists in the cost comparison of the context filtering either in each agent after the activation or in the environment during the scheduling process.

Example. In this paper, we consider a crisis management application where several emergency services must be coordinated in order to reduce the crisis effects. A crisis situation is a dynamic phenomenon defined by the initial situation, which depends on place and time, and by the impact on population and infrastructure. We focus our example on a specific point which is the victim evacuation. This task consists in coordinating two agents playing the role *medical porter*. The goal of the medical porters is to shift a victim to an emergency vehicle. This action requires one medical porter with the skill *medical monitoring* and one medical porter with the skill *victim handling*. The first skill allows the medical porter to monitor the victim and to inform the hospital of the evolution of the victim health. The second skill is necessary to handle the victim.

Each simulation component (agents, messages and objects) is situated on a grid. In this example, we consider that the victims belong to the set of the objects because they are not autonomous. The agents (medical porter) act in this environment and cooperate in order to evacuate victims. A medical porter can move *randomly* or towards *a given direction*, and it possesses only one skill. It can either *monitor* or *handle* a victim, but the two skills are required to evacuate the victim. The agents can communicate in order to find a partner with the complementary skill. Each agent has a field of perception that limits its perception of the environment. A medical porter is able to perform one of the following actions in a time cycle of simulation: 1) the action *move randomly*, 2) the action *move towards a location*, 3) the action *wait*, and 4) the action *evacuate a victim*.

We use this application along the paper to illustrate our study of the cost of the environment supporting interaction and simulation.

---

[1] http://education.mit.edu/starlogo/

3   Environment Modeling

To support multi-party communications and contextual activation, we propose to use the environment as a privileged intermediary to manage interaction (EASI, Environment as Active Support for Interaction) and simulation (EASS, Environment as Active Support for Simulation) process. The EASS model [1] embeds the EASI model [13]. The environment manages meta-informations on the MAS (agents, messages, context) and uses them to compute the agent context(s). In this section, we give the background elements to understand the assessment of the cost of the environment. More details about the models can be found in [13].

The environment model EASI is thus defined by $\langle \Omega, \mathcal{D}, P, \mathcal{F} \rangle$ with:

- $\Omega = \{\mathcal{A} \cup \mathcal{MSG} \cup \mathcal{O}\} = \{\omega_1, ..., \omega_m\}$ the set of entities (with $\mathcal{A}$ the set of agents, $\mathcal{MSG}$ the set of messages and $\mathcal{O}$ the set of objects, *i.e.* all entities that are not agents or messages),
- $\mathcal{D} = \{d_1, ..., d_m\}$ the set of domain descriptions of the properties,
- $P = \{p_1, ..., p_n\}$ the set of properties,
- $\mathcal{F} = \{f_1, ..., f_k\}$ the set of filters.

Entity. An *entity* $\omega_i \in \Omega$ is defined by $\langle e_r, e_d \rangle$ where e_r is a reference to an element of the MAS and e_d is the description of that element. An element of the MAS can be an agent, a message or an object. A reference is its physical address on the platform or other objects (such as URL, mailbox, *etc.*). The description e_d is a set of couples $\langle p_i, v_j \rangle$ where $p_i \in P$ and v_j is the value of the property for the entity. Any agent of the MAS has its own processing and knowledge settings. It is connected to the environment thanks to its description that the environment stores and updates. This description e_d is used for the routing of the informations to the agent with the reference e_r (EASI) or its activation (EASS). The agents can modify their description dynamically.

Property. A property $p_i \in P : \Omega \rightarrow d_j \cup \{unknown, null\}$ is a function whose description domain $d_j \in \mathcal{D}$ can be quantitative, qualitative or a finite set of data. The *unknown* value is used when the value of the property cannot be set, and *null* is used to state that the property is undefined in the given description. In order to simplify the notation, only the value of the description domain d_j is given to specify a property.

Filter. A filter identifies the entities according to their description (e_d) and realizes the interaction between the concrete objects (e_r). A filter $f_j \in \mathcal{F}$ is a tuple $f_j = \langle f_a, [f_m], [f_C], n_f \rangle$ with n_f the filter name. The assertion $f_a : \mathcal{A} \rightarrow \{true, false\}$ identifies the related agents, the assertion $f_m : \mathcal{MSG} \rightarrow \{true, false\}$ identifies the related messages, and $f_C : \mathcal{P}(\Omega) \rightarrow \{true, false\}$ is an optional set of assertions identifying other entities of the context. A filter is considered valid for any tuple $\langle agent, [message], [context] \rangle$ such as:

$$f_a(agent)[\wedge f_m(message)][\wedge f_C(context)] \text{ is true.}$$

The choice of the action specializes the filter. Every agent a whose description validates f_a is activated if there exists a set of entities in the *context* such that f_C is true. After its activation, the agent validates the action related to this context and executes it (or not). It is a communication filter if the action is to add the message m that satisfies f_m in the message box of the agent a. The agents can add and remove dynamically the filters. Thus, they adapt their relation with the environment according to their needs.

In a crisis simulation, the agents have observable properties: their identifier, which is unique, their location, their availability (*true/false* values) which specifies if an agent rescues a victim, their skill (*position/monitoring*), their field of vision, and their internal time. The messages have the following properties: their identifier that is unique, their sender identifier, the sending time, the message type (*'request'/'accept'*), their location which is the location of the sender when the message is put in the environment, the identifier of the victim to rescue and the skill required by the sender in the case of communication. The victims have four properties: their identifier, their location, their status (*stretcher/vehicle*) and "diagnosed". The victims to be evacuated must be diagnosed; this is expressed by the property *diagnosed* (*true/false* values).

In the simulation example, let f^e be a communication filter related to the following context: an agent is interested by a message if it is available (f_a), if the request message is close to the agent a and a has the requested skill (f_m) and if the victim has been diagnosed (f_c). In this example, f_c contains one entity which is a victim identified by the property id_v of the message. The request messages have information about the victim location and the skill of the medical porter.

The processing of the context is traditionally done by the agents, but in EASI and its extension EASS it is done by the environment. In this way, the computation can be done with information that is shared by the agents (the entities descriptions) and with a degree of mutualization. However, it implies the centralization of the computation and the management of the information update. Now, we propose to evaluate the cost of this centralization.

4   Theoretical Assessment

The theoretical assessment is the comparison between the processing by the environment of a communication filter for n_a agents and the processing by n_a agents of all the messages. In order to take into account the context dynamics, this theoretical assessment also studies the update process of the MAS. The theoretical assessment is based on two criteria. The first, noted $Cost_T$, is the number of tests performed during the filtering process and the second, noted $Cost_M$, is the number of resulting messages. The objective is to identify in which cases it is better to mediate the communication through the environment than to manage it in the agents.

Following the definition given in section 3, a filter f has tests related to one agent description (f_a), one message description (f_m) and a subset of entities (f_c). To simplify the explanation, we assume there is only one entity to match f_c and this entity cannot be an agent. This last assumption implies that there is no additional cost for the update process if the message processing is executed by the agents. The following generic filter illustrates our assessment, it is shown in a classical "if" structure:

Tests	N
$if \ \wedge_{i=1..n} \ T_i^m(?m)$	(1)
$and \ \wedge_{i=1..n} \ T_i^a(?a)$	(2)
$and \ \wedge_{i=1..n} \ T_i^{a,m}(?a, ?m)$	(3)
$and \ \wedge_{i=1..n} \ T_i^c(?c)$	(4)
$and \ \wedge_{i=1..n} \ T_i^{m,c}(?m, ?c)$	(5)
$and \ \wedge_{i=1..n} \ T_i^{a,c}(?a, ?c)$	(6)

For the filter example f^e: an agent $(?a)$ is interested by a request message $(?m)$ (1) if it is available (2), if this request message is close to the agent and the agent $?a$ has the skill specified in the message (3) and if the victim to evacuate $(?c)$ (5) is diagnosed (4). $T_1^a(?a)$ is the result of the first test on an agent related to its availability and $\wedge_{i=1..n}T_i^a(?a)$ are all the tests on the agent receiver. $T_1^{a,m}(?a, ?m)$ is the result of the first test on an agent and a message related to the requested skill and $\wedge_{i=1..n}T_i^a(?a, ?m)$ are all the tests on the receiver and the message.

Evaluation of the Number of Tests. A filter is composed of two types of tests. The first type is a comparison between the value of a property and a constant: the tests (1)(2)(4). The second type of test implies a matching process: the tests (3)(5)(6). For each of these types the cost in number of tests is related to the number of entities that have to be tested and on the result of the previous tests on the same set of entities. For example, the number of agents that is tested by $T_2^a(?a)$ is related to the number of agents that have validated the test $T_1^a(?a)$. Let t_i be the percentage of entities that validates the test i. Let $|C_i|$ be the number of tests and C_i be the number of evaluated tests of (i), the value of C_i is:

$$C_i = \begin{cases} 1 + \sum_{j=1}^{|C_i|} \prod_{k=1}^{j} t_k & \text{if } |C_i| > 1 \\ 1 & \text{if } |C_i| = 1 \\ 0 & \text{if } |C_i| = 0 \end{cases}$$

The cost of a filter is the sum of the costs of its tests (sequentially validated). The following table gives the cost of the generic filter if one message has to be processed by the environment (column E) or by the n_a agents (column A). There are two differences between these two processing: 1) the factorization of the tests if the processing is executed by E; 2) the implicit elimination of the agents if the processing is executed by the agents. If the processing is executed by E the tests are factorized, for example (1) is executed only once and is repeated for n_a agents. Each test (i) that is false implies an elimination of the agent that executes the processing. For example in (5) only the subset of agents where the test (4) is validated continue the processing. Following the hypothesis that the filtering and the shared information are the same, the matching done by the agents is more costly than by the environment for the tests (1)(4)(5) and are the same for (2)(3)(6). We conclude that the mediation by environment is always better than the broadcast when at least one agent is interested in the message. Nevertheless the implicit agent elimination implies that if the agents do not share all information some part of the filter evaluation can be less costly for the agents than for the environment. This parameter will be empirically tested in section 5.

N	E	A	tested entities
(1)	C_1	$n_a * C_1$	$n_m = 1$
(2)	$n_a * C_2$	$n_a * C_2$	all the agents in E and each agent individually (A)
(3)	$p_2 * n_a * C_3$	$p_2 * n_a * C_3$	p_2 : percentage of agents that have the requested state
(4)	$n_c * C_4$	$(p_2*p_3*n_a)*n_c*C_4$	n_c : number of entities that are tested in (4). p_3 : percentage of agents that are related to the message.
(5)	$p_4 * n_c * C_5$	$(p_2 * p_3 * n_a) * (p_4 * n_c) * C_5$	p_4 : percentage of entities related to the context (4)
(6)	$(p_2 * p_3 * n_a) * (p_5*p_4*n_c) * C_5$	$(p_2*p_3*n_a) * (p_5 * p_4 * n_c) * C_5$	p_5 : percentage of entities validating (5)

Moreover, the environment and the agents need the same information to process the filters and the result has to be balanced by the cost of the access to this information.

Evaluation of the Number of Messages. If the exchange of information is done via messages, there are two types of messages to take into account: 1) the messages related to the interaction in the MAS such as the request messages in our example f^e; 2) the messages related to the update process.

Firstly we evaluate the number of messages that are the result of the interaction in the MAS. Let n_m be the number of messages sent by the agents, let p_r be the average percentage of agents that have to receive the messages. If the matching process is executed by the environment then each message is sent to the environment, and the environment transmits it to the $p_r * n_a$ receivers. The total number of messages is therefore $n_m(1 + p_r * n_a)$. If the matching process is executed by the agents, then each message is sent to all the agents which locally execute the matching process. The total number of messages is therefore $n_m * n_a$. Except if p_r is close to 1 which means that the message is related to all the agents, the environment mediation is less costly. The more the selection is important, the more the mediation by the environment is beneficial.

Secondly we evaluate the number of messages that are the result of the update process of the MAS. To simplify, we consider only the update process of the descriptions of the agents and not of the context, although the principle is the same. In our example, it implies that the state of the victims is not updated (or that its cost is the same if it is computed by the environment or by the agents). When an agent is updated all its properties are updated in the environment. Remember that the filter example does not contain tests on the state of the agents to compute the context (f_c). Hence, the computation by the agents does not imply additional costs because we suppose the agents do not need to access updated information about the other agents, which is an underestimation of the cost of the local agent solution.

When the communication is mediated by the environment, the agents update their descriptions by putting in the environment new descriptions that replace the old ones. This action is associated to the sending of a message. One agent can make between 0 and n updates during a time period. Let $freq_a$ be the average frequency of the description update during a time period. There is therefore $freq_a * n_a$ update messages if the interaction is mediated by the environment during the reference time period.

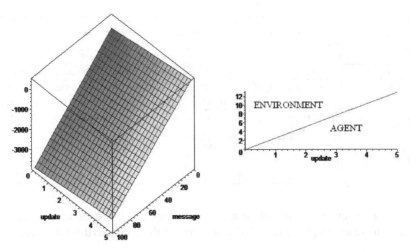

Fig. 1. Comparison of the cost (left) and comparison of the number of messages in function of the agent update frequency (right), with $p_r = 0.6$ and $n_a = 100$

To have the total cost according to the number of messages criteria we have to sum these costs. Let $Cost_M^E = n_m(1 + p_r n_a) + freq_a n_a$ be the total environment cost and $Cost_M^A = n_m n_a$ be the total agents cost.

To compare these two costs, we study the sign of the subtraction of the two costs $Cost_M = Cost_M^E - Cost_M^A$ in order to find when the environment mediation is more costly than the local computation in function of the number of messages and of the frequency of the update process. The resulting formula is $Cost_M = n_m(1 + n_a(p_r - 1)) + freq_a n_a$. Figure 1 (left) gives the plan corresponding to this function in the three dimensional space (message, agent update, cost) if the number of agents (n_a) and the average proportion of receiver (p_r) are respectively fixed to 100 and 0.6. These assessment parameters are unfavourable to the mediation by the environment: 1) the cost of the update process is not taken into account when the context is computed by the agents; 2) the agents are interested by more than half the messages. Nevertheless we can see that there are few cases in which the use of the environment is more costly than the local matching according to the number of messages criteria.

The function $Cost_M(n_m, freq_a) = 0$ enables to find the number of messages dedicated to the interaction in the MAS (m) that have to be mediated by the environment to compensate the cost of the update process ($freq_a$). The relation between these two parameters is $n_m = -(n_a/(1 + n_a(1 - p_r))) * freq_a$, the value of n_m is therefore inversely proportional to the proportion of the agents interested in the messages. Figure 1 (right) illustrates this proposition. The environment area represents the cases where it is better to use the environment than to mediate the messages in function of the number of messages. For instance, with $freq_a = 2$, if there is more than 5 messages related to the interaction in the MAS, then it is better to use the environment to mediate the communication even if that means that there are $freq_a n_a = 200$ messages related to the update process.

This study has been done in the worst case for the environment because the agents are not taken into account to evaluate the interaction context (f_c). If it were the case, there would be no modification of the update cost of the environment. However, when the matching is executed by agents, the number of messages in the update process would be increased by $freq_a\,(n_a - 1)$ (the cost of broadcasting the description updates), thus improving the comparative environment performance.

In this section the assessment has been done with the hypothesis that the agents have the same information to evaluate a filter. In the next section, we depict how these results relate to real experimentation where this hypothesis is not valid.

5   Empirical Assessment

In section 4, we have shown that using the environment to compute the contexts is less costly than computing them locally in the agents when the agents update frequency is not too high. In other words, the entities dynamics is the parameter that determines which solution is the best for a multi-agent system. This section consists in evaluating the cost of the context computing on a real example that is a simulation of the victim evacuation in the crisis situation.

A prototype of our ABS framework has been implemented as a plugin for the multi-agent platform Madkit [7]. This plugin is composed of an environment component with an API that enables the agents to add/retract/modify their descriptions and filters. We have chosen to implement the matching process within a Rules-Based System (RBS). The instantiation of the model into a RBS is straightforward: the descriptions are the facts of the rule engine, and the filters are its rules. Rule firing is based on the efficient RETE algorithm [6]. It is a network-based algorithm designed to speed the matching of patterns with data. RETE uses a static discrimination network, generated by the language compiler, that represents data dependencies between rule conditions.

We compare our model which encompasses contextual activation by the environment with a model which encompasses a classical activation process to evaluate the cost of supporting the simulation process through the environment. We have tested four simulation scenarios, using ten filters: seven filters for contextual activation (f_1 to f_7), two filters for communication (f_{recept} and f_{accept}). The filter $f_{classicalActivation}$ is used in the scenarios S_1 and S_3 in order to simulate a classical activation process. It activates each agent once by simulation cycle without taking into account the context. The other activation filter allows to activate an agent *medical porter* in a specific context to perform an action.

The scenarios are defined below. S_1 and S_2 are scenarios without agent communication, and S_3 and S_4 are scenarios with agent communication:

- $-S_1-\{f_{classical\ activation}\}$ + Local Agent Context Analysis (LACA),
- $-S_2-\{f_1, f_2, f_3, f_4, f_5\}$,
- $-S_3-\{f_{classical\ activation}\}$ + LACA + Broadcast,
- $-S_4-\{f_1, f_2, f_3, f_4, f_5, f_6, f_7\}$ + $\{f_{recept}, f_{accept}\}$

S_1 illustrates a classical scenario with an activation phase and a local agent context analysis; S_2 is a scenario with a contextual activation inside the environment. We describe the filters of S_2. The filter f_1 allows to activate the action *move randomly* that is triggered when a medical porter has no victim close to it (context *victim seeking*). The filters f_2 and f_3 correspond to the association of the action *move towards a location* with the contexts *closest victim discovery* (f_2) and *handled victim discovery* (f_3). The context *closest victim discovery* happens when an agent perceives the closest victim to evacuate in its perception field. The context *closest victim discovery* depicts the situation where an agent *medical porter* perceives a victim that is handled by another medical porter. The filter f_4 activates the action *wait* and triggers when a medical porter is alone close to a victim to evacuate (context *victim proximity*). The filter f_5 corresponds to the association between the action *evacuate victim* and the context *rescue victim*. This context appears when two medical porters with the complementary skill are close to a victim ready to be evacuated. This context is related to the following information: the availability of the porters, the location of the porter and the victim. The comparison between S_1 and S_2 enables to study the cost of the activation process.

S_3 and S_4 are the extensions of respectively the scenarios S_1 and S_2 with the communication filters f_{recept} and f_{accept}, and the associated activation filters f_6 and f_7. The filters f_{recept} and f_{accept} provide the support for the communication process: f_{recept} enables the reception of the request messages and f_{accept} enables the reception of their answer(s). When the agents communicate, the simulation has to manage two new agent actions that are activated by the filters f_6 and f_7. The filter f_6 is related to the action *contact*: an agent puts a request message in the environment and waits for an answer. The filter is activated when the agent is close to a victim to evacuate but does not have another agent nearby to evacuate this victim. The filter f_7 is related to the action *coordination*: an agent answers to the contact agent and moves towards the victim location. The context of this filter is the reception of a request message containing the location of the victim. The comparison between S_3 and S_4 enables to study the cost of the activation when the medical porters use communication protocol to contact another medical porter with the skill required.

Activation. We have run three series of simulations characterized by two parameters: number of agents *medical porters* and field of vision. For each series of simulations, we evaluate our model using the average run-time over 50 simulations with similar parameters. In the first group, we have experimented with 5 medical porters with the skill "medical monitoring", 5 medical porters with the skill "victim handling" and 5 victims, in the second group with 15 of each, and in the third group with 20 of each.

Figure 2 shows the result for the third experiment. As we can see, the scenario S_2 is faster than the classical scenario S_1, except for the lowest field of vision. For a field of vision from 8 to 30, S_2 run-time curves (dotted curves) are below S_1 run-time curves (solid curves). The increase of the field of vision improves the knowledge of the agents on the environment and should improve their efficiency: the probability for an agent to perceive victims and other agents increases. This is true with the use of the environment in the scenario S_2, a larger field improves the efficiency of the agents: as the field of vision increases from 5 to 30, run-time decreases from around $4000s$ to $800s$. But with the classical activation process in the scenario S_1, the improvement of victim perception barely offsets the cost of local context analysis. We can observe that in most cases, the cost of the context computing in each agent is more expensive in terms of simulation run-time than the cost of computing the context inside the environment. The results are similar for the first and second series of experimentations.

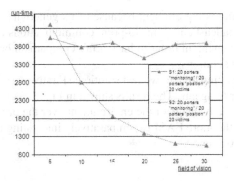

Fig. 2. Simulation run-time with the strategies S_1 and S_2 for the third group

Communication. In the following experimentation, we compare the use of broadcast and the use of the environment for the communication. In these tests, only the communication filters f_{recept} and f_{accept} of the scenario S_4 are evaluated and compared to the broadcast. We call *Broadcast* the experimentation with a matching process executed by the agents, *Environment* the experimentation with a matching process executed by the environment.

The agents send "request" messages. The agents interested in these messages are those which are close to the sender and have a different skill, if the victim can be moved. In *Broadcast*, the message is broadcasted to all the agents, while in *Environment* they are managed by f_{recept}. Those agents answer with either an accept or a reject message through an addressed message. In *Broadcast*, the message is sent to the agent via point-to-point communication, while in *Environment*, they are managed by f_{accept}.

Each data is an average on 50 simulations. Each simulation is composed of 2000 steps. We have run two series of simulations characterized by two parameters: number of agents and number of updates during the simulation.

Firstly, we have tried to use the RETE algorithm to implement the communication filter, and compared it to the broadcast and addressed messaging capabilities of MadKit. The following table sums up the simulation run-time in function of the messaging implementation: EASI with a RETE tree, MadKit Messaging support, *ad hoc* Environment and Broadcast.

Agent	RETE	Madkit	Environment	Broadcast
10	4624	1096	125	173
20	15215	4050	394	596
50	83785	26306	2377	4755

These results show that the cost of a RETE tree offsets the gains in the message treatment. This discrepancy with the previous results for activation can be explained by the cost of the addition and removal of the messages. In the activation part, the entities are the same all along the simulation, and only rule firing and description modifications take place.

This has lead us to use an *ad hoc* implementation of the environment for communication, which treats the calculations in the same way as described in section 4. However, the comparative results show that the MadKit implementation of the broadcast and addressed messaging is not efficient because of its genericity. Therefore, we also re-implemented the broadcast and point-to-point communications using the same data structures as the environment, in order to have a fair comparison between the environment and the broadcast solutions.

In the following, we compare the two last solutions. Firstly, we study the impact of a variation in the number of agents. The results shown in Fig. 3 feature 10 to 100 agents. The curve shows a clear advantage of the environment over the broadcast, and that the more agents there are, the more EASI is comparatively interesting (from 27.8% for 10 agents to 53.9% for 100 agents).

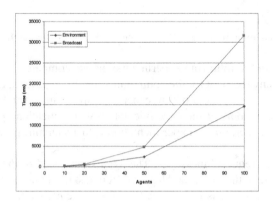

Fig. 3. Comparative results of Broadcast and EASI in function of the number of agents

In the second phase (Fig. 4), we study the impact of a variation in the number of updates, for 50 agents. During each update, all the agents move and in the case of the Environment implementation, they modify their description accordingly. The run-time gain of EASI over broadcast gets from 50.7% for 20 updates to 48.8% for 1000 updates. Therefore, the cost of the update mechanism is not significant in comparison with the difference in efficiency.

Another solution to reduce the cost of the matching process is to reduce the number of entities that are tested for each message. For example, if the reception of messages is conditioned by the skill of the agents, then we can cluster the agents in several communication environments according to this property value. This clustering should decrease the cost of the matching for two reasons: the number of agents in each environment decreases, and the filter can be simplified by removing the test on the skill property. To test this hypothesis, we have run experiments using several communication environments.

In these experimentation the following hypothesis have been done:

- We call $Environment3$ interaction through three environments, and the agents are grouped in function of their skill. The skill $driver$ has been added in order to have three skills.
- The agents send "request" messages. The agents interested in these messages are those which are close to the sender and have the same skill. In experimentation $Broadcast$, the messages are broadcasted to all the agents, while in $Environment3$ they are managed by f_{recept_2}. This filter is designed after the filter f_{recept}, without

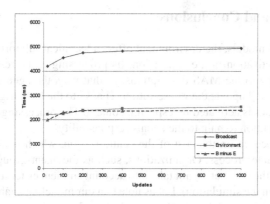

Fig. 4. Comparative results of Broadcast and EASI in function of the update process

the condition on the skill that becomes implicit : all the agents in a given environment share the same skill.

- An agent which has received a "request" answers with either an accept or a reject message through an addressed message in the same way than in the previous experimentation.

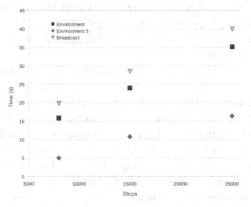

Fig. 5. Comparative results of *Broadcast*, *Environment* and *Environment3*

The results shown in Fig. 5 feature 60 agents, and each test was done 15 times. The number of steps of the simulation varies between 8000 and 25000. The receivers of request messages are selected over the position and skill properties. The results are consistent with the previous findings, the *Environment* being significantly better than the *Broadcast*. Furthermore, *Environment3* shows a 50% to 68% gain over *Environment*. This highlights the importance of the number of agents and of the filter complexity on the performance of the system. However, separating the agent descriptions can only be done when they are not used in the same filters.

These experimentations show a clear advantage to the environment over the broadcast, and this result is backed up by the theoretical results. It shows the efficiency of the context computing inside the environment in comparison with the classical approach where the context is computed locally in each agent.

6 Discussion and Conclusions

In the EASI/EASS models, the environment offers a technical support for the communication and the activation in the simulation. Its processing is the result of the filters triggering according to the MAS descriptions. In that way, the agents add or remove dynamically filters and update their description. Thanks to the filter triggering, an agent receives a message or is activated in a specific context. The processing of the messages and the action execution remain to the agents responsibility.

This paper is focused on the cost of the centralization, which is closely related to the cost of the update process. Organizations, such as the agent-group-role model [5] enable to decrease the cost but do not take into account ambient criteria like the location of the agents (e.g. in the simulation). Let us note that our modeling enables to reproduce a selection of percepts according to the organisation. Institutions (see e.g. [4]) generally do not share the same objective. The focus is on control and not on the filtering / matching process. However, the technical solutions may be close to ours.

The use of a shared knowledge has already been done within agents (for example Sycara's work, e.g. [14]) but in that case the update of the properties is not considered. In this paper, we focus on the cost of the update process that could limit the interest to centralize specific information. Furthermore, middle-agents are generally used only as a first step to find contacts, and not to manage all the communications. In our view, the environment is a facility, which can be used to facilitate the interaction, apply norms, or verify some rules related to the application design. These roles do not belong to the same design level as the agents.

Theoretically, we have shown that if the communication takes the context into account, there is no strictly dominant solution. It depends on the dynamicity of the multi-agent system, the number of agents and the average percentage of agents interested in each message. According to the number of tests criteria, we have shown that the environment is always better than the local context computation. According to the number of messages criteria, the result has to take into account the number of messages related to the MAS activity and the number of messages related to the update process. We have shown that the environment solution is generally better to mediate the communication of the MAS activity and that few messages to mediate are needed to compensate the cost of the update process.

To propose an empirical assessment of the cost of the environment, we have studied the run-time criterion in the crisis simulation example. We compare the cost of the local context analysis for each agent to a central and global control ensured by the environment, and the cost of communication. The main conclusion is that the environment cost is significantly lower than the local agent calculation of the context perception, except when there are very few agents.

In the future, we intend to investigate different ways to improve the environment performance. An ongoing effort concerns the theoretical evaluation of a RETE-based instantiation of the model. We also study how to take advantage of the filter and entity structures to speed up the matching process. The clustering of the agents in several environments is a perspective to improve the matching process. However, moving the agents from one environment to another dynamically according to their interaction needs is costly and therefore has to be taken into account.

References

1. Badeig, F., Balbo, F., Pinson, S.: Contextual activation for agent-based simulation. In: ECMS 2007, pp. 128–133 (2007)
2. Bousquet, F., Bakam, I., Proton, H., Page, C.L.: Cormas: Common-Pool Resources and Multi-Agent Systems. In: Pobil, A.P.D., Mira, J., Ali, M. (eds.) IEA/AIE 1998, Part II. LNCS, vol. 1416, pp. 826–837. Springer, Heidelberg (1998)
3. Branigan, H.: Perspectives on multi-party dialogue. Research on Language & Computation 4(2-3), 153–177 (2006)
4. Esteva, M., Rodriguez-Aguilar, J., Rosell, B., Arcos, J.: Ameli: An agent-based middleware for electronic institutions. In: Jennings, R., Sierra, C., Sonenberg, L., Tambe, M. (eds.) Proceedings of the third International Joint Conference on Autonomous Agents and Multi-Agent Systems (AAMAS 2004), pp. 236–243. ACM Press (2004)
5. Ferber, J., Gutknecht, O., Jonker, C., Muller, J., Treur, J.: Organization models and behavioural requirements specification for multi-agent systems (2002)
6. Forgy, C.L.: Rete: A fast algorithm for the many pattern/many object pattern match problem. Artificial Intelligence 19, 17–37 (1982)
7. Gutknecht, O., Ferber, J.: The MADKIT Agent Platform Architecture. In: Wagner, T.A., Rana, O.F. (eds.) AA-WS 2000. LNCS (LNAI), vol. 1887, pp. 48–55. Springer, Heidelberg (2001)
8. Kesaniemi, J., Katasonov, A., Terziyan, V.: An observation framework for multi-agent systems. In: International Conference on Autonomic and Autonomous Systems, pp. 336–341 (2009)
9. Laberge, D.: Perceptual Learning and Attention. In: Learning and Cognitive Processes, vol. 4, ch.5, pp. 237–273. W.K. Estes (1975)
10. Luke, S., Cioffi-Revilla, C., Panait, L., Sullivan, K., Balan, G.: Mason: A multiagent simulation environment. Simulation 81(7), 517–527 (2005)
11. Picco, G.P., Buschini, M.L.: Exploiting Transiently Shared Tuple Spaces for Location Transparent Code Mobility. In: Arbab, F., Talcott, C. (eds.) COORDINATION 2002. LNCS, vol. 2315, pp. 258–273. Springer, Heidelberg (2002)
12. Platon, E., Sabouret, N., Honiden, S.: Overhearing and Direct Interactions: Point of View of an Active Environment. In: Weyns, D., Van Dyke Parunak, H., Michel, F. (eds.) E4MAS 2005. LNCS (LNAI), vol. 3830, pp. 121–138. Springer, Heidelberg (2006)
13. Saunier, J., Balbo, F.: Regulated multi-party communications and context awareness through the environment. International Journal on Multi-Agent and Grid Systems 5(1), 75–91 (2009)
14. Sycara, K., Wong, H.: A taxonomy of middle-agents for the internet. In: ICMAS 2000: Proceedings of the Fourth International Conference on MultiAgent Systems (ICMAS 2000), pp. 465–466. IEEE Computer Society, Washington, DC, USA (2000)
15. Weiss, G. (ed.): Multiagent Systems: A Modern Approach to Distributed Artificial Intelligence. The MIT Press, Cambridge (1999)
16. Weyns, D., Omicini, A., Odell, J.: Environment as a first-class abstraction in multiagent systems. Autonomous Agents and Multi-Agent Systems 14(1) (2007)
17. Weyns, D., Steegmans, E., Holvoet, T.: Towards active perception in situated multi-agent systems. Special Issue of Journal on Applied Artificial Intelligence 18(9-10), 867–883 (2004)

Data Streams Classification: A Selective Ensemble with Adaptive Behavior

Valerio Grossi[1] and Franco Turini[2]

[1] Dept. of Pure and Applied Mathematics, University of Padova
via Trieste 63, 35121 Padova, Italy
[2] Dept. of Computer Science, University of Pisa
largo B. Pontecorvo 3, 56127 Pisa, Italy
grossi@math.unipd.it, turini@di.unipi.it

Abstract. Data streams classification represents an important and challenging task for a wide range of applications. The diffusion of new technologies, such as smartphones and sensor networks, related to communication services introduces new challenges in the analysis of streaming data. The latter requires the use of approaches that require little time and space to process a single item, providing an accurate representation of only relevant data characteristics for keeping track of concept drift. Based on these premises, this paper introduces a set of requirements related to the data streams classification proposing a new adaptive ensemble method. The outlined system employs two distinct structure, for managing both data aggregation and mining features. The latter are represented by a selective ensemble managed with an adaptive behavior. Our approach dynamically updates the threshold value for enabling the models directly involved in the classification step. The system is conceived to satisfy the proposed requirements even in the presence of concept drifting events. Finally, our method is compared with several existing systems employing both synthetic and real data.

1 Introduction

The constant and rapid diffusion of new technologies, such as smartphones, and sensor networks, related to communication and web services, and safety applications, has introduced new challenges in data management, analysis and mining. In these scenarios data arrives on-line, at a time-varying rate creating the so-called data stream phenomenon. Conventional knowledge discovery tools cannot manage this overwhelming volume of data. The unpredictable nature of data streams requires the use of new approaches, which involve at most one pass over the data, and try to keep track of time-evolving features, known as concept drift.

Ensemble approaches represent a valid solution for data streams classification [5]. In these methods, classification takes advantage of multiple classifiers, extracting new models from scratch and deleting the out-of-date ones continuously. In [20,19], it was stressed that the number of classifiers actually involved in the classification task cannot be constant through time. The cited works demonstrate that a selective ensemble which, based on current data distribution, dynamically calibrates the set of classifiers to use, provides a better performance than systems using a fixed set of classifiers constant

J. Filipe and A. Fred (Eds.): ICAART 2011, CCIS 271, pp. 208–223, 2013.

through time. In our former approach, the selection of the models involved in the classification step was chosen by a fixed activation threshold. This choice is the right solution if it is possible to study a-priori what is the best value to assign to the threshold. In many real environment, this information is unavailable, since the stream data behavior cannot be modeled. In several domains, such as intrusion detection, data distribution can remain stable for a long time, changing radically when an attack occurs.

This work presents an evolution of the system outlined in [20,19]. The new approach introduces a complete adaptive behavior in the management of the threshold required for the selection of the set of models actually involved in the classification. This work describes the adaptive approach for varying the value of the model activation threshold through time, influencing the overall behavior of the ensemble classifier, based on data change reaction. Our approach is explicitly explained with the use of binary attributes. This choice can be seen as a limitation, but it is worth observing that every nominal attribute can be easily transformed into a set of binary ones. The only inability is the direct treatment of numerical values. [14] represents a general approach to solve the on-line discretization of numerical attributes. The proposed method is particularly suitable in our context, since it proposes a discretization method based on two layers. The first layer summarizes data, while the second one constructs the final binning. The process of updating the first layer works on-line and requires a single scan over the data.

Paper Organization: Section 2 introduces our reference scenario, outlining some requirements that a system working on streaming environments should satisfy. Section 3 describes our approach in details, highlighting how the requirements introduced in Section 2.1 are verified by the proposed model. Furthermore, it present how our adaptive selection is implemented. Section 4 presents a comparative study to understand how the new adaptive approach guarantees a higher reliability of the system. In this section, our approach is compared with other well-know approaches available in the literature. Finally, Section 5 draws the conclusions and introduces some future works.

2   Data Streams Classification

Data streams represent a new challenge for the data mining community. In a stream scenario, traditional mining methods are further constrained by the unpredictable behavior of a large volume of data. The latter arrives on-line at variable rates, and once an element has been processed, it must be discarded or archived. In either cases, it cannot be easily retrieved. Mining systems have no control over data generation, and they must be capable of guaranteeing a near real-time response.

Definition 1. *A data stream is an infinite set of elements* $X = X_1, \ldots, X_j, \ldots$ *where each* $X_i \in X$ *has* $a + 1$ *dimensions,* $(x_i^1, \ldots x_i^a, y)$, *and where* $y \in \{\perp, 1, \ldots, C\}$, *and* $1, \ldots, C$ *identify the possible values in a class.*

A stream can be divided into two sets based on the availability of class label y. If value y is available in the record ($y \neq \perp$), it belongs to the training set. Otherwise the record represents an element to classify, and the true label will only be available after an unpredictable period of time.

Given Definition 1, the notion of *concept drift* can be easily defined. As reported in [23], a data stream can be divided into batches, namely $b_1, b_2, ..., b_n$. For each batch b_i, data is independently distributed w.r.t. a distribution $P_i()$. Depending on the amount and type of concept drift, $P_i()$ will differ from $P_{i+1}()$. A typical example is customers' buying preferences, which may change according to the day of the week, inflation rate and/or availability of alternatives. Two main types of concept drift are usually distinguished in the literature, i.e. *abrupt* and *gradual*. Abrupt changes imply a radical variation of data distribution from a given point in time, while gradual changes are characterized by a constant variation during a period of time. The concept drifting phenomenon involves data expiration directly, forcing stream mining systems to be continuously updated to keep track of changes. This implies making time-critical decisions for huge volumes of high-speed streaming data.

2.1 Requirements

As introduced in Section 2, the stream features influence the development of a data streams classifier radically. A set of requirements must be taken into account before proposing a new approach. These needs highlight several implementation decisions inserted in our approach.

Since data streams can be potentially unbounded in size, and data arrives at unpredictable rates, there are rigid constraints on time and memory required by a system through time:

Req. 1: the time required for processing every single stream element must be constant, which implies that every data sample can be analyzed almost only once.
Req. 2: the memory needed to store all the statistics required by the system must be constant in time, and it cannot be related to the number of elements analyzed.
Req. 3: the system must be capable of updating their structures readily, working within a limited time span, and guaranteeing an acceptable level of reliability.

Given Definition 1, the elements to classify can arrive in every moment during the data flow.

Req. 4: the system must be able to classify unseen elements every time during its computation.
Req. 5: the system should be able to manage a set of models that do not necessarily include contiguous ones, i.e. classifiers extracted using adjacent parts of the stream.

2.2 Related Work

Mining data streams has rapidly become an important and challenging research field. As proposed in [12], the available solutions can be classified into *data-based* and *task-based* ones. In the former approaches a data stream is transformed into an approximate smaller-size representation, while task-based techniques employ methods from computational theory to achieve time and space efficient solutions. *Aggregation* [1,2,3], *sampling* [10] or *summarized data structure*, such as histograms [21,17], are popular example of data-based solutions. On the contrary, *approximation algorithms* such as those introduced in [15,10] are examples of task-based techniques.

In the context of data streams classification, two main approaches can be outlined, namely *instance selection* and *ensemble learning*. Very Fast Decision Trees (VFDT) [9] with its improvements [22,14,27] for concept drifting reaction and numerical attributes managing represent examples of instance selection methods. In particular, the Hoeffding bound guarantees that the split attribute chosen using n examples, is the same with high probability as the one that would be chosen using an infinite set of examples. Last et al. [8] propose another strategy using an info-fuzzy technique to adjust the size of a data window. Ensemble learning employs multiple classifiers, extracting new models from scratch and deleting the out-of-date ones continuously. On-line approaches for bagging and boosting are available in [26,7,5]. Different methods are available in [30,34,29,25,11], where an ensemble of weighted-classifiers, including an adaptive genetic programming boosting, as in [11], is employed to cope with concept drifting. None of the two techniques can be assumed to be more appropriate than the other. [5] provides a comparison between different techniques not only in terms of accuracy, but also including computational features, such as memory and time required by each system. By contrast, our approach proposes an ensemble learning that differs from the cited methods since it is designed to concurrently manage different sliding windows, enabling the use of a set of classifiers not necessarily contiguous and constant in time.

3 Adaptive Selective Ensemble

A detailed description of our system is available in [19,18]. In the following subsections, we introduce only the main concepts of our approach highlighting the relations between the requirements outlined in Section 2.1 and the aggregate structures introduced. The proposed structures are primarily conceived to capture evolving data features, and guarantee data reduction at the same time. Ensuring a good trade-off between data reduction, and a powerful representation of all the evolving data factors is a non-trivial task.

3.1 The Snapshot

The snapshot definition implies the naïve Bayes classifier directly. In our model, the streaming training set is partitioned into chunks. Each data chunk is transformed into an approximate more compact form, called *snapshot*.

Definition 2 (Snapshot). *Given a data chunk of k elements, with A attributes and C class values, a snapshot computes the distribution of the values of attribute $a \in A$ with class value c, considering the last k elements arrived:*

$$S_k : C \times A \mapsto freq(a,k,c), \ \forall a \in A, c \in C$$

The following properties are directly derived from Definition 2.

Property 1. Given a stream with C class values and A attributes, a snapshot is a set of $C \times A$ tuples.

x y z	class value
0 1 1	no
0 1 0	yes
0 1 1	no
1 1 1	yes
1 0 0	yes
1 0 1	no
1 1 1	yes
0 0 0	ind
0 0 1	yes
0 0 0	ind

$$\Rightarrow S_{10} = \begin{Vmatrix} (x,2,3) & (y,2,3) & (z,3,2) \\ (x,2,1) & (y,2,1) & (z,0,3) \\ (x,2,0) & (y,2,0) & (z,2,0) \end{Vmatrix} \begin{matrix} yes \\ no \\ ind \end{matrix}$$

(b) The resulting snapshot.

(a) A stream chunk of 10 elements.

Fig. 1. From data stream (a) to snapshot (b)

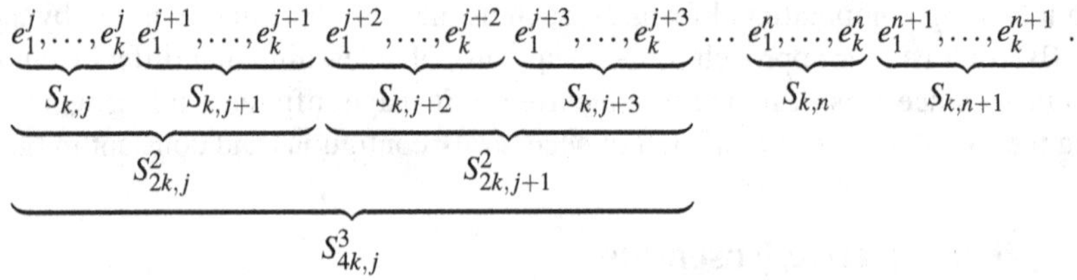

$$\underbrace{e_1^j,\dots,e_k^j}_{S_{k,j}} \underbrace{e_1^{j+1},\dots,e_k^{j+1}}_{S_{k,j+1}} \underbrace{e_1^{j+2},\dots,e_k^{j+2}}_{S_{k,j+2}} \underbrace{e_1^{j+3},\dots,e_k^{j+3}}_{S_{k,j+3}} \cdots \underbrace{e_1^n,\dots,e_k^n}_{S_{k,n}} \underbrace{e_1^{n+1},\dots,e_k^{n+1}}_{S_{k,n+1}} \cdots$$

$$\underbrace{\phantom{S_{k,j}\ S_{k,j+1}}}_{S_{2k,j}^2} \underbrace{\phantom{S_{k,j+2}\ S_{k,j+3}}}_{S_{2k,j+1}^2}$$

$$\underbrace{\phantom{S_{2k,j}^2 \ \ \ S_{2k,j+1}^2}}_{S_{4k,j}^3}$$

Fig. 2. Snapshots and their order

Property 2. Building a snapshot S_k requires k accesses to the data stream. Every element is accessed only once. Computing a snapshot is linear to the k number of elements.

Figure 1 shows an example of snapshot creation. The latter implies only a single access to every stream element. A snapshot is built incrementally accessing the data one by one, and updating the respective counters. Properties 1 and 2 guarantee that a snapshot requires a constant time and memory space, satisfying Requirements 1 and 2.

Snapshots of Higher Order. The only concept of snapshot is not sufficient to guarantee all the features needed for data managing and drift reaction. The concept of high-order snapshot is necessary to maximize data availability for the mining task guaranteeing only one data access.

Definition 3 (High-order Snapshot). *Given an order value $i > 0$, a high-order snapshot, is obtained by summing h snapshots of $i - 1$ order:*

$$S_{h \times k}^i = \sum_{j=1}^h S_{k,j}^{i-1}$$

$$\sum_{j=1}^h \left[freq_j(a,k,c) \right]^{i-1}, \forall a \in A, c \in C$$

where, given a class value c and an attribute a, $\left[freq_j(a,k,c)\right]^{i-1}$ *refers to the distribution of the values of attribute a with class value c of the j-th snapshot of order i − 1.*

Figure 2 shows the relation between snapshots and their order. The aim is to employ a set of snapshots created directly from the stream to build new ones, representing increasingly larger data windows, simply by summing the frequencies of their elements.

A high-order snapshot satisfies Property 1, since it has the same structure of a basic one. Moreover, it further verifies Requirements 3, since the creation of a new high-order snapshot is linear in the number of attributes and class values. The creation of high-order snapshots does not imply any loss of information. This aspect guarantees that a set of different size sliding windows is simultaneously managed by accessing data stream only once, enabling the approach to consider every window as computed directly from the stream.

From a snapshot, or a high-order one, the system extracts an approximated decision tree, or employs the snapshot as naïve Bayes classifier directly.

3.2 The Frame

Snapshots are stored to maximize the number of elements for training classifiers. A model mined from a small set of elements tends to be less accurate than the one extracted from a large data set. If this observation is obvious in "traditional" mining contexts, where training sets are accurately built to maximize the model reliability, in a stream environment this is not necessarily true. Due to concept drifting, a model extracted from a large set of data can be less accurate than the one mined from a small training set. The large data set can include mainly out-of-date concepts.

Snapshots are then stored and managed, based on their order, in a structure called *Frame*. The order of a snapshot defines its level of time granularity. Conceptually similar to *Pyramidal Time Frame* introduced by Aggarwal et al. in [1] and inherited by *logarithmic tilted-time window*, our structure sorts snapshots based on the number of elements from which a snapshot was created.

Definition 4 (Frame). *Given a level value i, and a level capacity j, a frame is a function that, given a pair of indexes (x, y) returns a snapshot of order x and position y:*

$$F_{i,j} : (x, y) \mapsto Snapshot_{x,y}$$

where: $x \in \{0, \ldots, i-1\}$ *and* $y \in \{0, \ldots, j-1\}$.

As shown in Figure 3, level 1 contains snapshots created directly from the stream. Upper levels use the snapshots of the layer immediately lower to create a new one. The maximum number of snapshots available in the frame is constant in time and is defined by the number of levels and the level capacity. For each layer, the snapshot are stored with FIFO policy. The frame memory occupation is constant in time and is linear with the number of snapshots storable in the structure.

3.3 Ensemble Management

The concepts introduced in Section 3.1 and 3.2 are employed to define and manage an ensemble of classifiers. The selective ensemble management is defined by as a four phase approach:

1. For each snapshot S_i^j, a triple (C_i, w_i, b_i) representing the classifier, its weight and the classifier enabling variable b_i is extracted from S_i^j.
2. Since data distribution can change through time, the models currently in the structure are re-weighed with the new data distribution, using a test set of complete data taken from the last portion of the stream directly.
3. Given a level i, every time a new classifier is generated, the system decides if the new model must be inserted in the ensemble based on new data distribution. Apart from the lowest level, where the new one is inserted in any case, the system selects the k_i most promising models, based on the current weight associated to the classifiers, to classify the new data distribution from $k_i + 1$ models correctly.
4. Finally, a set of *active models* is selected, setting the boolean value b_i associated with a C_i as *true*. The set of *active models* is selected based on the value of an activation threshold θ. All the classifiers that differ at most θ to the best classifier with the highest weight are enabled.

The defined approach satisfies Requirements 4 and 5, since it can classify a new instance every time it is required, and can employ a set of not necessarily contiguous classifiers, since it is not necessarily true that every classifier generated through time enters the ensemble, but even in that case it can be disabled.

Figure 4 shows the overall organization of our system. For each level in the frame structure we have a corresponding level in the ensemble. The subdivision of the data aggregation task from the mining aspects makes our approach suitable in distributed/parallel environments as well. One or more components can be employed to manage the concepts of snapshot and frame, while another can manage the ensemble classifier.

level			
1	$S_{k,j+2}$	$S_{k,j+1}$	$S_{k,j}$
2	$S_{2k,l+2}^2$	$S_{2k,l+1}^2$	$S_{2k,l}^2$
3	$S_{4k,g+2}^3$	$S_{4k,g+1}^3$	$S_{4k,g}^3$
4		$\cdots$	
$\cdots$			
i	$S_{2^{i-1}k,n+2}^i$	$S_{2^{i-1}k,n+1}^i$	$S_{2^{i-1}k,n}^i$

Fig. 3. The frame structure

3.4 Adaptive Behavior

As proposed in Section 3.3 our approach has two key factors influencing its behavior, the weight measure to employ and the selection of the θ value. If in the literature, several weight measures, mainly related to classifier accuracy, are available and guarantee a

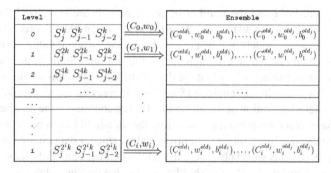

Fig. 4. The overall system architecture

Activation Threshold Algorithm.

```
1:  if (oneModel() = true) then
2:      θ ← 0.00; activation threshold initialized
3:      oldActModels ← 1
4:      return
5:  end if
6:  actModels ← getActiveModel(θ)
7:  if (actModels > oldActModels) then
8:      θ ← θ + 0.01; increment threshold value
9:  else if (actModels < oldActModels) then
10:     θ ← θ div 2; decrement threshold value
11: end if
12: oldActModels ← actModels
13: return
```

Fig. 5. Pseudo-code of the activation threshold algorithm

good reliability of the system, the θ threshold represents the real key factor for the quality of our approach.

In our experiments [18], we noticed that the reliability of the system is heavily influenced by the θ value. As we shall present in Section 4.3, independently from the data set employed, activation values which are too high (or too small) decrease the predictive power of the ensemble. On the one hand, in case of relatively stable data, small activation threshold values limit the use of large sets of classifiers. On the contrary, large threshold values damage the selective ensemble in the case of concept drifts. In the cited experiments, the θ value was fixed by the user and it did not change through time. Only our experience and the experimental results drove the selection of the right value.

In this work we introduce an adaptive approach for varying the value of the activation threshold through time, thus influencing the overall behavior of the entire system, based on data change reaction. The basic idea of the adaptive method is similar to the additive-increase / multiplicative-decrease algorithm adopted by TCP Internet protocol for managing the transfer rate value used in TCP congestion avoidance.

The pseudo-code of the method for managing the activation threshold is proposed in Figure 5. The algorithm is quite simple. When the first model is inserted in the structure, the activation threshold and the number of active models are initialized (Steps 1-5). Successively, every time a new model is inserted in the ensemble, the procedure at Step 6 computes how many models will be activated with the current θ value. If the number of models potentially activatable is higher than the old one (Step 7), the threshold is increased. This situation happens when the data distribution remains stable and the new inserted model is immediately enabled (Step 7-8). Increasing the threshold value, we can obtain a better exploitation of the ensemble. On the contrary, if the number of active models decreases from the previous invocation, the threshold has to be decreased. It is useless and dangerous to maintain the current value, since a data change might be in progress (Step 9-10). It is worth observing that, if the number of models does not change between the two invocations, the threshold does not change, since there is no evidence of model improving or data change.

From a computational point of view the algorithm does not introduce appreciable overhead. Only the *getActiveModel()* procedure requires to access the ensemble structure. If we consider n as the number of classifiers storable in the ensemble, the complexity of the algorithm is linear in $O(n)$.

The experimental section demonstrates that our system is no more heavily influenced by θ value, since it changes automatically, adapting it to data distribution.

4   Comparative Experimental Evaluation

4.1   Data Sets

Several synthetic data sets and a real one were introduced in our experiments. This kind of data enables an exhaustive investigation about the reliability of the systems involving different scenarios. The data behavior can be described exactly, characterizing the number of concept drifts, the rate between a change to another and the number of irrelevant attributes, or the percentage of noisy data.

LED24: Proposed by Breiman et al. in [6], this generator creates data for a display with 7 LEDs. In addition to the 7 necessary attributes, 17 irrelevant boolean attributes with random values are added, and 10 percent of noise is introduced, to make the solution of the problem harder. This type of data generates only stable data sets.

Stagger: Introduced by Schlimmer and Granger in [28], this problem consists of three attributes, namely colour ∈ {*green, blue, red*}, shape ∈ {*triangle, circle, rectangle*}, and size ∈ {*small, medium, large*}, and a class y ∈ {0,1}. In its original formulation, the training set includes 120 instances and consists of three target concepts occurring every 40 instances. The first set of data is labeled according to the concept color = *red* ∧ size = *small*, while the others include color = *green* ∨ shape = *circle* and size = *medium* ∨ size = *large*. For each training instance, a test set of 100 elements is randomly generated according to the current concept.

cHyper: Introduced in [7], a data set is generated by using a n-dimensional unit hypercube, and an example x is a vector of n-dimensions $x_i \in [0,1]$. The class boundary is a hyper-sphere of radius r and center c. Concept drifting is simulated by changing the c position by a value Δ in a random direction. This data set generator

Table 1. Description of data sets

dataSet	#inst	#attrs	#irrAttrs	#classes	%noise	#drifts
LED24$_{a/b/test}$	10k / 100k / 25k	24	17	10	10%	none
Hyper$_{a/b/test}$	10k / 100k / 25k	15	0	2	0	none
Stagger$_{a/test}$	1200 / 120k	9	0	2	0	3 (every 400) / (every 40k)
Stagger$_{b/test}$	12k / 1200k	9	0	2	0	3 (every 4k) / (every 400k)
cHyper$_{a/b/c}$	10k / 100k / 1000k	15	0	2	10%	20 (every 500) / (every 5k) / (every 50k)
cHyper$_{test}$	250k	15	0	2	10%	20 (every 12.5k)
Cyclic	600k	25	0	2	5%	15×4 (every 10k)
Cyclic$_{test}$	150k	25	0	2	5%	15×4 (every 2.5k)

introduces noise by randomly flipping the label of a tuple with a given probability. Two additional data sets, namely Hyper and Cyclic are generated using this approach. Hyper does not consider any drifts, while Cyclic proposes the problem of periodic recurring concepts.

KddCup99: this real data set concerns the significant problem of automatic and real-time detection of cyber attacks [31]. The data includes a series of network connections collected from two weeks of LAN network traffic. Each record can either correspond to a normal connection or an intrusive one. Each connection is represented by 42 attributes (34 numerical), such as the *duration* of the connection, the number of *bytes transmitted*, and the *type of protocol* used, e.g. tcp, udp. The data contains 23 training attack types, that can be further aggregated into four categories, namely *DOS, R2L,U2R*, and *Probing*. Due to its instable nature, KddCup99 is largely employed to evaluate several data streams classification systems, including [3,16]

The features of the data sets actually employed are reported in Table 1. The stable LED24 and Hyper are useful for testing whether the mechanism for change reaction has implications for the reliability of the systems. The evolving data sets test different features of a stream classification system. The Stagger problem verifies, if all the systems can cope with concept drift, without considering any problem dimensionality. Then, the problem of learning in the presence of concept drifting is evaluated with the other data sets, also considering a huge quantity of data with cHyper.

4.2   Systems

Different popular stream ensemble methods are introduced in our experiments. All the systems expect the data streams to be divided into chunks based on a well-defined value. All the approaches are implemented in Java 1.6 with MOA [32] and WEKA libraries [33] for the implementation of the basic learners and employ complete non-approximate data for the mining task.

Fix: This approach is the simplest one. It considers a fixed set of classifiers, managed as a FIFO queue. Every classifier is unconditionally inserted in the ensemble, removing the oldest one, when the ensemble is full.

SEA: A complete description and evaluation of this system can be found in [30]. In this case classifiers are not deleted indiscriminately. Their management is based on a weight measure related to model reliability. This method represents a special case of our selective ensemble, where only one level is defined.

DWM: This system is introduced in [24,25]. The approach implemented here considers a set of data as input to the algorithm, and a batch classifier as the basic one. A weight management is introduced, but differently from SEA, every classifier has a weight associated with it, when it is created. Every time the classifier makes a mistake, its weight decreases.

Oza: This system implements the online bagging method of Oza and Russell [26] with the addition of the ADWIN technique [5] as a change detector and as estimator of the weight of the boosting method.

Single: This approach employs an incremental single model with EDDM [13,4] techniques for drift detection. Both Oza and Single were tested using ASHoeffdingTree and naïve Bayes models available in MOA.

4.3 Results

All the experiments were run on a PC with Intel E8200 DualCore with 4Gb of RAM, employing Linux Fedora 10 (kernel 2.6.27) as operating system. Our experiments consider a frame with 8 levels of capacity 3. Every high-order snapshot is built by adding 2 snapshots. This frame size is large enough to consider snapshots that represent big portions of data at higher-levels. For each level, an ensemble of 8 classifiers was used. The tests were conducted comparing the use of the naïve Bayes (NB), and the decision tree (DT) as base classifiers. In all the cases, we compare our Selective Ensemble (SE) (with fixed model activation threshold set to 0.1 and 0.25) with our Adaptive Selection Ensemble ASE. For each data generator, a collection of 100 training sets (and corresponding test sets) are randomly generated with respect to the features outlined in Table 1. Every system is run, and the average accuracy and 95% of interval confidence are reported. Each test consists of a set of 100 observations. All the statistics reported are computed according to the results obtained.

Results with Stable Data Sets. The results obtained with stable data sets confirm that the drift detection approach provided by each system does not heavily influence its overall accuracy. With LED24 and Hyper problems, all the systems reach a quite accurate result. Table 2 reports the results obtained with Hyper data sets using the naïve Bayes approach. These results can be compared with the ones provided in Table 3 in Section 4.3, where the concept drifting problem is added to the same type of data.

It is worth observing that there are no significant differences between the results obtained by SE approach, varying the model activation threshold. The new ASE approach provides a result in line with the best ones. The adaptive behavior mechanism does not negatively influence the reliability of the system in the case of stable data streams. On the contrary, the new approach enables a better ensemble exploitation.

Moreover, Table 2 highlights that Single model requires a large quantity of data to provide a good performance. Finally, Fix_{64} and SEA_{64} provide good results that, compared with the ones obtained by the same systems analyzing the cHyper and Cyclic

Table 2. Results using naïve Bayes with the Hyper problem

| | Hyper_a / Hyper_b | | |
	avg	std dev	conf
ASE	92.74 / 93.93	1.92 / 2.88	0.38 / 0.49
$SE_{0.1}$	92.72 / 93.92	2.20 / 2.52	0.43 / 0.50
$SE_{0.25}$	92.70 / 93.91	2.22 / 2.53	0.43 / 0.50
Fix_{64}	91.82 / 92.35	2.54 / 2.98	0.50 / 0.58
SEA_{64}	92.97 / 94.44	1.80 / 1.64	0.50 / 0.32
DWM_{64}	91.82 / 92.76	2.12 / 2.74	0.42 / 0.54
Oza_{64}	92.39 / 93.73	2.30 / 0.40	0.45 / 0.08
Single	90.04 / 92.68	3.25 / 2.82	0.64 / 0.55

problems, demonstrate that these kinds of approaches guarantee appreciable results only with a quite stable phenomenon. They do not provide a fast reaction to concept drift, since the number of models involved in the classification task is constant in time, and when a drift occurs, they have to change a large part of the models, before classifying new concepts correctly.

Table 3. Overall results with the cHyper problem

| | cHyper_a / cHyper_b / cHyper_c - decision tree | | |
	avg	std dev	conf
ASE	83.58 / 88.72 / 93.19	0.51 / 0.40 / 0.28	0.10 / 0.08 / 0.06
$SE_{0.1}$	84.05 / 89.43 / 93.09	0.49 / 0.40 / 0.32	0.10 / 0.08 / 0.06
$SE_{0.25}$	78.42 / 86.10 / 91.86	0.86 / 0.35 / 0.23	0.17 / 0.07 / 0.23
Fix_{64}	70.26 / 82.02 / 90.62	2.58 / 1.23 / 0.13	0.51 / 0.24 / 0.13
SEA_{64}	70.26 / 82.14 / 90.04	2.58 / 1.10 / 0.14	0.51 / 0.22 / 0.14
DWM_{64}	77.75 / 85.18 / 92.65	1.94 / 0.60 / 0.14	0.38 / 0.04 / 0.14
Oza_{64}	81.99 / 89.60 / 92.40	0.97 / 0.37 / 0.25	0.19 / 0.07 / 0.25
Single	81.50 / 87.85 / 89.99	1.60 / 0.70 / 0.34	0.31 / 0.14 / 0.34

| | cHyper_a / cHyper_b / cHyper_c - naïve Bayes | | |
	avg	std dev	conf
ASE	87.52 / 92.23 / 95.94	0.38 / 0.43 / 0.33	0.09 / 0.08 / 0.06
$SE_{0.1}$	87.62 / 92.62 / 95.98	0.42 / 0.43 / 0.47	0.08 / 0.09 / 0.09
$SE_{0.25}$	79.90 / 86.80 / 92.14	0.83 / 0.40 / 0.22	0.16 / 0.08 / 0.22
Fix_{64}	73.72 / 83.69 / 94.16	2.60 / 1.35 / 0.40	0.51 / 0.26 / 0.40
SEA_{64}	73.72 / 84.23 / 94.78	2.60 / 1.27 / 0.31	0.51 / 0.25 / 0.31
DWM_{64}	85.93 / 92.18 / 95.63	1.76 / 0.18 / 0.38	0.35 / 0.04 / 0.38
Oza_{64}	80.01 / 87.31 / 89.78	1.23 / 0.54 / 0.56	0.24 / 0.11 / 0.56
Single	81.25 / 89.47 / 93.34	2.02 / 0.87 / 0.84	0.40 / 0.17 / 0.84

Results with Evolving Data Sets. Table 3 reports the overall results obtained analyzing the cHyper problem, considering both decision tree and naïve Bayes models. Differently from the results obtained with stable data sets, the active model threshold influences the overall results. Varying the value from 0.1 to 0.25, and especially considering cHyper_a and cHyper_b, SE system presents a difference even larger than 6% between the two values. On the contrary, our ASE approach provides an accuracy in line with the best one, even considering standard deviation. This demostrates that, without knowing the ideal threshold value for model activation, our ASE approach represents the right solution to the different situations involved in a stream scenario, and simulated by the three cases of the cHyper problem. As stated in the previous section, it is worth observing the poor performances of Fix and SEA in the case of evolving data. These obsevations are further validated by the results obtained with the Stagger problem, that essentially follow the ones proposed in Table 3.

Finally, Table 4 outlines the resources required by the systems. The memory requirements were tested using NetBeans 6.8 Profiler. We can state that Single requires less memory than ensemble methods, which need a quantity of memory that is essentially linear with respect to the number of classifiers stored in the ensemble. The different nature of the two classes of systems influences this value. The average memory required by our system is slightly higher than the others, since our system manages two different structures, as suggested at the end of Section 3.3. The run time behavior confirms this trend. In this case the drift detection approach influences the execution time of a method. Let us compare the bagging method Oza with respect to DWM, SEA_{64} and ASE. These tests highlight that incremental single model systems are faster than ensemble ones, since they have to update only one model. On the contrary, considering the accuracy, single model systems rarely provide best average values. Finally, Oza guarantees an appreciable reliability with every data set, but its execution time is definitely higher than the others.

Table 4. $cHyper_c$ time and memory required

	decision tree		naïve Bayes	
	avg used heap (KB)	run time (sec.)	avg used heap (KB)	run time (sec.)
ASE	9276	82.40	7572	27.42
SE	9233	80.80	7894	27.45
Fix_{64}	8507	47.54	5317	23.82
SEA_{64}	7980	152.07	5371	97.76
DWM_{64}	5111	77.56	5137	21.21
Oza_{64}	10047	393.93	6664	290.24
Single	5683	11.54	5399	8.26

Figure 6a shows the results obtained considering the Cyclic problem. The latter are presented considering the naïve Bayes approach and analyzing different rates between the chunk size and the elements to classify. As shown in Figure 6a, even in this case, our ASE approach is in-line with the $SE_{0.1}$ and better than the others. Since this problem presents recurring concepts, our approach can exploit the selective ensemble better than the others, since some models which are currently out of context are not deleted by the system, but simply disabled. If a concept becomes newly valid, the model can be reactivated. This behavior is still valid, even in the case of the adaptive approach.

We conclude this section, proposing the results obtained analyzing the KddCup99 problem, and considering the decision tree approach. In this case, only an execution is run considering the whole data set. As shown is Figure 6b, the approaches employing an advanced method to keep track of concept drift propose an accuracy in line with the ones obtained by Aggarwal et al. in [3]. Even in this case, ASE proposes a performance comparable with $SE_{0.1}$, showing that the adaptive behavior guarantees a good level of reliability. The run time requirements needed for analysing KddCup99 dataset are in line with the ones proposed in Table 4.

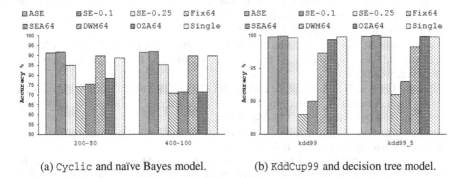

(a) Cyclic and naïve Bayes model.   (b) KddCup99 and decision tree model.

Fig. 6. Average accuracy with Cyclic ans KddCup99 problems

5   Conclusions

Starting from the requirements constrained by the unpredictable nature of streaming data, this paper proposed an adaptive selective ensemble approach for data streams classification. The aim of this work is represented by a new adaptive behavior for an ensemble model selection approach. The new feature enables the system to automatically adapt the active model threshold to the current stream status. The idea is not providing a fixed value of the threshold set up experimentally, but letting its value automatically adapt to the data flow changes. When data are quite stable, the system can use a large part of the ensemble. On the contrary, when data changes the threshold, it has to be reduced to disable the not up-to-date models. The preliminary results show that, with respect to the use of a fixed threshold, our adaptive algorithm provides a slightly worse performance than the ones using the best value of the threshold. Unfortunately, the choice of the best value is not always feasible, and if a wrong selection is made, the system loses its precision. Our adaptive approach does not require any assumption about active model values and displays good adaptation to the different scenarios. This work represents a first step to guarantee a system completely adaptable to the different streaming factors. As future works, our aims are to test our adaptive model in a real stream application with real data. Moreover, we are currently studying the introduction of runtime monitoring tools for automatically adapting our system, e.g varying the number of frame levels, or the models available for each layer, dynamically considering memory consumption and time response constraints.

References

1. Aggarwal, C.C., Han, J., Wang, J., Yu, P.: A framework for clustering evolving data streams. In: Proceedings of the 2003 International Conference on Very Large Data Bases (VLDB 2003), Berlin, Germany, pp. 81–92 (2003)
2. Aggarwal, C.C., Han, J., Wang, J., Yu, P.: A framework for projected clustering of high dimensional data streams. In: Proceedings of the 2004 International Conference on Very Large Data Bases (VLDB 2004), Toronto, Canada, pp. 852–863 (2004)

3. Aggarwal, C.C., Han, J., Wang, J., Yu, P.: On demand classification of data streams. In: Proceedings of the 10th International Conference on Knowledge Discovery and Data Mining (KDD 2004), Seattle, WA, pp. 503–508 (2004)
4. Baena-Garcia, M., del Campo-Avila, J., Fidalgo, R., Bifet, A., Ravalda, R., Morales-Bueno, R.: Early drift detection method. In: International Workshop on Knowledge Discovery from Data Streams (2006)
5. Bifet, A., Holmes, G., Pfahringer, B., Kirby, R., Gavaldá, R.: New ensemble methods for evolving data streams. In: Proceedings of the 15th International Conference on Knowledge Discovery and Data Mining, pp. 139–148 (2009)
6. Breiman, L., Friedman, J., Olshen, R., Stone, C.: Classification and Regression Trees. Wadsworth International Group, Belmont (1984)
7. Chu, F., Zaniolo, C.: Fast and Light Boosting for Adaptive Mining of Data Streams. In: Dai, H., Srikant, R., Zhang, C. (eds.) PAKDD 2004. LNCS (LNAI), vol. 3056, pp. 282–292. Springer, Heidelberg (2004)
8. Cohen, L., Avrahami, G., Last, M., Kandel, A.: Info-fuzzy algorithms for mining dynamic data streams. Applied Soft Computing 8(4), 1283–1294 (2008)
9. Domingos, P., Hulten, G.: Mining high-speed data streams. In: Proceedings of the 6th International Conference on Knowledge Discovery and Data Mining (KDD 2000), Boston, MA, pp. 71–80 (2000)
10. Domingos, P., Hulten, G.: A general method for scaling up machine learning algorithms and its application to clustering. In: Proceedings of the 18th International Conference on Machine Learning (ICML 2001), Williamstown, MA, pp. 106–113 (2001)
11. Folino, G., Pizzuti, C., Spezzano, G.: Mining Distributed Evolving Data Streams using Fractal GP Ensembles. In: Ebner, M., O'Neill, M., Ekárt, A., Vanneschi, L., Esparcia-Alcázar, A.I. (eds.) EuroGP 2007. LNCS, vol. 4445, pp. 160–169. Springer, Heidelberg (2007)
12. Gaber, M.M., Zaslavsky, A., Krishnaswamy, S.: Mining data streams: a review. ACM SIGMOD Records 34(2), 18–26 (2005)
13. Gama, J., Medas, P., Castillo, G., Rodrigues, P.: Learning with drift detection. In: SBIA Brazilian Symposium on Artificial Intelligence, pp. 286–295 (2004)
14. Gama, J., Pinto, C.: Discretization from data streams: applications to histograms and data mining. In: Proceedings of the 2006 ACM Symposium on Applied Computing (SAC 2006), Dijon, France, pp. 662–667 (2006)
15. Gama, J., Fernandes, R., Rocha, R.: Decision trees for mining data streams. Intelligent Data Analysis 10(1), 23–45 (2006)
16. Gao, J., Fan, W., Han, J., Yu, P.S.: On appropriate assumptions to mine data streams: Analysis and practice. In: Proceedings of the 7th IEEE International Conference on Data Mining (ICDM 2007), Omaha, NE, pp. 143–152 (2007)
17. Gilbert, A., Guha, S., Indyk, P., Kotidis, Y., Muthukrishnan, S., Strauss, M.: Fast, small-space algorithms for approximate histogram maintenance. In: Proceedings of the 2002 Annual ACM Symposium on Theory of Computing (STOC 2002), Montreal, Quebec, Canada, pp. 389–398 (2002)
18. Grossi, V.: A New Framework for Data Streams Classification. Ph.D. thesis, Supervisor Prof. Franco Turini, University of Pisa (2009), http://etd.adm.unipi.it/theses/available/etd-11242009-124601/
19. Grossi, V., Turini, F.: Stream mining: a novel architecture for ensemble based classification. Accepted as full paper by Internl. Journ. of Knowl. and Inform. Sys., forthcoming, draft (2011), www.di.unipi.it/~vgrossi
20. Grossi, V., Turini, F.: A new selective ensemble approach for data streams classification. In: Proceedings of the 2010 International Conference in Artificial Intelligence and Applications (AIA 2010), Innsbruck, Austria, pp. 339–346 (2010)

21. Guha, S., Koudas, N., Shim, K.: Data-streams and histograms. In: Proceedings of the 2001 Annual ACM Symposium on Theory of Computing (STOC 2001), Heraklion, Crete, Greece, pp. 471–475 (2001)

22. Hulten, G., Spencer, L., Domingos, P.: Mining time changing data streams. In: Proceedings of the 7th International Conference on Knowledge Discovery and Data Mining (KDD 2001), San Francisco, CA, pp. 97–106 (2001)

23. Klinkenberg, R.: Learning drifting concepts: Example selection vs. example weighting. Intelligent Data Analysis 8, 281–300 (2004)

24. Kolter, J.Z., Maloof, M.A.: Using additive expert ensembles to cope with concept drift. In: Proceedings of the 22nd International Conference on Machine learning (ICML 2005), Bonn, Germany, pp. 449–456 (2005)

25. Kolter, J.Z., Maloof, M.A.: Dynamic weighted majority: An ensemble method for drifting concepts. Journal of Machine Learning Research 8, 2755–2790 (2007)

26. Oza, N.C., Russell, S.: Online bagging and boosting. In: Proceedings of 8th International Workshop on Artificial Intelligence and Statistics (AISTATS 2001), Key West, FL, pp. 105–112 (2001)

27. Pfahringer, B., Holmes, G., Kirkby, R.: Handling Numeric Attributes in Hoeffding Trees. In: Washio, T., Suzuki, E., Ting, K.M., Inokuchi, A. (eds.) PAKDD 2008. LNCS (LNAI), vol. 5012, pp. 296–307. Springer, Heidelberg (2008)

28. Schlimmer, J.C., Granger, R.H.: Beyond incremental processing: Tracking concept drift. In: Proceedings of the 5th National Conference on Artificial Intelligence, Menlo Park, CA, pp. 502–507 (1986)

29. Scholz, M., Klinkenberg, R.: An ensemble classifier for drifting concepts. In: Proceeding of 2nd International Workshop on Knowledge Discovery from Data Streams, in Conjunction with ECML-PKDD 2005, Porto, Portugal, pp. 53–64 (2005)

30. Street, W.N., Kim, Y.: A streaming ensemble algorithm (SEA) for large-scale classification. In: Proceedings of the 7th International Conference on Knowledge Discovery and Data Mining (KDD 2001), San Francisco, CA, pp. 377–382 (2001)

31. The UCI KDD: University of California: KDD Cup 1999 Data,
http://kdd.ics.uci.edu/databases/kddcup99/kddcup99.html

32. The University of Waikato: MOA: Massive Online Analysis (August 2009),
http://www.cs.waikato.ac.nz/ml/moa

33. The University of Waikato: Weka 3: Data Mining Software in Java, Version 3.6,
http://www.cs.waikato.ac.nz/ml/weka

34. Wang, H., Fan, W., Yu, P.S., Han, J.: Mining concept-drifting data streams using ensemble classifiers. In: Proceedings of the 9th International Conference on Knowledge Discovery and Data Mining (KDD 2003), Washington, DC, pp. 226–235 (2003)

On kNN Classification
and Local Feature Based Similarity Functions

Giuseppe Amato and Fabrizio Falchi

ISTI-CNR, via G. Moruzzi 1, 56124 Pisa, Italy
{giuseppe.amato,fabrizio.falchi}@isti.cnr.it
http://www.isti.cnr.it

Abstract. In this paper we consider the problem of image content recognition and we address it by using local features and kNN based classification strategies. Specifically, we define a number of image similarity functions relying on local features comparing their performance when used with a kNN classifier. Furthermore, we compare the whole image similarity approach with a novel two steps kNN based classification strategy that first assigns a label to each local feature in the document to be classified and then uses this information to assign a label to the whole image. We perform our experiments solving the task of recognizing landmarks in photos.

Keywords: Image classification, Recognition, Landmarks, Pattern recognition, Machine learning, Local features.

1 Introduction

Image content recognition is a very important issue that is being studied by many scientists worldwide. In fact, with the explosion of the digital photography, during the last decade, the amount of digital pictures available on-line and off-line has extremely increased. However, many of these pictures remain unannotated and are stored with generic names on personal computers and on on-line services. Currently, there are no tools and effective technologies to help users in searching for pictures by real content, when they are not explicitly annotated. Therefore, it is becoming more and more difficult for users to retrieve even their own pictures.

A picture contains a lot of implicit conceptual information that is not yet possible to exploit entirely and effectively. Automatically content based image recognition opens up opportunities for new advanced applications. For instance, pictures themselves might be used as queries on the web. An example in this direction is the service "Google Goggles" [11] recently launched by Google, that allows you to obtain information about a monument through your smartphone using this paradigm.

Note that, even if many smartphones and cameras are equipped with a GPS and a compass, the geo-reference obtained with this is not enough to infer what the user is actually aiming at. Content analysis of the picture is still needed to determine more precisely the user query or the annotation to be associated with a picture. A promising approach toward image content recognition is the use of classification techniques to associate images with classes (labels) according to their content. For instance, if an

J. Filipe and A. Fred (Eds.): ICAART 2011, CCIS 271, pp. 224–239, 2013.

image contains a car, it might be automatically associated with the class *car* (labelled with the label *car*).

In this paper we study the problem of image content recognition by using SIFT [14] and SURF [5] local features, to represent image visual content, and kNN based classifiers to decide about the presence of conceptual content. In more details we will define 20 different functions that measure similarity between images. These functions are defined using various options and combinations of local feature matching and similarities. Some of them also take into consideration geometric properties of the matching local features. These functions are used in combination of a standard Single-label Distance-Weighted kNN algorithm. In addition we also propose a new classification algorithm that extend the traditional kNN classifiers by making direct use of similarity between local features, rather than similarity between entire images. We will see that the similarity functions that also make use of geometric considerations offer a better performance than the others. However, the new kNN based classifier that exploit directly the similarity between local features has an higher performance even without using geometric information.

The paper is organized as follows. In Section 3 we briefly introduce local features. In Section 4 we present various iamge similarity features relying on local features to be used with a kNN classification algorithm. Section 5 propose a novel classification approach. Finally, Sections 6 and 7 presents the experimental results. An earlier version of this paper has been presented at the Third International Conference on Agents and Artificial Intelligence [1].

2 Related Work

The first approach to recognizing location from mobile devices using image-based web search was presented in [17]. Two image matching metrics were used: energy spectrum and wavelet decompositions. Local features were not tested. In the last few years the problem of recognizing landmarks have received growing attention by the research community. In [16] methods for placing photos uploaded to Flickr on the World map was presented. In the proposed approach the images were represented by vectors of features of the tags, and visual keywords derived from a vector quantization of the SIFT descriptors. In [13] a combination of context- and content-based tools were used to generate representative sets of images for location-driven features and landmarks. Visual information is combined with the textual metadata while we are only considering content-based classification. In [19], Google presented its approach to building a web-scale landmark recognition engine. Most of the work reported was used to implement the Google Goggles service [11]. The approach makes use of the SIFT feature. The recognition is based on best matching image searching, while our novel approach is based on local features classification. In [7] a survey on mobile landmark recognition for information retrieval is given. Classification methods reported as previously presented in the literature include SVM, Adaboost, Bayesian model, HMM, GMM. The kNN based approach which is the main focus of this paper is not reported in that survey. In [9], various MPEG-7 descriptors have been used to build kNN classifier committees. However local features were not considered.

In [6] the effectiveness of NN image classifiers has been proved and an innovative approach based on Image-to-Class distance that is similar in spirit to our approach has been proposed.

3 Local Features

The approach described in this paper focuses on the use of image local features. Specifically, we performed our tests using the SIFT [14] and SURF [5] local features. In this section, we briefly describe both of them.

The Scale Invariant Feature Transformation (SIFT) [14] is a representation of the low level image content that is based on a transformation of the image data into scale-invariant coordinates relative to local features. Local feature are low level descriptions of keypoints in an image. Keypoints are interest points in an image that are invariant to scale and orientation. Keypoints are selected by choosing the most stable points from a set of candidate location. Each keypoint in an image is associated with one or more orientations, based on local image gradients. Image matching is performed by comparing the description of the keypoints in images. For both detecting keypoints and extracting the SIFT features we used the public available software developed by David Lowe[1].

The basic idea of Speeded Up Robust Features (SURF) [5] is quite similar to SIFT. SURF detects some keypoints in an image and describes these keypoints using orientation information. However, the SURF definition uses a new method for both detection of keypoints and their description that is much faster still guaranteeing a performance comparable or even better than SIFT. Specifically, keypoint detection relies on a technique based on a approximation of the Hessian Matrix. The descriptor of a keypoint is built considering the distortion of Haar-wavelet responses around the keypoint itself. For both, detecting keypoints and extracting the SURF features, we used the public available noncommercial software developed by the authors [2].

4 Image Similarity Based Classifier

In this section we discuss how traditional kNN classification algorithms can be applied to the task of classifying images described by local features, as for instance SIFT or SURF. In particular, we define 20 image similarity measures based on local features description. These will be later on compared to the new classification strategy that we propose in Section 5.

4.1 Single-Label Distance-Weighted kNN

Given a set of documents D and a predefined set of *classes* (also known as *labels*, or *categories*) $C = \{c_1, \ldots, c_m\}$, *single-label document classification* (SLC) [8] is the task of automatically approximating, or estimating, an unknown *target function* $\Phi : D \to C$, that describes how documents ought to be classified, by means of a function $\hat{\Phi} : D \to C$, called the *classifier*, such that $\hat{\Phi}$ is an approximation of Φ.

---

[1] http://people.cs.ubc.ca/~lowe/

[2] http://www.vision.ee.ethz.ch/~surf

A popular SLC classification technique is the *Single-label distance-weighted kNN*. Given a training set Tr containing various examples for each class c_i, it assigns a label to a document in two steps. Given a document d_x (an image for example) to be classified, it first executes a kNN search between the objects of the *training set*. The result of such operation is a list $\chi^k(d_x)$ of labelled documents d_i belonging to the *training set* ordered with respect to the decreasing values of the similarity $s(d_x, d_i)$ between d_x and d_i. The label assigned to the document d_x by the classifier is the class $c_j \in C$ that maximizes the sum of the similarity between d_x and the documents d_i, labelled c_j, in the kNN results list $\chi^k(d_x)$

Therefore, first a score $z(d_x, c_i)$ for each label is computed for any label $c_i \in C$:

$$z(d_x, c_j) = \sum_{d_i \in \chi^k(d_x)\,:\,\Phi(d_i)=c_j} s(d_x, d_i) \,.$$

Then, the class that obtains the maximum score is chosen:

$$\hat{\Phi}^s(d_x) = \arg\max_{c_j \in C} z(d_x, c_j) \,.$$

It is also convenient to express a degree of confidence on the answer of the classifier. For the *Single-label distance-weighted kNN* classifier described here we defined the confidence as 1 minus the ratio between the *score* obtained by the second-best label and the best label, i.e,

$$\nu_{doc}(\hat{\Phi}^s, d_x) = 1 - \frac{\arg\max\limits_{c_j \in C - \hat{\Phi}^s(d_x)} z(d_x, c_j)}{\arg\max\limits_{c_j \in C} z(d_x, c_j)} \,.$$

This classification confidence can be used to decide whether or not the predicted label has an high probability to be correct.

4.2   Image Similarity

In order the kNN search step to be executed, a similarity function between images should be defined. Global features, generally, are defined along with a similarity (or a distance) function. Therefore, similarity between images, is computed as the similarity between the corresponding global features. On the other hand, a single image has several local features. Therefore, computing the similarity between two images requires combining somehow the similarities between their numerous local features.

In the following we define a function for computing similarity between images on the basis of their local features that is derived from the work presented in [14]. In the experiments, at the end of this paper, we will compare the performance of the similarity function, when used with the *single-label distance-weighted kNN* classification technique, against the local feature based classification algorithm proposed in Section 5.

Local Feature Similarity. The Computer Vision literature related to local features, generally uses the notion of distance, rather than that of similarity. However in most cases a similarity function $s()$ can be easily derived from a distance function $d()$. For both SIFT and SURF the Euclidean distance is typically used as measure of dissimilarity between two features [14,5].

Let $d(p_1, p_2) \in [0, 1]$ be the normalized distance between two local features p_1 and p_2. We can define the similarity as:

$$s(p_1, p_2) = 1 - d(p_1, p_2)$$

Obviously $0 \leq s(p_1, p_2) \leq 1$ for any p_1 and p_2.

Local Features Matching. A useful aspect that is often used when dealing with local features is the concept of local feature matching. In [14], a distance ratio matching scheme was proposed that has also been adopted by [5] and many others. Let's consider a local feature p_x belonging to an image d_x (i.e. $p_x \in d_x$) and an image d_y. First, the point $p_y \in d_y$ closest to p_x (in the remainder $NN_1(p_x, d_y)$) is selected as candidate match. Then, the distance ratio $\sigma(p_x, d_y) \in [0, 1]$ of closest to second-closest neighbors of p_x in d_y is considered. The distance ratio is defined as:

$$\sigma(p_x, d_y) = \frac{d(p_x, NN_1(p_x, d_y))}{d(p_x, NN_2(p_x, d_y))}$$

Finally, p_x and $NN_1(p_x, d_y)$ are considered matching if the distance ratio $\sigma(p_x, d_y)$ is smaller than a given threshold. Thus, a function of matching between $p_x \in d_x$ and an image d_y is defined as:

$$m(p_x, d_y) = \begin{cases} 1 \text{ if } \sigma(p_x, d_y) < c \\ 0 \text{ otherwise} \end{cases}$$

In [14], $c = 0.8$ was proposed reporting that this threshold allows to eliminate 90% of the false matches while discarding less than 5% of the correct matches. In Section 7 we report an experimental evaluation of classification effectiveness varying c that confirms the results obtained by Lowe. Please note, that this parameter will be used in defining the image similarity measure used as a baseline and in one of our proposed local feature based classifiers.

For Computer Vision applications, the distance ratio described above is used for selecting good candidate matches. More sophisticated algorithms are then used to select actual matches from the selected ones considering geometric information as scale, orientation and coordinates of the interest points. In most of the cases a Hough transform [3] is used to search for keys that agree upon a particular model pose. To avoid the problem of boundary effects in hashing, each match is hashed into the 2 closest bins giving a total of 16 entries for each hypothesis in the hash table. This method has been proposed for SIFT [14] and is very similar to the weak geometry consistency check used in [12].

Thus, we define the set $M_h(d_x, d_y)$ as the matching points in the most populated entry in the Hash table containing the Hough transform of the matches in d_y obtained using the distance ratio criteria.

4.3   Similarity Measures

In this section, we define 5 different image similarity measures approaches and 4 different versions of each of them for a total of 20 measures.

1-NN Similarity Average – s^1. The simplest similarity measure only consider the closest neighbor for each $p_x \in d_x$ and its distance from the query point p_x. The similarity between two documents d_x and d_y can be defined as the average similarity between the local features in d_x and their closest neighbors in d_y. Thus, we define the *1-NN Similarity Average* as (for simplicity, we indicate the number of local features in an image d_x as $|d_x|$):

$$s^1(d_x, d_y) = \frac{1}{|d_x|} \sum_{p_x \in d_x} \max_{p_y \in d_y} (s(p_x, p_y))$$

Percentage of Matches – s^m. A reasonable measure of similarity between two image d_x and d_y is the percentage of local features in d_x that have a match in d_y. Using the distance ratio criterion described in 4.2 for individuating matches, we define the *Percentage of Matches* similarity function s^m as follows:

$$s^m(d_x, d_y) = \frac{1}{|d_x|} \sum_{p_x \in d_x} m(p_x, d_y)$$

where $m(p_x, d_y)$ is 1 if p_x has a match in d_y and 0 otherwise (see Sec. 4.2).

Distance Ratio Average – s^σ. The matching function $m(p_x, d_y)$ used in the *Percentage of Matches* similarity function is based on the ratio between closest to second-closest neighbors for filtering candidate matches as proposed in [14] and reported in Section 4.2. However, this distance ratio value can be used directly to define a *Distance Ratio Average* function between two images d_x and d_y as follows:

$$s^\sigma(d_x, d_y) = \frac{1}{|d_x|} \sum_{p_x \in d_x} \sigma(p_x, d_y)$$

Please note that function does not require a distance ratio c threshold.

Hough Transform Matches Percentage – s^h. As mentioned in Section 4.2, an Hough transform is often used to search for keys that agree upon a particular model pose. The Hough transform can be used to define a *Hough Transform Matches Percentage*:

$$s^h(d_x, d_y) = \frac{|M_h(d_x, d_y)|}{|d_x|}$$

where $M_h(d_x, d_y)$ is the subset of matches voting for the most voted pose. For the experiments, we used the same parameters proposed in [14], i.e. bin size of 30 degrees for orientation, a factor of 2 for scale, and 0.25 times the maximum model dimension for location.

Managing the Asymmetry. All the proposed similarity functions are not symmetric, i.e., $s(d_x, d_y) = s(d_y, d_x)$ does not hold. Consider the case in which the set of local features belonging to d_x is a subset of the ones belonging to d_y. In this case the similarity $s(d_x, d_y)$ is 1 while the same does not hold for $s(d_y, d_x)$.

In searching for images similar to d_x, it is not clear in advance whether $s(d_x, d_y)$ or $s(d_y, d_x)$ would be a better similarity measure for the recognition task. Thus, we tested various combinations.

Given an image d_{Te} belonging to Te (i.e., an image that we want to automatically classify), and an image d_{Tr} belonging to Tr (i.e., an image for which the class label is known in advance) we define various versions of the similarities defined before:

- $s^{Te}(d_{Te}, d_{Tr}) = s(d_{Te}, d_{Tr})$ – is the canonical approach which tries to find points in the test image that are similar to the ones in the training one.
- $s^{Tr}(d_{Te}, d_{Tr}) = s(d_{Tr}, d_{Te})$ – is the inverse approach which uses the points in training documents as queries.
- $s^{or}(d_{Te}, d_{Tr}) = \max(s(d_{Te}, d_{Tr}), s(d_{Tr}, d_{Te}))$ – is the fuzzy *or* of s^{Te} and s^{Tr}. This considers equivalent two images if any of the two is a crop of the other.
- $s^{and}(d_{Te}, d_{Tr}) = \min(s(d_{Te}, d_{Tr}), s(d_{Tr}, d_{Te}))$ – is the fuzzy *and* of s^{Te} and s^{Tr}. This never considers equivalent two images if any of the two is a crop of the other.
- $s^{avg}(d_{Te}, d_{Tr}) = (s(d_{Te}, d_{Tr}) + s(d_{Tr}, d_{Te}))/2$ – is the mean of s^{Te} and s^{Tr}.

Thus, we have defined 5 versions of our 4 similarity measures for a total of 20 similarity measures that will be denoted as $s^{m,Te}$, $s^{m,Tr}$, $s^{m,or}$, ..., $s^{h,Te}$, etc.

5   Local Feature Based Image Classifier

In the previous section, we considered the classification of an image d_x as a process of retrieving the most similar ones in the *training set* Tr and then applying a kNN classification technique in order to predict the class of d_x.

In this section, we propose a new approach that first assigns a label to each local feature of an image. The label of the image is then assigned by analyzing the labels and confidences of its local features. This approach has the advantage that any access method for similarity search in metric spaces [18] can be used to speed-up classification. The proposed *Local Feature Based Image Classifiers* classify an image d_x in two steps:

1. first each local feature p_x belonging to d_x is classified considering the local features of the images in Tr;
2. second the whole image is classified considering the class assigned to each local feature and the confidence of the classification.

Note that classifying individually the local features, before assigning the label to an image, we might loose the implicit dependency between interest points of an image. However, surprisingly, we will see that this method offers better effectiveness than the baseline approach. In other words we are able to improve at the same time both efficiency and effectiveness.

In the following, we assume that the label of each local feature p_x, belonging to images in the training set Tr, is the label assigned to the image it belongs to (i.e., d_x). Following the notation used in Section 4,

$$\forall p_x \in d_x, \ \forall d_x \in Tr \ , \Phi(p_x) = \Phi(d_x).$$

In other words, we assume that the local features generated over interest points of the images in the training set can be labeled as the image they belong to. Note that the noise introduced by this label propagation from the whole image to the local features can be managed by the local features classifier. In fact, we will see that when very similar training local features are assigned to different classes, a local feature close to them is classified with a low confidence. The experimental results reported in Section 7 confirm the validity of this assumption.

As we said before, given $p_x \in d_x$, a classifier $\hat{\Phi}$ returns both a class $\hat{\Phi}(p_x) = c_i \in C$ to which it believes p_x to belong *and* a numerical value $\nu(\hat{\Phi}, p_x)$ that represents the confidence that $\hat{\Phi}$ has in its decision. High values of ν correspond to high confidence.

5.1   Local Feature Classifier

Among all the possible approach for assigning a label to a interest point, the simplest is to consider the label of its closest neighbor in Tr. The confidence value can be evaluated using the idea of the distance ratio discussed in Section 4.2.

We thus define a local feature based classifier $\hat{\Phi}^m(p_x)$ that assign a candidate label $\hat{\Phi}^m(p_x)$ as the one of the nearest neighbor in Tr closest to p_x (i.e., $NN_1(p_x, Tr)$):

$$\hat{\Phi}^m(p_x) = \Phi(NN_1(p_x, Tr))$$

The confidence here plays the role of a matching function, where the idea of the distance ratio is used to decide if the candidate label is a good match:

$$\nu(\hat{\Phi}^m, p_x) = \begin{cases} 1 \text{ if } \dot{\sigma}(p_x, t_r) < c \\ 0 \text{ otherwise} \end{cases}$$

The distance ratio $\dot{\sigma}$ here is computed considering the nearest local feature to p_x and the closest local feature that has a label different than the nearest local feature. This idea follows the suggestion given by Lowe in [14], that whenever there are multiple training images of the same object, then the second-closest neighbor to consider for the distance ratio evaluation should be the closest neighbor that is known to come from a different object than the first. Following this intuition, we define the distance ratio $\dot{\sigma}$ as:

$$\dot{\sigma}(p_x, T_r) = \frac{d(p_x, NN_1(p_x, Tr))}{d(p_x, NN_2^*(p_x, Tr))}$$

where $NN_2^*(p_x, Tr)$ is the closest neighbor that is known to be labeled differently than the first as suggested in [14].

The parameter c used in the definition of the confidence is the equivalent of the one used in [14] and [5]. We will see in Section 7 that $c = 0.8$ proposed in [14] by Lowe is able to guarantee good effectiveness. It is worth to note that c is the only parameter to be set for this classifier considering that the similarity search performed over the local features in Tr does not require a parameter k to be set.

5.2   Whole Image Classification

As we said before, the local feature based feature classification is composed of two steps (see Section 5). In previous section we have dealt with the issue of classifying the local feature of an image. Now, in this section, we discuss the second phase of the local feature based classification of images. In particular we consider the classification of the whole image given the label $\hat{\Phi}(p_x)$ and the confidence $\nu(\hat{\Phi}, p_x)$ assigned to its local features $p_x \in d_x$ during the first phase.

To this aim, we use a confidence-rated majority vote approach. We first compute a score $z(p_x, c_i)$ for each label $c_i \in C$. The score is the sum of the confidence obtained for the local features predicted as c_i. Formally,

$$z(d_x, c_i) = \sum_{p_x \in d_x, \hat{\Phi}(p_x) = c_i} \nu(\hat{\Phi}, p_x) \, .$$

Then, the label that obtains the maximum score is chosen:

$$\hat{\Phi}(d_x) = \arg\max_{c_j \in C} z(d_x, c_j) \, .$$

As measure of confidence for the classification of the whole image we use ratio between the predicted and the second best class:

$$\nu_{img}(\hat{\Phi}, d_x) = 1 - \frac{\arg\max_{c_j \in C - \hat{\Phi}(p_x)} z(d_x, c_j)}{\arg\max_{c_i \in C} z(d_x, c_i)} \, .$$

This whole image classification confidence can be used to decide whether or not the predicted label has an high probability to be correct. In the experimental results Section 7 we will show that the proposed confidence is reasonable.

6   Evaluation Settings

For evaluating the various classifiers we need at least: a data set, an interest points detector, a local feature extractor, some performance measures. In the following, we present all the evaluation setting we used for the experimentation.

6.1   The Dataset

The dataset that we used for our tests is composed of 1,227 photos of landmarks located in Pisa and was used also in [2]. The photos have been crawled from Flickr, the well known on-line photo service. The dataset we built is publicly available. The IDs of the photos used for these experiments together with the assigned label and extracted features can be downloaded from [10]. In the following we list the classes that we used and the number of photos belonging to each class. In Figure 1 we reported an example for each class that are: *Leaning Tower* (119 photos); *Duomo* (130 photos); *Battistero*

Fig. 1. Example images taken from the dataset

(104 photos); *Camposanto Monumentale (exterior)* (46 photos); *Camposanto Monumentale (field)* (113 photos); *Camposanto Monumentale (portico)* (138 photos); *Chiesa della Spina* (112 photos); *Palazzo della Carovana* (101 photos); *Palazzo dell'Orologio* (92 photos); *Guelph tower* (71 photos); *Basilica of San Piero* (48 photos); *Certosa* (53 photos).

In order to build and evaluating a classifier for these classes, we divided the dataset in a *training set* (Tr) consisting of 226 photos (approximately 20% of the dataset) and a *test set* (Te) consisting of 921 (approximately 80% of the dataset). The image resolution used for feature extraction is the standard resolution used by Flickr i.e., maximum between width and height equal to 500 pixels.

The total number of local features extracted by the SIFT and SURF detectors were about 1,000,000 and 500,000 respectively.

6.2 Performance Measures

For evaluating the effectiveness of the classifiers in classifying the documents of the *test set* we use the micro-averaged *accuracy* and micro- and macro-averaged *precision, recall* and F_1.

Micro-averaged values are calculated by constructing a global contingency table and then calculating the measures using these sums. In contrast macro-averaged scores are calculated by first calculating each measure for each category and then taking the average of these. In most of the cases we reported the micro-averaged values for each measure.

Precision is defined as the ratio between correctly predicted and the overall predicted documents for a specific class. *Recall* is the ratio between correctly predicted and the overall actual documents for a specific class. F_1 is the harmonic mean of *precision* and *recall*.

Note that for the *single-label* classification task, micro-averaged *accuracy* is defined as the number of documents correctly classified divided by the total number of documents in the *test set* and it is equivalent to the micro-averaged *precision, recall* and F_1 scores.

7   Experimental Results

In this section we report the experimental results obtained for all the 20 image similarity based and local feature based classifiers. For the image similarity based classifier results are reported for each similarity measure defined in Section 4.3. We also show that the proposed measure of confidence can be used to improve effectiveness on classified images accepting a small percentage of not classified objects.

7.1   Image Similarity Based Classifiers

In Table 1, *Accuracy* and macro averaged F_1 of the image similarity based classifiers for the 20 similarity functions defined in Section 4 are reported. Note that the *single-label distance-weighted kNN* technique has a parameter k that determines the number of closest neighbors retrieved in order to classify a given image (see Section 4). This parameter should be set during the training phase and is kept fixed during the test phase. However, in our experiments we decided to report the result obtained ranging k between 1 and 100. For simplicity, in Table 1, we report the best performance obtained and the k for which it was obtained. Moreover, we report the performance obtained for $k = 1$ which is a particular case in which the kNN classifier simply consider the closest image.

Let's first consider the approach used for managing the asymmetry of the distance functions discussed in Section 4.3. The best approach for all the similarity functions using both SIFT and SURF features is the fuzzy *and*, i.e., $s^{*,and}$. The more traditional approach $s^{*,Te}$ is the second best in most of the cases. On the contrary, $s^{*,Tr}$ always offers the worst performance. In other words, the best results were obtained when the similarity between two images is computed as the minimum of the similarity obtained considering as query in turn the test image local features and the training images. The result is the same both when using SIFT and SURF.

The *Hough Transform Matches Percentage* (s^h) similarity function is the best choice for both SIFT and SURF for all the 5 versions for managing the asymmetry. The geometric information considered by this function allows to obtain significantly better performance in particular for SURF.

The second best is *Distance Ratio Average* (s^σ) which only considers the distance ratio as matching criterion. Please note that s^σ does not require a distance ratio threshold (c) because it weights every match considering the distance ratio value. Moreover, s^σ performs sightly better than *Percentage of Matches* (s^m) which requires the threshold c to be set.

The results obtained by the *1-NN Similarity Average* (s^1) function show that considering just the distance between a local features and its closest neighbors gives worst performance than considering the distance ratio s^σ. In other words, the similarity between a local feature and its closest neighbor is meaningful only if compared to the other nearest neighbors, which is exactly what the distance ratio does.

Regarding the parameter k it is interesting to note that the k value for which the best performance was obtained for each similarity measure is typically much higher for SURF than SIFT. In other words, the test image closest neighbors in the training set are more relevant using SIFT than using SURF.

Table 1. Image similarity based classifier ($\hat{\Phi}^s$) performance obtained using various image similarity functions

similarity function			s^1 - Avg 1-NN					s^m - Perc. of Matches					s^o - Avg Sim. Ratio					s^h - Hough Transform				
		version	Te	Tr	or	and	avg	Te	Tr	or	and	avg	Te	Tr	or	and	avg	Te	Tr	or	and	avg
Best	Acc	SIFT	.75	.52	.55	.85	.82	.88	.80	.81	.90	.88	.89	.80	.81	.91	.89	.92	.88	.88	.93	.91
		SURF	.79	.70	.73	.80	.82	.85	.73	.76	.88	.86	.82	.73	.75	.87	.84	.89	.76	.79	.92	.86
	F_1	SIFT	.72	.55	.56	.84	.84	.86	.80	.80	.89	.86	.87	.80	.81	.91	.88	.90	.87	.86	.93	.90
		SURF	.76	.67	.70	.78	.80	.83	.70	.74	.87	.84	.81	.68	.73	.86	.82	.87	.74	.77	.89	.85
$k=1$	Acc	SIFT	.73	.52	.55	.85	.82	.88	.78	.80	.90	.88	.89	.78	.80	.91	.88	.91	.87	.87	.93	.91
		SURF	.79	.63	.67	.80	.82	.81	.60	.62	.86	.79	.81	.63	.64	.84	.76	.87	.66	.68	.90	.81
	F_1	SIFT	.72	.55	.53	.84	.84	.86	.78	.80	.89	.86	.87	.79	.80	.90	.87	.90	.86	.86	.92	.90
		SURF	.76	.63	.67	.78	.80	.79	.65	.65	.84	.78	.80	.67	.67	.83	.77	.85	.68	.70	.89	.81
Best k	Acc	SIFT	*9*	*1*	*1*	*1*	*1*	*1*	*7*	*4*	*2*	*3*	*1*	*5*	*5*	*3*	*5*	*2*	*3*	*9*	*2*	*1*
		SURF	*3*	*6*	*8*	*1*	*1*	*20*	*28*	*42*	*14*	*20*	*8*	*23*	*17*	*11*	*14*	*21*	*35*	*39*	*11*	*18*
	F_1	SIFT	*1*	*1*	*1*	*1*	*1*	*1*	*7*	*4*	*3*	*3*	*1*	*5*	*5*	*3*	*5*	*2*	*8*	*5*	*9*	*9*
		SURF	*1*	*6*	*3*	*1*	*1*	*18*	*28*	*19*	*23*	*20*	*8*	*5*	*17*	*11*	*14*	*21*	*14*	*30*	*3*	*28*

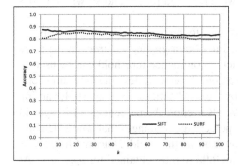

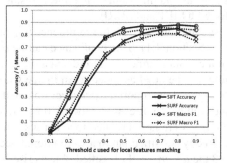

Fig. 2. Accuracy obtained by both SIFT and SURF for various k using the $s^{m,Te}$ similarity function with the image similarity based classifier

Fig. 3. Accuracy and Macro F_1 for various matching thresholds c, obtained by the image similarity based classifier ($\hat{\Phi}^s$) using the $s^{m,Tr}$ similarity and SIFT

This is more evident in Figure 2 where we report the *accuracy* obtained for k between 1 and 100 by both SIFT and SURF using the $s^{m,Te}$ similarity function. SIFT obtains the best performance for smaller values of k with respect to SURF. Moreover, SIFT performance is generally higher than SURF. It is interesting to note that performance obtained for $k = 1$ is typically just slightly worst than that of the best k. Thus, $k = 1$ gives very good performance even if a better k could be selected during a learning phase.

Two of the similarity measures proposed in Section 4.3 require a parameter to be set. In particular, the similarity measures *Percentage of Matches* (s^m) and *Hough Transform Matches Percentage* (s^h) use the matching function defined in Section 4.2 that requires a threshold for the distance ratio threshold (c) to be fixed in advance.

Table 2. Accuracy and Macro F_1 for the local feature based classifiers $\hat{\Phi}^m$ and for the kNN classifiers based on the various image similarity measures proposed for best k and related to the *and* version

	classifier	$\hat{\Phi}^m$	$\hat{\Phi}^s$			
	similarity		$s^{1,\,and}$	$s^{m,\,and}$	$s^{\sigma,\,and}$	$s^{h,\,and}$
Accuracy	SIFT	.94	.85	.90	.91	.93
	SURF	.93	.80	.88	.87	.92
F_1 Macro	SIFT	.94	.84	.89	.91	.93
	SURF	.91	.78	.87	.86	.89

In Figure 3 we report the performance obtained by using the *Percentage of Matches* classifier, i.e., the image similarity based classifier $\hat{\Phi}^s$ using the similarity measure s^m. For each distance ratio threshold c we report the best result obtained for k between 0 and 100. As mentioned in Section 4.2, in the paper where SIFT [14] was presented, Lowe suggested to use 0.8 as distance ratio threshold (c). The results confirm that the threshold proposed in [14] is the best for both SIFT and SURF and that the algorithm is stable around this values. In Table 1, results were reported for s^m and s^h with $c = 0.8$ for both SIFT and SURF.

Let us now consider the confidence ν_{doc} assigned to the predicted label of each image (see Section 4.1). This confidence can be used to obtain greater *accuracy* at the price of a certain number of false dismissals. In fact, a confidence threshold can be used to filter all the label assigned to an image with a confidence ν_{doc} less than the threshold. In Figure 4 we report the *accuracy* obtained by the $s^{h,and}$ measure using SIFT, varying the confidence threshold between 0 and 1. We also report the percentage of images in Te that were not classified together with the percentage of images that where actually correctly classified but that were filtered because of the threshold. Note that for $\nu_{doc} = 0.3$ the *accuracy* of classified objects rise from 0.93 to 0.99 obtained for $\nu_{doc} = 0$. At the same time the percentage of correctly predicted images that are filtered (i.e., the classifier does not assign a label because of the low confidence threshold ν_{doc}) is less than 10%.

This prove that the measure of confidence defined is meaningful. However, the best confidence threshold to be used depends on the task. Sometimes it could be better to try to *guess* the class of an image even if we are not sure, while in other cases it might be better to assign a label only if the classification has an high confidence.

7.2 Local Feature Based Classifier

In this section we compare the performance of the image similarity based classifiers using the 20 similarity measures defined in Section 4.3 with the local feature based classifier defined in 5.

In Table 2, we report *accuracy* and macro-averaged F_1 obtained by the *Local Feature Based Image Classifier* ($\hat{\Phi}^m$) using both SIFT and SURF together with the results obtained by the image similarity based approach ($\hat{\Phi}^s$) for the various similarity measures. Considering that in the previous section we showed that the fuzzy *and* approach

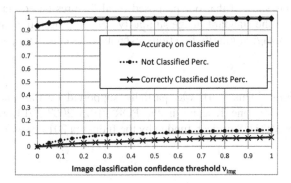

Fig. 4. Results obtained by the image similarity based classifier for similarity $s^{h,and}$ using SIFT, for various classification confidence thresholds (ν_{img})

performs better than the other, we only report the result obtained for the *and* version of each measures and for the best k.

The first observation is that the *Local Feature Based Image Classifier* ($\hat{\Phi}^m$) performs significantly better then any *Image Similarity Based Classifier*. In particular $\hat{\Phi}^m$ performs better then $s^{h,and}$, even if no geometric consistency checks are performed by $\hat{\Phi}^m$ while matches in $s^{h,and}$ are filtered making use of the Hough transform.

Even if in this paper we did not consider the computational cost of classification, we can make some simple observations. In fact, it is worth saying that the local feature based classifier is less critical from this point of view. First, because closest neighbors of local features in the test image are searched once for all in the Tr and not every time for each image of Tr. Second, because it is possible to leverage on global spatial index for all the features in Tr, to support efficient k nearest neighbors searching. In fact, the similarity function between two local features is the Euclidean distance, which is a metric. Thus, it could be efficiently indexed by using a metric data structures [18,15,4].

Regarding the local features used and the computational cost, we underline that the number of local features detected by the SIFT extractor is twice that detected by SURF. Thus, on one hand SIFT has better performance while on the other hand SURF is more efficient.

8 Conclusions

In this paper we addressed the problem of image content recognition using local features and kNN based classification techniques. We defined 20 similarity functions and compared their performance on a image content landmarks recognition task. We found that a two-way comparison of two images based on fuzzy *and* allows better performance than the standard approach that compares a query image with the ones in a training set. Moreover, we showed that the similarity functions relying on matching of local features that makes use of geometric constrains perform slightly better than the others.

Finally, we defined a novel kNN classifier that first assigns a label to each local feature of an image and then labels the whole image by considering the labels and the confidences assigned to its local features.

The experiments showed that our proposed local features based classification approach outperforms the standard image similarity kNN approach in combination with any of the defined image similarity functions, even the ones considering geometric constrains.

Acknowledgements. This work was partially supported by the VISITO Tuscany project, funded by Regione Toscana, in the POR FESR 2007-2013 program, action line 1.1.d, and the MOTUS project, funded by the Industria 2015 program.

References

1. Amato, G., Falchi, F.: Local feature based image similarity functions for kNN classfication. In: Proceedings of the 3rd International Conference on Agents and Artificial Intelligence (ICAART 2011), vol. 1, pp. 157–166. SciTe Press (2011)
2. Amato, G., Falchi, F., Bolettieri, P.: Recognizing landmarks using automated classification techniques: an evaluation of various visual features. In: Proceeding of the Second Interantional Conference on Advances in Multimedia (MMEDIA 2010), Athens, Greece, June 13-19, pp. 78–83. IEEE Computer Society (2010)
3. Ballard, D.H.: Generalizing the hough transform to detect arbitrary shapes. Pattern Recognition 13(2), 111–122 (1981)
4. Batko, M., Novak, D., Falchi, F., Zezula, P.: Scalability comparison of peer-to-peer similarity search structures. Future Generation Comp. Syst. 24(8), 834–848 (2008)
5. Bay, H., Tuytelaars, T., Van Gool, L.: SURF: Speeded Up Robust Features. In: Leonardis, A., Bischof, H., Pinz, A. (eds.) ECCV 2006, Part I. LNCS, vol. 3951, pp. 404–417. Springer, Heidelberg (2006)
6. Boiman, O., Shechtman, E., Irani, M.: In defense of nearest-neighbor based image classification. In: CVPR (2008)
7. Chen, T., Wu, K., Yap, K.H., Li, Z., Tsai, F.S.: A survey on mobile landmark recognition for information retrieval. In: MDM 2009: Proc. of the Tenth International Conference on Mobile Data Management, pp. 625–630. IEEE (2009)
8. Dudani, S.: The distance-weighted k-nearest-neighbour rule. IEEE Transactions on Systems, Man and Cybernetics SMC-6(4), 325–327 (1975)
9. Fagni, T., Falchi, F., Sebastiani, F.: Image classification via adaptive ensembles of descriptor-specific classifiers. Pattern Recognition and Image Analysis 20, 21–28 (2010), http://dx.doi.org/10.1134/S1054661810010025
10. Falchi, F.: Pisa landmarks dataset (2011), http://www.fabriziofalchi.it/pisaDataset/ (last accessed on March 3, 2011)
11. Google: Google Goggles (2011), http://www.google.com/mobile/goggles/ (last accessed on March 3, 2011)
12. Jégou, H., Douze, M., Schmid, C.: Improving bag-of-features for large scale image search. Int. J. Comput. Vision 87(3), 316–336 (2010)
13. Kennedy, L.S., Naaman, M.: Generating diverse and representative image search results for landmarks. In: WWW 2008: Proceeding of the 17th International Conference on World Wide Web, pp. 297–306. ACM Press, New York (2008)
14. Lowe, D.G.: Distinctive image features from scale-invariant keypoints. International Journal of Computer Vision 60(2), 91–110 (2004)

15. Samet, H.: Foundations of Multidimensional and Metric Data Structures. Computer Graphics and Geometric Modeling. Morgan Kaufmann Publishers Inc., San Francisco (2005)
16. Serdyukov, P., Murdock, V., van Zwol, R.: Placing flickr photos on a map. In: Allan, J., Aslam, J.A., Sanderson, M., Zhai, C., Zobel, J. (eds.) SIGIR, pp. 484–491. ACM (2009)
17. Yeh, T., Tollmar, K., Darrell, T.: Searching the web with mobile images for location recognition. In: CVPR (2), pp. 76–81 (2004)
18. Zezula, P., Amato, G., Dohnal, V., Batko, M.: Similarity Search: The Metric Space Approach. In: Advances in Database Systems, vol. 32. Springer, Heidelberg (2006)
19. Zheng, Y., Song, M.Z., Adam, Y., Buddemeier, H., Bissacco, U., Brucher, A., Chua, F., Neven, T.S., Tour, H.: The world: Building a web-scale landmark recognition engine. In: CVPR, pp. 1085–1092. IEEE (2009)

Web Service Composition Plans in OWL-S

Eva Ziaka, Dimitris Vrakas, and Nick Bassiliades

Department of Informatics, Aristotle University of Thessaloniki, Thessaloniki, Greece
{evziaka,dvrakas,nbassili}@csd.auth.gr

Abstract. One of the main visions of Semantic Web has been the ability of software agents to compose atomic web services in order to facilitate the automation of complex tasks. One of the approaches used in the past in order to automatically construct composite web services has been AI planning. The most important advantage of this approach is its dynamic character that reduces the interference of the user. Although there have been various attempts to utilize planning algorithms and systems in the composition process, there has been little work in the field of converting web service composition plans in OWL-S. This paper studies the use of two well established standards in expressing plans and composite web services, namely the Planning Domain Definition Language (PDDL) and the Ontology Web Language for Services (OWL-S) and suggests a method for translating the produced PDDL plans of any planning system to OWL-S descriptions of the final composite web services. The result is a totally new web service that can later be discovered and invoked or even take part in a new composition.

Keywords: Web services composition, AI planning, Semantic web services, OWL-S, PDDL.

1   Introduction

Nowadays, many different systems all over the globe can communicate with each other through the Internet. The need for supporting interoperability of web applications so that they can be used by all platforms, no matter their implementation, has led to web services technology and a new, web-service-oriented way of programming. This new technology is based on open protocols, such as the XML and the well known HTTP transfer protocol.

There is often the need to execute more complex tasks that simple web services do not have the potential to complete on their own. In such cases, simple web services must cooperate so as to combine their functionalities to create a new complex web service that will hold the desirable functionality. Semantic information about all the available atomic web services is very important for their cooperation in web services composition field, so as to be able to understand the meaning of their inputs and outputs and to match them to achieve cooperation.

During the past decade a large number of approaches for composing web services have been proposed, some fully automated, other partially automated, whereas a lot of them are even completely manual. A promising way that aims at fully automated web

J. Filipe and A. Fred (Eds.): ICAART 2011, CCIS 271, pp. 240–254, 2013.

services composition is the use of AI planning technology. Each web service is represented as a planning operator and the desired composite service's inputs and outputs form the initial state and the goals respectively. The plans that arise are encoded in languages such as PDDL [5] that describe the actions, that is the web services, that must be executed and the order of their execution.

The contribution of this paper focuses on the automatic translation of the plans, expressed in PDDL, to OWL-S descriptions [10] that take advantage of the OWL-S control constructs and facilitate the automatic invocation of the composite service. Specifically, information from the PDDL descriptions of the domain, the composition problem, and the plan is used to create a functional representation of the composition. This representation describes with a specific syntax the way each atomic web service is connected to each other in order to produce the final output. In a second phase, this functional representation is utilized to generate the OWL-S descriptions of the new composite web service.

In terms of functionality, the method described in this paper is merely based on the PDDL descriptions of the planning operators and does not explicitly deal with semantic information of the initial atomic services. Therefore, it can be applied to compositions arising from both syntactic and semantic matching of inputs and outputs of the atomic services. However, since the final expression will be encoded in OWL-S language, we will use the notion of semantic web service throughout the rest of the paper.

In the sections to follow, the relative research field is explored. The suggested technique is analyzed in detail and some conclusions along with future directions are given. Specifically, the rest of the paper is organized as follows:

In section 2, the field of automated web services composition using AI Planning techniques is presented and some studies on the field are exposed. In section 3, the developed method for translating the PDDL plans to OWL-S descriptions is analyzed. This section is divided into two sub-sections, reflecting the two phases of the method; in the first sub-section, the algorithm that creates the functional representation describing the composition is presented, whereas in the second sub-section, the method for converting this representation to OWL-S description is described. Finally, in the last section, conclusions of the research so far are given along with some ideas on how the developed algorithms and the web services composition procedure could be enhanced.

2   Related Work

The process of automated web services composition by the point of view of planning has been studied extensively. The most important advantage of this approach is the dynamic character that is offered to the composition process, which reduces a lot the interference of the user.

One of the most known systems in the field of web services composition via planning is SHOP2 (Simple Hierarchical Ordered Planner), [15]. It is based on HTN planning (Hierarchical Task Network) methods [14]. One basic difference between SHOP2 and the other HTN systems is that it locates all the actions of the plan in the same order that they will be later executed. In this way, the current state of the system in every step of the planning procedure is known and inference mechanisms or heuristic techniques can be used to augment the effectiveness and the efficiency of the whole process.

The functionality of SHOP2 consists of three basic steps. In the first one, the domain is constructed by the process OWL-S files of the available web services. The atomic services are represented by operators and methods for analyzing the complex services to simpler ones are constructed. In the second step, the composition problem is transformed to planning problem. This is realized by describing the problem as an abstract composite process that need decomposition with the use of methods so as to obtain simple processes that refer to web services. In the third step, the problem is solved by decomposing the tasks and creating the plan, i.e. is the description of the final composite service.

Another technique, analyzed in [12], is based on situation calculus, where the states are not considered as instances of the environment but as sequences of actions that were executed in the past and resulted to this state. This technique uses also the language Golog (alGOL in LOGic), which is based on logic and the problems that are encoded in it can be solved by methods that use logic. For the appropriate representation of the planning problem in Golog, the language was extended so as to be able to contain constraints on the composition process defined by the user. These constraints in essence reflect the desired outputs. The OWL-S descriptions are used as requirements of the processes that must be executed and also as descriptions of the actions that are provided by the web services. The composition problem is transformed into the problem of finding the appropriate Golog program that when executed, all the defined constraints will be satisfied. In the solution process, intelligent agents are used whom knowledge base contains the preconditions and the results of the services, encoded in situation calculus terms. The available web services correspond to operators, primitive or composite. The role of the agents is the inference on the web services, in order to discover, execute and compose them.

A different and quite simple web services composition method is presented in [18]. It is based on regression in a state space. The algorithms belonging to this category start from exploring the goals that must be succeeded and seek for the actions that lead from the goals to the initial state. The method proposed introduces a new structure called SLM (Semantic Links Matrix) and is a table containing the values of semantic relevance between the parameters of the web services. For the construction of this table, the process models and the relative ontologies of the atomic services are used. Generally, the SLM structure groups the candidate web services based on their semantic relevance and in the same time provides information on their quality characteristics so as to ease the choice among them. The algorithm begins from the goals, but because of the SLM structure it does not need to calculate the previous states. In the step of locating the actions that satisfy the current goals, all the services that have a positive value in the relevance function are considered as candidates. The best service is chosen based on the QoS characteristics. The process continues until it reaches the initial state.

Another approach described in [17] uses model checking techniques for producing the plan. The algorithm consists of four steps. In the first step, the goal and the initial states are defined. In the second step, the model of the process on which the checks will be running is extracted. The web services that could be used for the domain are automatically detected and the state space where the solution is searched is constructed. Information on the services is retrieved by the ontologies and is inserted to the model. In the third step, the search algorithm in the plan space is executed and some plans that satisfy the goals are collected. In the fourth and last step, the best plan is chosen and is converted in a composite web service, encoded in BPEL.

A system which was developed recently and is analyzed in [6] is the system PORSCE. The approach is based on transforming the web services composition problem to a planning problem. The straight forward mapping of these two fields is exploited and the OWL-S descriptions of the available web services are used to construct PDDL plan files. The initial state is derived by the data given as input to the final web service by the user, whereas the goals are reflected by the desired outputs. The operators of the problem correspond to the available atomic web services that can be used. Their preconditions are mapped to the inputs of the services and theirs results to the outputs. Simultaneously, the ontologies that are connected to the types of the parameters of the available web services are used so as for the semantics of the concepts to be provided. The system starts by representing the composition problem with planning terms. Then, a solution to the problem is provided by an external planner, such as LPG-td [3], [4] or JPlan [8], according to the user's selection. Finally, the quality of the produced plan is measured based on some quality measures selected by the user at the beginning of the process and the results are provided to the user. There is also the possibility of replacing instantly some of the web services in the plan with other relevant, as they are discovered during the planning process.

Another approach that exploits the similarities between the AI planning and semantic web services composition research fields is the OWLS-Xplan [9]. This system uses the OWL-S descriptions of the available web services, the relevant OWL ontologies that define the types of the parameters in the descriptions and a planning query as input. After some preprocessing of the above data and the execution of the Xplan planning algorithm, the result is a plan describing the sequence of composed services that satisfies the goals.

The OWLS-Xplan approach consists of two basic parts. The first one is an OWLS2PDDL converter which converts the OWL-S descriptions along with the OWL ontologies to the equivalent PDDL domain and problem of the composition. Specifically, the conversion results to descriptions of the domain and problem in a XML dialect of PDDL (developed by the authors), referred to as PDDXML, that simplifies parsing, reading and communicating the descriptions using SOAP. An atomic operator is directly related to a service profile as they both provide a general description of their instances, actions and web services, respectively. A complex action can be linked to a service model that describes how simpler actions should cooperate to result to the composite one. Finally, the methods used in HTN planning are related to composite web services and may be used by the planner as a hierarchical task network during the planning process.

The second part of OWLS-Xplan is the developed heuristic hybrid Xplan AI planner that combines the benefits of the action-based FF-planner [7] with HTN planning. Xplan always finds a solution, if it exists in the state space, over the space of possible plans, in contrast to HTN approaches. It combines guided local search with graph planning and a simple form of hierarchical task networks to produce a plan. Also, a re-planning component is included to improve flexibility is cases changes happen in the world during planning, a property well needed in semantic web services composition field.

The solution analyzed in [16] also translates the composition problem to PDDL descriptions and suggests that in this way an appropriate planner could be found each time according to the problem so as to provide an improved solution. The paper presents a three step technique for the creation of a composite web service with the first step being the translation of the OWL-S descriptions and OWL ontologies to PDDL domain and problem descriptions; the second one is the creation of a plan that solves the problem with the execution of a planner; the third one is the translation of the plan to a new OWL-S description of the resulting composite web service. However, the paper focuses only on the first step of the procedure. Some assumptions are made to ease the translation function, such as considering that each atomic process has either effects or outputs but not both simultaneously. Also, the authors of the paper do not deal with OWL-S process models that have composite process using Repeat-While and Repeat-Until or Any-Order and Split-Join constructs. The algorithm proposed, deals separately with the OWL-S process model, the atomic and simple processes, the sequence, if-then-else, choice and split processes and with the OWL-S target service description to create the domain and problem descriptions. The process of choosing the appropriate planner for each problem and the translation of the plan to OWL-S description of the new service are not elaborated in the paper.

The aforementioned methods tackle the problem of web services composition using a variety of fully or partially automated techniques. However, they don't deal with the task of expressing the resulting composite service in OWL-S, taking advantage of the supported control constructs.

3   Translating PDDL to OWL-S

This section analyzes the method for translating a composite web service expressed in the PDDL language to the corresponding OWL-S description. The translation completes in two phases. The first one concerns the extraction of all the required information from the plan for the creation of a composite web service's functional representation. The second is about the conversion of this representation to an OWL-S description of the resulting composite web service.

3.1   Constructing the Composite WS

The first step in the creation of an OWL-S description based on data derived from a PDDL plan is the manipulation of these data and their conversion to a composite web service functional representation. This representation refers to the available simple or atomic web services and the order in which they should be executed and is structured using the OWL-S control constructs *sequence*, *split* and *split-join*.

In the following algorithm the functional representation of a composite web service C is represented as a predicate $f(a_0,a_1,...,a_n)$, where f is the control construct used to describe the composition structure and $a_0,a_1,...,a_n$ stand for the simple web services that participate in the composition. Each a_i could be another composite service or, in a simpler case, an atomic process, which is represented as $atomic(a_i)$.

The developed algorithm consists of three general steps, as shown in **Fig. 1**. The first step concerns the parsing of the files associated with the composition planning problem and the extraction of all the information needed in the next steps. In the second step, a web service composition graph is created. The nodes of the graph are the actions of the plan and the edges are the links that express the order constraints among the actions. The creation of the graph is based on the information collected from the previous step. Finally, in the last step, the composite web service functional representation is formed using the ordering constraints that are extracted from the composition graph. In the following paragraphs, these three steps are described in more detail.

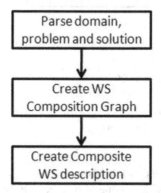

Fig. 1. Converting a PDDL plan to a composite web service functional representation

The initial available information is derived from the PDDL domain and problem files of the composition problem. For the parsing of these files, an external library, called PDDL4J, [13] is used. The types of information that are required by the translation process are the following: a) the name of the operator, b) the parameters list, c) the preconditions list, d) the effects list, e) the initial state and f) the goals of the problem. Finally, the resulting plan is parsed in order to extract information concerning the actions of the plan. Exploiting the syntax of this file, information on the actions used can easily be extracted. The data that are needed in the later steps of the algorithm involve the timestamp of each action, which is the time step when the action will be executed and the name, parameters and duration of it. The actions are read in the order that they are presented in the plan, so the procedure keeps track of this order.

When all these data are retrieved, the procedure continues combining them so as to create objects representing the steps of the plan. Every step contains the name of the action that will be executed, the parameters with which the action is called, the timestamp and duration of the action, the operator from which the action is derived, the substitution imposed on the operation, the list of preconditions that must hold for the action to be executed and the list of the effects, the facts that will change due to the execution of the action.

The second step creates the web service composition graph. The nodes list is identical to the list of actions of the plan. In essence, the contribution of this step is the computation of the edges, that is, the links between the actions. The general idea is to traverse all the actions and locate cases where one precondition of an action matches one effect of another. This ought to happen in theory because of the causal links that are present among the actions of the plan, which imply that the preconditions of the later actions will appear as effects of other previous actions. An order constraint link is then created between the two actions.

Algorithm 1 (Graph): Computes the web services composition graph
Inputs: P = {a₀,a₁,...,aₙ}, the plan
Output: G = (P,E), web services composition graph

```
E = ∅
for i = n down to 1
    for each c ∈ prec(ai)
        for j = i-1 down to 0
            for each p∈ add(aj)
                if (c = p)
                    E = E U {(aj,ai)}
return G = (P,E)
```

The algorithm that discovers such kinds of links is called *Graph* and starts from the last elements of the action list. Each one of its preconditions is then examined so as to discover a previous action in the plan that produces this fact. This means to discover an action that contains this fact in its effect list. So, another loop is needed to access all the previous possible producers of this imminent link. When such a previous action is found, a link is created among the two actions. This link illustrates an order constrain and ensures that the action that produces the fact will be executed before the one that consumes it in its preconditions list.

A simple example of the above procedure is depicted in **Fig. 2**. In this example there are two actions in the plan, the actions *Drop Ball B* with which a robot puts down the ball B and the action *Grab Ball A* that results in a state where the robot is holding the ball A. The algorithm examines first the action *Grab Ball A* and loops on its preconditions. In this case there is only one precondition, declaring that for executing this action, the robot's gripper must be free. So, somewhere in the plan there should be an action that realizes this fact. Exploring the previous actions of the plan, the algorithm confronts the action *Drop Ball B* and matches the fact under consideration with the second result of this action. Automatically, an order constraint link is created between the two actions meaning that the robot should definitely perform first the action *Drop Ball B* so as to be able then to perform the action *Grab Ball A*.

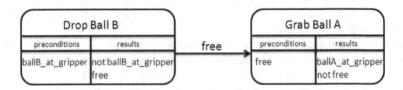

Fig. 2. Example on discovering links

When all the edges and the corresponding order constraints are discovered in the plan, the procedure can continue and exploit these relationships in order to construct a composite web service functional representation that illustrates in a more formal way how the actions of the composite service relate to each other. This representation is built upon the control constructs that OWL-S uses to describe the different possible connections between web services. In the algorithm we use three basic control constructs: *sequence*, *split* and *split-join*. The control *sequence* declares that all its members should be executed in the exact order they appear. The control *split* is used to describe cases of parallel execution of web services. The last control, *split-join*, describes the case where a split occurs in the plan and the parallel executions connect again in a next step in one web service. It is important that the web services that happen to be last in the parallel executions, have to synchronize their outputs to supply the web service following the connecting point with the sufficient inputs.

The general algorithm that constructs the composite web service's functional representation consists of 2 basic steps, presented in Algorithm 2 (*Basic*) and Algorithm 3 (*Join*). Before the execution of these algorithms, a manipulation of the data gathered so far is needed. First, the order constraints list is reduced by removing all the constraints not needed. Then the algorithm *Basic* is called, locates the web services that will be invoked first and creates functional representations of the sub-compositions that start from these services. All these representations are then added to an empty split control. Up to this point, the first version of the requested functional representation is ready. But some refinement steps should be performed in order to provide a more concise representation. So, next in the developed algorithm, a process named *Join* takes place and simplifies the functional representation by replacing *split* controls with *split-join* where possible. The generated functional representation of algorithm *Basic* contains null expressions and unnecessary controls, such as a split control with only one parameter. In the following paragraphs a more detailed description of the translation procedure is provided.

The output of *Graph* algorithm may contain some unnecessary ordering constraints, so the first step is about locating such constraints and removing them from the set. Unnecessary constraints are the ones that can be implied by others, so there is no need for their existence in the set. One order constraint A can be inferred by others if there exists another constraint B with the same left part as A and a constraint C whose left part is identical to the right part of constraint B and its right part is identical to the right part of constraint A. An example will clarify more the above situation. Let the set {$A<C$, $A<B$, $B<C$} be the set of constraints of the composition problem. Examining the need of existence of the first order constraint, which is interpreted as 'the web service A must be executed before the execution of the service C', the constraint $A<B$ has the same web service at the left part. The process continues by exploring the set for constraints that have service B in the left part, because this is the right part of the constraint $A<B$. Such a constraint exists and is the third of the set. Also, this constraint has identical right part with the first constraint that is examined in the process. This means that the constraint $A<C$ is unnecessary because it can be inferred by the constraints $A<B$ and $B<C$, so it is removed from the set.

The next procedure that takes place is the *Basic* procedure, shown in Algorithm 2. The first step of this algorithm is the location of the so called 'clear' services, the web services that are executed first in the plan. The main characteristic of these services is that they are not consumers in any causal link, which means that there is no need for another web service to be executed before them. Such services can be located by searching for the existence of each web service in the plan as a right part of an order constraint. If this search returns no results, then the service can be marked as "clear". For example, having the set of web services {A, B, C} and the order constraints {A<B, B<C} it can be easily inferred that only the service A is clear, because it does not appear as a right member of any order constraint. For each clear web service, the construction of sub-representations of the desired composition takes place. In essence, the relationship among a clear web service and all its children, all the services that can be executed after the completion of the clear service, is revealed.

Algorithm 2 (Basic): Computes an initial composite service with Sequence and Split constructs
Inputs: G = (V,E), the web service graph
Output: C, a composite service

```
// R is the set of root nodes in G
set R ← {r∈V: ∀x ∈ V, (x→r)∉E }
if |R| = 0 then return NULL
if |R| = 1 then
    set G'← the tree in G with r∈R
             as the root
    return sequence(r, Basic(G'-{r}))
set c ← {}
for each r in R
    set G' ← the tree in G with r∈R
             as the root
    set c ← c ∪ Basic(G'-{r})
return Join(split(c))
```

In the next steps of the algorithm *Basic*, the number of clear services is examined. In the trivial case, where there are no such services, a null value is returned. If there is only one clear service, then the only representation that can be constructed is a simple sequence of the clear service and the composition of the child. So in this point, the algorithm calls recursively itself with the rest of the graph as a parameter. This is because the expression beginning from the clear service must contain all the information about the expressions that can be built from the children of this service.

If there are more than one clear services, then an empty composite web service is created and for every clear service the *Basic* procedure is invoked having as parameter the *Graph* without the service in question. All the returned functional representations are then added to a *split* control. The resulting *split* expression is simplified by an algorithm that will be analyzed later in the paper. A short example is given to clarify the procedure. Suppose there are a clear service A and two children B and C. The functional representation returned from the algorithm, in terms of control constructs, will be *seq(A,split(Basic(B),Basic(C)))*. Supposing that there are no other web services in the plan, the final result will be *seq(A,split(B,C))*.

Next, the composition representation that resulted from the clear services (algorithm *Basic*) is simplified by the algorithm *Join* (Algorithm 3). The main function of this algorithm is to replace the *split* controls with *split-join*, wherever this is possible. In every step, two parameters of the functional representation are examined for the existence of a common part. If one such part is found, it is removed from both the parameters and the results are added to a new *split-join* relationship. Finally, a new *sequence* control is created, the *split-join* is added as the first parameter and the common part is added as a second parameter.

Algorithm 3 (Join): Replaces split with split-join where possible in a composite service
Inputs: $C=f(a_0,a_1,...,a_n)$, a composite service with sequence and split constructs
Output: C, a composite service with sequence, split and split-join constructs

---

```
do
    for each (aᵢ,aⱼ): i,j in [0,n]
        Set L(aᵢ,aⱼ) = 0
        if aᵢ = aᵢ'∪k,  aⱼ = aⱼ'∪k then
            L(aᵢ,aⱼ) = |k|
    (aₓ, a_y) = arg max (L(aᵢ,aⱼ))
                       (aᵢ,aⱼ)
    L_xy = max(L_ij)
    if L_xy > 0 then
        Let f_ax(a_x0,a_x1,...,a_xn) the
            construct containing k in a_x
        Let f_ay(a_y0,a_y1,...,a_yn) the
            construct containing k in a_y
        k₁=k₂=k
        if f_ax = split then
            k₁ = f_ax(a_x0,a_x1,...,a_xn)
        if f_ay = split then
            k₂ = f_ay(a_y0,a_y1,...,a_yn)
        C= C-{a_x,a_y}
        C=C∪seq(s+j(a_x',a_y'),s(k₁,k₂))
while L_xy > 0
return C
```

For each pair (a_i,a_j) of parameters, the size of their common part is stored in the structure $L(a_i,a_j)$. The size of x is expressed as $|x|$ and refers to the number of simple web services that take part in the functional representation of x. When all the pairs are traversed, the one with the largest common part is selected, that is the pair (a_x,a_y). If the size is a positive number, then the next step checks whether the common part is in a *split* control in the two parameters of the selected pair. If so, the *split* expression must not be divided instead it should be completely removed.

Since this procedure is performed twice, once for every parameter of the couple, the results are two new common parts that should be removed respectively from the parameters. This is realized in parameters a_x' and a_y'. The resulting expressions are added as members of the *split-join* control, symbolized as '*s+j*', which in turn is added as a parameter of the sequence control. Then, the common parts are combined in a *split* control, symbolized as '*s*' and the result becomes the second parameter of the *sequence* control. Finally, this new *sequence* representation replaces the two parameters in the initial composite web service, a_x, a_y. All the previous steps are repeated for the altered composite web service C until no common part exists between its' parameters. Then, C is returned, as was formed from the procedure and represents a composition having *sequence*, *split* and *split-join* control constructs that functionally represents the data flow among the participating simple web services.

After the completion of *Join*, the null parameters of the functional representation created so far are cleared and the pointless control constructs are removed, e.g. the expression *split(A)* becomes *A*. Finally, the duplicate references to control constructs are eliminated This means, that the expression *seq(seq(A,B),C)* is transformed to the equivalent one *seq(A,B,C)*.

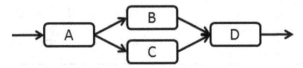

Fig. 3. Composition example

A short example of the whole procedure is given to clarify its workings. In **Fig. 3** a web services composition plan is depicted in a graphical way. The clear service is only the service A, so the result of the *Basic* algorithm, before calling the algorithm *Join*, will be *seq(A,split(seq(B,D),seq(C,D)))*. The *Join* algorithm will notice that the parameters *seq(B,D)* and *seq(C,D)* have the service D as a common part, so the *split* control construct can be replaced by a *split-join* one. By removing the common part from each parameter, the results are the representations *seq(B,null)* and *seq(C,null)* and they are added as parameters in a new *split-join* control. Since the common part is not in a *split* expression in none of the two parameters, the resulting common part is just the service D and the new *sequence* representation is constructed as follows: *seq(split-join(seq(B,null),seq(C,null)),D)*. This representation replaces the *split* of the initial expression and the result is the representation *seq(A,seq(split-join(seq(B,null),seq(C,null)),D))*.

After the completion of the clearing algorithm the functional representation is transformed to *seq(A,seq(split-join(B,C),D))* which finally becomes *seq(A,split-join(B,C),D)* at the last step, which is an accurate functional representation of the composition.

3.2   Creating OWL-S Descriptions

Up to this point, a functional representation has been constructed that supplies suffi-cient information on the data flow of the composition. But, for the procedure to be complete so as to provide the user with a new semantic web service ready for execu-tion, the OWL-S description has to be constructed. This is done based on this repre-sentation. The descriptions that are constructed by the algorithm are the process and the profile descriptions. The OWL-S API, which can be found at [11], was used for their creation. This OWL-S API is a JAVA library providing functions that facilitate the creation of OWL-S descriptions.

First, the process file is created by the algorithm 4, *OWLSProcess*. The algorithm takes as input parameter the composite web service representation C, as formed by the previous algorithms and discerns two cases. If *C* is an atomic service, then the appropriate parts of the OWL-S process description are created that describe the service along with its inputs and outputs. Specifically, for every input of the atomic service, an input element is created by calling the *InputElement* function of the OWL-S API. All the input elements are gathered in a list which is then set as the value of the *hasInput* field of the OWL-S process description. The same steps are followed for the creation of the output list which is the value of the *hasOutput* field in the description.

Algorithm 4 (OWLSProcess): Creates the OWL-S process description
Inputs: $C = f(a_0, a_1, .. a_n)$
Output: The OWL-S process description of C

```
if f = atomic then
    A = OWLSAPI.AtomicProcessElement
    LI = LO = {}
    for each pi ∈ prec(a0)
        ki = OWLSAPI.InputElement(pi)
        LI = LI + {ki}
    OWLSAPI.hasInput(LI)
    for each oi ∈ add(a0)
        mi = OWLSAPI.OutputElement(oi)
        LO = LO + {mi}
    OWLSAPI.hasOutput(LO)
    PE = OWLSAPI.PerformElement
    return PE.add(A)
else
    CC = OWLSAPI.ControlContruct(f)
    CC.add(CLO(C))
    return CC
```

If *C* is not just an atomic service, but instead a composite one, then the appropriate control construct element is created (*seq, split, split-join*) according to *f* and the algorithm *CLO* is called to create the list of the services that takes part in this element. Then, this list is added to the control construct element and this is the object that the *OWLSProcess* algorithm returns. In fact, this object contains all the information about the OWL-S process description of *C*.

Algorithm 5 (CLO): Creates the List Object containing the atomic services of the composite one
Inputs: $C = f(a_0, a_1, .. a_n)$
Output: LO: the List Object

```
if n = 0 then
    return null
LO = OWLSAPI.ListObjectElement
LO.First = OWLSProcess(a0)
LO.Rest = CLO(f(a1, a2, …, an))
return LO
```

The algorithm *CLO* has as input a composite web service functional representation, which is in essence a functional representation with OWL-S control construct connecting the participants services, and creates using the OWL-S API a *List Object* element with the atomic services as parameters. The *list object* is a structure with *First* and *Rest* parts and could be described by an expression like: $First(a_0, Rest(First(a_1), Rest(\ldots)))$.

In *CLO* algorithm, the first parameter of the expression is examined and the *OWLSProcess* algorithm is called for this. The result becomes the head of the constructing list, because it is the service or the composition of services that will be executed first. Then, the *CLO* algorithm is called recursively for *C'*, the composite web service C with a_0 omitted. The result of this call is set as the *Rest* part. Finally, the constructed list object is returned.

The last step in converting the composite web service functional representation to OWL-S description is the creation of the profile description. Here, the composite web service is treated as an atomic service with specific inputs and outputs. The construction of this description is merely based on the methods provided by the OWL-S API's functions.

4   Conclusions and Future Work

Web services are playing an important role in the web applications development field, with which many different systems through the globe can communicate and exchange data using the World Wide Web. Users that need a specific functionality can retrieve the desired web services from the UDDI registries and use them to create the output they are looking for.

SOA architecture has contributed to the rapid and easy web applications development, using as units the web services and combining them to create new, complex services of advanced functionality that can serve even as complete business models. The composition methods studied in this paper differ on user's involvement level. Some initial solutions, of limited autonomy, use workflows and leave the details regarding the location the appropriate services execution and their order to the user. In some more creative solutions, the user doesn't have to find the exact services that will be used, but just provides a description of them. The discovering of services that match with the descriptions and the execution of the resulting workflow are automatically performed without the intervention of the user.

In later studies, the autonomy of the composition procedure is increased. Semantic information concerning the web services is used to describe in a semantic level their functionality. Languages such as OWL-S are used for this purpose. In this way, concept matching becomes possible and so is the check whether two or more services can cooperate. The semantic information is used also by automatic web services composition via planning methods, which are examined in this paper. The composition problem is treated as a planning problem and solved by algorithms of the field.

The result is a plan encoded in planning languages, such as PDDL+ that describes the services that will be used for the composition and the way in which they will be combined to create the desired composite web service. But, for this final service to be available to other users too and to be published in a UDDI registry as an atomic web service and take part to possible future compositions, semantic description of the service have to be created.

The contribution of this paper focuses on converting the PDDL+ plans that constitute the composite web service to OWL-S descriptions of the new web service. Information extracted from the domain of the composition problem is used to construct a composite web service functional representation that describes sufficiently the composition. Then, this representation is used to create the OWL-S profile description of the composite web service, containing information on its inputs and outputs. Also, the OWL-S process description is constructed, that analyzes the way the atomic services are used for the production of the final composite web service.

As for future plans, a complete system could be developed as an extension to the already existing automatic web services composition systems, taking advantage of the algorithms proposed by this paper to construct new semantic web services and publish them in UDDI registries so as to be available to everyone who could be seeking such functionality. In this way, an integrated solution to the composition problem would be provided. Already developed solutions could be used to this direction, such as the system SiTra described in [2], which transforms the OWL-S description of a web service to BPEL, the execution language for web services.

Also, the possibility of creating the grounding OWL-S descriptions of the composite web service could be explored. In this description, the exact data flow among the atomic services will be described and the result will be an even more automated solution. So far, our approach provides the order and the way of the execution of the services taking part in the composition. However, the information of which output is offered as input to the next service is not provided from the OWL-S descriptions of the composite service. This procedure is left to the system that tries to execute the resulting service. It is obvious that by providing this kind of information through the grounding description, the development of systems that execute complex services is greatly simplified.

Moreover, characteristics concerning the quality could be considered for the composite web service. In case there is such data in the semantic descriptions of the atomic web services, procedures that take advantage of them could be developed to construct the quality characteristics of the resulting composite service.

Finally, we aim at integrating web service composition via planning into a decision support system for industrial risk reduction, which represents risk case studies via domain dependent ontologies including the mechanism for building up the risk as a composition of simple physical processes [1].

Acknowledgements. This work has been supported by the Project "Integrated European Industrial Risk Reduction System (IRIS)" (7th Framework Programme, Theme: 4 – NMP, FP7-NMP-2007-LARGE-1, CP-IP 213968-2).

References

1. Angelides, D., Xenidis, Y.: Fuzzy vs. Probabilistic Methods for Risk Assessment of Coastal Areas. In: Linkov, I., Kiker, G.A., Wenning, R.J. (eds.) Environmental Security in Harbors and Coastal Areas: Management using Comparative Risk Assessment and Multi-Criteria Decision Analysis. NATO Security through Science Series (Series C: Environmental Security), pp. 251–266. Springer, Heidelberg (2007) ISBN: 978-1-4020-5801-1

2. Bordbar, B., Howells, G., Evans, M., Staikopoulos, A.: Model Transformation from OWL-S to BPEL Via SiTra. In: Akehurst, D.H., Vogel, R., Paige, R.F. (eds.) ECMDA-FA. LNCS, vol. 4530, pp. 43–58. Springer, Heidelberg (2007)

3. Gerevini, A., Saetti, A., Serina, I.: LPG-TD: a Fully Automated Planner for PDDL2.2 Domains (short paper). In: 14th Int. Conference on Automated Planning and Scheduling (ICAPS 2004), booklet of the system demo section, Whistler, Canada (2004)

4. Gerevini, A., Saetti, A., Serina, I.: LPG-td a planning system (2005), http://zeus.ing.unibs.it/lpg/

5. Ghalab, M., Howe, A., Knoblock, C., McDermott, D., Ram, A., Veloso, M., Weld, D., Wilkins, D.: PDDL – the Planning Domain Definition Language. Technical report. Yale University, New Haven, CT (1998)

6. Hatzi, O., Meditskos, G., Vrakas, D., Bassiliades, N., Anagnostopoulos, D., Vlahavas, I.: Semantic Web Service Composition using Planning and Ontology Concept Relevance with PORSCE II. In: Proceeding of the 2009 Web Intelligence and Intelligent Agent Technology, Milan, Italy, pp. 418–421 (2009)

7. Hoffman, J., Nebel, B.: The FF Planning System: Fast Plan Generation Through Heuristic Search. Journal of Artificial Intelligence Research 14, 253–301 (2001)

8. JPlan: Java Graphplan Implementation, http://sourceforge.net/projects/jplan

9. Klusch, M., Gerber, A., Schmidt, M.: Semantic Web Service Composition Planning with OWLS-XPlan. In: Proceedings of the AAAI Fall Symposium on Semantic Web and Agents. AAAI Press, Arlington (2005)

10. Martin, D., Burstein, M., Lassila, O., McIlraith, S., Narayanan, S., Paolucci M., Parsia, B., Payne, T., Sirin, E., Srinivasan,N., Sycara, K.: OWL-S: Semantic Markup for Web Services (2004), http://www.daml.org/services/owl-s/1.1/

11. OWL-S API, http://www.daml.ri.cmu.edu/owlsapi/

12. Peer, J.: Web Service Composition as AI Planning – a Survey. Technical report. University of St. Gallen (2005)

13. Pellier, D.: PDDL4J (2008), http://sourceforge.net/projects/pddl4j

14. Sacerdoti, E.: The nonlinear nature of plans. In: Proc. of the International Joint Conference on AI, pp. 206–214 (1975)

15. Sirin, E., Parsia, B., Wu, D., Hendler, J., Nau, D.: HTN planning for web service composition using SHOP. Journal of Web Semantics 1(4), 377–396 (2004)

16. Yang, B., Qin, Z.: Composing semantic web services with PDDL. Inform. Technol. J. 9, 48–54 (2009)

17. Yu, H.Q., Reiff-Marganiec, S.: Semantic Web Services Composition via Planning as Model Checking. Technical Report. CS-06-003, University of Leicester (2006)

18. Zhang, P., Huang, B., Sun, Y.: Automatic Web services composition based on SLM. In: Workshop on Semantic Web and Ontology, SWON 2008 (2008)

Convergence Classification and Replication Prediction for Simulation Studies

Andreas D. Lattner[1], Tjorben Bogon[1,2], and Ingo J. Timm[2]

[1] Information Systems and Simulation, Institute of Computer Science
Goethe University Frankfurt, P.O. Box 11 19 32, 60054 Frankfurt, Germany
[2] Business Informatics I, University of Trier, D-54286 Trier, Germany

Abstract. Providing assistance systems for simulation studies can support the user by performing monotonous tasks and keeping track of relevant results. In this paper we present approaches to estimate, if – and when – statistically significant results are expected for certain investigations. This information can be used to control simulation runs or to provide information to the user for interaction. The first approach is used to classify if significance is expected to occur for given samples and the second approach estimates the needed replications until significance is expected be reached. For an initial evaluation of the approaches, experiments are performed on samples drawn from normal distributions.

Keywords: Significance estimation, Simulation control, Statistical tests, Machine learning.

1 Introduction

Nowadays, simulation is widely used in order to evaluate system changes, to perform parameter optimization of systems, or to compare existing alternatives. A clear advantage of simulation is that costs or damages on real systems can be avoided while investigating effects of changes or testing newly planned systems. Simulation is used in various domains, e.g., for marine container terminal planning (Berth Planning and Quay Resources Assignment Problem; [10]), multi-location transshipment problems [3], and clinical resource planning [15].

If complex systems with many parameters are modeled, simulation studies can consist of a large number of single simulation runs and a rather structured and disciplined evaluation has to be performed in order to avoid getting lost in the vast of result data. A support for the non-creative, monotonous tasks in simulation is desirable.

In this work, we present one aspect of the current research project *AssistSim* addressing a support for the performance of simulation studies. The project aims at supporting planning and execution of simulation studies including simulation system control and an automated analysis of intermediate simulation results. In this paper we present an approach to significance estimation in order to estimate, if – and when – statistically significant results are expected for certain investigations. The approach itself can also be applied in other situations, i.e., beyond simulation – for any task where two samples should be compared and where preliminary samples should be used for estimation how many further examples might be needed in order to satisfy certain statistical properties.

J. Filipe and A. Fred (Eds.): ICAART 2011, CCIS 271, pp. 255–268, 2013.

The paper is structured as follows: In Section 2, we discuss some approaches related to ours. The context of the work and the framework of automated operation and control of simulation experiments is presented in Section 3. In Section 4 we introduce our approach to significance estimation. Experimental results are presented in Section 5. A conclusion as well as ideas for further works are discussed in Section 6.

2   Related Work

The automation of (simulation) experiments as well as the application of data mining approaches to experimental settings and results has been addressed by various researchers. Explora is a knowledge discovery assistant system for multipattern and multistrategy discovery (e.g., [8,9]). Klösgen lists four analysis tasks that can be aimed at in such a setting [8]: single-variant analysis (e.g., influence of predefined factors on output variables), comparison of variants, analysis of whole space of variants, and optimization. Klösgen reports that the discovery approach "can constitute a valuable approach also in an area where the analyst has already a lot of knowledge on the domain". Referring to Klösgen three paradigms are fundamental in order to support data exploration: search, visualization, and navigation, and KDD should combine these three paradigms in a semi-automatic process [9]. The Explora system "constructs hierarchical spaces of hypotheses, organizes and controls the search for interesting instances in these spaces, verifies and evaluates the instances in data, and supports the presentation and management of the discovery findings" [9, p. 250]. Different facets of interestingness are also discussed in this paper: evidence, redundancy, usefulness, novelty, simplicity, and generality. The application of Explora to simulation experiments in practical political planning is presented in [8].

King et al. [6] address the "automation of science"; they present the development of the robot scientist "Adam" who autonomously generates functional genomics hypotheses and tests these hypotheses using laboratory automation. An ontology and logical language has been developed to describe the research performed by the robot. The automated conclusions have been confirmed through manually performed experiments. In earlier work, King et al. present genomic hypothesis generation with their "robot scientist" [7]. Experiments and hypothesis generation are performed in a loop where experimental results are evaluated and machine learning (with access to background knowledge) is applied. The output of this step is used in order to select experiments for the next cycle.

Huber et al. apply decision tree learning (ID3) in order to extract knowledge from simulation runs in model optimization [5]. They set up a classification task where the relation between input and output of simulation runs is learned. The result of the learning phase is a decision tree indicating which attributes are important and what attribute values lead to "good" or "bad" behavior. In their paper, they apply the approach to find the range of configuration and workload parameters to optimize the performance for a multiprocessor system. Referring to Huber et al. this qualitative information of the system behavior can be helpful for interpretation of the optimization results.

Burl et al. [2] present an approach to automated knowledge discovery from simulators. They address the "landscape characterization problem" with the aim to identify regions in the parameter space which lead to a certain output behavior. Their approach is based on support vector machines (SVM) and active learning, i.e., they aim at an intelligent selection of new points in the parameter space in order to maximize "the amount of new information obtained" [2, p. 83]. As applications they use asteroid collision simulation and simulation of the Earth's magnetosphere. They report an increase of the efficiency over standard gridding ($2\times$ to $6\times$).

Hoad et al. [4] introduce an algorithm for the automated selection of the number of replications for discrete-event simulation in order to achieve a certain accuracy for simulation output measures taking into account confidence intervals. They apply the approach to different statistical distributions and to a set of simulation models. The authors report that the algorithm is effective in selecting the needed number replications in order to cover the expected mean at a given level of precision.

Similar to some of the related approaches, we apply machine learning in combination with simulation. In this work, machine learning is not used to discover knowledge from simulation results but to learn a classifier for the estimation of statistical properties. In our approach, we take into account statistical tests and the development of their results for the control of simulation runs.

3   Control of Simulation Experiments

In this section, we briefly describe the project context of the approach presented in this paper. The goal of the associated project *AssistSim* is the provision of support functionalities for the performance of simulation studies. Assistance is intended for planning, execution, and analysis of simulation studies. The first aspect – planning assistance – aims at capturing relevant information for a simulation study, e.g., identification of the objects of investigation including parameters as well as their domains, and selection of measurements and target functions. Details about this aspect are planned to be published in a separate paper by our project partners.

The aim of the second aspect – the execution assistance – is the automated operation and control of the simulation system, i.e., the automated execution of simulation runs. This phase is partially connected with the analysis assistance as simulation control depends on intermediate results of simulation runs. However, in the current project, we restrict the analysis assistance to a relevant set of functions for simulation control. A thoroughly designed analysis assistance for the investigation of a large result set of simulation studies is planned to be part of a follow-up project.

The essential task of the simulation execution assistance is the systematic execution of the different settings of the planned experiments. It is distinguished between three different kinds of simulation studies:

1. Exploration: The parameter space has to be explored and interesting findings should be captured.
2. Optimization: Parameter configurations which are expected to lead to good results w.r.t. a target functions should be identified.
3. Comparison: Two or more parameter configurations of a simulation model (or different simulation models) should be compared identifying the best one or ranking the variants w.r.t. a target function.

Using a straight-forward approach, exploration studies can be performed by testing all possible parameter configurations. In the case of continuous variables, a step size for discretization or a selection of parameter values to be investigated has to be performed. Optimization studies can be performed by coupling optimization methods (e.g., meta-heuristics for stochastic combinatorial optimization [1]). For comparison studies, different approaches in the fields of ranking, selection, and multiple comparisons have been introduced (e.g., [16]).

In our work, we focus on discrete-event simulation where various random variables can influence simulation runs. In production scenarios, for instance, randomness can affect the delivery times of parts, duration of processes, and breakdowns of machines. Thus, multiple runs of the same simulation model with identical parameter configurations but different seed values for the random number generators usually leads to varying simulation runs and consequently, to different results of the corresponding observed measurements (e.g., manufacturing output). Technically, this situation can be described as a stochastic process with a (usually unknown) probability distribution and expected value for the target function. Having this situation in mind, a meaningful simulation study has to perform multiple runs of the same simulation setting (i.e., model and parameter configuration) with different random number seed values in order to draw conclusions about configurations' qualities. This multiple runs of the same parameter settings are called replications.

The number of replications and their results are highly relevant for computation of statistical evidence. Depending on these results, mean values and confidence intervals of measurement variables can be computed or statistical tests can be applied in order to check if experimental data supports the hypothesis that one variant leads to better results than another. Obviously, if more replications are performed, a higher confidence w.r.t. the statistical results will be received. However, complex simulation models can lead to costly execution times for single simulation runs and a large parameter space might prohibit performing a large number of replications for each parameter configuration.

The approach presented here aims at the estimation if certain statistical results are expected to be generated and when this could be the case, i.e., how many replications are expected to be needed in order to satisfy certain statistical properties. In this work, we focus on situations where two different variants should be compared by a statistical test. A similar approach could be developed for an estimation when a confidence interval of a measurement is expected to be accurate enough for the expert performing the simulation study.

4   Significance Estimation

In this section, we present our approaches to significance estimation. For initial studies, we have abstracted from simulation runs and use probability distributions and randomly drawn samples of these distributions for a first investigation how data can look like. We assume that observed measurement variables of different simulation runs also underly certain distributions. Using well-known probability distributions allows for structured investigations of our approaches where we can easily generate samples from distributions with known properties. Evaluations with data generated by simulation models can

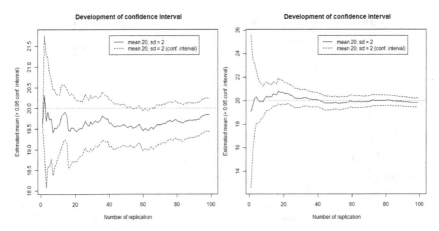

Fig. 1. Two examples for confidence interval development for growing number of samples (normal distribution with mean 20 and standard deviation 2)

be more difficult as the real underlying distribution is not known and if two simulation model variants are compared, it is not clear from the beginning if the distributions of their results differ. In the next subsection we present an analysis of statistical properties before we actually introduce our approaches to significance estimation, namely convergence classification and replication prediction.

4.1 Analysis of Statistical Properties

If we take a look at different successively drawn samples of distributions, we can see an interesting development of values. Figure 1 shows two developments of values from the same distribution (normal distribution with mean 20 and standard deviation 2). The solid blue line shows the estimated mean value using a specific number of sample values. The dashed light blue line shows the confidence interval. It is known that we need four times as many samples in order to halve the size of the confidence interval (e.g., [11]). It can be seen that in one case the mean of the sample is below the actual expected value of the distribution (left part of Figure 1) while in the other case, the line comes close to the actual expected value rather quickly (right part of Figure 1).

Figure 2 shows the development curves of p values of performed t-tests on varying sample sizes. In these graphs, we can see two curves: One where the compared samples are actually drawn from different distributions (blue line; mean 20, stdev 2 vs. mean 21, stdev 3) and another where both compared samples are drawn from the same distribution (dashed red line; mean 20, stdev 2). The two distributions have been selected to have a good overlap in the values on purpose in order to take a look at samples where the difference is not obvious after drawing a few examples. Interestingly, it can be seen (e.g., in the right part of Figure 2) that for these distributions in some cases the graphs can be hardly distinguished (for less than 100 samples for each distribution).

Additionally to the graphs comparing two single samples, the average p values of 100 runs are plotted in Figure 3. As it can be seen, the p values of identical distributions

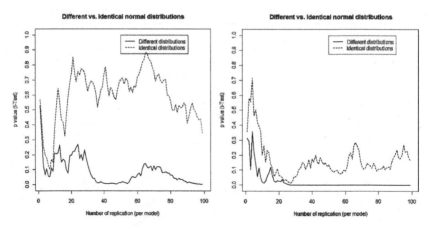

Fig. 2. Two examples for p values of t-test development for growing number of samples (normal distributions with mean 20, stdev 2 vs. mean 21, stdev 3)

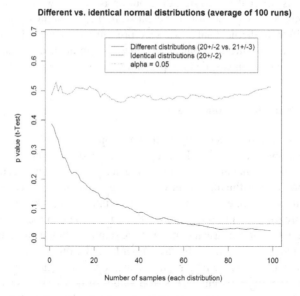

Fig. 3. p value of t-test development for growing number of samples (normal distributions with mean 20, stdev 2 vs. mean 21, stdev 3 - Average of 100 runs

(dashed red line) are close by 0.5 while the p values of the different distributions (solid blue line) move towards the x-axis.

In this study, we focus on the comparisons of two different distributions and leave out the single sample case where only one measurement variable of one variant is taken into account. The following two sections describe two approaches to significance estimation.

4.2 Convergence Classification

Convergence classification aims at estimating if it can be shown that samples from one distribution are better on average (e.g., if it can be shown by a statistical test that the mean is greater than the mean of another distribution). The basic idea is to observe the development of p values while the number of samples is increasing. We have set up the convergence estimation as a classification task. A classifier is trained using a set of positive and negative examples (different distribution vs. identical distribution). This classifier can later be used in order to classify unseen p value series.

In our current implementation, we extract five straight-forward features which are used for classification and have a target attribute with two possible outcomes:

- p_{min}: The minimal p value observed so far.
- p_{man}: The maximal p value observed so far.
- p_{avg}: The average of all observed p values.
- p_{last}: The last known p value (taking into account the whole samples).
- p_{grad}: The "gradient" of the p value development, taking into account first and last known p value in relation to the number of samples.
- $class$: Different or same distribution (diff/same).

In order to train the classifier, we apply the C4.5 algorithm for decision tree learning [13]. In our work, we have integrated the WEKA machine learning program and have used the J4.8 implementation of C4.5 [17].

4.3 Replication Prediction

While significance estimation only aims at the classification if a significant statistical result is expected, the replication prediction task has the goal to estimate the number of needed replications in order to reach the significant result with a statistical test. Thus, in this case we are facing a numeric prediction task.

Various prediction methods could be applied to the data, e.g., from the field of time series prediction. For our initial experiments we decided to apply regression to the known series of p values in order to estimate the subsequent development. Therefore, we use the R project implementation of the nonlinear least squares method (*NLS*) [14].

In order to fit a function to the provided data, we let the regression identify the coefficients a and b of the following formula: $f(x) = \frac{1}{a+bx}$.

The prediction of the number of necessary replications is done by computing the interception point of the curve with the desired significance level α. Equalizing the function with α and solving it for x leads to the predicted number of replications: $x = \frac{1}{\alpha b} - \frac{a}{b}$.

Figure 4 shows the development of p values as well as the regression curve which has been generated from the first 30 p values.

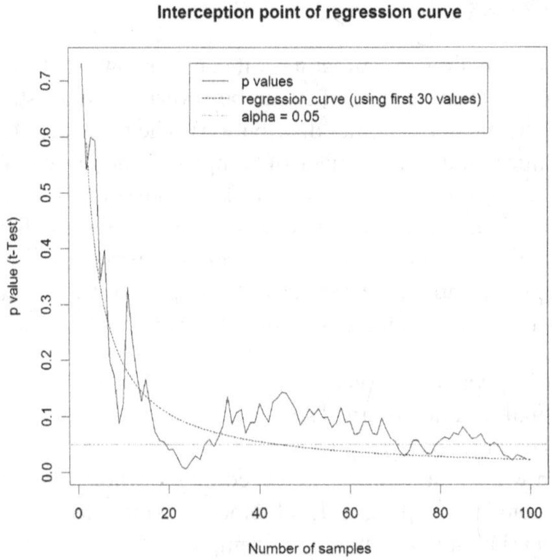

Fig. 4. Replication prediction using nonlinear regression

5   Evaluation

The evaluation consists of three parts. In the first part, the significance classification is applied to distributions with fixed mean and standard deviation. The second part applies the significance classification to randomly generated distributions. In the third part, we apply the replication prediction to fixed distributions.

5.1   Significance Classification for Fixed Distributions

In the first experiment series, we apply the significance classification approach to samples drawn from different distributions with fixed mean and standard deviation. Altogether, we set up three different distribution pairs which are evaluated. In our evaluation, we investigate the classifier accuracy for varying numbers of p values $(5, 10, \ldots, 95)$ taken into account for training and classification. For each distribution pair, ten independent runs are performed where 500 training and 500 testing examples (50% same, 50% different distributions) are generated.

Table 1 shows a summary of the results indicating the average accuracy of the approach as well as the accuracy if simple comparison of the last p value with the α threshold is performed, i.e., if $p_{last} < \alpha$, it will be classified to *diff*, otherwise to *same*. Additionally, for each number of p values we perform a statistical significance test comparing the accuracies of the classifier vs. the α-threshold approach (ten accuracy values each) and capture the corresponding p values of the test. Significant results are emphasized with bold letters. The accuracies for the second distribution pair ($\mu_1 = 20$, $sd_1 = 2$ vs. $\mu_2 = 22$, $sd_2 = 2$) is shown in Figure 5.

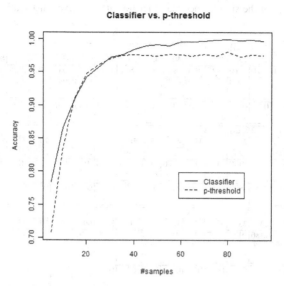

Fig. 5. Accuracies of significance classifier and p-threshold (normal distributions with mean 20, stdev 2 vs. mean 22, stdev 2)

The following tree is an example for a trained classifier with 30 provided p values:

```
currP <= 0.129937
|    currP <= 0.033323: diff (158.0/6.0)
|    currP > 0.033323
|    |    avg <= 0.099929: diff (16.0)
|    |    avg > 0.099929
|    |    |    totgrad <= 0.000215: diff (63.0/21.0)
|    |    |    totgrad > 0.000215: same (4.0)
currP > 0.129937: same (259.0/40.0)
```

The results of these experiments indicate an advantage of the trained classifier in comparison to the threshold-based method in many cases. Significant ($\alpha = 0.05$) differences in the accuracies can be observed for 5-40 p values in the first setting. In the second setting (where the mean difference is greater), for 5 and 10 as well as from 40 - 95 better results can be achieved using the classifier. In the third setting (even greater difference between means), the classifier is better for 5 and the settings with 20 or more p values.

5.2 Significance Classification for Random Distributions

In a second test, we do not use distributions with fixed mean and standard deviation values, but randomly generated distributions. The generation of the random distributions works as follows:

- Select a random mean value for the first distribution: $\mu_1 \in [50, 500]$.
- Randomly select a standard deviation value for the first distribution $sd_1 \in [0, 0.3\mu_1]$.

Table 1. Accuracies of the significance classifier for different fixed distributions (0^* indicates p values < 0.001)

$\mu_1 = 20, sd_1 = 2, \mu_2 = 21, sd_2 = 2$																		
Appr. 5	10	15	20	25	30	35	40	45	50	55	60	65	70	75	80	85	90	95
Sign.Cl..655	.698	.751	.770	.810	.825	.836	.855	.865	.895	.898	.909	.915	.928	.929	.935	.944	.940	.952
p thresh. .570	.628	.660	.716	.763	.783	.810	.839	.857	.887	.891	.903	.922	.930	.933	.940	.947	.946	.955
p (t-test) 0^*	0^*	0^*	0^*	0^*	0^*	.001	.004	.151	.171	.129	.185	.845	.636	.783	.914	.688	.908	.755
$\mu_1 = 20, sd_1 = 2, \mu_2 = 22, sd_2 = 2$																		
Appr. 5	10	15	20	25	30	35	40	45	50	55	60	65	70	75	80	85	90	95
Sign.Cl..784	.865	.910	.942	.957	.972	.975	.984	.989	.991	.989	.995	.996	.996	.998	.999	.998	.998	.997
p thresh. .708	.836	.911	.947	.960	.970	.974	.976	.974	.974	.976	.975	.974	.976	.973	.980	.973	.976	.975
p (t-test) 0^*	0^*	.561	.825	.807	.341	.408	.021	0^*	0^*	0^*	0^*	0^*	0^*	0^*	0^*	0^*	0^*	0^*
$\mu_1 = 20, sd_1 = 2, \mu_2 = 23, sd_2 = 2$																		
Appr. 5	10	15	20	25	30	35	40	45	50	55	60	65	70	75	80	85	90	95
Sign.Cl..893	.958	.972	.986	.994	.997	.997	.997	.997	.999	.999	.999	1.0	1.0	1.0	1.0	1.0	1.0	1.0
p thresh. .865	.959	.970	.970	.97	.975	.977	.976	.973	.974	.980	.975	.975	.976	.975	.972	.974	.976	.974
p (t-test) 0^*	.653	.418	0^*	0^*	0^*	0^*	0^*	0^*	0^*	0^*	0^*	0^*	0^*	0^*	0^*	0^*	0^*	0^*

Table 2. Accuracies of the significance classifier for randomly generated distributions (0^* indicates p values < 0.001)

Appr. 5	10	15	20	25	30	35	40	45	50	55	60	65	70	75	80	85	90	95
Sign.Cl..563	.601	.655	.680	.701	.717	.733	.730	.745	.761	.775	.780	.778	.777	.793	.808	.801	.790	.806
p thresh. .539	.564	.580	.599	.612	.615	.626	.627	.635	.634	.646	.642	.649	.652	.653	.666	.654	.654	.658
p (t-test) .014	.002	0^*	0^*	0^*	0^*	0^*	0^*	0^*	0^*	0^*	0^*	0^*	0^*	0^*	0^*	0^*	0^*	0^*

- Generate random mean value for the second distribution within the standard deviation of the first one: $\mu_2 \in [\mu_1 - sd_1, \mu_1 + sd_1]$.
- Randomly select a standard deviation value for the second distribution: $sd_2 \in [0, 2sd_1]$.

Instead of drawing random samples from the same distribution, in this experiment series for each training and testing example, the distributions are generated randomly. Thus, more general classifiers are trained taking into account various different distributions. Once again, ten independent runs with 500 training and 500 testing examples are performed. The results (average accuracies and p value of t-tests) of these experiments are presented in Table 2. A graph comparing the significance classifier with the p-threshold method is shown in Figure 6. The results indicate better results of the classifier for all tested numbers of p values. In some cases an accuracy difference with approximately 15 percent points occurs in these experiments.

5.3   Replication Prediction for Fixed Distributions

The third part of the evaluation addresses the replication prediction. Additionally to the approach presented in Section 4.3, we use a statistical power analysis in order to

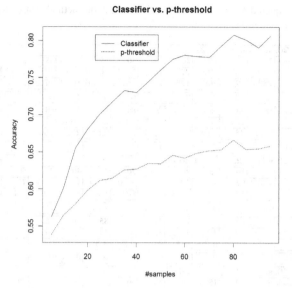

Fig. 6. Accuracies of significance classifier and p-threshold for different number of used p values for randomly generated distributions

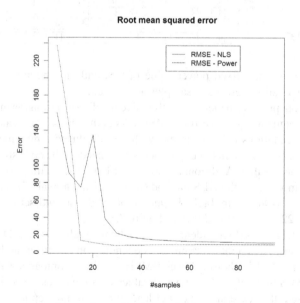

Fig. 7. Root mean squared error for replication prediction (normal distributions with mean 20, stdev 2 vs. mean 22, stdev 2)

estimate the needed sample size (e.g., [12]). We use the implementation of R Project (*power.t.test*) with the estimated mean difference of the corresponding number of

Table 3. Root mean squared errors of the replication prediction methods for different fixed distributions (mean values of 100 runs)

$\mu_1 = 20$, $sd_1 = 2$, $\mu_2 = 21$, $sd_2 = 2$									
Approach	5	10	15	20	25	30	35	40	45	50
NLS	179.7	157.1	192.1	77.7	153.5	55.3	102.6	62.0	145.2	149.1
Power	282.1	223.5	166.9	125.2	102.4	167.8	155.6	103.6	97.9	27.6
#invalid	25	13	9	9	7	6	5	6	6	5
Approach	55	60	65	70	75	80	85	90	95	
NLS	153.1	101.9	129.3	100.0	80.7	70.9	65.0	60.6	57.1	
Power	96.1	38.3	26.8	24.9	25.5	25.3	24.8	25.7	26.3	
#invalid	3	3	2	2	1	1	1	1	1	

$\mu_1 = 20$, $sd_1 = 2$, $\mu_2 = 22$, $sd_2 = 2$									
Approach	5	10	15	20	25	30	35	40	45	50
NLS	160.1	91.1	74.8	134.5	38.5	22.0	18.0	15.9	14.7	13.8
Power	237.4	145.0	13.8	11.3	9.2	8.0	8.5	8.5	8.8	9.1
#invalid	13	8	3	2	2	2	1	1	1	0
Approach	55	60	65	70	75	80	85	90	95	
NLS	13.3	12.8	12.5	12.2	12.0	11.8	11.7	11.5	11.4	
Power	9.2	9.2	9.2	9.3	9.2	9.3	9.3	9.4	9.4	
#invalid	0	0	0	0	0	0	0	0	0	

$\mu_1 = 20$, $sd_1 = 2$, $\mu_2 = 23$, $sd_2 = 2$									
Approach	5	10	15	20	25	30	35	40	45	50
NLS	115.4	103.3	8.1	6.3	5.6	5.2	4.9	4.8	4.6	4.6
Power	42.5	4.1	3.1	3.1	3.3	3.5	3.5	3.6	3.6	3.6
#invalid	6	5	5	5	5	4	4	4	4	4
Approach	55	60	65	70	75	80	85	90	95	
NLS	4.5	4.4	4.4	4.4	4.3	4.3	4.3	4.2	4.2	
Power	3.6	3.6	3.7	3.7	3.7	3.7	3.7	3.7	3.7	
#invalid	4	4	4	4	4	4	4	4	4	

sample sizes, $\alpha = 0.05$, a fixed power value of 0.8, and the one-sided test setting. The result is an estimation how many samples are needed.

We apply the both prediction methods to the same fixed distributions as in Section 5.1 and capture the root mean squared error (RMSE). As both methods generate unrealistic high replication estimations in some cases, we have introduced a maximal threshold. Whenever this threshold (1000 in our experiments) is exceeded, the corresponding value is set to the threshold value. Additionally, we count how many times no interception point could be computed for the NLS method (marked with "#invalid"). The results of these experiments are shown in Table 3. One graph of the second setting ($\mu_1 = 20$, $sd_1 = 2$ vs. $\mu_2 = 22$, $sd_2 = 2$) is shown in Figure 7.

The experimental results do not identify one of the methods as better. Depending on the number of p values taken into account and depending on the different distributions, one or the other method leads to a lower RMSE. A direct comparison is not really possible, as the NLS method leads to invalid values in some cases. Especially, if only few values are used, the regression does not lead to a valid interception point (25 out 100 for the first setting and 5 p values). For the first two distribution pairs (those with a higher overlap) and low numbers of p values (5 and 10), the NLS method leads to better mean error of the 100 performed runs. Early prediction results are of special interest as it allows for an early intervention (of the system or user).

6 Conclusions

In this paper, we have addressed the estimation of statistical properties. We have presented two approaches: one for classification if a development of observed p values is expected to lead to a statistical significant result and another one for the prediction of needed sample sizes, also by taking into account previous samples.

The comparison of the significance classifier with a threshold-based classification leads to significantly better results in most cases. Especially in the experiments with randomly generated distributions, a better performance could be observed. For samples where the mean values of the distributions are not too close, high classification accuracies (almost 90%) can be reached even if only five p values are used.

The experiments with the replication prediction do not exhibit that clear results. The power-based predictor leads to lower average error rates for the setting with a greater difference of the mean values as well as in the cases where many p values are used. In some settings, the regression-based approach leads to better results, e.g., if only 5 or 10 p values are used for the closer distribution pairs.

It should be at least mentioned that the approaches presented here – multiple statistical tests with increasing sample sizes – are violating regular statistical procedures where the setting should be clear before experiments are performed and multiple tests with the same data should be avoided or at least taken into account by using adapted significance levels. For exploration-based studies such approaches might be acceptable in order to filter out certain variants or if one is aware of the statistical statement.

The current significance classifier uses a rather small set of straight-forward features. It would be interesting to investigate if further features can lead to an improvement of the classifier's accuracy. The prediction of the needed number of replications has not been addressed deeply within this study. In this case, an investigation of further statistical or time series prediction methods should be performed. Further experiments are needed in order to make statements in what situations adequate results are expected. Another topic for future work is the application of the approaches to simulation systems. In this context, relevant research questions are how the approaches perform if other distributions (than normal distributions) are present and what the underlying distributions of certain observation variables of simulation models are.

Acknowledgements. The content of this paper is a partial result of the AssistSim project (Hessen Agentur Project No.: 185/09-15) which is funded by the European Union (European Regional Development Fund - ERDF) as well as the German State Hesse in context of the *Hessen ModellProjekte*. We would like to thank our AssistSim project partners for interesting discussions on the automation of simulation experiments.

References

1. Bianchi, L., Dorigo, M., Gambardella, L.M., Gutjahr, W.J.: A survey on metaheuristics for stochastic combinatorial optimization. Natural Computing: An International Journal 8(2), 239–287 (2009)

2. Burl, M.C., DeCoste, D., Enke, B.L., Mazzoni, D., Merline, W.J., Scharenbroich, L.: Automated knowledge discovery from simulators. In: Ghosh, J., Lambert, D., Skillicorn, D.B., Srivastava, J. (eds.) Proceedings of the Sixth SIAM International Conference on Data Mining, Bethesda, MD, USA, April 20-22 (2006)
3. Ekren, B.Y., Heragu, S.S.: Simulation based optimization of multi-location transshipment problem with capacitated transportation. In: WSC 2008: Proceedings of the 40th Conference on Winter Simulation, pp. 2632–2638 (2008)
4. Hoad, K., Robinson, S., Davies, R.: Automated selection of the number of replications for a discrete-event simulation. Journal of the Operational Research Society (October 2009), http://dx.doi.org/10.1057/jors.2009.121
5. Huber, K.P., Syrjakow, M., Szczerbicka, H.: Extracting knowledge supports model optimization. In: Proceedings of the International Simulation Technology Conference, SIMTEC 1993, San Francisco, pp. 237–242 (November 1993)
6. King, R.D., Rowland, J., Oliver, S.G., Young, M., Aubrey, W., Byrne, E., Liakata, M., Markham, M., Pir, P., Soldatova, L.N., Sparkes, A., Whelan, K.E., Clare, A.: The automation of science. Science 324(5923), 85–89 (2009)
7. King, R.D., Whelan, K.E., Jones, F.M., Reiser, P.G.K., Bryant, C.H., Muggleton, S.H., Kell, D.B., Oliver, S.G.: Functional genomic hypothesis generation and experimentation by a robot scientist. Nature 427, 247–252 (2004)
8. Klösgen, W.: Exploration of simulation experiments by discovery. In: AAAI 1994 Workshop on Knowledge Discovery in Databases (KDD 1994), Technical Report WS-94-03. pp. 251–262. The AAAI Press, Menlo Park (1994)
9. Klösgen, W.: Explora: A multipattern and multistrategy discovery assistant. In: Fayyad, U.M., Piatetsky-Shapiro, G., Uthurusamy, R. (eds.) Advances in Knowledge Discovery and Data Mining, pp. 249–271. AAAI Press, Menlo Park (1996)
10. Laganá, D., Legato, P., Pisacane, O., Vocaturo, F.: Solving simulation optimization problems on grid computing systems. Parallel Comput. 32(9), 688–700 (2006)
11. Law, A.M.: Simulation Modeling & Analysis, 4th edn. McGraw-Hill (2007)
12. Park, H.M.: Hypothesis testing and statistical power of a test. Working paper. The university information technology services (UITS), Center for Statistical and Mathematical Computing, Indiana University (2008)
13. Quinlan, J.R.: C4.5 - Programs for Machine Learning. Morgan Kaufmann Publishers, Inc. (1993)
14. R Development Core Team: R: A Language and Environment for Statistical Computing. R Foundation for Statistical Computing, Vienna, Austria (2010), http://www.R-project.org; ISBN 3-900051-07-0
15. Swisher, J.R., Jacobson, S.H.: Evaluating the design of a family practice healthcare clinic using discrete-event simulation. Health Care Management Science 5(2), 75–88 (2002)
16. Swisher, J.R., Jacobson, S.H., Yücesan, E.: Discrete-event simulation optimization using ranking, selection, and multiple comparison procedures: A survey. ACM Trans. Model. Comput. Simul. 13(2), 134–154 (2003)
17. Witten, I.H., Frank, E.: Data Mining: Practical machine learning tools and techniques, 2nd edn. Morgan Kaufmann, San Francisco (2005)

Part II

Agents

The Life of Concepts: An ABM of Conceptual Drift in Social Groups

Enrique Canessa[1], Sergio Chaigneau[2], and Ariel Quezada[3]

[1] Facultad de Ingeniería y Ciencias & Centro de Investigación de la Cognición
Universidad Adolfo Ibáñez, Balmaceda 1625, Recreo, Viña del Mar, Chile
[2] Escuela de Psicología & Centro de la Investigación de la Cognición
Universidad Adolfo Ibáñez, Diagonal Las Torres 2640, Peñalolén, Santiago, Chile
[3] Escuela de Psicología & Centro de la Investigación de la Cognición
Universidad Adolfo Ibáñez, Balmaceda 1625, Recreo, Viña del Mar, Chile
{ecanessa,sergio.chaigneau,ariel.quezada}@uai.cl

Abstract. Based on the premise that conceptual agreement (i.e., feeling that we share an idea with others) is always inferential, we develop an ABM that models the conditions under which a concept will gain or loose strength in the minds of individuals. The ABM is based on simple assumptions, generally consistent with psychological and philosophical analyses on the subject. We assume that different members of a population have slightly different versions of one similar conceptualization, that inferred agreement may be true or illusory, and that a concept that promotes agreement (true or illusory) increases its strength. Our analyses (simulated experiments and probability models) test the influence of several variables on the fate of a concept (i.e., whether it strengthens or weakens in the minds of individuals), and support the conclusion that the most important parameters are the probabilities of true and illusory agreement afforded by the concept.

Keywords: Agent-based modelling, Shared meaning, Conceptual content, Markov chain.

1 Introduction

Concepts appear to have a life-cycle in the cultures in which they exist. Concepts are born at a certain point in time, spread or not through culture, and die out. Our view here is that the fate of concepts in culture depends on their usefulness, and that a concept is useful when it generates episodes of shared meaning, thus allowing social cohesion and the coordination of behaviour. Given that meaning is something that happens in individual minds, how is it possible that people agree about a meaning? Psychological inquiry often assumes that meaning is shared by resorting to direct reference, i.e., by pointing to the referred object, rather than by describing it [1], [2], [3], [4], [5], [6], [7], [8] and [9]. Though this approach may work for concrete objects, it does not solve the problem of how people agree about the meaning of diffuse objects (abstract entities like, e.g., democracy, womanhood, happiness). Direct

J. Filipe and A. Fred (Eds.): ICAART 2011, CCIS 271, pp. 271–286, 2013.

reference does not apply for these objects because they lack clear spatio-temporal limits, thus preventing the use of direct reference in interactions. Furthermore, everyday concepts like those illustrated above are notoriously ill-defined; making shared meaning even more mysterious [10]. In our current work, we hold the view that shared meaning is possible because meaning is conventional, i.e., there is a limited set of meanings that apply to a given situation [11], [12], [13]. Constraining the number of concepts that apply on a given occasion, makes agreement a tractable problem. However, even if a group of people has developed conceptual conventions, the likely case is that each person instantiates a somewhat different version of those concepts (e.g., people may conceptualize "leadership" in slightly different ways). Furthermore, even if a group of people has conventions about more or less dichotomous concepts (e.g., "cowardice" and "courage"), a person could still be wrong about which one is being deployed by someone else at a given moment (e.g., if someone says "suicide", she may be thinking of "cowardice" while I may be thinking of "courage"). Consequently, an individual can never know for sure whether someone else agrees or not with his conceptualization of a given event (even when being explicit). Agreement is a probabilistic inference [14].

The ABM we report here focuses on two probabilities that represent the above mentioned inference. First, the probability of *true agreement* (symbolized by $p(a1)$), which stands for the probability that two agents (an *observer* and an *actor*) agree on something given that they instantiate different versions of the same concept (i.e., the "leadership" example above). Second, the probability of *illusory agreement* (symbolized by $p(a2)$), which stands for the probability that observer and actor agree, given that they instantiate different concepts altogether (i.e., the "courage" or "cowardice" example above).

2   Conceptual Description of the ABM

Our current ABM represents a social group which has a set of conventional conceptual states that, for ease of exposition, we will call the *focal set*. These states can represent different versions of the same concept (e.g., different versions of "leadership"; or a set of closely related concepts, such as "miserly", "stingy", "scrooge"). Our $p(a1)$ probability reflects the degree of overlap between the different versions in the focal set (greater overlap implies greater probability of true agreement). Our $p(a2)$ probability reflects the degree of contrast against alternative conceptualizations (lower contrast implies greater probability of illusory agreement). The system models the dynamical trajectories of concepts as they become increasingly or decreasingly relevant for agents depending on their capacity to generate agreement of any type.

In each simulation run, agents act as observers and actors. Observers seek evidence that actors share their concept. Actors have a certain probability that they will or will not act according to the focal set concept. If they act according to the focal set concept, that specific interaction has a probability $p(a1)$ of providing observers evidence of a shared concept. If actors don't act according to the focal set concept (i.e., they act according to the contrast concept), that specific interaction has a probability $p(a2)$ of providing observers evidence of a shared concept.

We make some very simple and quite generally accepted assumptions about our agents' psychology. If the observer witnesses evidence (blind to whether it is true or illusory agreement), then his own conceptual state increases its relevance in his mind (i.e., our cognitive assumption; c.f., [15], [16], [17]), will be more likely to guide his behaviour in the future (i.e., our motivational assumption; c.f., [18]), and the observer will want to interact with that particular agent again in the future (i.e., our social assumption; c.f., [19]).

3   ABM Implementation

The theory presented in section 2 is implemented in an agent-based model (ABM). In summary, the ABM represents how concepts spread and get stronger (or weaker) in a social group, by observing the behaviour of other members. In the ABM, each individual is an agent (actor, A), which acts according to its concept with probability equal to the strength of the concept. That behaviour is observed by another agent (observer, O), and that changes the strength of its concept. In general, if the observed behaviour agrees with the behaviour expected from O's concept, then O's concept strengthens. Conversely, if the observed behaviour differs from what is expected from that concept, then O's concept weakens. Concurrently, the agents begin to interact more frequently with those that have strengthened their concepts. In the following paragraphs we describe the details of the ABM.

In the social group that the ABM represents, one can set the number of members that belong to the group. Each agent can have one of five different related concepts or versions of the same concept and each of the concepts or versions is represented by a number in the [0, 1] interval, labelled the *coefficient* of the concept. This coefficient determines the probability that an agent behaves according to the given concept. The initial values of the coefficients are sampled from a normal distribution with a mean and standard deviation, which can be set. The model checks that the assigned coefficients will always remain in the [0, 1] interval.

Agents modify the strength of their concept's coefficients by observing the behaviour of other agents. Every time they see behaviour consistent with their concepts, the corresponding coefficients are incremented by 0.02. On the other hand, if the observed behaviour is not consistent with their concepts, the corresponding coefficients are decremented by 0.02. The model makes sure that the coefficients always remain inside the [0, 1] interval. Thus, when an agent sees that another agent acts according to its concept, it is more probable that the agent will act according to its own concept in the future. These actions spread concepts throughout the group.

Agents develop interaction preferences as they observe each other. Specifically, agents will tend to interact more frequently with agents who have confirmed their concepts in previous interactions, and indirectly, they will be less likely to interact with those that have not confirmed their concepts. This aspect of the ABM limits the diffusion of concepts, given that it imposes certain heterogeneity to the diffusion speed of the concepts. It could even cause the weakening of some concepts among certain members of the group. To simulate this aspect of our theory, each agent has an interaction probability with the rest of the agents. Taking into account computational restrictions, those probabilities take only discrete values (0.08, 0.11, 0.17, 0.26 and 0.38, probabilities which increase by approximately 50% between successive values). At the start of a simulation run, all the agents are assigned a probability equal to 0.08, which means that an agent will randomly interact with any other agent. Then, as the run advances, if agent A confirms O's concept, agent O will increase its interaction probability with A to the immediately larger value. For example, if agent A's interaction probability was 0.08, then agent A will increase that probability to 0.11.

A last aspect incorporated in the ABM is that in a social group, it might exist more than one version of a concept. Thus, the model allows setting the number of versions that will be present in a group between 1 and 5. Each version will be assigned to a number of agents equal to the total number of agents in a group divided by the number of versions.

Each agent O determines whether its concept will strengthen or weaken according to the following rules:

a) If A acts according to its own concept in the focal set, and A's conceptual content completely coincides with O's conceptual content, then O's concept will strengthen with probability equal to 1.

b) If A acts according to its own concept in the focal set, and that concept is a version of the same concept in O's focal set (but not identical), then O's concept will strengthen with probability equal to $p(a1)$ and will weaken with probability equal to $1 - p(a1)$.

c) If A does not act according to its concept in the focal set (i.e., acts according to a contrasting concept), and the contrasting concept overlaps somewhat with the O's concept in the focal set, then O's concept will strengthen with probability equal to $p(a2)$ and will weaken with probability equal to $1 - p(a2)$.

d) If A does not act according to its concept in the focal set (i.e., acts according to a contrasting concept), and A's conceptual content completely coincides with O's conceptual content, then O's concept will weaken with probability equal to 1.

Finally, each simulation cycle or step of the ABM is composed of the following actions:

i) From the set of all agents, randomly select without replacement an observer agent (O).

ii) O selects one actor agent (A), according to the interaction probabilities that O has for the rest of the agents.

iii) A behaves according to its concept with probability equal to the value of the coefficient of the concept that it has.

iv) O observes that behaviour and modifies its coefficient of the concept, according to the rules that were previously described.

v) Repeat steps i) through iv) until all agents have been observers.

We acknowledge that this description may not provide the reader with a complete understanding of our ABM. Space restrictions preclude providing greater detail. In lieu, the ABM is available as a zip file. The interested reader can download it from http://www.uai.cl/images/stories/CentrosInvetigacion/CINCO/CAT_1_English.zip. To run it, you will need first to install Netlogo version 4.0.4 (http://ccl.northwestern.edu/netlogo/). Once on the ABM interface, do the following steps to run a simulation: (1) Input the simulation parameters, be it with the slider controls or typing the desired $p(a1)$ and $p(a2)$ values in the appropriate windows. (2) Press SETUP. (3) Press SIMULATE. (4) If you want to pause a run, simply press SIMULATE (you will need to press it again to resume).

4   Preliminary Results

Once the ABM was implemented and verified, we carried out several runs to assess the dynamics of the coefficients of the concepts that emerged. After gaining some insights into the dynamics of those coefficients and how different combination of parameters changed those dynamics, we performed experiments fixing the value of some parameters as follows: number of agents = 100; number of versions of a concept (or number of related concepts) = 5; initial value of coefficients of concepts = 0.5; and changing the value of $p(a1)$ and $p(a2)$ between 0.1 and 0.95. According to the combination of values for $p(a1)$ and $p(a2)$, three different dynamics emerged.

4.1   Convergence to Zero

When we set small values for $p(a1)$ and $p(a2)$, for example $p(a1) = 0.1$ and $p(a2) = 0.2$, then the coefficients rapidly decrease and get close to zero, remaining at that value. This can be seen in Figure 1, where we plotted the mean value of the coefficient of each version of the concept (c1 through c5) over simulation steps. The mean value of each coefficient is calculated by averaging the individual value of the coefficient of the agents that have each version of the concept.

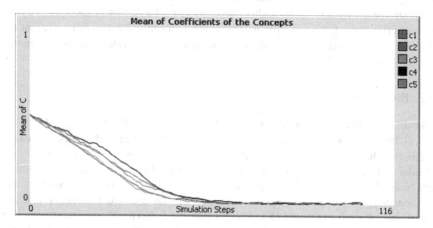

Fig. 1. Mean of coefficients of concepts (c1 through c5), for $p(a1) = 0.1$ and $p(a2) = 0.2$

That happens because the probability that O observes a behaviour consistent with its concept is very low, since $p(a1)$ and $p(a2)$ are small. Thus, in general, concepts tend to weaken, which in turn makes it more probable that the coefficients will keep decreasing throughout the run. Conceptually, this is equivalent to concepts that are not useful to generate agreement in a social group, and rapidly die out.

4.2   Convergence to One

When both $p(a1)$ and $p(a2)$ are set to large values, for example $p(a1) = 0.8$ and $p(a2) = 0.9$, then the coefficients quickly increase and take values close to 1.0. This can be seen in Figure 2.

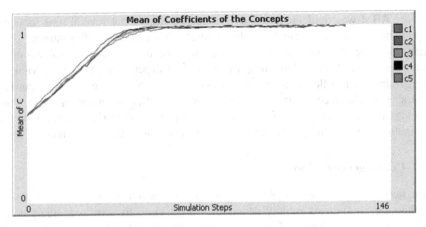

Fig. 2. Mean of coefficients of concepts (c1 through c5), for $p(a1) = 0.8$ and $p(a2) = 0.9$

Contrary to what happened in 4.1, in this situation $p(a1)$ and $p(a2)$ are large, thus favouring that O observes a behaviour consistent with its concept, which will strengthen the coefficient. In turn, this makes more probable that in successive cycles, all the coefficients of the concepts will increase. Conceptually, this is equivalent to a group of related concepts synergistically increasing their relevance by promoting agreement in culture.

4.3   Bifurcation

Using different combinations for $p(a1)$ and $p(a2)$, such as (0.20, 0.80), (0.60, 0.40), (0.80, 0.16), we saw that some concepts tended to strengthen and others to weaken. We labelled this type of dynamic a "bifurcation", which is shown in Figure 3.

Under this condition, a relatively large value of $p(a1)$ or $p(a2)$, but not of both of them, will promote that on each run some O's observe behaviours consistent with their concepts. However, since $p(a1)$ or $p(a2)$ will have a small value, it will also happen that on each run some O's will not observe behaviours consistent with their concepts. Thus, some versions of the concept will strengthen and others weaken in

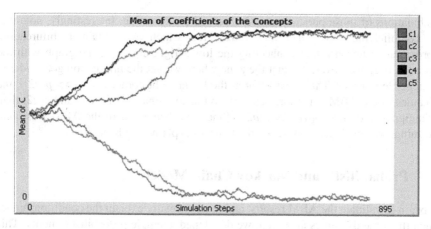

Fig. 3. Mean of coefficients of concepts (c1 through c5), for $p(a1) = 0.2$ and $p(a2) = 0.8$

agents' minds. Conceptually, this is equivalent to a group of related concepts that have a weakly contrasting (i.e., somewhat overlapping) conceptual alternative, such as might be the case of concepts of *male* versus *female* gender, and concepts of *liberal* versus *conservative* political views. Concepts like these tend to become polarized in large social groups, just as occurs in our model's bifurcations.

4.4 Map of Dynamics

Since we realized the significant influence of $p(a1)$ and $p(a2)$ on the type of dynamics that emerged, we ran simulations using more combinations for these two variables. The types of dynamics of the coefficients that emerged are presented in Figure 4.

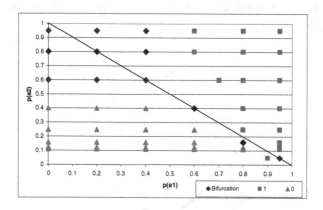

Fig. 4. Dynamics that emerge for coefficients of concepts, according to values set to $p(a1)$ and $p(a2)$ (bifurcation, 1 = convergence to 1, 0 = convergence to 0)

We confirmed that for small values of $p(a1)$ and $p(a2)$, we obtained convergence to zero; for large values of $p(a1)$ and $p(a2)$, we saw convergence to one; and for other combinations of those parameters, we observed a bifurcation. Interestingly, note that the combinations for $p(a1)$ and $p(a2)$, for which we obtain a bifurcation, approximately lie on a line connecting the lower right corner of the graph with the upper left one. Moreover, see that the zone where we get the bifurcation gets wider at the upper left corner. That means that at the lower right corner $(p(a1) \gg p(a2))$, the dynamics of the ABM gets more sensitive to the combination of $p(a1)$ and $p(a2)$ than at the opposite corner $(p(a1) \ll p(a2))$. Since that behaviour of the ABM was quite intriguing, we developed another model to try to explain such behaviour.

5   Probabilistic and Markov Chain Model

To begin to validate the ABM results, and more formally explain the conditions under which the three dynamics appeared, we developed a simple probabilistic model. This initial model justified why the bifurcation emerged when the values for $p(a1)$ and $p(a2)$ roughly lie on a line connecting the lower right corner of the graph with the upper left one, as shown in Figure 4.

5.1   Simple Conditional Probability Model

To explain the three different types of dynamics that emerge from the ABM, we use a simple conditional probability model to calculate an initial probability that a concept will strengthen (p_{if}). If p_{if} is small, then most probably, the coefficient, which represents the concept, will decrease. On the other hand, if p_{if} is large, the coefficient will increase. If p_{if} is about 0.5, then we obtain the ideal situation under which a bifurcation might occur, i.e. each coefficient will have a 50% chance of decreasing and a 50% chance of increasing, thus making it possible that about half of them will diminish and half of them will augment. Figure 5 shows a conditional probability tree that helps calculate p_{if}.

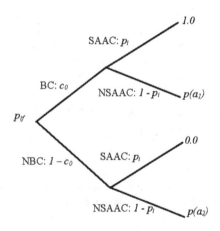

Fig. 5. Conditional probability tree for calculating p_{if}

In this model, p_{if} depends on whether an agent A (actor) behaves according to its concept (event BC), which has probability equal to the initial value of the coefficient that we set (c_0), or not (event NBC, with probability $1 - c_0$). Then, if A acts according to its concept, then there is a p_i probability that A and O share all their conceptual content (event SACC), and a $1- p_i$ probability that they don't share it all (i.e., each has a different version of the same concept, event NSAAC). If they share all their conceptual content (with probability pi), then it is certain that O will strengthen its concept's coefficient. If they share versions of the same concept (with probability $1-p_i$), then it is less than certain ($p(a1)$) that O will strengthen its concept.

On the other hand, if A does not behave according to its concept (event NBC), and A and O share all their conceptual content (event SAAC, with probability p_i), then O's concept will certainly weaken. Alternatively, if A and O do not share all their conceptual content (event NSAAC), then it might happen that A provides O with some portion of the conceptual content, and thus O's concept might strengthen with probability $p(a2)$.

Solving the probability tree of Figure 5 for p_{if}, we obtain:

$$P_{if} = c_0[p_i + p_{a1}(1 - p_i)] + (1 - c_0)(1 - p_i)p_{a2}. \tag{1}$$

In (1), remember that p_i corresponds to the probability that agents share all their conceptual content, i.e. that they have the same version of a concept. Thus, we can calculate p_i for the beginning of a run. In such initial condition, we will have N/V agents with the same version of a concept, where N equals the total number of agents and V is the number of different versions of a concept. Then, the initial probability that agent O will interact with an agent A that has the same version of the concept will be equal to the number of other agents that have the same version as O has (without counting O), divided by the total number of agents (without counting O):

$$p_i = \frac{\frac{N}{V} - 1}{N - 1}. \tag{2}$$

For the value of the parameters used in the runs, $N = 100$ and $V = 5$, so that $p_i = 19/99 = 0.1919$.

Now, if we set $p_{if} = 0.5$ in (1), i.e. the ideal condition for obtaining a bifurcation, and establish $c_0 = 0.5$ (the value we used in our simulation runs), we can get equation (3), which states the ideal condition for $p(a1)$ and $p(a2)$ for getting a bifurcation.

$$P_{a1} + P_{a2} = 1.0. \tag{3}$$

Note that (3) does not contain p_i, which means that that condition applies for any value of p_i. Equation (3) corresponds to a line with an intercept with the y axis ($p(a2)$ axis) equal to 1.0 and slope equal to -1.0, which coincides with the line depicted in Figure 4 that represents the combinations of $p(a1)$ and $p(a2)$ where we obtained a bifurcation. Now, if the sum of $p(a1)$ and $p(a2)$ is bigger than 1.0, we obtain a parallel line to (3), but located above (3). In that case, p_{if} is larger than 0.5, and thus most probably the coefficients will converge to one. This coincides with the region of combinations for $p(a1)$ and $p(a2)$, shown on Figure 4, where the coefficients converge to one. To see that, we can rewrite (1), replacing $c_0 = 0.5$:

$$p_{a1} + p_{a2} = \frac{2\,p_{if} - p_i}{1 - p_i}. \tag{4}$$

If we replace in (4) p_i for its value 0.1919 and, for example, set $p_{if} = 0.6$, we obtain $p(a1) + p(a2) = 1.248$. On the other hand, if the sum of $p(a1)$ and $p(a2)$ is smaller than 1.0, we also obtain a parallel line to (3) but located below (3). In this case, p_{if} will be smaller than 0.5, and thus we will get a convergence to zero of the coefficients. For example, if we put $p_{if} = 0.4$ in (4), we get the line $p(a1) + p(a2) = 0.753$. That line is located within the region of combinations of $p(a1)$ and $p(a2)$ shown in Figure 4, where we obtain that dynamic.

However, from Figure 4, we can also see that on the upper left corner of the graph, the line $p(a1) + p(a2) = 1.0$ does not represent all the combinations of $p(a1)$ and $p(a2)$ where the ABM exhibits the bifurcation. Thus, the simple probability model only partially explains the empirical results.

5.2   Markov Chain Model

Since the model in 5.1 calculates p_{if} only for the initial state of a simulation run, it cannot fully capture the dynamical nature of the ABM. Remember that the concepts' coefficients change during a run, as well as the interaction probabilities among agents. In the ABM, that means that c_0 and p_i will change as the simulation run advances. Thus, we need to build a model that captures that dynamical aspect of the ABM. To do so, we use a simple Markov chain, with four states, as described in Table 1.

Table 1. State transition probability matrix of the Markov chain

	S_{t+1} ($j = 0$)	W_{t+1} ($j = 1$)
S_t ($i = 0$)	p_{if}^{+}	$1 - p_{if}^{+}$
W_t ($i = 1$)	p_{if}^{-}	$1 - p_{if}^{-}$

Table 1 indicates that if a concept strengthens (state S_t ($i = 0$)), then the probability that it will increase in the next step (state S_{t+1} ($j = 0$)) is p_{if}^{+}, and that it will weaken is $1 - p_{if}^{+}$ (state W_{t+1} ($j = 1$)). On the other hand, if a concept weakens (state W_t ($i = 1$)), then the probability that it will strengthen in the next step (state S_{t+1} ($j = 0$)) is p_{if}^{-} and that it will weaken (state W_{t+1} ($j = 1$)) is $1 - p_{if}^{-}$. In the ABM, each of those p_{if} has a meaning. From expression (1), we know that p_{if} depends on c and p_i, which change during a simulation run. The description of the ABM states that if a concept strengthens, then its coefficient will increase by a certain Δc, and the same will happen with p_i, which will increase by Δp_i. If the concept weakens, then the coefficient c will decrease by Δc, but p_i will remain the same. Therefore, using those facts, we can write the following expressions for p_{if}^{+} and p_{if}^{-}:

$$p_{if}^+ = p_{if}(c^+, p_i^+) \text{ with } c^+ = c_0 + \Delta c, \; p_i^+ = p_i + \Delta p_i. \tag{5}$$

$$p_{if}^- = p_{if}(c^-, p_i^-) \text{ with } c^- = c_0 - \Delta c, \; p_i^- = p_i. \tag{6}$$

Then, replacing (5) and (6) in (1), we can write the explicit equations for p_{if}^+ and p_{if}^-:

$$p_{if}^+ = (c_0 + \Delta c)[(p_i + \Delta p_i) + p_{a1}(1 - p_i - \Delta p_i)]$$
$$+ (1 - c_0 - \Delta c)(1 - p_i - \Delta p_i) p_{a2}. \tag{7}$$

$$p_{if}^- = (c_0 - \Delta c)[p_i + p_{a1}(1 - p_i)]$$
$$+ (1 - c_0 + \Delta c)(1 - p_i) p_{a2}. \tag{8}$$

Now, if we apply the properties of an ergodic Markov chain (c.f. [20]), we can compute a long-run probability that a concept will strengthen (Π_0):

$$\Pi_0 = \frac{p_{if}^-}{1 + p_{if}^- - p_{if}^+}. \tag{9}$$

Since (9) is written in terms of p_{if}^+ and p_{if}^-, which in turn are given by (7) and (8), it would be rather cumbersome to write an explicit equation for (9) in terms of c, Δc, p_i, Δp_i, $p(a_1)$ and $p(a_2)$. Thus, we prefer to use the following definitions and write a simpler expression for Π_0:

$$
\begin{aligned}
a &= (c_0 - \Delta c) p_i & e &= (c_0 + \Delta c)(p_i + \Delta p_i) \\
b &= (c_0 - \Delta c)(1 - p_i) & f &= (c_0 + \Delta c)(1 - p_i - \Delta p_i) \\
d &= (1 - c_0 + \Delta c)(1 - p_i) & g &= (1 - c_0 - \Delta c)(1 - p_i - \Delta p_i)
\end{aligned} \tag{10}
$$

Then, using (10), we can write:

$$\Pi_0 = \frac{a + b \, p_{a1} + d \, p_{a2}}{1 + a + b \, p_{a1} + d \, p_{a2} - e - f \, p_{a1} - g \, p_{a2}}. \tag{11}$$

Expression (11) can be rearranged so that it looks similar to equation (4), i.e. represents a line that states the relationship that must exist between $p(a1)$ and $p(a2)$ for a given Π_0:

$$p_{a2} = \frac{b + \Pi_0 f - \Pi_0 b}{\Pi_0 d - \Pi_0 g - d} \, p_{a1} + \frac{a + \Pi_0 - \Pi_0 a + \Pi_0 e}{\Pi_0 d - \Pi_0 g - d}. \tag{12}$$

Note that (12) is an equation of a line with a slope equal to the expression located to the left of $p(a1)$ and an intercept with the y axis ($p(a2)$ axis) equal to the far right hand expression. If we compare (12) with (4), we can see that the slope in (4) does not change, depending on the values that $p(a1)$, $p(a2)$ and p_{if} take; but in (12) the slope changes (remember that p_{if} in (4) is equivalent to Π_0 in (12)). Moreover, if we set in (10), $c_0 = 0.5$, $p_i = 0.1919$, $\Delta c = 0.45$ and $\Delta p_i = 0.05$; and put the values a through g, defined in (10), in (12), we can get a family of lines that represents the condition that

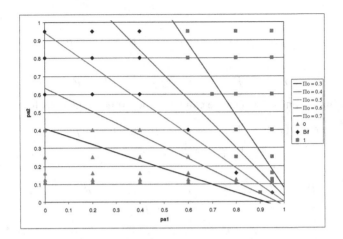

Fig. 6. Dynamics that emerge for coefficients of concepts, according to values set to $p(a1)$ and $p(a2)$(bifurcation, 1 = convergence to 1, 0 = convergence to 0) and lines for different values of Π_0 according to the Markov chain model

must meet $p(a1)$ and $p(a2)$ for obtaining different values of Π_0. Figure 6 shows the same graph presented in Figure 4, but displaying lines for $\Pi_0 = 0.3$ to 0.7 according to the Markov chain model that corresponds to expression (12).

From Figure 6, we can see that the line for $\Pi_0 = 0.5$ (the ideal condition for getting a bifurcation) approximately coincides with a line equal to the one we calculated for the simple probabilistic model (see expression (3) and Figure 4). The other lines for $\Pi_0 = 0.6$ and 0.7 are located in the region where the ABM exhibits the convergence to one dynamic and the lines for $\Pi_0 = 0.3$ and 0.4 lie in the region where the convergence to zero dynamic emerges. Thus, we can see that the Markov chain model represents fairly well the conditions under which the ABM exhibits the three different dynamics. Moreover, note that the lines tend to converge toward the lower right corner of the graph, where $p(a1) \gg p(a2)$ and tend to diverge toward the upper left corner, where $p(a1) \ll p(a2)$. This means, that the region where we get the bifurcation and which separates the areas where we obtain the convergence to one and zero, gets narrower when $p(a1) \gg p(a2)$ and wider when $p(a1) \ll p(a2)$. That characteristic was the one that the probabilistic model described in 5.1 was not able to capture.

5.3   Sensitivity of Models to Changes in Values of Some Parameters

By analyzing the ABM's and its associated probabilistic models' sensitivity to different parameters, we are able to derive predictions for "real world" situations. Although the Markov chain model presented in 5.2 better explains the dynamical properties of the ABM than the probabilistic model described in 5.1, the latter model is easier to analyze from a substantive point of view. Thus, based on expression (1), we will compute the sensitivity of that model to changes in values of some parameters. Here, we will present only two results of such analyses. To do so, we use (1) and take the partial derivatives of p_{if} with respect to $p(a1)$ and $p(a2)$:

$$\frac{\partial p_{if}}{\partial p_{a1}} = c_0(1 - p_i).$$ (13)

$$\frac{\partial p_{if}}{\partial p_{a2}} = (1 - c_0)(1 - p_i).$$ (14)

Additionally, since p_i appears in (13) and (14) and that variable depends on the number of agents and concepts (N, V see (2)), we can express (13) and (14) in terms of N and V. Moreover, given that for reasonably large values of N, p_i tends to $1/V$, we will analyze (13) and (14) taking into consideration that $p_i \approx 1/V$.

From (13) and (14) we can see that the sensitivity of p_{if} with respect to $p(a1)$ and $p(a2)$ is always positive (remember that $0 \le c_0$, $p_i \le 1$), i.e. the larger $p(a1)$ and $p(a2)$, the larger p_{if}. Now, the larger the number of concepts a group has (V), the smaller p_i will be and the more influential $p(a1)$ and $p(a2)$ will be on the value that takes p_{if}. That means that for groups with a large set of related concepts (or many different versions of the same concept), the probability of true and illusory agreement ($p(a1)$ and $p(a2)$) will greatly influence p_{if}. The significance of that influence will also be determined by the value of c_0. Note that for large values of c_0, $p(a1)$ will have a larger influence on p_{if} than $p(a2)$ and vice-versa. Thus, for groups with many concepts, the degree of agreement, either true or illusory, and the initial strength of each concept will dictate whether each concept strengthens or weakens, and eventually disappears.

Several "real world" situations could conform to the dynamics described above. As an illustration, imagine a social group that has an abstract concept, such as *conservative*. Presumably, people would have many different versions of such concept (i.e., a small p_i), with some people, e.g., considering that conservative is a view about economics, while others considering that it is a view about values, and so on. Imagine, furthermore, that this concept's relevance in that society is moderate, in the sense that it does not persistently determine people's actions (i.e., $c_0 \ne 1$). For concepts like this, our sensitivity analyses predict that their fate as a cultural phenomenon will depend mainly on their capacity to generate agreement.

Imagine, furthermore, that *conservative* has *liberal* as a weakly contrasting alternative concept (liberal is weakly contrasting to conservative because it does not clearly divide political opinion in two sharply contrasting clusters). Our sensitivity analyses predict that the fate of this pair will depend on agreement, regardless of whether it is true ($p(a1)$) or illusory ($p(a2)$). Additionally, as discussed in 4.3 above, these conditions promote bifurcations akin to social polarization.

Perhaps, an even more interesting situation arises in groups that have a small number of concepts or versions of them. In that case, p_i will be large, and thus $p(a1)$ and $p(a2)$ will not have a large influence on p_{if} (i.e., the degree of true and illusory agreement will not have a large influence on the fate of the concepts). Examining Figure 5, we can see that in the above mentioned situation, the fate of each concept will be predominantly dictated by its initial strength c_0, i.e., an initially rather strong concept will disseminate throughout the group and become stronger, and an initially quite weak concept will die out. Note that since in this situation, agreement of any type is almost irrelevant, that implies that a concept may spread even if people do not share the same meaning of it.

Again, a "real world" situation that could conform to these conditions is the following. Imagine a social group in which an authority (moral, political, or other) pushes an oversimplified concept (e.g., a slogan), and creates the conditions to make it relevant (e.g., punishes dissent). As occurs with commands, slogans may leave little room for alternative interpretations (i.e., p_i is large), which, by equations (13) and (14) implies that agreement ceases to be the predominant force that drives that concept's path. In other words, if an authority presents a very simple idea that allows little room for alternative interpretations, and succeeds in making it relevant in people's minds (i.e., makes c_0 sufficiently large), that condition will be sufficient to strengthen the concept and disseminate it throughout the social group, regardless of whether its meaning is shared or not.

6   Conclusions

In the work we report here, we use our ABM to develop a complex theory about the dynamics of shared meaning in social groups. This use of ABMs is not new, and has been advocated by [21]. Our ABM embodies some very simple rules of interaction, in keeping with Axelrod's KISS principle [22]. However, the ABM's dynamics are not simple, as attested by the expanded region of combinations of $p(a1)$ and $p(a2)$ in Figure 4, where bifurcations emerge.

Our theory development approach to Agent Based Modeling led us to formalize the dynamics through increasingly refined probabilistic models. Not only is this currently allowing us to recursively improve our ABM, but it also allowed us to clearly link the conceptual and mathematical formulations of our theory (respectively, sections 1 and 2, and section 5), and to gain a more general and clear understanding of the ABM's dynamics.

It is true that our model is, at this point, purely theoretical, and that it requires data to support it. However, we incorporated into the ABM generally accepted psychological theory, and as our sensitivity analyses in 5.3 show, the ABM makes intuitively correct predictions that were not built into it in an ad hoc fashion. These two aspects, we think, are at least evidence of the ABM's face validity. Currently, we are working on designing experiments to gather data from subjects to validate our model. Pilot tests of one of the designed experiments suggest that the concepts of true and illusory agreement, as developed in our theory and included in the ABM, indeed hold. Thus, we would be very disappointed if future work shows that the validity of our model is only illusory.

Regarding the application of the ABM and related theory about concepts, we will use it to explain and understand psychological and sociological phenomena, such as the ones we already discussed in section 5.3. (e.g., how can culture be defined in relation to concepts, what happens with minorities that hold certain conceptual views, can group agreement/disagreement be predicted). Additionally, we think that for Computer Science, our work could be applied to enhance the capability of search engines. In particular, note that the user of an engine needs to communicate his/her needs for specific information to the engine. Thus, user and engine have to create a shared conceptual meaning of the search terms and information needs. The stronger the shared conceptual meaning created, the more appropriate the information that the search engine will provide to the user. Hence, our inquiry into conceptual and shared meaning could give insights to the development of better communication tools between users and search engines.

References

1. Brennan, S.E., Clark, H.H.: Conceptual pacts and lexical choice in conversation. J. Exp. Psychol. Learn. 22, 1482–1493 (1996)
2. Brown-Schmidt, S., Tanenhaus, M.K.: Real-time investigation of referential domains in unscripted conversation: A targeted language game approach. Cognitive Sci. 32, 643–684 (2008)
3. Carpenter, M., Nagell, K., Tomasello, M.: Social cognition, joint attention, and communicative competence from 9 to 15 months of age. Monographs of the Society for Research in Child Development 63(4), 1–143 (1998)
4. Clark, H.H., Krych, M.A.: Speaking while monitoring addressees for understanding. J. Mem. Lang. 50(1), 62–81 (2004)
5. Galantucci, B., Sebanz, N.: Joint action: current perspectives. Topics in Cognitive Science 1, 255–259 (2009)
6. Garrod, S., Anderson, A.: Saying what you mean in dialogue: A study in conceptual co-ordination. Cognition 27, 181–218 (1987)
7. Moses, L.J., Baldwin, D.A., Rosicky, J.G., Tidball, G.: Evidence of referential understanding in the emotions domain at 12 and 18 months. Child Dev. 72, 718–735 (2001)
8. Richardson, D.C., Dale, R., Tomlinson, J.M.: Conversation, gaze coordination, and beliefs about visual context. Cognitive Sci. 33, 1468–1482 (2009)
9. Tomasello, M.: Joint attention as social cognition. In: Moore, C., Dunham, P.J. (eds.) Joint Attention: its Origins and Role in Development, pp. 103–130. Lawrence Erlbaum, Hillsdale (1995)
10. Rosch, E., Mervis, C.B.: Family resemblances: Studies in the internal structure of categories. Cognitive Psychol. 7, 573–605 (1975)
11. Lewis, D.: Convention: A Philosophical Study. Harvard University Press, Cambridge (1969)
12. Lewis, D.: Languages and Language. In: Gunderson, K. (ed.) Minnesota Studies in the Philosophy of Science, vol. VII, pp. 3–35. University of Minnesota Press, Minneapolis (1975)
13. Millikan, R.G.: Language: A biological model. Oxford University Press, Oxford (2005)
14. Chaigneau, S.E., Canessa, E., Gaete, J.: Conceptual Agreement Theory. Manuscript submitted to New Ideas Psychol. (2010)
15. Evans, J.B.T.: Dual-processing accounts of reasoning, judgment, and social cognition. Annu. Rev. Psychol. 59, 255–278 (2008)

16. Brewer, M.B.: A dual process model of impression formation. In: Srull, T.K., Wyer Jr., R.S. (eds.) Advances in Social Cognition, pp. 1–36. Lawrence Erlbaum Associates, Hillsdale (1988)

17. Lenton, A.P., Blair, I.V., Hastie, R.: Illusions of gender: Stereotypes evoke false memories. J. Exp. SOC Psychol. 37(1), 3–14 (2001)

18. Rudman, L.A., Phelan, J.E.: Backlash effects for disconfirming gender stereotypes in organizations. Res. Organ. Behav. 28, 61–79 (2008)

19. Nickerson, R.: Confirmation bias: A ubiquitous phenomenon in many guises. Rev. Gen. Psychol. 2, 175–220 (1998)

20. Ross, S.M.: A First Course in Probability, 5th edn. Prentice Hall, New Jersey (1998)

21. Ilgen, D.R., Hulin, C.L.: Computational modeling of behavior in organizations. American Psychological Association, Washington D.C (2000)

22. Axelrod, R.: Advancing the Art of Simulation in the Social Sciences. In: Conte, R., Hegselmann, R., Terna, P. (eds.) Simulating Social Phenomena. Springer, Berlin (1997)

Improving File Sharing Experience with Incentive Based Coalitions

M.V. Belmonte, M. Díaz, and A. Reyna

E.T.S.I. Informática, Bulevar Louis Pasteur, N.35
Universidad de Málaga, 29071 Málaga, Spain
{mavi,mdr,reyna}@lcc.uma.es

Abstract. The natural selfish behavior of P2P system users, has given rise to the appearance of freeriding. These users download but don't contribute to the system, leading to a degradation of the system, and the user experiences. It becomes mandatory to find proper mechanisms to incentive cooperation among users in these systems. In this paper we provide an incentive based coalitions to improve the system welfare and users experiences. A peer that participates in a coalition lends "bandwidth" to other peers of the coalition, in exchange for utility and consequently a far better experience. Taking concepts from game theory we provide a solid ground upon which we build our mechanism . Simulated experiments support the approach, showing how encouraging participation stops freeriding and therefore improve the system performance and the user experience.

Keywords. P2P, Coalition, Game theory, File sharing, Free-riding, Incentive mechanisms.

1 Introduction

In general, a P2P content distribution system creates a distributed storage medium that allows the publishing, searching and retrieval of files by members of the network [11]. Traditionally, the main problem of the P2P systems is limited to file search. However, the efficient download of content and the fairness in the bandwidth contribution is also an important aim in the design goal of these kind of systems. The early P2P systems (Gnutella, Kazaa,...) lack mechanisms for fairness in bandwidth usage. For this reason, these systems suffer from free-loaders, peers that consume many more resources or contents (bandwidth) than they contribute. In [6] and [17] empirical studies have observed this behaviour in Napster, Gnutella or even eDonkey.

One of the reasons for this problem is that the mechanisms used for downloading and sharing in the P2P systems, do not take the selfish behaviour of the peers into account at the design stage. P2P system users act rationally trying to maximise the benefits obtained from using the system's shared resources [3]. Therefore, it is important to find mechanisms that provide incentives and encourage cooperation. One possible solution could be to use an economic framework that provides incentives. In this sense game theory may be a good tool on which to model the interactions between peers in a P2P file sharing system. The idea is to define "the rules of the game" so that the system

J. Filipe and A. Fred (Eds.): ICAART 2011, CCIS 271, pp. 287–301, 2013.

as a whole exhibits good behaviour when self-interested nodes pursue self-interested strategies ([7]).

Our approach proposes the application of a coalition formation scheme based on game theory to P2P file sharing systems (in [1] we presented an early version of this work). The main idea of the coalition formation scheme is the fact that peers which contribute more get a better experience. We define a "responsiveness bonus" that reflects the peer's overall contribution to the system, and we use the game theory utility concept to calculate it. It is possible to form a coalition among peers with a re-distribution of the number of bytes to be transferred. A peer that participates in a coalition lends "bandwidth" to other peers of the coalition, in exchange for utility; and this utility will increase its responsiveness bonus. The coalition formation scheme rewards the peers with a higher responsiveness bonus (therefore giving them greater bandwidth to download files), and penalises the ones that only consume resources, decreasing their responsiveness bonus and consequently their bandwidth.

The proposed incentive mechanism encourages cooperative behaviour between the peers preventing the free-riding problem. Using the game theory concept of "core", the peers forming the coalition get in return a fair utility in relation to the bandwidth they supply (achieving fairness in bandwidth sharing); And in addition, it allows the formation of coalitions of peers that help each other in downloading files, increasing the performance of the P2P network.

2   Downloading with Coalitions

In this section, we describe the model of the environment in which the system is deployed and the mechanism of coalition formation among peers. We firstly describe a simplified situation, illustrating the advantages of forming coalitions for P2P downloads and the way of computing and dividing the utility or profit obtained by peer that participates in the coalition. Secondly, and in more general terms, we describe the coalition formation process and how the data and the bandwidth are distributed among the coalition members.

2.1   P2P Network Type

For our work, we have selected a P2P system with a partially centralised architecture and an unstructured network. The first characteristic is related to the degree of centralisation of the peer's network, and the second with the fact that the network is created in a non-deterministic way. When a peer wants to download a file, it directs its request to a supernode and this searches the file in its index (a supernode is a peer that acts as a central index for files shared for a subpart of the peer network). When the file is located, supernode sends to the "requester" peer an indexed result with the set of nodes that store the requested file. Then, the requester peer opens a direct connection with one or more peers that hold the requested file, and proceeds to download it.

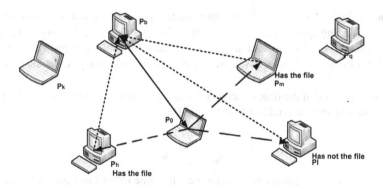

Fig. 1. The coalition formation model

2.2   Coalition Formation Model

Coalition formation is an important mechanism for cooperation in Multi-Agent Systems (MAS). In order to be used by autonomous agents, a coalition formation mechanism must solve the following issues: i) maximise the agents' profit or utility. For every coalition S, coalitional value $V(S)$ must be computed, i.e., the total utility obtained by S as a whole ii) divide the total utility among agents in a fair and stable way, so that the agents in the coalition are not motivated to abandon it. For every coalition S and every agent $i \in S$, payment configuration $x(i)$ must be computed, i.e., the share of $V(S)$ that is assigned to the agent i. iii) do this within a reasonable amount of time and using a reasonable amount of computational efforts. Our coalition formation model allows cooperation to take place between autonomous, rational and self-interested agents in a class of superadditive task oriented domains [15]. Each agent has the necessary resources to carry out its own task, however it is possible to form a coalition between agents with a new re-distribution of the task that may allow them to obtain benefits. The proposed model guarantees an optimum task allocation and a stable payoff division. Furthermore, computational complexity problems are solved.

In this section the coalition formation scheme is applied with the goal of improving the performance of P2P file exchange systems. In this case, the central idea is based on sharing the task of downloading a file among a set of peers forming a coalition. From the point of view of the peer that wants to download the file there is a clear advantage, since the total download time is reduced. From the point of view of uploading peers, for each one the task of transferring the file is alleviated, since it is divided between the members of the coalition.

Let us consider the simplified situation illustrated in figure 1. In this scenario, peer p_b, asks p_a for a file Z. This peer p_a forms a coalition S with three other peers p_h, p_l and p_m to transfer that file. In P2P file sharing systems, every node p_m has an upload b_i^{in} and download b_i^{out} bandwidth dedicated to file sharing. Usually these bandwidths are user defined and indicate maximum values, and b_i^{in} is much lower than b_i^{out}. This simplified scenario can be generalised, we can consider that the downloader peer splits its bandwidth in order to perform simultaneous downloads, determining the b_i^{out} dedicated for each download. The model is valid for both scenarios.

In general, there will be an initial uploading agent p_0 (in the figure p_a) and a set of n additional uploading agents, $p_1, \ldots, p_n$ (in the figure p_h, p_l and p_m), all of which have the file that has to be downloaded and they dedicate their upload capacity b_i^{in} to this transfer. Let us call $size(Z) = T$ the size of the file to download. Let us also assume that $\sum_0^n b_i^{in} \leq b_b^{out}$.

Then an estimation of the time necessary for the transfer is given by the ratio between the size of the file and the coalition bandwidth(1):

$$t_S = \frac{T}{b_0^{in} + \sum_1^n b_i^{in}} \tag{1}$$

On the other hand, if the coalition is not formed, the time for just an uploading agent p_0 is given by (2):

$$t_0 = \frac{T}{b_0^{in}} \tag{2}$$

Therefore the value obtained by the coalition S can be defined as Δt, the difference between (2) and (1), as shown in (3):

$$\Delta t = \frac{T}{b_0^{in}} - \frac{T}{b_0^{in} + \sum_1^n b_i^{in}} = \frac{T}{b_0^{in}} \frac{\sum_1^n b_i^{in}}{\sum_0^n b_i^{in}} = t_0 \frac{\sum_1^n b_i^{in}}{\sum_0^n b_i^{in}} \tag{3}$$

Of course, if $p_0 \notin S$, $V(S) = 0$. To sum up, the coalitional value for every coalition is given by (4):

$$V(S) = \begin{cases} t_0 \frac{\sum_1^n b_i^{in}}{\sum_0^n b_i^{in}} & \text{if } p_0 \in S \\ 0 & \text{if } p_0 \notin S \end{cases} \tag{4}$$

Now we address the following problem, to define a stable payoff division of $V(S)$ between agents, i. e., given a partition of the set of all agents into different coalitions, to assign an amount $x(i)$ to every agent i. The problem is to distribute the utility in a fair and stable way, so that the agents in the coalition are not motivated to abandon it. Game theory provides different concepts (core, kernel, Shapley value, etc.) for the stability of coalitions [4]. The core is the simplest to define. A payment configuration belongs to the core if there is no other coalition that can improve on the payoffs of all of its members.

Formally, let N be the set of all agents, and let us denote $x(S) = \sum_{i \in S} x(i)$. The payoff division lies inside the core iff the following holds: i) for all sets of agents $S \subseteq N$, $V(S) \leq x(S)$ (group rationality); and ii) $V(N) = x(N)$ (global rationality). The existence of the core is not guaranteed in the general case, in a given situation the core may be empty. However, we will show for our case a payoff division that lies inside the core. We will expand on the ideas presented in [15].

The proposed payoff division scheme is calculated by means of the marginal profit concept. Thus, the payment to each agent is given by the marginal profit according to the resources that describe that agent, and multiplied by the value of the resources. Since the concept of marginal profit is really that of partial derivative, the payment vector x will be computed as follows: for the original uploading agent, p_0, there are 2 parameters

(t_0 and b_0^{in}), hence its payment is be given by $t_0 \frac{\partial V}{\partial t_0} + b_0^{in} \frac{\partial V}{\partial b_0^{in}}$. For the remaining agents, there is only one parameter (b_i^{in}), hence their payments are $b_i^{in} \frac{\partial V}{\partial b_i^{in}}$. Finally, we obtain the expressions shown in equations (5):

$$
x_i = \begin{cases} t_0 \frac{(\sum_1^n b_i^{in})^2}{(\sum_0^n b_i^{in})^2} & \text{if } i = 0 \\ t_0 \frac{b_0^{out} b_i^{in}}{(\sum_0^n b_i^{in})^2} & \text{if } i \neq 0 \end{cases} \tag{5}
$$

In appendix A we prove that the equations (5) define a payoff division that lies inside the core.

Finally, we will show that the computations can be done within a reasonable amount of time and using a reasonable amount of computational efforts. It is obvious that the above schema provides a set of explicit formulae that compute payments in time linear with the number of peers.

2.3 Data and Bandwidth Distribution Model

Following on from the example, let us suppose that the peer p_b asks p_0 (p_a in the figure) to download the file Z, and p_0 decides to initiate a coalition to download the file (figure 1) Then, p_0 carries out the following steps:

1. To set the coalition size. If the sum of the upload bandwidth, b_i^{in}, of all the interested peers (let us suppose this value is equal to n) joined with p_0, is lower than the download bandwidth of p_b, b_b^{out}, we are in the trivial case (equation 6). In this case, all the interested peers joined with p_0 will form the coalition. So, $| S | = N$:

$$
(\sum_{i \in N} b_i^{in}) \leq b_b^{out} \tag{6}
$$

Conversely if the expression (6) is false, we must distribute the bandwidth of p_b, b_b^{out} between all the interested peers[1].

For this, it is necessary to distribute the b_b^{out} of p_b among all of them. This can be done by "the progressive filling algorithm" [10]. Let us suppose w_i is the assigned bandwidth for the interested peers. The algorithm initialises the bandwidth of all the interested peers to 0, $w_i = 0$, $\forall i \in S$. Then, it increases all the bandwidths at the *same rate*, until one or several peers hit their limits, $w_i = b_i^{in}$, $\forall i \in S$. Once the bandwidth assigned to one peer, p_i, reaches its limit, it is taken out of the process. The algorithm will continue to increase the bandwidth of the remaining peers at the *same rate*. The algorithm will finish when all the peers reach their limits, or when the bandwidth of p_b, b_b^{out} is wasted.

This algorithm provides the *max-min fairness* [9]. A bandwidth allocation is max-min fair if and only if an increase of b_x^{in} within its domain of feasible allocation is at the cost of decreasing some other b_y^{in}. So, it gives the peer with the smallest bidding value the largest feasible bandwidth.

---

[1] If there are many interested peers, a maximum size of coalition is established in order to avoid an undue partitioning of b^{out}. This value depends on the b^{out} and the file size.

2. To split the file size Z, $size(Z) = T$, between the coalition members, $p_i \in S$.

   (a) We estimate the minimum amount of time needed to transfer Z as a function of the known bandwidth limits. Following this, the minimum amount of time can be estimated as follows:

$$t_i = T / \sum_{i \in S} \min(b_i^{in}, b_i^{out}) \tag{7}$$

This estimated time is the same initially for all the peers in the coalition S^2.

   (b) Once we know the estimation of the time for each peer, we can carry out a partitioning of the file taking into account the capacities of each peer. Every peer will have to transfer a number of bytes $b_i^{in} * t_i$. The file is divided into blocks of this size that are assigned to the peers.

3. To inform each coalition member of the size of the block to be transferred. p_0 communicates to each peer member of S the number of bytes to be transferred.

3   The Incentive Mechanism

As we have already mentioned our incentive mechanism is based on providing a better quality of service to the peers that participate in the coalitions. In order to achieve this, we define a *Responsiveness Bonus*, Rb, for every peer. This value reflects the peer's overall contribution to the system (i.e. how much work it has carried out for the other peers in the system). In accordance with the above model, in the proposed payoff division each peer obtains a utility which is proportional to the resources that it supplies. Therefore, the peer p_i that supplies a greater bandwidth (uploading peers) will obtain a greater utility, and this utility will increase its Rb. Conversely, the value of Rb_i should be reduced when p_i acts as a downloading peer and does not contribute. We consider that an auditing authority is responsible for storing and updating the Rb, using proper methods to control the concurrency.

So the value of Rb_i will be calculated as a heuristic function of x_i that can be adjusted with data from the real system behaviour or from simulation results. This uses the x_i values obtained by the uploading peers of the coalition as uploading points, Up_i. For the downloading peer of the coalition its downloading points, Dp_i, are calculated as the average of the utility obtained by the uploading peers of the coalition. Each peer p_i accumulates Dp_i and Up_i points by adding the points obtained in each coalition formation process in which it participates[3].

$$Up_i = Up_i + x_i \tag{8}$$

---

[2] b_i^{out} will be used to download files from the P2P sharing file system, in the case p_i does not have the file previously and it works for another node in the coalition (it will be 0 if the peer had the file).

[3] Since a new peer that joins the coalition formation system will have its uploading and downloading points set to 0, we allow the peers to download a minimum amount set to a parameter $MinDownload$.

$$Dp_i = Dp_i + \sum_{s \in S} x_s / \mid S \mid \tag{9}$$

Rb_i is a value included in the interval $[0..1]$. The correction of the bandwidth is only applied to the download bandwidth $Rb_i b_i^{out}$ (it makes no sense to correct the upload bandwidth, because we would be decreasing the upload capacity of the collaborative peers). Initially the Rb_i of the peers (uploading/downloading) is 1^4. A higher responsiveness bonus (Rb_i closer to 1) will mean that p_i will be able to fill all its reserved bandwidth, since it can add more peers to the coalition in order to complete its bandwidth, reducing the download time. Otherwise, an Rb_i closer to 0 will limit the possibility of adding peers to the coalition (in fact, in some cases it will avoid creating any coalition for the download). In this way, our incentive mechanism promotes cooperation taking into account the selfish behaviour of the peers.

$$Rb_i(Up_i, Dp_i, Fs_i) =$$

$$\begin{cases} 1 & \text{if } (Up_i - Dp_i) \geq 0 \\ 0 & \text{if } (Up_i - Dp_i) < 0 \wedge Up_i = 0 \wedge Fs_i = 0 \\ 1 & \text{if } (Up_i - Dp_i) < 0 \wedge Up_i = 0 \wedge Fs_i > 0 \\ \frac{Up_i \cdot \gamma}{Dp_i} & \text{if } (Up_i - Dp_i) < 0 \wedge Up_i > 0 \end{cases} \tag{10}$$

The equation 10 computes Rb_i in relation to Dp_i, Up_i, γ and Fs_i (the number of files shared by peer i).

If $Up_i - Dp_i \geq 0$, it means that the peer is contributing to the system more than it is consuming from it, and so $Rb_i = 1$. If, $Up_i = 0$, the peer has not contributed anything to the system and, if, in addition, the number of shared files is 0 obviously the peer is a free-rider and its Rb_i must be 0. Conversely, if the peer has not contributed to the system, but the number of shared files is not 0, it means that the peer wants to contribute to the system but its shared files have still not been downloaded by other peers; So its Rb_i must stay at 1. Finally, if the peer has contributed to the system, but less than what it has been consuming from it, its Rb_i will be proportional to $Up_i / Dp_i{}^5$. The variable γ allows us to regulate this formula in order to increase or decrease the proportional relation between the benefit, Up_i, and the penalty, Dp_i.

4 Performance Evaluation

In this section we describe the simulations we performed and the corresponding results. In order to simulate our coalition formation model for P2P file sharing, we have defined and implemented a generic P2P simulator for service oriented networks. The simulation tool is presented in detail in [16]. Additionally, it should be noted that we are dealing with situations which are different from traditional system simulators, since, we are also trying to model the user behaviour. For this reason, and in order to model the user as close to reality as possible, the peers are classified in three categories according to their

---

[4] Otherwise their bandwidths would be reduced from the beginning, and the download times of the files would be higher (compared to the scheme without coalitions).
[5] This value is always < 1.

behaviour: free-rider, adaptive and collaborative. We will first describe how the user behaviour has been modeled, and then the simulation results.

4.1   Modelling the User Behaviour

Free-riding is a consequence of selfish user behaviour in file sharing systems. In the case that we want to study actions to take in order to improve cooperation, modifying user's behaviour, a key step would be the modelling of users that are going to take part in the simulation. A realistic simulation should include at least three kinds of users(behaviours):

1. **Free-Rider.** Represents the selfish peer which only downloads files and rejects all the incoming file requests.
2. **Collaborative.** Represents collaborative behaviour. These peers always try to maximise the system's performance, so they offer all their available b_i^{in} and accept all incoming file requests until their bandwidth is full.
3. **Adaptive.** Represents intelligent behaviour, and so, is adapted to the evolution of peer welfare. These users accept download requests as long as they are interested in downloading a file, that is as long as they benefit. When the number of target files is 0 the b_i^{in} will be 0. Otherwise, all the entire available b_i^{in} will be offered so that a high Rb is maintained and all the target files can be downloaded. In case of multiple requests, the b_i^{in} will be divided between all the requests, taking into account the Rb of the requester peers.

Finally, all of them have a limit of download tries to avoid them repeatedly asking for the same file.

4.2   Experimental Results

We have run simulations of a P2P network of 1000 peers for 2000 units of simulated time (steps). All peers have the same bandwidth capabilities, 1024 kbs for downloads and 512 kbs for uploads. We have defined a collection of files of different sizes, a random number of copies of these files are delivered through the peers at the start of the simulation. The minimum number of copies for a file is 5 and the maximum is the half part of the number of peers that forms the network (500 for our experiments). This means that peers have a random number of initially stored files, between 0 and the whole collection of files. The objective of the simulation is that every peer manages to get the whole file collection, by this we mean, to download the files that are not initially stored. Depending on the peer's behaviour it will face this objective in different ways. File sizes range from 10000KB to 90000 KB.

To measure the impact of the Adaptive users on the system the experiments have been run with two different populations. The first one without Adaptive users, and the same population for Free Riders and Collaborative users (50% Free Riders, 50% Collaborative and 0% Adaptive users), we will hereafter refer to this as Population 1. And the second one with the same population for Adaptive and Collaborative users (40% Free Riders, 30% Collaborative and 30% Adaptive users), hereafter refer to as Population 2. In addition, the simulations have been run with two different incentive policies:

Table 1. No Coalition Mechanism Downloaded Bytes

No Coalition	Population 1	Population 2
Free Rider	157 Gb	123 Gb
Collaborative	156 Gb	92 Gb
Adaptive		91Gb
total	313Gb	306 Gb

No Coalitions (NC), where no incentive mechanism is considered and Coalitions (C), which implements our proposal. After repeating the simulation experiments 100 times we take the average to give the results. To compare how the incentive mechanisms and the user behaviours affect the P2P system, two main metrics have been considered: Downloaded Bytes and Average Time. In addition, we analyse the Work Progress; This measure shows how the simulations evolve towards the final target.

Number of Downloaded Bytes. Figure 2 shows the evolution of the downloaded bytes distribution for No Coalitions mechanism for experiments run with Population 1 on the left and Population 2 on the right. Similar figures for Coalitions in Figure 3. Downloaded bytes can be interpreted as the benefit obtained from the system. Next, we analyse these results for each of the different populations.

In Population 1 all work is done by the collaborative users, since free-riders do not collaborate. Figure 2 left shows the evolution of the distribution of the downloaded bytes is round 50 % for both users during the simulation. This means, the collaborative users do all the work, and the benefits are shared equally with the free riders. However, when we run the Coalition mechanism, Figure 3 left, free riders are stopped, the percentage of bytes downloaded by free riders drastically decreases after the first 100 steps of simulation. This demonstrates how the coalition formation prevents free-riders from obtaining more bytes as simulation time advances, and so from fully using the system's resources.

When Adaptive users are simulated, this is Population 2, distribution of downloaded bytes are affected as shown in figures 2 right and 3 right. With respect to collaborative and adaptive users, both are 30 % of the population, they do all the work and share more or less equally the benefits with free riders. In figure 3 right the evolution of the distribution of the downloaded bytes shows how free riders are again stopped, as in Population 1, and this means that the benefit is shared between the collaborative and adaptive users, these are those that are uploading files. In addition, collaborative users increase the percentage of downloaded bytes during the simulation; However, adaptive users first increase and after decrease the percentage. This is due to the behaviour of adaptive users, which are penalised when they are not sharing enough.

Tables 1 and 4.2 summarise the downloaded bytes per populations and per behaviour. In both tables it can be observed that the bytes downloaded by free riders are slightly reduced in Population 2 with respect to Population 1, this is because there are 10% less users in this population, the average bytes per user is very similar. With Population 1 it can also be observed that the coalition mechanism reduces the total bytes downloaded to 50,54%, with respect to No Coalitions, but 83,61% of this reduction is due to the Free Riders detection. This shows again how the algorithm prevents free-riders from

Table 2. Coalition Mechanism Downloaded Bytes

Coalitions	Population 1	Population 2
Free Rider	24 Gb	20 Gb
Collaborative	130 Gb	82 Gb
Adaptive		84 Gb
total	154 Gb	186 Gb

abusing as simulation time advances, and from stressing the system resources; and this leads to a more healthy system.

The Coalition Algorithm () reacts to the inclusion of Adaptive users by increasing by 20% the total amount of downloaded bytes compared to Population 1, this means that they benefit the whole system. In addition, comparing Coalitions and No Coalitions with Population 2 (when Adaptive users are simulated), the results are better than with the Population 1. In this case, the total amount is reduced by 38,78% where 83,15% is due to the Free Rider's detection.

Note that when using the second population there are fewer Free Riders and Collaborative users in the simulation. This affects the data shared and demanded in the system,and it justifies the smaller amount of downloaded bytes.

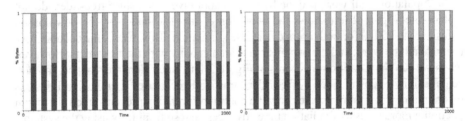

Fig. 2. Evolution of the Downloaded Bytes Distribution (% Bytes vs Time) for No Coalitions Mechanism , Population 1 (left) and Population 2 (right). Free Rider in red, Adaptive in blue and green for Collaborative users.

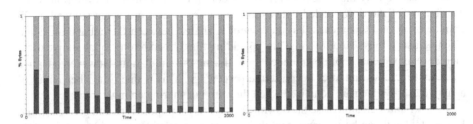

Fig. 3. Evolution of the Downloaded Bytes Distribution (% Bytes vs Time) for Coalitions Mechanism, Population 1 (left) and Population 2 (right). Free Rider in red, Adaptive in blue and green for Collaborative users.

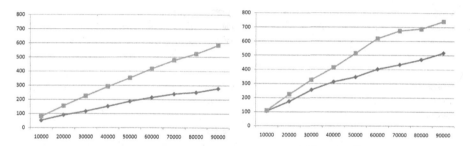

Fig. 4. Average Download Time (Time vs Bytes), Population 1 (left) and Population 2 (right)

Average Time. In addition to the total amount of bytes downloaded during the simulation, the time spent on each download is also a significant measure of the system performance. In Figure 4 the average time for each file for both algorithms is compared (squares for No coalitions and diamonds for Coalitions).

Population 1: When Adaptive users are not considered, Figure 4 left, the best download times are the ones obtained with Coalitions. For the smallest files, the times for No Coalitions and Coalitions are quite similar then, the higher the file size is, the greater the time difference. As expected, the benefit of using Coalition is increased as the file sizes grows.

Population 2: When Adaptive users are introduced average download time is increased, this is because the system is more stressed. Adaptive users implement a selfish behaviour, but they have to share in order to obtain benefits and they are capable of simultaneous downloads. All of this increases the download time. In Figure 4 right No Coalitions and Coalitions show a very similar slope and smaller values for Coalition mechanism, which also stops Free Riders. In this way, the system benefits without penalizing user's downloading times.

Work Progress. Our simulations have been modelled with a final objective, this is, that every node stores the whole file collection available in the system. Work progress allows us to study how the different configurations satisfy this final objective. In addition, it helps us to analyse the difference between the two different populations used in the experiment. To study the evolution of the Work Progress we have monitored the evolution of the bytes shared and the bytes demanded. As in economic equilibrium these lines are crossed, meaning that the supply and the demand are balanced. This is an indication of the System's health.

Figure 5 shows the sharing and demand for Coalitions in the two studied populations. The offer, initial stored files in the system is slightly bigger for the second population, this is observed where the offer (solid line) crosses the Y axis. This is due to the decrease in the number of Free Riders in Population 2,which justifies what was introduced in section 4.2, the experiments run with Population 2 get a smaller number of downloaded bytes.

Note that the evolution of the offer and the demand are significantly different when Adaptive users are introduced (population 2). On one hand, the slope of the offer is much smaller on Population 2, this is because of the sharing policy of Adaptive users,

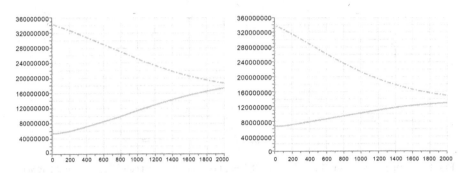

Fig. 5. System's Sharing vs Demand, Population 1 (left) and Population 2 (right). Coalitions. Bytes vs Time, solid line represents the offer and dotted-dashed the demand.

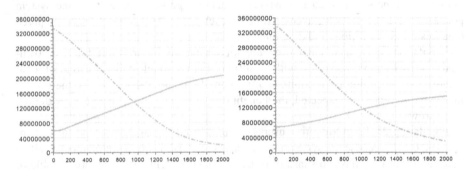

Fig. 6. System's Sharing vs Demand, Population 1 (left) and Population 2 (right). No Coalitions. Bytes vs Time, solid line represents the offer and dotted-dashed the demand.

which only share if they have outstanding downloads. On the other hand, the demand is not negatively affected by the decrease of the offer in Population 2, it is even better. The initial demand is similar, but the result is 10% better (demand is 10% lower) in Population 2. Once again it is shown that Adaptive users benefit the whole System when using the Coalition algorithm.

When analysing the sharing and demand in No Coalitions, Figure 6, we found that the offer behaves similarly to Coalitions; adaptive users cause a reduction of the slope of the offer. This is common to all the algorithms, and it justifies that some of them decrease the total amount of downloaded bytes. However, the demand seems not to be affected by adaptive users, that is to say, the introduction of adaptive users does not benefit the system when using this algorithm.

Comparing algorithms, seems that No Coalitions manage to get a better results. However, the reason for the fall of the demand on Coalitions is because the detection of Free Riders. Numerically, The Coalitions algorithm in Population 1 reaches 54% of Work Progress and in Population 2 44%, considering that Free Riders are 50% in Population 1 and 40% in Population 2, and subtracting their demands, the work progress is near to 4% to finishing in both populations. There is a proportional relationship between population size and demand.

5   Related Work

Reviewing the bibliography, several approaches have been proposed to combat the free riding problem. Karakaya [2] et al. have categorised them into three main types: firstly, incentive mechanisms based on monetary payments: one party offering a service to another is remunerated and inversely, resources consumed must be remunerated or paid for. Secondly, mechanisms based on reputation: it keeps information about the peer reputation, and peers with a good reputation are offered better services. Thirdly, incentive mechanisms based on differential services or reciprocity-based: peers that contribute more get a better quality of service [14] [5] [8] [12] [13]. Our approach could be included in this category, although its foundation is different and innovative.

Although some of the above approaches [5] [8] [13] are based on differential services, they do not promote a cooperative behaviour between peers that improves the download performance in the P2P System. And, in addition they do not achieve fair service differentiation between peers.

Those remaining, more similar to our approach, propose incentive mechanisms that encourage collaboration among peers. For example, 2Fast [12] is based on creating groups of peers that collaborate in downloading a file. However, compared to our proposal, it does not enforce fairness among the collector and helper peers, and in addition it is not specified how the helper may reclaim its contributed bandwidth in the future. Bit torrent [14] is also based on collaboration between peers. Its "tit-for-tat" policy of data sharing works right when the peers show a reciprocal interest in a particular file. However, in bit-torrent, the peers' download bandwidth is limited to their upload capacity, thereby reducing the achievable download performance. However, in our approach, the system's download capacity is not reduced to its upload capacity; And, using the Rb does not force a "mutual reciprocity" mechanism (like "tit-for-tat"); and thus the bandwidth contributed by a peer can be used in later downloads.

EMule [18] also promotes cooperation between peers. It uses a credit system to reward frequent uploaders and alleviates the free-riding problem. However, credits are exchanged between two specific peers, so content trading can happen only between peers that have mutual interests, and in addition it does not enforce fairness in bandwidth sharing.

Finally, the work of Ma et al. [10], also provides service differentiation based on the amount of services that each node has provided to a P2P community, and it uses a game theoretic framework. However, while we use a cooperative approach that proposes coalition formation, they propose a mechanism that makes different requesting users bid for resources, creating a dynamic competitive game.

6   Conclusions

In this paper we have presented a coalition formation based incentive mechanism for P2P file sharing systems. It is based on game theory and takes into account the rational and self-interested behaviour of the peers. In [1], the initial idea of applying this model to this problem was presented. Now, we have formally demonstrated the fairness of the model using game theory and, more concretely, the concept of "core". In addition, we have modelled user behaviour and defined the coalition formation model in order to perform simulations, using the simulator presented in [16]. The paper also includes some analysis of simulations.

Our approach allows any peer with idle bandwidth to participate in a coalition, uploading files for other peers in exchange for utility, and consequently greater download bandwidth; And in addition, it provides, using the "core", a fair utility to the peers forming the coalition in relation to the bandwidth they supply. To achieve this, a *Responsiveness Bonus* that reflects the peer's overall contribution to the system is defined, and the game theory utility concept is used to calculate it.

The simulation results have shown that in relation to downloaded bytes, the coalition mechanism prevents free-riders from obtaining more bytes as simulation time increases. In addition, it reacts to the inclusion of adaptive users increasing by 20% the total amount of downloaded bytes, so they benefit the system. In relation to download time, coalitions are capable of getting the best average download times and stopping free riders at the same time. Finally, we the analysis of Work Progress have shown that equilibrium between offer and demand is better when using our mechanism. This helps to keep the systems healthy.

Finally, we are working on the simulation of other approaches in order to be able to compare our results with existing proposals. And we plan to generalise the proposed coalition formation algorithm in order to include Quality of Service information. Our idea is to form coalitions in such a way that they are able to provide or guarantee QoS in different aspects depending on the service or application, i.e. real time constraints or fault tolerance.

References

1. Belmonte, M.V., Conejo, R., Pérez-de-la-Cruz, J.L., Triguero, F.: Coalitions among Intelligent Agents: A Tractable Case. Computational Intelligence. An International Journal 22(1), 52–68 (2006)
2. Karakaya, M., Korpooglu, I., Ulusoy, O.: Free Riding in Peer-to-Peer Networks. In: IEEE Internet Computing, pp. 92–98. IEEE Press (2009)
3. Golle, P., Leyton-Brown, K., Mironov, I., Lillibridge, M.: FIncentives for Sharing in Peer-to-Peer Networks. In: ACM Conference on Electronic Commerce. ACM Press (2001)
4. Kahan, J. Rapaport, A.: Theories of Coalition Formation. Lawrence Erlbaum Associates Publishers (1984)
5. Ramaswamy, L., Liu, L.: A New Challenge to Peer-to-Peer File Sharing Systems. In: 6th Hawaii International Conference on System Sciences (2003)
6. Sariou, S., Gummadi, P.K., Gribble, S.D.: A measurement study of peer-to-peer file sharing systems. In: Proceedings of Multimedia Computing and Networking (2002)
7. Shneidman, J., Parkes, D.: Rationality and Self-Interest in Peer to Peer Networks. In: Kaashoek, M.F., Stoica, I. (eds.) IPTPS 2003. LNCS, vol. 2735, pp. 139–148. Springer, Heidelberg (2003)
8. Karakaya, M., Korpooglu, I., Ulusoy, O.: Countrecting free riding in Peer-to-Peer networks. Computer Networks 52, 675–694 (2008)
9. Bertsekas, D. Gallager, R.: Data Networks. Prentice-Hall (1992)
10. Ma, R.T.B., Lee, S.C.M., Luis, J.C.S., Yau, D.K.Y.: Ncentive and Service Differentiation in P2P Networks: A Game Theoretic Approach. IEEE/ACM Transactions on Networking 14(5), 978–991 (2006)

11. Androutsellis-Theotokis, S., Spinellis, D.: A survey of Peer-to-Peer Content Distributing Technologies. ACM Computing Surveys 36(4), 335–371 (2004)
12. Garbacki, P., Iosuo, A., Epema, D., Van Steen, M.: 2Fast:Collaboartive Downloads in P2P Networks. In: Sixth International Conference on Peer-to-Peer Computing, P2P 2006 (2006)
13. Mekouar, L. Iraqi, Y. Boutaba, R.: Handling Free Riders in Peer-to-Peer Systems. In: Sixth International Conference on Peer-to-Peer Computing, P2P 2006 (2006)
14. Cohen, B.: Incentives build robutness in Bit Torrent. In: Proceedings of the First Workshop on Economics of Peer-to-Peer Systems (2003)
15. Belmonte, M.V., Conejo, R., Díaz, M., Pérez-de-la-Cruz, J.L.: Coalition Formation in P2P File Sharing Systems. In: Marín, R., Onaindía, E., Bugarín, A., Santos, J. (eds.) CAEPIA 2005. LNCS (LNAI), vol. 4177, pp. 153–162. Springer, Heidelberg (2006)
16. Belmonte, M.V., Díaz, M., Pérez-de-la-Cruz, J.L., Reyna, A.: File sharing service over a generic P2P simulator. In: Proceedings of Semantic, Knowledge and Grid, SKG 2007, pp. 370–373. IEEE Computer Society, USA (2007)
17. Handurukande, S.B., Kermarrec, A., Le Fessant, F., Massoulié, L., Patarin, S.: Peer sharing behaviour in the eDonkey network, and implications for the design of server-less file sharing systems. In: 1st ACM Sigops/Eurosys European Conference on Computer Systems, EuroSys 2006 (2006)
18. The e-Mule web site, http://www.emule.net

Appendix

We must prove that equations (5) define a payoff division that lies inside the core.

Group Rationality. First note that always $x_i \geq 0$. Moreover, a coalition S such that $p_0 \notin S$ has $V(S) = 0$. So the only thing to prove is that, for every $P \subseteq N - \{p_0\}$, the coalition $S = \{p_0\} \cup P$ has a coalitional value $V(S)$ such that $x(0) + \sum_{i \in P} x(i) \geq V(S)$. Let us define $Q = N - P - \{b_0\}$, $p = \sum_{b_i \in P} b_i^{in}$, $q = \sum_{b_i \in Q} b_i^{in}$. Then

$$\sum_{i \in P} x(i) = t_0 \frac{(p+q)^2 + b_0^{in} p}{(b_0^{in} + p + q)^2} \tag{11}$$

and

$$V(S) = t_0 \frac{p}{b_0^{in} + p} \tag{12}$$

The difference between (11) and (12) is

$$t_0 b_0^{in} q \geq 0 \tag{13}$$

Therefore we have proved group rationality.

Global Rationality. Note that the coalitional value as a function $V = v(t_0, b_0^{in}, b_1^{in}, \ldots, b_n^{in})$ (equation 4) is homogeneous of degree 1. Therefore,

$$\sum_0^n x(i) = t_0 \frac{\partial V}{\partial t_0} + b_0^{in} \frac{\partial V}{\partial b_0^{in}} + \ldots + b_n^{in} \frac{\partial V}{\partial b_n^{in}} = V \tag{14}$$

Basics of Intersubjectivity Dynamics: Model of Synchrony Emergence When Dialogue Partners Understand Each Other

Ken Prepin and Catherine Pelachaud

LTCI/TSI, Telecom-ParisTech/CNRS, 37-39 rue Dareau, 75014, Paris, France
{ken.prepin,catherine.pelachaud}@telecom-paristech.fr

Abstract. Since Condon's annotations of videotaped interactions in 1966, an increasing amount of studies points the crucial role of non-verbal behaviours in communication. Among others, synchrony between interactants is claimed to be an evidence of the interaction quality: to give to humans a feeling of natural dialogue, agents must be able to react on appropriate time. Recent dynamical models propose that synchrony emerges from the coupling between interactants. We propose here, and test in simulation, a model of verbal communication which links the mutual understanding of dialogue partners to the emergence of synchrony between their non-verbal behaviours: if interactants understand each other, synchrony emerges; if they do not understand, synchrony is disrupted. In addition to propose and test a model explaining the link between synchrony and interaction quality (synchrony accounts for mutual understanding and good interaction, di-synchrony accounts for misunderstanding) our tests point the fact that synchronisation and di-synchronisation emerging from mutual understanding are fast phenomenons: agents have a quick answer to whether they understand each other or not.

1 Introduction

When we design agents capable of being involved in verbal exchange, with humans or with other agents, it is clear that the interaction cannot be reduced to speech. When an interaction takes place between two partners, it comes with many non-verbal behaviours that are often described by their type such as smiles, gaze at the other, speech pauses, head nod, head shake, raise eyebrows, mimicry of posture and so on [12,27]. But another aspect of these non-verbal behaviours is their timing according to the partner's behaviours.

In 1966, Condon and Ogston's annotations of interactions have suggested that there are temporal correlations between the behaviours of two person engaged in a discussion [4]: micro analysis of discussion videotaped conduces Condon to define in 1976 the notions of auto-synchrony (synchrony between the different modalities of an individual) and hetero-synchrony (synchrony between partners).

Since Condon et al.'s findings, synchronisation between interactants has been investigated in both behavioural studies and cerebral activity studies. These studies tend to show that when people interact together, their synchronisation is tightly linked to

J. Filipe and A. Fred (Eds.): ICAART 2011, CCIS 271, pp. 302–318, 2013.

the quality of their communication: they synchronise if they managed to exchange and share information; synchronisation is directly linked to their friendship, affiliation and mutual satisfaction of expectations.

- In developmental psychology, generations of protocols have been created, from the "still face" [26] to the "double video" [16,18], in order to stress the crucial role of synchronisation during mother-infant interactions.

- Behavioural and cerebral imaging studies show that oblivious synchrony and mimics of facial expressions [2,5] are involved in the emergence of a shared emotion as in emotion contagion [11].

- In social psychology, in teacher-student interaction or in group interactions, synchrony between behaviours occurring during verbal communication has been shown to reflect the rapports (relationship and intersubjectivity) within the groups or the dyads [8,13].

- The very same results have been found for human-machine interactions: on one hand synchrony of non-verbal behaviour improves the comfort of the human and her/his feeling of sharing with the machine (either a robot or a virtual agent) [22] and on the other hand, the human spontaneously synchronises during interaction with a machine when her/his expectations are satisfied by the machine [24].

In the case of non-verbal interactions, the phenomenon of synchronisation between two partners has recently been investigated as a phenomenon emerging from the dynamical coupling of interactants: that is to say a phenomenon whose description and dynamics are not explicited in each of the partners but appear when the interactants are put together and when the new dynamical system they form is more complex and richer than the simple sum of partners dynamics.

In mother-infant interactions via the "double-video" design cited above, synchrony is shown to emerge from the mutual engagement of mother and infant in the interaction [15,18]. In adult-adult interactions mediated by a technological device, synchrony and coupling between partners has been shown to emerge from the mutual attempt to interact with the other in both behavioral studies [1] and cerebral activity studies [7].

These descriptions of synchrony as emerging from the coupling between interactants, are consistent with the fact cited before, that synchrony reflects the quality of the interaction. Given interactants, both the quality of their interaction and the degree of their coupling are tightly linked to the amount of information they exchange and share: high coupling involves both synchrony and good quality interaction; synchrony and quality of the interaction are covarying indices of the interaction. That makes the synchrony parameter particularly crucial: on one hand it carries dyadic information, concerning the quality of the ongoing interaction; on the other hand it can be retrieved by each partner of the interaction, comparing its own actions to its perceptions of the other [24].

The emergence of synchrony during non-verbal interaction has been modelled by both robotics implementation [23] and virtual agent coupling [19].

- In the robotic experiment, two robots controlled by neural oscillators are coupled together by the way of their mutual influence: turn-taking and synchrony emerge [23].

- In the virtual agent experiment, Evolutionary Robotics was used to design a dyad of agents able to favour cross-perception situation; the result obtained is a dyad of agents with oscillatory behaviours which share a stable state of both cross perception and synchrony [19].

The stability of these states of cross-perception and synchrony is a direct consequence of the reciprocal influence between the agents.

We have seen there that literature stresses two main results concerning synchrony. First, synchrony of non-verbal behaviours during verbal-interactions is a necessary element for a good interaction to take place: synchrony reflects the quality of the interaction. Second, synchrony has been described and modelled as a phenomenon emerging from the dynamical coupling between agents during non-verbal interactions. In this paper, we propose to conciliate these two results in a model of synchrony emergence during verbal interactions.

We propose and test in simulation a model of verbal communication which links the emergence of synchrony of non-verbal behaviours to the level of shared information between interactants: if partners understand each other, synchrony will arise, and conversely if they do not understand each other enough, synchrony could not arise. By constructing this model of agents able to interact as humans do, on the basis of psychology, neuro-imaging and modelisation results, that are both the understanding of humans and the believability of artifacts (e.g. virtual humans) which are assessed.

In Sect.2 we describe the architecture principle and show how a level of understanding can be linked to non-verbal behaviours. In Sect.3, we test this architecture, i.e. we test in simulation a dyad of architectures which interact together. We characterise the conditions of emergence of coupling and synchrony between the two virtual agents. Finally, in Sect.4, we discuss these results and their outcomes.

2   Model Principle

We propose a model accounting for the emergence of synchrony depending directly on a shared level of understanding between agents. This model is based on the four next properties of humans' interactions:

P1. To emit or receive a discourse modify the internal state of the agent [25].

P2. Non-verbal behaviours reflect the internal states [14].

P3. Humans are particularly sensitive to synchrony, as a cue of the interaction quality and and the mutual understanding between participants [6,22,24].

P4. Synchrony can be modelled as a phenomenon emerging from the dynamical coupling of agents [23,19,1]

The model of agent we propose in the present section is implemented in Sect.3 as a Neural Network (NN). Groups of neurons are vectors of variables represented by capital letters (e.g. $V_{Input} \in [-1,1]^n$ and $S \in [-1,1]^m$) and the weights matrices which modulate the links between these groups are represented by lower case letters (e.g. $u \in [-1,1]^{m \times n}$): we obtain equations such as $u \cdot V_{Input} = S$. For sake of simplicity, in both the description of the model principle (this section) and in its implementation and tests (Sect.3) groups of neurons and weights matrices are reduced to single numerical variables ($\in [-1,1]$).

In the next two subsections, we model the two first properties, P1 and P2. We describe how the non-verbal behaviour can be linked to a level of mutual understanding.

Then, in the subsections 2.3 and 2.4, we describe how this will give to a dyad of agents coupling capabilities. That constitute the modelling of the third and fourth properties, P3 and P4.

2.1 Speak and Listen Modifies Internal State

Let us consider a dyad of agents, Agent1 and Agent2. Each agent's state is represented by one single variable, S_1 for Agent1 and S_2 for Agent2 ($\in [-1, 1]$). Now, let us consider the speech produced by each agent, the verbal signal $V_{Act\ i}$ ($\in \{0, 1\}$), and the speech heard by each agent, the perceived signal $V_{Per\ i}$ ($\in \{0, 1\}$).

P1 claims that each agent, either listener or speaker, has its internal state S_i modified by verbal signals: the listener's internal state is modified by what it hears, and the speaker's internal state is modified by what it says. Two "level of understanding", the weights u_i and u'_i, are defined for each agent of the dyad. u_i modulates the perceived verbal signal $V_{Per\ i}$, and u'_i modulates the produced verbal signal $V_{Act\ i}$ (see fig.1). To

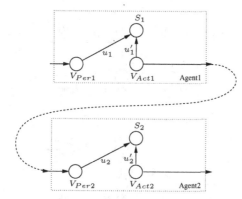

Fig. 1. Verbal perception, $V_{Per\ i}$, and verbal action, $V_{Act\ i}$, both influence the internal state S_i. These influences depend respectively on the level of understanding u_i and u'_i.

model interaction in more natural settings these u_i parameters should be influenced by many variables, such as the context of the interaction (discussion topic, relation-ship between interactants), the agents moods and personalities. However in the present model we combine all these parameters in the single variable u_i ($\in [-1, 1]$). The choice of the values of u_1 and u_2 is arbitrary near 0.01: it enables a well balanced sampling of the oscillators' activations, the period last around 100 time steps; the other parameters of the architecture are chosen depending on this one so as not to modify the whole systems dynamics.

If t is the time we have the following equations:

$$\begin{cases} S_1(t+1) = S_1(t) + u_1 V_{Per1}(t+1) + u'_1 V_{Act1}(t+1) \\ S_2(t+1) = S_2(t) + u_2 V_{Per2}(t+1) + u'_2 V_{Act2}(t+1) \end{cases} \tag{1}$$

Assuming that communication is ideal, i.e. $V_{Per i} = V_{Act j}$, and that Agent1 is the only one to speak, i.e. $V_{Act2} = V_{Per1} = 0$, the system of equations 1 gives:

$$\begin{cases} S_1(t+1) = S_1(t) + u'_1 V_{Act1}(t+1) \\ S_2(t+1) = S_2(t) + u_2 V_{Act1}(t+1) \end{cases} \tag{2}$$

This first property P1 is crucial in our model, as it links together the agents' internal states: each one is modified by speech depending on its own parameter u_i. In the present model, we assume that for a given agent, understanding of its productions and of its perceptions are similar: for Agent i, $u_i = u_i'$.

2.2 Non-verbal Behaviours Reflect Internal State

The second property P2, claims that "non-verbal behaviours reflect internal state". That is to say, agent's arousal, mood, satisfaction, awareness, are made visible thanks to facial expressions, gaze, phatics, backchannel, prosody, gestures, speech pauses. To make visible the internal properties of Agent i, a non-verbal signal, $NV_{Act\ i}$, is triggered depending on its internal state, S_i. When S_i reaches the threshold β, the agent produces non-verbal behaviours with th_β the threshold function (see Fig.2):

$$NV_{Act\ i}(t) = th_\beta(S_i(t)) \tag{3}$$

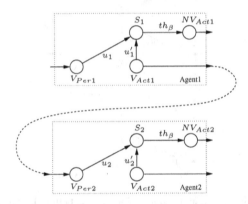

Fig. 2. Each agent produces non-verbal behaviours $NV_{Act\ i}$ when S_i reaches the threshold β. $NV_{Act\ i}$ depends on how much the internal state S_i has been influenced by what has been said.

We suggest here that pitch accents, pauses, head nods, changes of facial expressions and other non-verbal cues are, for a certain part, produced by agents when a particularly important idea arises, when the explanation reach a certain point, when an idea or a concept starts to be outlined. We assume that the phenomenon is similar in both speaker and listener, it is driven by the evolution of what is wanted to be expressed in one case and it is driven by what is heard in the other case. If speaker and listener understand each other, these peaks of arousal and understanding should co-occur: they appear to be temporally linked. These peaks will be the bases of entrainment for intentional coordination between partners. And then this coordination could be seen as a marker of interaction quality.

Considering these two first points, that is to say, equations 2 and 3 we have the following system of equations :

$$\begin{cases} NV_{Act1}(t_1) = th_\beta(\sum_{t_0}^{t_1} u_1 V_{Act1}(t)) \\ NV_{Act2}(t_1) = th_\beta(\sum_{t_0}^{t_1} u_2 V_{Act1}(t)) \end{cases} \tag{4}$$

If an agent is enough influenced by what is said, it produces non-verbal signals. And if $u_1 = u_2$ then $NV_{Act1} = NV_{Act2}$, agents' non-verbal behaviours may be synchronised, where as if u_1 and u_2 are too different, agents will not be able to synchronise.

2.3 Sensitivity to Synchrony

To account for the property P3, "sensitivity of human to synchrony", we use the fact that sensitivity to synchrony can be modelled by simple model of mutual reinforcement of the perception-action coupling [1,19]. In addition to the influence from speech (either during its perception or its production), each agent's internal state S_i is influenced by the non-verbal behaviour it perceives from the other $NV_{Act\,j}$, modulated by sensitivity to non-verbal signal σ (see fig.3).

The internal state of each agent is modified by both what it understand of the speech

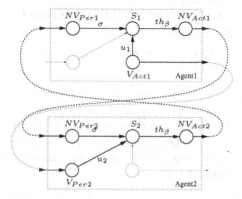

Fig. 3. Agent1's internal state, S_1, is influenced by both its own understanding of what it is saying $u_1 \cdot V_{Act1}$ and the non-verbal behaviour of Agent2, $\sigma \cdot NV_{Act2}$. Agent2's internal state, $_2$, is influenced by its own understanding of what Agent1 says $u_2 \cdot V_{Act1}$ and the non-verbal behaviour of Agent1, $\sigma \cdot NV_{Act1}$.

and what it sees from the non-verbal behaviour of the other:

$$\begin{cases} S_1(t+1) = S_1(t) + u_1 V_{Act1}(t+1) + \sigma NV_{Act2}(t) \\ S_2(t+1) = S_2(t) + u_2 V_{Act1}(t+1) + \sigma NV_{Act1}(t) \end{cases} \tag{5}$$

This last equation will favour the synchronisation by increasing the reciprocal influence when agents' internal state reach together a high level.

2.4 Coupling between Dynamical Systems

How to enable agents involved in a verbal interaction, to be as much synchronised as they share information? To enable synchrony to emerge between the two agents, we used the fact that synchronisation can be modelled as a phenomenon emerging from the dynamical coupling within the dyad [23]: on one hand agents must have internal dynamics which control their behaviour; on the other hand, they must be influenced by the other's behaviours.

In the previous subsections, we proposed a dyad of agent which mutually influence. If we replace the non-verbal behaviours of agents by their internal states in the system of equations 5, it gives:

$$\begin{cases} S_1(t+1) = S_1(t) + u_1 V_{Act1}(t+1) + \sigma th_\beta(S_2(t)) \\ S_2(t+1) = S_2(t) + u_2 V_{Act1}(t+1) + \sigma th_\beta(S_1(t)) \end{cases} \tag{6}$$

To enable coupling to occur, the agents should also be dynamical systems: systems which state evolves along time by themselves. The internal state of the agents S_i produces behaviours and is influenced by the other agent's behaviour. To ensure internal dynamics, we made this internal state a relaxation oscillator, which increases linearly and decreases rapidly when it reaches the threshold 0.95 (Fig.5 shows an example of the signals obtained). By oscillating , the internal states agents will not only influence each other but also be able to correlate one with the other [23].

Here, two cases are interesting.

When the internal states of both agents are under the threshold triggering non-verbal behaviours, β, the system of equation 6 becomes:

$$\begin{cases} S_1(t+1) = S_1(t) + u_1 V_{Act1}(t+1) \\ S_2(t+1) = S_2(t) + u_2 V_{Act1}(t+1) \end{cases} \tag{7}$$

The two agents are almost independent, they are only influenced by the speech of Agent1 and each one produces its own oscillating dynamic. That could be the case if two tired people (high β) speak about a not so interesting subject (u_i are low): they are made apathic by the conversation, they do not express anything.

The second interesting case is when both agents' internal states are above the threshold β. The system of equation 6 becomes:

$$\begin{cases} S_1(t+1) = S_1(t) + u_1 V_{Act1}(t+1) + \sigma S_2(t) \\ S_2(t+1) = S_2(t) + u_2 V_{Act1}(t+1) + \sigma S_1(t) \end{cases} \tag{8}$$

In this case agents are not anymore independent, they influence each other depending on the way they understand speech. If we push the recursivity of these equations one step further we obtain:

$$\begin{cases} S_1(t+1) = S_1(t) + u_1 V_{Act1}(t+1) + \sigma(S_2(t-1) + u_2 V_{Act1}(t) + \sigma S_1(t-1)) \\ S_2(t+1) = S_2(t) + u_1 V_{Act1}(t+1) + \sigma(S_1(t-1) + u_1 V_{Act1}(t) + \sigma S_2(t-1)) \end{cases} \tag{9}$$

And now we see the effect of coupling, that is to say that agents are not only influenced by the state of the other but they are influenced by their own state, mediated by the other: the non-verbal behaviours of the other becomes their own biofeedback [17]. When the threshold β is overtaken, the reciprocal influence is recursive and becomes exponential: the dynamics of S_1 and S_2 are not any more independent, they are influenced in their phases and frequencies [21,23].

3 Test of the Model

We tested this model by implementing a dyad of agent as a neuronal network in the neuronal network simulator Leto/Prometheus (developed in the ETIS lab. by Gaussier et al. [9,10]), and by studying its emerging dynamics with different sets of parameters.

3.1 Implementation

We implemented the model on the neural networks simulator Leto/Prometheus. Leto/Prometheus simulates the dynamics of neural networks by an update of the whole network at each time step. We use groups of neurons with one neuron, and non-modifiable links between groups. The schema of Fig.4 show this implementation.

The internal states of agents, S_i, are relaxation oscillators: the re-entering link of

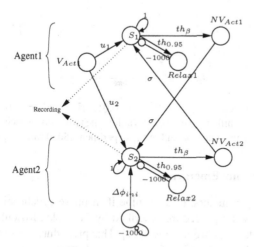

Fig. 4. Implementation of the two agents. The couples $(S_1; Relax1)$ and $(S_2; Relax2)$ are relaxation oscillators. The parameters which will be tested are the following: β, the threshold which controls the non-verbal production; u_1 and u_2 which control the agents' level of sharing; $\Delta\phi_{ini}$, the initial phase-shift between agents.

weight 1 makes the neuron behave as a capacity, and the $Relax$ neuron which fires when a 0.95 threshold is reached, inhibits S_i and makes it relax (see Fig.5 for an example of the activation obtained).

V_{Act1}, Agent1's verbal production, is a neuron of constant activity 1. This neuron feeds the oscillators of both agents, weighted by their level of understanding u_1 and u_2. The values of u_1 and u_2 are near 0.01: it enables a well balanced sampling of the oscillators' activations, the period last around 100 time steps.

In addition to agent understanding u_1 and u_2, three other parameters are modifiable in this implementation:

- The threshold β which controls the triggering of non-verbal signal.
- The sensitivity of agent's internal state to non-verbal signal σ which weights $NV_{Act\,i}$.

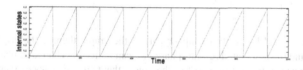

Fig. 5. Activations of the internal state $S_1(t)$ for $u_1 = 0.01$

These two parameters β and σ directly control the amount of non-verbal influence between the agents: they must be high enough to enable coupling, for instance reducing initial phase-shift between oscillators or compensating phase deviation when $u_1 \neq u_2$.
- The initial phase shift $\Delta\phi_{ini}$, which makes agents start with a phase shift between $S_1(t_{ini})$ and $S_2(t_{ini})$ at the beginning of each test of the architecture.

Finally, the variables recorded during these tests are the internal states of both agents, $S_1(t)$ and $S_2(t)$ (see Fig.6 for an example).

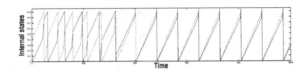

Fig. 6. Activations recorded for $u_1 = 0.01$, $u_2 = 0.011$, $\beta = 0.85$, $\sigma = 0.05$ and $\Delta\phi_{ini} = 0.4$. Despite the initial phase shift and the phase deviation, the two agents synchronise. This is a stable state of the dyad, it remains until the end of the experiment (5000 time steps).

3.2   Test of Synchrony Emergence

For a given set of parameters, to determine if in-phase synchronisation occurred between agents, we used a procedure described by Pikovsky, Rosenblum and Kurths in their reference book "Synchronisation" [21]. This procedure consists in comparing the phases of two signals to determine if they are synchronous or not.

First we used the fact that relaxation oscillators can be characterised by their peaks. There is a peak at time t_k when $S_i(t_k) \geq 0.9\beta$ and $S_i(t_k + 1) = 0$. Then, we used the fact that phase can be rebuilt from these peaks [21]. We assign to the time t_k the values of the phase $\phi(t_k) = 2\pi k$, and for every instants of time $t_k < t < t_k + 1$ determine the phase as a linear interpolation between these values (see fig.7):

$$\phi(t) = 2\pi k + 2\pi \frac{t - t_k}{t_{k+1} - t_k} \tag{10}$$

After that, when the phases of signals are obtained, we consider their difference modulo 2π (see fig.8). Horizontal plateaus in this graph reflect periods of constant phase-shift

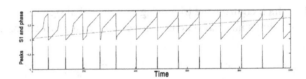

Fig. 7. Signal, Peaks and Phase. In the upper part of the graph, there is the original signal S_1 (shown in fig.6) and the associated re-built phase (we can notice the change of phase slope when synchronisation occurs). In the lower part of the graph, there are the peaks extracted from S_1 in order to re-build the phase.

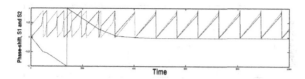

Fig. 8. Signals of two agents and their associated phase-shift $\Delta_{\phi_1,\phi_2}(t)$. When agents synchronise with each other, their phase-shift remains constant and near zero.

between signals, i.e. synchronisation. Horizontal plate aux near zero reflect periods of synchronisation and co-occurrence of non-verbal signals.

Finally, for each 5000 time steps simulation, we define that in-phase synchronisation occurs if the phase-shift becomes near zero at a time t_{synch}, smaller than 3000, and remains constant until the end. We defined the synchronisation speed as $SynchSpeed = (3000 - t_{synch})/3000$. If in-phase synchronisation is immediate $SynchSpeed = 1$; if in-phase synchronisation occurs at time step 3000 $SynchSpeed = 0$; and if in-phase synchronisation do not occurs $SynchSpeed < 0$.

3.3 Test of Architecture Parameters

We tested different parameters of this model, first to show the direct link existing between emergence of synchrony and level of sharing between interactants, and second to characterise the different properties of this model.

To show the direct link existing between emergence of synchrony and level of sharing between interactants, we fixed u_1 to 0.01 and made u_2 vary between 0 and 0.02, that is to say the shared understanding of the two agents differs between 0 and 100%. Notice here the importance to test synchronisation when $u_2 = 0$: if synchronisation occurs when $u_2 = 0$, i.e. when Agent2 does not perceived the speech of Agent1, that means that agents synchronise every time just thank to non-verbal signal of Agent1; in that case, synchrony is not any more an in dice of the interaction quality, the influence of non-verbal signals (linked to β and σ) is too high.

To evaluate the influence of the amount of non-verbal signal exchanged, we made the threshold β vary between 0 and 0.95.

To evaluate the influence of the sensitivity to non-verbal signal, we made the sensitivity σ vary between 0 and 0.09.

Finally, to evaluate the abilities of such a dyad of agents to re-synchronise after an induced phase-shift or after a misunderstanding, we made the initial phase shift $\Delta\phi_{ini}$ vary between 0 and π.

Shared Understanding Influence. When the two agents are synchronous in phase ($\Delta\phi_{ini} = 0$), we tested which of the u_2 values keep agents synchronised or make them disynchronise. For fixed $\beta = 0.7$, $\sigma = 0.05$ and $\Delta\phi_{ini} = 0$, u_2 varies between 0 and 0.02. The following graph of Fig.9 shows the associated disynchronisation speed.

When the difference between u_1 and u_2 is to high, no synchronisation can occur since even when synchrony is forced at the beginning of the experiment, agent disynchronise.

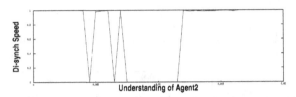

Fig. 9. Di-synchronisation speed of the dyad, depending on the Agent2 understanding u_2. u_2 varies from left to right between 0 and 0.02. A null disynchronisation speed means that synchronisation has been maintained until the end of the experiment. A disynchronisation speed 1 is for a dis-synchronisation occurring at the very beginning of the experiment.

Influence of Amount of Non-verbal Signals. The coupling and synchronisation capabilities of the dyad of agents, may directly depend on the amount of non-verbal signals they exchange: among other, the ability to compensate a difference of understanding may be improved by an increase of non-verbal signals exchanged. We tested this effect by calculating disynchronisation speeds as just above, making u_2 vary between 0 and 0.02 and the threshold β varying between 0 and 0.9 ($\sigma = 0.05$). We obtained the 3D graph of Fig.10.

When $\beta = 0.9$, that is to say when very few non-verbal signals are exchanged, synchrony maintains only when the two agents have equal level of understanding, $u_1 = u_2 = 0.01$. For other values, the influence of the threshold β is not so clear: the dyad does not resist better to disynchronisation when $\beta < 0.5$ than when $6 \leq \beta \leq 8$. This effect, or this absence of effect, may be due to the fact that the more β decreases, the less accurate in time the non-verbal signals are: if β is low, non-verbal signals are emit earlier before the peaks of S_i activation and on a larger time window, they are not enough precise in time to maintain synchrony. We chosen $\beta = 0.7$, i.e. the mean of its best performances values.

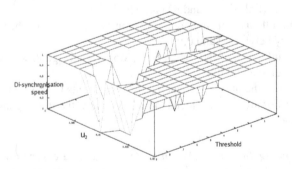

Fig. 10. Di-synchronisation speed of the dyad, depending on the Agent2 understanding u_2 and the threshold β ($\sigma = 0.5$). u_2 varies between 0 and 0.02. β varies from 0.9 to 0, in the sens of non-verbal signals increase. When the d i-synchronisation speed value is null, synchronisation has been maintained until the end of the experiment. A disynchronisation speed 1 is for a disynchronisation occurring at very beginning of the experiment.

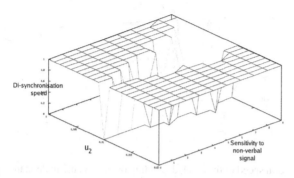

Fig. 11. Di-synchronisation speed of the dyad, depending on the Agent2 understanding u_2 and the sensitivity σ ($\beta = 0.7$). u_2 varies between 0 and 0.02. σ varies from 0 to 0.09. When the d i-synchronisation speed value is null, synchronisation has been maintained until the end of the experiment. A disynchronisation speed 1 is for a disynchronisation occurring at the very beginning of the experiment.

Sensitivity to Non-verbal Signals. Another way to modify the influence of non-verbal signals on coupling and synchronisation properties of the dyad, is to modify the sensitivity to the perceived non-verbal signal, σ. We tested this effect by calculating disynchronisation speeds as previously, making u_2 vary between 0 and 0.02 and the sensitivity σ varying between 0 and 0.09 ($\beta = 0.07$). We obtained the 3D graph of Fig.11.

Sensitivity to non-verbal signal σ have a direct effect on agents to stay synchronous even with different understandings: the higher is sensitivity σ, the more resistant to difference between u_i the synchronisation capability of the dyad is. The effect of σ is important despite its low value ($\sigma < 0.1$) due to the high number of non-verbal signal exchanged: when Agent i's internal state S_i reaches the threshold β, it produces the non-verbal signals $NV_{Act\ i}$ at every time step until S_i relaxes. That can last between 0 and 20 time steps for each oscillation period. The effect of σ is multiplied by this number of steps.

It is important to notice here that the σ effect on the dyad resistance to u_i differences, has a counter-part. This counter-part is the fact that when σ increase and make the dyad more resistant to disynchronisation, it also makes the synchronisation of the dyad less related to mutual understanding. For instance, when $\sigma \geq 0.7$, agents stay synchronous even when Agent2 do not understand anything, $u_2 = 0$. To balance these two effects, facilitation of synchronisation and decrease of synchrony significance, we chosen a default value of $\sigma = 0.05$.

Re-synchronisation Capability. Given a value of Agent2 understanding u_2, we tested the ability of the dyad Agent1-Agent2 to re-synchronise after a phase shift. We made the initial phase-shift $\Delta\phi_{ini}$ vary between 0 and π for every values of u_2 and calculated the speed of synchronisation if any. The 3D graph of Fig.12 shows the synchronisation speed for each couple $(u_2; \Delta\phi_{ini})$.

The initial phase-shift between S_1 and S_2 does not appear to affect the synchronisation capacities of the dyad. With the chosen $\sigma = 0.05$ and $\beta = 0.7$, when the agents'

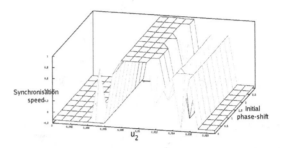

Fig. 12. Synchronisation speed of the dyad, depending on the Agent2 understanding u_2 and initial phase-shift $\Delta\phi_{ini}$ ($\sigma = 0.05$ and $\beta = 0.7$). u_2 varies between 0 and 0.02. $\Delta\phi_{ini}$ varies from 0 to π. When the synchronisation speed value is null, the dyad did not synchronised until the end of the experiment. A synchronisation speed 1 is for a synchronisation occurring at the very beginning of the experiment.

levels of understanding u_1 and u_2 do not differ more than 15% of each other, they synchronise systematically and very quickly: for instance they synchronise even when they start in anti-phase ($\Delta\phi_{ini} = \pi$). And conversely, when the levels of understanding u_1 and u_2 are more than 15% different, synchronisation is no more immediate.

4   Discussion

We proposed and tested a model which links emergence of synchrony between dialogue partners to their level of shared understanding. This model assesses both the understanding of humans and the believability of artifacts (e.g. virtual humans). When two interactants have similar understanding of what the speaker says, their non-verbal behaviours appear synchronous. Conversely, when the two partners have different understanding of what is being said, they disynchronise. This model is implemented as a dynamical coupling between two talking agents: on one hand, each agent proposes its own dynamics; on the other hand, each agent is influenced by its perception of the other. These are the two minimal conditions enabling coupling. What makes this model particular is that the internal dynamics of agents are generated by the meaning exchanged through speech. It links the dynamical side of interaction to the formal side of speech.

We tested this model in simulation, and showed that synchrony effectively emerges between agents when they have close level of understanding. We noticed a clear effect of the level of understanding on the capacity of the agents to both remain synchronous and re-synchronise: agents disynchronise if the level of shared understanding is lower than 85% (with our parameters) and conversely agents synchronise if the level of shared understanding is higher than 85%. These results tend to prove that, considering that synchrony between agents is an indice of good interaction and shared understanding, the reciprocal property is true too; that is disynchrony accounts for misunderstanding.

We have shown that agents remain synchronous depends on both their shared understanding (the ratio between u_1 and u_2) and their sensitivity to non-verbal behaviour (σ in our implementation). The more sensitive to non-verbal behaviours are the agents, the more resistant to disynchronisation is the dyad and the easier is the synchronisation.

An important counter-part of this easier synchronisation is that it makes synchrony less representative of shared understanding: agents or people with very different levels of understanding will be able to synchronise; if sensitivity to non-verbal behaviour is too high, the dyadic parameter of synchrony is not a cue of shared understanding. By contrast, the facility agents trigger non-verbal behaviours when their internal states are high (threshold β) does not appear to change the synchronisation properties of the dyad: the higher number of exchanged non-verbal signals seems to be compensated by their associated decrease of precision.

In addition to the effect of shared understanding on the stability of synchrony between agents, we have tested the effect of shared understanding on the capacity of the dyad to re-synchronise. For instance, during a dialogue, synchrony can be broken by the use of new concept by the speaker. That may result in lowering the level of shared understanding below the 85% necessary for remaining synchronous. Synchrony can also be disrupted by an external event which can introduce a phase-shift between interactants. Given fixed sensitivity to non-verbal behaviour (σ) and facility to trigger non-verbal behaviours (β), we tested how quickly the dyad can re-synchronise after a phase-shift. The shared level of understanding necessary to enable re-synchronisation appeared to be the same as the one under which agents disynchronise.

Two crucial points must be noticed here. First, when agents' understanding do not differ more than 15% (shared understanding higher than 85%), agents synchronise systematically whatever the phase-shift is, and when agent's understanding differ more than 15% they disynchronise. Second, both synchronisation and disynchronisation of agents are very quick, lasting about one oscillation of the agents' internal states. Synchronisation and disynchronisation are very quick effects of respectively misunderstanding and shared understanding: agents involved in an interaction do not have to wait to see synchrony appears when they understand each other, they have a fast answer to whether they understand each other or not.

The 5000 time steps length of our tests allowed us to test the stability of synchrony or disynchrony after their occurrence; however it is clearly not a natural situation. Synchrony in natural interaction is a varying phenomenon involving multiple synchronisation and disynchronisation phases: the level of shared understanding varies along the interaction. In fact disynchrony may be quite informative for the dyad as its detection enables agents to adapt one another. In natural interactions, synchrony occurring after disynchrony shows that agents share understanding whereas they did not before: they have benefited from the interaction and exchanged information.

As a consequence, the mean level of shared understanding necessary for good interaction to take place between persons in natural context would be much more reasonable: the 85% of shared understanding occurs in phases of particularly good interaction and its is not a hard constraint on the whole dialogue; this very high level necessary for synchronisation should be divided by the ratio of synchrony vs disynchrony phases present in natural interaction. For instance we can imagine that a level of shared understanding higher than 85% would occur when people involved in a discussion have just reached an agreement. By contrast, when the level of shared understanding stays all along the dialogue far under 85%, the dyad would be more like two strangers trying to talk together, or a professional talking with technical words to a naive listener.

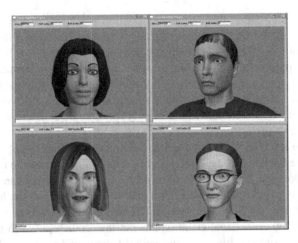

Fig. 13. Greta, Obadia, Poppy and Prudence. They are four agents implemented on the open source system Greta. Each one has its own personality and level of understanding. When interacting together, different levels of non-verbal synchrony should appear between the agents of this group.

Our model has been tested and its principle has been validated in agent-agent context. To go a step farther, in "wild world" situations involving humans, two elements must be added: Understanding of language during interaction with human; Recognition of non-verbal behaviours of human users. In the near future, we will adapt the present neural architecture to the open source virtual agent Greta [20]. The system Greta enables one to generate multi-modal (verbal and non-verbal) behaviours online and with accurate timing. The verbal signals will be modelled as elements of "small-talk" and the non-verbal signal will be modelled as, pitch accents, pauses, head nods, head shakes and facial expressions. To test the real impact of such a model on human perception of interaction, we will perform perceptive evaluation: we aim to simulate a group of virtual agents dialoguing with each other (see fig.4). Each agent will have its own personality and level of understanding of what being said. This will lead to pattern of synchronisation and disynchronisation. Among other, agents which share understanding should display inter-synchrony pattern [3]. Finally, human observers should clearly fill which agent is sharing understanding with which other agent.

In conclusion, we can notice that, in addition to the two main results of this study —"disynchrony accounts for misunderstanding" and "synchronisation and disynchronisation are very quick phenomenons"— another result is the model itself. It proposes a link between synchrony and inter-subjectivity by the use of dynamical system coupling: synchrony and dynamical coupling emerge together when agents mutually understand each other; as a consequence synchrony account for good interaction.

We believe, this model is a start to answer the issues of what is the part of dynamical coupling between agents involved in verbal interaction? What is the part of emerging dynamics in the communication of meanings and intentions? And moreover, how these two parts can co-exist and feed each other?

Acknowledgements. This work has been partially financed by the European Project NoE SSPNet (Social Signal Processing Network). Nothing could have been done without the Leto/Prometheus NN simulator, lent by the Philippe Gaussier's team (ETIS lab, Cergy-Pontoise, France).

References

1. Auvray, M., Lenay, C., Stewart, J.: Perceptual interactions in a minimalist virtual environment. New Ideas in Psychology 27, 32–47 (2009)
2. Chammat, M., Foucher, A., Nadel, J., Dubal, S.: Reading sadness beyond human faces. Brain Research (2010) (in Press, accepted, manuscript)
3. Condon, W.S.: An analysis of behavioral organisation. Sign Language Studies 13, 285–318 (1976)
4. Condon, W.S., Ogston, W.D.: Sound film analysis of normal and pathological behavior patterns. Journal of Nervous and Mental Disease 143, 338–347 (1966)
5. Dubal, S., Jouvent, A.F.R., Nadel, J.: Human brain spots emotion in non humanoid robots. Social Cognitive and Affective Neuroscience (2010) (in press)
6. Ducan, S.: Some signals and rules for taking speaking turns in conversations. Journal of Personality and Social Psychology 23(2), 283–292 (1972)
7. Dumas, G., Nadel, J., Soussignan, R., Martinerie, J., Garnero, L.: Inter-brain synchonization during social interaction. PLoS One 5(8), e12166 (2010)
8. Bernieri, F.J.: Coordinated movement and rapport in teacher-student interactions. Journal of Nonverbal Behavior 12(2), 120–138 (1988)
9. Gaussier, P., Cocquerez, J.: Neural networks for complex scene recognition :simulation of a visual system with several cortical areas. In: IJCNN, Baltimore, pp. 233–259 (1992)
10. Gaussier, P., Zrehen, S.: Avoiding the world model trap: An acting robot does not need to be so smart. Journal of Robotics and Computer-Integrated Manufacturing 11(4), 279–286 (1994)
11. Hatfield, E., Cacioppo, J.L., Rapson, R.L.: Emotional contagion. Current Directions in Psychological Sciences 2, 96–99 (1993)
12. Kendon, A.: Conducting Interaction: Patterns of Behavior in Focused Encounters. Cambridge University Press, Cambridge (1990)
13. LaFrance, M.: Nonverbal synchrony and rapport: Analysis by the cross-lag panel technique. Social Psychology Quarterly 42(1), 66–70 (1979)
14. Matsumoto, D., Willingham, B.: Spontaneous facial expressions of emotion in congenitally and non-congenitally blind individuals. Journal of Personality and Social Psychology 96(1), 1–10 (2009)
15. Mertan, B., Nadel, J., Leveau, H.: The effect of adult presence on communicative behaviour among toddlers. In: New Perspective in Early Communicative Development. Routledge, London (1993)
16. Murray, L., Trevarthen, C.: Emotional regulation of interactions vetween two-month-olds and their mothers. In: Social Perception in Infants, pp. 101–125 (1985)
17. Nadel, J.: Imitation and imitation recognition: their functional role in preverbal infants and nonverbal children with autism, pp. 42–62. Cambridge University Press, UK (2002)
18. Nadel, J., Tremblay-Leveau, H.: Early perception of social contingencies and interpersonal intentionality: dyadic and triadic paradigms. In: Early Social Cognition, pp. 189–212. Lawrence Erlbaum Associates (1999)
19. Paolo, E.A.D., Rohde, M., Iizuka, H.: Sensitivity to social contingency or stability of interaction? modelling the dynamics of perceptual crossing. New Ideas in Psychology 26, 278–294 (2008)

20. Pelachaud, C.: Modelling multimodal expression of emotion in a virtual agent. Philosophical Transactions of Royal Society. Biological Science 364, 3539–3548 (2009)

21. Pikovsky, A., Rosenblum, M., Kurths, J.: Synchronization: A Universal Concept in Nonlinear Sciences. Cambridge University Press, Cambridge (2001)

22. Poggi, I., Pelachaud, C.: Emotional Meaning and Expression in Animated Faces. In: Paiva, A.C.R. (ed.) IWAI 1999. LNCS, vol. 1814, pp. 182–195. Springer, Heidelberg (2000)

23. Prepin, K., Revel, A.: Human-machine interaction as a model of machine-machine interaction: how to make machines interact as humans do. Advanced Robotics 21(15), 1709–1723 (2007)

24. Prepin, K., Gaussier, P.: How an Agent Can Detect and use Synchrony Parameter of its Own Interaction with a Human? In: Esposito, A., Campbell, N., Vogel, C., Hussain, A., Nijholt, A. (eds.) Second COST 2102. LNCS, vol. 5967, pp. 50–65. Springer, Heidelberg (2010)

25. Scherer, K., Delplanque, S.: Emotions, signal processing, and behaviour. Firmenich, Geneva (2009)

26. Tronick, E., Als, H., Adamson, L., Wise, S., Brazelton, T.: The infants' response to entrapment between contradictory messages in face-to-face interactions. Journal of the American Academy of Child Psychiatry (Psychiatrics) 17, 1–13 (1978)

27. Yngve, V.H.: On getting a word in edgewise, pp. 567–578 (April 1970)

Stability and Optimality in Matching Problems with Weighted Preferences

Maria Silvia Pini[1], Francesca Rossi[1],
Kristen Brent Venable[1], and Toby Walsh[2]

[1]Department of Pure and Applied Mathematics, University of Padova
Padova, Italy
[2]NICTA and UNSW, Sydney, Australia
{mpini,frossi,kvenable}@math.unipd.it,
Toby.Walsh@nicta.com.au

Abstract. The stable marriage problem is a well-known problem of matching men to women so that no man and woman, who are not married to each other, both prefer each other. Such a problem has a wide variety of practical applications, ranging from matching resident doctors to hospitals, to matching students to schools or more generally to any two-sided market. In the classical stable marriage problem, both men and women express a strict preference order over the members of the other sex, in a qualitative way. Here we consider stable marriage problems with weighted preferences: each man (resp., woman) provides a score for each woman (resp., man). Such problems are more expressive than the classical stable marriage problems. Moreover, in some real-life situations it is more natural to express scores (to model, for example, profits or costs) rather than a qualitative preference ordering. In this context, we define new notions of stability and optimality, and we provide algorithms to find marriages which are stable and/or optimal according to these notions. While expressivity greatly increases by adopting weighted preferences, we show that in most cases the desired solutions can be found by adapting existing algorithms for the classical stable marriage problem. We also investigate manipulation issues in our framework. More precisely, we adapt the classical notion of manipulation to our context and we study if the procedures which return the new kinds of stable marriages are manipulable.

Keywords: Stable marriages, Weighted preferences.

1 Introduction

The stable marriage problem (SM) [9] is a well-known problem of matching the elements of two sets. It is called the *stable marriage* problem since the standard formulation is in terms of men and women, and the matching is interpreted in terms of a set of marriages. Given n men and n women, where each person expresses a strict ordering over the members of the opposite sex, the problem is to match the men to the women so that there are no two people of opposite sex who would both rather be matched with each other than their current partners. If there are no such people, all the marriages are said to be *stable*. In [4] Gale and Shapley proved that it is always possible

J. Filipe and A. Fred (Eds.): ICAART 2011, CCIS 271, pp. 319–333, 2013.

to find a matching that makes all marriages stable, and provided a polynomial time algorithm which can be used to find one of two extreme stable marriages, the so-called *male-optimal* or *female-optimal* solutions. The Gale-Shapley algorithm has been used in many real-life scenarios [20], such as in matching hospitals to resident doctors [19], medical students to hospitals, sailors to ships [13], primary school students to secondary schools [21], as well as in market trading.

In the classical stable marriage problem, both men and women express a strict preference order over the members of the other sex in a qualitative way. Here we consider stable marriage problems with weighted preferences. In such problems each man (resp., woman) provides a score for each woman (resp., man). Stable marriage problems with weighted preferences are interesting since they are more expressive than the classical stable marriage problems, since in classical stable marriage problem a man (resp., a woman) cannot express how much he (resp., she) prefers a certain woman (resp., man). Moreover, they are useful in some real-life situations where it is more natural to express scores, that can model notions such as profit or cost, rather than a qualitative preference ordering. In this context, we define new notions of stability and optimality, we compare such notions with the classical ones, and we show algorithms to find marriages which are stable and/or optimal according to these notions. While expressivity increases by adopting weighted preferences, we show that in most cases the desired solutions can be found by adapting existing algorithms for the classical stable marriage problem. We also investigate manipulation issues in our framework. More precisely, we adapt the classical notion of manipulation to our context and we show that in some cases the procedures which return the new kinds of stable marriage are manipulable.

Stable marriage problems with weighted preferences have been studied also in [8,12]. However, they solve these problems by looking at the stable marriages that maximize the sum of the weights of the married pairs, where the weights depend on the specific criteria used to find an optimal solution, that can be minimum regret criterion [8], the egalitarian criterion [12] or the Lex criteria [12]. Therefore, they consider as stable the same marriages that are stable when we don't consider the weights. We instead use the weights to define new notions of stability that may lead to stable marriages that are different from the classical case. They may rely on the difference of weights that a person gives to two different people of the other sex, or by the strength of the link of the pairs (man,woman), i.e., how much a person of the pair wants to be married with the other person of the pair. The classical definition of stability for stable marriage problems with weighted preferences has been considered also in [2] that has used a semiring-based soft constraint approach [3] to model and solve these problems.

The paper is organized as follows. In Section 2 we give the basic notions of classical stable marriage problems, stable marriage problems with partially ordered preferences and stable marriage problems with weighted preferences (SMWs). In Section 3 we introduce a new notion of stability, called α-stability for SMWs, which depends on the difference of scores that every person gives to two different people of the other sex, and we compare it with the classical notion of stability. Moreover, we give a new notion of optimality, called lex-optimality, to discriminate among the new stable marriages, which depends on a voting rule. We show that there is a unique optimal stable marriage and we give an algorithm to find it. In Section 4 we introduce other notions of stability for SMWs that are based on the strength of the link of the pairs (man,woman), we compare them with the classical stability notion, and we show how to find marriages that are stable according to these notions with the highest global link. In Section 5 we analyze manipulation issues in our context. In Section 6 we summarize the results contained in the paper, and we give some hints for future work.

A preliminary version of this paper has been presented in [16].

2 Background

We now give some basic notions on classical stable marriage problems, stable marriage problems with partial orders, and stable marriage problems with weighted preferences.

2.1 Stable Marriage Problems

A *stable marriage problem* (SM) [9] of size n is the problem of finding a stable marriage between n men and n women. Such men and women each have a preference ordering over the members of the other sex. A marriage is a one-to-one correspondence between men and women. Given a marriage M, a man m, and a woman w, the pair (m, w) is a *blocking pair* for M if m prefers w to his partner in M and w prefers m to her partner in M. A marriage is said to be *stable* if it does not contain blocking pairs.

The sequence of all preference orderings of men and women is usually called a *profile*. In the case of classical stable marriage problem (SM), a profile is a sequence of strict total orders.

Given a SM P, there may be many stable marriages for P. However, it is interesting to know that there is always at least one stable marriage.

Given an SM P, a *feasible partner* for a man m (resp., a woman w) is a woman w (resp., a man m) such that there is a stable marriage for P where m and w are married.

The set of all stable marriages for an SM forms a lattice, where a stable marriage M_1 dominates another stable marriage M_2 if men are happier (that is, are married to more or equally preferred women) in M_1 w.r.t. M_2. The top of this lattice is the stable marriage where men are most satisfied, and it is usually called the *male-optimal* stable marriage. Conversely, the bottom is the stable marriage where men's preferences are least satisfied (and women are happiest, so it is usually called the *female-optimal* stable marriage). Thus, a stable marriage is male-optimal iff every man is paired with his highest ranked feasible partner.

The *Gale-Shapley* (GS) *algorithm* [4] is a well-known algorithm to solve the SM problem. At the start of the algorithm, each person is free and becomes engaged during the execution of the algorithm. Once a woman is engaged, she never becomes free again (although to whom she is engaged may change), but men can alternate between being free and being engaged. The following step is iterated until all men are engaged: choose a free man m, and let m propose to the most preferred woman w on his preference list, such that w has not already rejected m. If w is free, then w and m become engaged. If w is engaged to man m', then she rejects the man (m or m') that she least prefers, and becomes, or remains, engaged to the other man. The rejected man becomes, or remains, free. When all men are engaged, the engaged pairs form the male optimal stable matching. It is female optimal, of course, if the roles of male and female participants in the algorithm were interchanged.

This algorithm needs a number of steps that, in the worst case, is quadratic in n (that is, the number of men), and it guarantees that, if the number of men and women coincide, and all participants express a strict order over all the members of the other group, everyone gets married, and the returned matching is stable.

Example 1. Assume $n = 2$. Let $\{w_1, w_2\}$ and $\{m_1, m_2\}$ be respectively the set of women and men. The following sequence of strict total orders defines a profile:

- $m_1 : w_1 > w_2$ (i.e., man m_1 prefers woman w_1 to woman w_2),
- $m_2 : w_1 > w_2$,
- $w_1 : m_2 > m_1$,
- $w_2 : m_1 > m_2$.

For this profile, the male-optimal solution is $\{(m_1, w_2), (m_2, w_1)\}$. For this specific profile the female-optimal stable marriage coincides with the male-optimal one.   □

2.2   Stable Marriage Problems with Partially Ordered Preferences

In SMs, each preference ordering is a strict total order over the members of the other sex. More general notions of SMs allow preference orderings to be partial [14,11,10,7,6]. This allows for the modelling of both indifference (via ties) and incomparability (via absence of ordering) between members of the other sex. In this context, a stable marriage problem is defined by a sequence of $2n$ partial orders, n over the men and n over the women. We will denote with SMP a stable marriage problem with such partially ordered preferences.

Given an SMP, we will sometimes use the notion of a *linearization* of such a problem, which is obtained by linearizing the preference orderings of the profile in a way that is compatible with the given partial orders.

A marriage M for an SMP is said to be *weakly-stable* if it does not contain blocking pairs. Given a man m and a woman w, the pair (m, w) is a blocking pair if m and w are not married to each other in M and each one *strictly prefers* the other to his/her current partner.

A weakly stable marriage M dominates a weakly stable marriage M' iff for every man m, $M(m) \geq M'(m)$ or $M(m) \bowtie M'(m)$ ($\bowtie$ means incomparable) and there is a man m' s.t. $M(m') > M'(m')$. Notice that there may be more than one undominated weakly stable marriage for an SMP.

Example 2. Let $\{w_1, w_2\}$ and $\{m_1, m_2\}$ be respectively the set of women and men. An instance of an SMP is the following:

- $m_1 : w_1 >\bowtie w_2$,
- $m_2 : w_1 > w_2$,
- $w_1 : m_1 \bowtie m_2$,
- $w_2 : m_1 > m_2$.

For this instance, both $M_1 = \{(m_1, w_2), (m_2, w_1)\}$ and $M_2 = \{(m_1, w_1), (m_2, w_2)\}$ are weakly stable marriages and M_1 dominates M_2.   □

2.3 Stable Marriage Problems with Weighted Preferences

In classical stable marriage problems, men and women express only qualitative preferences over the members of the other sex. For every pair of women (resp., men), every man (resp., woman) states only that he (resp., she) prefers a woman (resp., a man) more than another one. However, he (resp., she) cannot express how much he (resp., she) prefers such a woman (resp., a man). This is nonetheless possible in stable marriage problems with weighted preferences.

A *stable marriage problem with weighted preferences* (SMW) [12] is a classical SM where every man/woman gives also a numerical preference value for every member of the other sex, that represents how much he/she prefers such a person. Such preference values are natural numbers and higher preference values denote a more preferred item. Given a man m and a woman w, the *preference value* for man m (resp., woman w) of woman w (resp., man m) will be denoted by $p(m, w)$ (resp., $p(w, m)$).

Example 3. Let $\{w_1, w_2\}$ and $\{m_1, m_2\}$ be respectively the set of women and men. An instance of an SMW is the following:

- $m_1 : w_1^{[9]} > w_2^{[1]}$ (i.e., man m_1 prefers woman w_1 to woman w_2, and he prefers w_1 with value 9 and w_2 with value 1),
- $m_2 : w_1^{[3]} > w_2^{[2]}$,
- $w_1 : m_2^{[2]} > m_1^{[1]}$,
- $w_2 : m_1^{[3]} > m_2^{[1]}$.

The numbers written into the round brackets identify the preference values.   □

In [12] they consider stable marriage problems with weighted preferences by looking at the stable marriage that maximizes the sum of the preference values. Therefore, they use the classical definition of stability and they use preference values only when they have to look for the optimal solution. We want, instead, to use preference values also to define new notions of stability and optimality.

We will introduce new notions of stability and optimality that are based on the weighted preferences expressed by the agents and we will show how to find them by adapting the classical Gale-Shapley algorithm [4] for SMs described in Section 2.

3 α-Stability

A simple generalization of the classical notion of stability requires that there are not two people that prefer *with at least degree* α (where α is a natural number) to be married to each other rather than to their current partners.

Definition 1 (α-stability). *Let us consider a natural number α with $\alpha \geq 1$. Given a marriage M, a man m, and a woman w, the pair (m, w) is an α-blocking pair for M if the following conditions hold:*

- *m prefers w to his partner in M, say w', by at least α (i.e., $p(m, w) - p(m, w') \geq \alpha$),*

- *w prefers m to her partner in M, say m', by at least α (i.e., $p(w, m) - p(w, m') \geq \alpha$).*

A marriage is α-stable if it does not contain α-blocking pairs. A man m (resp., woman w) is α-feasible for woman w (resp., man m) if m is married with w in some α-stable marriage.

3.1   Relations with Classical Stability Notions

Given an SMW P, let us denote with $c(P)$, the classical SM problem obtained from P by considering only the preference orderings induced by the preference values of P.

Example 4. Let us consider the SMW, P, shown in Example 3. The stable marriage problem $c(P)$ is shown in Example 1.   □.

If α is equal to 1, then the α-stable marriages of P coincide with the stable marriages of $c(P)$. However, in general, α-stability allows us to have more marriages that are stable according to this definition, since we have a more relaxed notion of blocking pair. In fact, a pair (m, w) is an α-blocking if both m and w prefer each other to their current partner *by at least* α and thus pairs (m', w') where m' and w' prefer each other to their current partner of less than α are not considered α-blocking pairs.

The fact that α-stability leads to a larger number of stable marriages w.r.t. the classical case is important to allow new stable marriages where some men, for example the most popular ones or the least popular ones, may be married with partners better than all the feasible ones according to the classical notion of stability.

Given an SMW P, let us denote with $I_\alpha(P)$ the set of the α-stable marriages of P and with $I(c(P))$ the set of the stable marriages of $c(P)$. We have the following results.

Proposition 1. *Given an SMW P, and a natural number α with $\alpha \geq 1$,*

- *if $\alpha = 1$, $I_\alpha(P) = I(c(P))$;*
- *if $\alpha > 1$, $I_\alpha(P) \supseteq I(c(P))$.*

Given an SMP P, the set of α-stable marriages of P contains the set of stable marriages of $c(P)$, since the α-blocking pairs of P are a subset of the blocking pairs of $c(P)$.

Let us denote with $\alpha(P)$ the stable marriage with incomparable pairs obtained from an SMW P by setting as incomparable every pair of people that don't differ for at least α, and with $I_w(\alpha(P))$ the set of the weakly stable marriages of $\alpha(P)$. It is possible to show that the set of the weakly stable marriages of $\alpha(P)$ coincides with the set of the α-stable marriages of P.

Theorem 1. *Given an SMW P, $I_\alpha(P) = I_w(\alpha(P))$.*

Proof. We first show that $I_\alpha(P) \subseteq I_w(\alpha(P))$. Assume that a marriage $M \notin I_w(\alpha(P))$, we now show that $M \notin I_\alpha(P)$. If $M \notin I_w(\alpha(P))$, then there is a pair (man,woman), say (m, w), in $\alpha(P)$ such that m prefers w to his partner in M, say w', and w prefers m to her partner in M, say m'. By definition of $\alpha(P)$, this means that m prefers w to w' by at least degree α and w prefers m to m' by at least degree α in P, and so $M \notin I_\alpha(P)$.

Similarly, we can show that $I_\alpha(P) \supseteq I_w(\alpha(P))$. In fact, if $M \notin I_\alpha(P)$, then there is a pair (man,woman), say (m, w), in P such that m prefers w to w' by at least degree α and w prefers m to m' by at least degree α. By definition of $\alpha(P)$, this means that m prefers w to w' and w prefers m to m' in $\alpha(P)$ and so $M \notin I_w(\alpha(P))$, i.e., M is not a weakly stable marriage for $\alpha(P)$. $\square$

This means that, given an SMW P, every algorithm that is able to find a weakly stable marriage for $\alpha(P)$ provides an α-stable marriage for P.

Example 5. Assume that α is 2. Let us consider the following instance of an SMW, say P.

- $m_1 : w_1^{[3]} > w_2^{[2]}$
- $m_2 : w_1^{[4]} > w_2^{[2]}$,
- $w_1 : m_1^{[8]} > m_2^{[5]}$,
- $w_2 : m_1^{[3]} > m_2^{[1]}$.

The SMP $\alpha(P)$ is the following:

- $m_1 : w_1 \bowtie w_2$,
- $m_2 : w_1 > w_2$,
- $w_1 : m_1 > m_2$,
- $w_2 : m_1 > m_2$.

The set of the α-stable marriages of P, that coincides with the set of the weakly stable marriages of $\alpha(P)$, by Theorem 1, contains the following marriages: $M_1 = \{(m_1, w_1), (m_2, w_2)\}$ and $M_2 = \{(m_1, w_2), (m_2, w_1)\}$. $\square$

On the other hand, not all stable marriage problems with partially ordered preferences can be expressed as stable marriage problems with weighted preferences such that the stable marriages in the two problems coincide. More precisely, given any SMP problem P, we would like to be able to generate a corresponding SMW problem P' and a value α such that, in P', the weights of elements ordered in P differ more than α, while those of elements that are incomparable in P differ less than α. Consider for example the case of a partial order over six elements, defined as follows: $x_1 > x_2 > x_3 > x_4 > x_5$ and $x_1 > y > x_5$. Then there is no way to choose a value α and a linearization of the partial order such that the weights of x_i and x_j differ for at least α, for any i,j between 1 and 5, while at the same time the weight of y and each of the x_i's differ for less than α.

3.2 Dominance and Lex-Male-Optimality

We recall that in SMPs a weakly-stable marriage dominates another weakly-stable marriage if men are happier (or equally happy) and there is at least a man that is strictly happier. The same holds for α-stable marriages. As in SMPs there may be more than one undominated weakly-stable marriage, in SMWs there may be more than one undominated α-stable marriage.

Definition 2 (Dominance). *Given two α-stable marriages, say M and M', M dominates M' if every man is married in M to more or equally preferred woman than in M' and there is at least one man in M married to a more preferred woman than in M'.*

Example 6. Let us consider the SMW shown in Example 5. We recall that α is 2 and that the α-stable marriages of this problem are $M_1 = \{(m_1, w_1), (m_2, w_2)\}$ and $M_2 = \{(m_1, w_2), (m_2, w_1)\}$. It is possible to see that:

- M_2 does not dominate M_1 since, for m_1, $M_1(m_1) > M_2(m_1)$ and
- M_1 does not dominate M_2 since, for m_2, $M_2(m_2) > M_1(m_2)$.   □

We now discriminate among the α-stable marriages of an SMW, by considering the preference values given by women and men to order pairs that differ for less than α.

In this paper we will consider a marriage optimal when the most popular men are as happy as possible and they are married with their most popular best α-feasible women. However, we could also consider a marriage optimal when the least popular men are as happy as possible and they are married with their most popular best α-feasible women.

To compute a strict ordering on the men where the most popular men (resp., the most popular women) are ranked first, we follow a reasoning similar to the one considered in [15,17], that is, we apply a voting rule [1] to the preferences given by the women (resp., by the men). More precisely, such a voting rule takes in input the preference values given by the women over the men (resp., given by the men over the women) and returns a strict total order over the men (resp., women).

Definition 3 (Lex-male-Optimal). *Consider an SMW P, a natural number α, and a voting rule r. Let us denote with o_m (resp., o_w) the strict total order over the men (resp., over the women) computed by applying r to the preference values that the women give to the men (resp., the men give to the women). An α-stable marriage M is lex-male-optimal w.r.t. o_m and o_w, if, for every other α-stable marriage M', the following conditions hold:*

- *there is a man m_i such that $M(m_i) \succ_{o_w} M'(m_i)$,*
- *for every man $m_j \prec_{o_m} m_i$, $M(m_j) \leq M'(m_j)$.*

Proposition 2. *Given an SMW P, a strict total ordering o_m (resp., o_w) over the men (resp., women),*

- *there is a unique lex-male-optimal α-stable marriage w.r.t. o_m and o_w, say L.*
- *L may be different from the male-optimal stable marriage of $c(P)$;*
- *if $\alpha(P)$ has a unique undominated weakly stable marriage, say L', then L coincides with L', otherwise L is one of the undominated weakly stable marriages of $\alpha(P)$.*

Example 7. Let us consider the SMW, P, shown in Example 5. We have shown previously that this problem has two α-weakly stable marriages that are undominated. We now want to discriminate among them by considering the lex-male-optimality notion. Let us consider as voting rule the rule that takes in input the preference values given by the women over the men (resp., by the men over the women) and returns a strict preference ordering over the men (resp., women). This preference ordering is induced by the overall score that each man (resp., woman) receives: men (women) that receive higher overall scores are more preferred. The overall score of a man m (resp., woman w), say $s(m)$ (resp., $s(w)$), is computed by summing all the preference values that the women give to him (the men give to her). If two candidates receive the same overall score, we use a tie-breaking rule to order them. If we apply this voting rule to the preference values given by the women in P, then we obtain

- $s(m_1) = 8 + 3 = 11$,
- $s(m_2) = 5 + 1 = 6$,

and thus the ordering o_m is such that $m_1 \succ_{o_m} m_2$. If we apply the same voting rule to the preference values given by the men in P,

- $s(w_1) = 3 + 4 = 7$,
- $s(w_2) = 2 + 2 = 4$,

and thus the ordering o_w is such that $w_1 \succ_{o_w} w_2$. The lex-male-optimal α-stable marriage w.r.t. o_m and o_w is the marriage $M_1 = \{(m_1, w_1), (m_2, w_2)\}$.   □

3.3 Finding the Lex-Male-Optimal α-Stable Marriage

It is possible to find optimal α-stable marriages by adapting the GS-algorithm for classical stable marriage problems [4].

Given an SMW P and a natural number α, by Theorem 1, to find an α-stable marriage it is sufficient to find a weakly stable marriage of $\alpha(P)$. This can be done by applying the GS algorithm to any linearization of $\alpha(P)$.

Given an SMW P, a natural number α, and two orderings o_m and o_w over men and women computed by applying a voting rule to P as described in Definition 3, it is possible to find the α-stable marriage that is lex-male-optimal w.r.t o_m and o_w by applying the GS algorithm to the linearization of $\alpha(P)$ where we order incomparable pairs, i.e., the pairs that differ for less than α in P, in accordance with the orderings o_m and o_w.

Algorithm 1. Lex-male-α-stable-GS

Input: P: an SMW, α: a natural number, r: a voting rule
Output: μ: a marriage
$o_m \leftarrow$ the strict total order over the men obtained by applying r to the preference values given by the women over the men
$o_w \leftarrow$: the strict total order over the women obtained by applying r to the preference values given by the men over the women
$P' \leftarrow$ the linearization of $\alpha(P)$ obtained by ordering incomparable pairs of $\alpha(P)$ in accordance with o_m and o_w;
$\mu \leftarrow$ the marriage obtained by applying the GS algorithm to P';
return μ

Proposition 3. *Given an SMW P, a natural number α, o_m (resp., o_w) an ordering over the men (resp., women), algorithm* Lex-male-α-stable-GS *returns the lex-male-optimal α-stable marriage of P w.r.t. o_m and o_w.*

4   Stability Notions Relying on Links

Until now we have generalized the classical notion of stability by considering separately the preferences of the men and the preferences of the women. We now intend to define new notions of stability that take into account simultaneously the preferences of the men and the women. Such a new notion will depend on the strength of the link of the married people, i.e., how much a man and a woman want to be married with each other. This is useful to obtain a new notion of stable marriage, that looks at the happiness of the pairs (man,woman) rather than at the happiness of the members of a single sex.

A way to define the strength of the link of two people is the following.

Definition 4 (Link Additive-strength). *Given a man m and a woman w, the* link additive-strength *of the pair (m, w), denoted by $la(m, w)$, is the value obtained by summing the preference value that m gives to w and the preference value that w gives to m, i.e., $la(m, w) = p(m, w) + p(w, m)$. Given a marriage M, the* additive-link *of M, denoted by $la(M)$, is the sum of the links of all its pairs, i.e., $\sum_{\{(m,w)\in M\}} la(m, w)$.*

Notice that we can use other operators beside the sum to define the link strength, such as, for example, the maximum or the product.

We now give a notion of stability that exploit the definition of the link additive-strength given above.

Definition 5 (Link-additive-Stability). *Given a marriage M, a man m, and a woman w, the pair (m, w) is a* link-additive-blocking pair *for M if the following conditions hold:*

- $la(m, w) > la(m', w)$,
- $la(m, w) > la(m, w')$,

where m' is the partner of w in M and w' is the partner of m in M. A marriage is link-additive-stable *if it does not contain link-additive-blocking pairs.*

Example 8. Let $\{w_1, w_2\}$ and $\{m_1, m_2\}$ be, respectively, the set of women and men. Consider the following instance of an SMW, P:

- $m_1 : w_1^{[30]} > w_2^{[3]}$,
- $m_2 : w_1^{[4]} > w_2^{[3]}$,
- $w_1 : m_2^{[6]} > m_1^{[5]}$,
- $w_2 : m_1^{[10]} > m_2^{[2]}$.

In this example there is a unique link-additive-stable marriage, that is $M_1 = \{(m_1, w_1), (m_2, w_2)\}$, which has additive-link $la(M_1) = 35 + 5 = 40$. Notice that such a marriage has an additive-link higher than the male-optimal stable marriage of $c(P)$ that is $M_2 = \{(m_1, w_2), (m_2, w_1)\}$ which has additive-link $la(M_2) = 13 + 10 = 23$.   □

The strength of the link of a pair (man,woman), and thus the notion of link stability, can be also defined by considering the maximum operator instead of the sum operator.

Definition 6 (Link Maximal-strength). *Given a man m and a woman w, the* link maximal-strength *of the pair (m, w), denoted by $lm(m, w)$, is the value obtained by taking the maximum between the preference value that m gives to w and the preference value that w gives to m, i.e., $lm(m, w) = max(p(m, w), p(w, m))$. Given a marriage M, the* maximal-link *of M, denoted by $lm(M)$, is the maximum of the links of all its pairs, i.e., $max_{\{(m,w)\in M\}} lm(m, w)$.*

Definition 7 (Link-maximal-Stability). *Given a marriage M, a man m, and a woman w, the pair (m, w) is a link-maximal-blocking pair for M if the following conditions hold:*

- $lm(m, w) > lm(m', w)$,
- $lm(m, w) > lm(m, w')$,

where m' is the partner of w in M and w' is the partner of m in M. A marriage is link-maximal-stable *if it does not contain link-maximal-blocking pairs.*

4.1 Relations with Other Stability Notions

Given an SMW P, let us denote with $Linka(P)$ (resp., $Linkm(P)$) the stable marriage problem with ties obtained from P by changing every preference value that a person x gives to a person y with the value $la(x, y)$ (resp., $lm(x, y)$), by changing the preference rankings accordingly, and by considering only these new preference rankings.

Let us denote with $I_{la}(P)$ (resp., $I_{lm}(P)$) the set of the link-additive-stable marriages (resp., link-maximal-stable marriages) of P and with $I_w(Linka(P))$ (resp., $I_w(Linkm(P))$) the set of the weakly stable marriages of $Linka(P)$ (resp., $Linkm(P)$). It is possible to show that these two sets coincide.

Theorem 2. *Given an SMW P, $I_{la}(P) = I_w(Linka(P))$ and $I_{lm}(P) = I_w(Linkm(P))$.*

Proof. Let us consider a marriage M. We first show that if $M \in I_w(Linka(P))$ then $M \in I_{la}(P)$. If $M \notin I_{la}(P)$, there is a pair (m, w) that is a link-additive-blocking pair, i.e., $la(m, w) > la(m, w')$ and $la(m, w) > la(m', w)$, where w' (resp., m') is the partner of m (resp., w) in M. Since $la(m, w) > la(m, w')$, m prefers w to w' in the problem $Linka(P)$, and, since $la(m, w) > la(m', w)$, w prefers m to m' in the problem $Linka(P)$. Hence (m, w) is a blocking pair for the problem $Linka(P)$. Therefore, $M \notin I_w(Linka(P))$.

We now show that if $M \in I_{la}(P)$ then $M \in I_w(Linka(P))$. If $M \notin I_w(Linka(P))$, there is a pair (m, w) that is a blocking pair for $I_w(Linka(P))$, i.e., m prefers w to w' in the problem $Linka(P)$, and w prefers m to m' in the problem $Linka(P)$. By definition of the problem $Linka(P)$, $la(m, w) > la(m, w')$ and $la(m, w) > la(m', w)$. Therefore, (m, w) is a link-additive-blocking pair for the problem P. Hence, $M \notin I_{la}(P)$.

It is possible to show similarly that $I_{lm}(P) = I_w(Linkm(P))$. $\qquad\square$

When no preference ordering changes in $Linka(P)$ (resp., $Linkm(P)$) w.r.t. P, then the link-additive-stable (resp., link-maximal-stable) marriages of P coincide with the stable marriages of $c(P)$.

Proposition 4. *Given an SMW P, if $Linka(P) = c(P)$ $(Linkm(P) = c(P))$, then $I_{la}(P) = I(c(P))$ (resp., $I_{lm}(P) = I(c(P))$).*

If there are no ties in $Linka(P)$ (resp., $Linkm(P)$), then there is a unique link-additive-stable marriage (resp., link-maximal-stable marriage) with the highest link.

Proposition 5. *Given an SMW P, if $Linka(P)$ (resp., $Linkm(P)$) has no ties, then there is a unique link-additive-stable (resp., link-maximal-stable) marriage with the highest link.*

If we consider the definition of link-maximal-stability, it is possible to define a class of SMWs where there is a unique link-maximal-stable marriage with the highest link.

Proposition 6. *In an SMW P where the preference values are all different, there is a unique link-maximal-stable marriage with the highest link.*

4.2   Finding Link-Additive-Stable and Link-Maximal-Stable Marriages with the Highest Link

We now show that for some classes of preferences it is possible to find optimal link-additive-stable marriages and link-maximal-stable marriages of an SMW by adapting algorithm GS, which is usually used to find the male-optimal stable marriage in classical stable marriage problems.

By Proposition 2, we know that the set of the link-additive-stable (resp., link-maximal-stable) marriages of an SMW P coincides with the set of the weakly stable marriages of the SMP $Linka(P)$ (resp., $Linkm(P)$). Therefore, to find a link-additive-stable (resp., link-maximal-stable) marriage, we can simply apply algorithm GS to a linearization of $Linka(P)$ (resp., $Linkm(P)$).

Proposition 7. *Given an SMW P, the marriage returned by algorithm* link-additive-stable-GS *(link-maximal-stable-GS) over P, say M, is link-additive-stable (resp., link-maximal-stable). Moreover, if there are not ties in $Linka(P)$ (resp., $Linkm(P)$), M is link-additive-stable (resp., link-maximal-stable) and it has the highest link.*

Algorithm 2. Link-additive-stable-GS (resp., link-maximal-stable-GS)

Input: P: an SMW
Output: μ: a marriage
$P' \leftarrow Linka(P)$ (resp., $Linkm(P)$);
$P'' \leftarrow$ a linearization of P';
$\mu \leftarrow$ the marriage obtained by applying GS algorithm to P'';
return μ

When there are no ties in $Linka(P)$ (resp., $Linkm(P)$), the marriage returned by algorithm *link-additive-stable-GS* (resp., *link-maximal-stable-GS*) is male-optimal w.r.t. the profile with links. Such a marriage may be different from the classical male-optimal stable marriage of $c(P)$, since it considers the happiness of the men reordered according to their links with the women, rather than according their single preferences.

This holds, for example, when we assume to have an SMW with preference values that are all different and we consider the notion of link-maximal-stability.

Proposition 8. *Given an SMW P where the preference values are all different, the marriage returned by algorithm* link-maximal-stable-GS *algorithm over P is link-maximal-stable and it has the highest link.*

5   Manipulation

In [18] Roth has shown that, when there are at least three men and three women, every stable marriage procedure is manipulable, i.e., there is a profile where an agent, mis-reporting his preferences, obtains a stable marriage which is better than or equal to the one obtained by telling the truth. In stable marriage problems, agents can manipulate in two ways: by changing the preference ordering [18], or by truncating the preference list [5].

We now would like to check if agents have additional ways of manipulating in our context, by changing only the preference weights, while preserving the preference ordering and not truncating the preference list.

In the following, we will call a *w-manipulation attempt* by an agent a the mis-reporting of the weights in a's preferences which preserves the true preference ordering. Also, we will say that a w-manipulation attempt of an agent a is *successful* if the resulting marriage for a is better than or equal to the marriage obtained by using the true preference weights of a, and that a stable marriage procedure is *w-manipulable* if there is a profile with a successful w-manipulation attempt for an agent.

We will show that every stable marriage procedure which returns an α-stable marriage, a link-additive, or a link-maximal stable marriage, is w-manipulable.

Theorem 3. *Let α be any natural number > 1. Every procedure which returns an α-stable marriage is w-manipulable.*

Proof. Let $\{w_1, w_2, w_3\}$ and $\{m_1, m_2, m_3\}$ be, respectively, the set of women and men. Consider the following instance of an SMW, say P, $\{m_1 : w_2^{[x+2\alpha]} > w_1^{[x+\alpha]} > w_3^{[x]}, m_2 : w_1^{[x+2\alpha]} > w_2^{[x+\alpha]} > w_3^{[x]}, m_3 : w_1^{[x+2\alpha]} > w_2^{[x+\alpha]} > w_3^{[x]}, w_1 : m_1^{[x+\alpha+1]} > m_2^{[x+\alpha]} > m_3^{[x]}, w_2 : m_3^{[x+2\alpha]} > m_1^{[x+\alpha]} > m_2^{[x]}, w_3 : m_1^{[x+2\alpha]} > m_2^{[x+\alpha]} > m_3^{[x]}\}$. P has two α-stable marriages: $M_1 = \{(m_1, w_1), (m_2, w_3), (m_3, w_2)\}$ and $M_2 = \{(m_1, w_3), (m_2, w_1), (m_3, w_2)\}$. Assume that w_1 mis-reports her preferences as follows: $w_1 : m_1^{[x+2\alpha]} > m_2^{[x+\alpha]} > m_3^{[x]}$, i.e., assume that she changes the weight given to m_1 from $x + \alpha + 1$ to $x + 2\alpha$. Let us denote with P' the resulting problem. P' has a unique α-stable marriage, that is M_1, which is the best α-stable marriage for w_1 in P. Therefore, it is convenient for w_1 to change her weights to get a better or equal result w.r.t the one obtained by telling the truth. Also, since P' has a unique α-stable marriage, every procedure which returns an α-stable marriage returns such a marriage. Thus, every procedure is w-manipulable.   □

Theorem 4. *Every stable marriage procedure that returns a link-additive stable marriage is w-manipulable.*

Proof. Let $\{w_1, w_2\}$ and $\{m_1, m_2\}$ be, respectively, the set of women and men. Consider the following instance of an SMW, say P: $\{m_1 : w_2^{[6]} > w_1^{[4]}, m_2 : w_2^{[5]} > w_1^{[2]}, w_1 : m_1^{[3]} > m_2^{[2]}, w_2 : m_1^{[5]} > m_2^{[2]}\}$. P has a unique link-additive stable marriage, which is $M_1 = \{(m_1, w_2), (m_2, w_1)\}$. Assume that w_1 mis-reports her preferences as follows: $w_1 : m_1^{[100]} > m_2^{[2]}$, i.e., she changes the weight given to m_1 from 3 to 100. Then, in the new problem, that we call P', there is a unique link-additive stable marriage, i.e., $M_2 = \{(m_1, w_1), (m_2, w_2)\}$, which is better for w_1 in P. Since in P' there is a unique link-additive-stable marriage, every procedure which returns a link-additive stable marriage will return it. Thus, every procedure is w-manipulable. □

A similar result can be shown also for stable marriage procedures that return a link-maximal stable marriage.

Summarizing, even if we don't allow the agents to modify their preference ordering or to truncate their preference list, they can manipulate simply by changing the values of their weights. Moreover, it is possible to see that some ideas to prevent manipulation, such as to assign equal weight to all top alternatives and to put a bound over the relative weights of two consecutive elements in every ordering, are ineffective to avoid w-manipulation.

6   Conclusions and Future Work

In this paper we have considered stable marriage problems with weighted preferences, where both men and women can express a score over the members of the other sex. In particular, we have introduced new stability and optimality notions for such problems and we have compared them with the classical ones for stable marriage problems with totally or partially ordered preferences. Also, we have provided algorithms to find marriages that are optimal and stable according to these new notions by adapting the Gale-Shapley algorithm. Moreover, we have investigated manipulation issues in our context.

We have also considered an optimality notion (that is, lex-male-optimality) that exploits a voting rule to linearize the partial orders. We intend to study if this use of voting rules within stable marriage problems may have other benefits. In particular, we want to investigate if the procedure defined to find such an optimality notion inherits the properties of the voting rule with respect to manipulation: we intend to check whether, if the voting rule is NP-hard to manipulate, then also the procedure on SMW that exploits such a rule is NP-hard to manipulate. This would allow us to transfer several existing results on manipulation complexity, which have been obtained for voting rules, to the context of procedures to solve stable marriage problems with weighted preferences.

Acknowledgements. This work has been partially supported by the MIUR PRIN 20089 M932N project "Innovative and multi-disciplinary approaches for constraint and preference reasoning".

References

1. Arrow, K.J., Sen, A.K., Suzumura, K.: Handbook of Social Choice and Welfare. North Holland, Elsevier (2002)
2. Bistarelli, S., Foley, S., O'Sullivan, B., Santini, F.: From Marriages to Coalitions: A Soft CSP Approach. In: Oddi, A., Fages, F., Rossi, F. (eds.) CSCLP 2008. LNCS, vol. 5655, pp. 1–15. Springer, Heidelberg (2009)
3. Bistarelli, S., Montanari, U., Rossi, F.: Semiring-based constraint solving and optimization. Journal of the ACM 44(2), 201–236 (1997)
4. Gale, D., Shapley, L.S.: College admissions and the stability of marriage. Amer. Math. Monthly 69, 9–14 (1962)
5. Gale, D., Sotomayor, M.: Semiring-based constraint solving and optimization. American Mathematical Monthly 92, 261–268 (1985)
6. Gelain, M., Pini, M.S., Rossi, F., Venable, K.B., Walsh, T.: Male Optimal and Unique Stable Marriages with Partially Ordered Preferences. In: Guttmann, C., Dignum, F., Georgeff, M. (eds.) CARE 2009 / 2010. LNCS, vol. 6066, pp. 44–55. Springer, Heidelberg (2011)
7. Gelain, M., Pini, M.S., Rossi, F., Venable, K.B., Walsh, T.: Male optimality and uniqueness in stable marriage problems with partial orders - Extended abstract. In: AAMAS 2010 (2010)
8. Gusfield, D.: Three fast algorithms for four problems in stable marriage. SIAM J. Comput. 16(1), 111–128 (1987)
9. Gusfield, D., Irving, R.W.: The Stable Marriage Problem: Structure and Algorithms. MIT Press, Boston (1989)
10. Halldorsson, M., Irving, R.W., Iwama, K., Manlove, D., Miyazaki, S., Morita, Y., Scott, S.: Approximability results for stable marriage problems with ties. Theor. Comput. Sci. 306(1-3), 431–447 (2003)
11. Irving, R.W.: Stable marriage and indifference. Discrete Applied Mathematics 48, 261–272 (1994)
12. Irving, R.W., Leather, P., Gusfield, D.: An efficient algorithm for the "optimal" stable marriage. J. ACM 34(3), 532–543 (1987)
13. Liebowitz, J., Simien, J.: Computational efficiencies for multi-agents: a look at a multi-agent system for sailor assignment. Electonic government: an International Journal 2(4), 384–402 (2005)
14. Manlove, D.: The structure of stable marriage with indifference. Discrete Applied Mathematics 122(1-3), 167–181 (2002)
15. Pini, M.S., Rossi, F., Venable, K.B., Walsh, T.: Manipulation and gender neutrality in stable marriage procedures. In: Proc. AAMAS 2009, vol. 1, pp. 665–672 (2009)
16. Pini, M.S., Rossi, F., Venable, K.B., Walsh, T.: Stable marriage problems with quantitative preferences. In: Informal Proc. of COMSOC 2010 - Third International Workshop on Computational Social Choice (2010)
17. Pini, M.S., Rossi, F., Venable, K.B., Walsh, T.: Manipulation complexity and gender neutrality in stable marriage procedures. Journal of Autonomous Agents and Multi-Agent Systems 22(1), 183–199 (2011)
18. Roth, A.E.: The economics of matching: Stability and incentives. Mathematics of Operations Research 7, 617–628 (1982)
19. Roth, A.E.: The evolution of the labor market for medical interns and residents: a case study in game theory. Journal of Political Economy 92, 991–1016 (1984)
20. Roth, A.E.: Deferred acceptance algorithms: History, theory, practice, and open questions. International Journal of Game Theory, Special Issue in Honor of David Gale on his 85th birthday 36, 537–569 (2008)
21. Teo, C.-P., Sethuraman, J., Tan, W.-P.: Gale-shapley stable marriage problem revisited: Strategic issues and applications. Manage. Sci. 47(9), 1252–1267 (2001)

A Conditional Game-Theoretic Approach to Cooperative Multiagent Systems Design

Wynn Stirling

Electrical and Computer Engineering Department, Brigham Young University
Provo, Utah 84602, U.S.A.
wynn_stirling@byu.edu
http://www.ee.byu.edu/faculty/wynns/

Abstract. Neoclassical game theory focuses exclusively on individual preferences, which are more naturally attuned to competitive, rather than cooperative,
decision scenarios. Conditional game theory differs from classical theory in two
fundamental ways. First, it involves a utility structure that permits agents to define their preferences conditioned on the preferences of other agents, and second,
it accommodates a notion of group rationality as well as individual rationality.
The resulting framework permits a notion of group preferences to be defined, and
permits solution concepts that account for both individual and group interests.

Keywords: Group Rationality, Cooperation, Negotiation.

1 Introduction

Game theory provides a mathematical framework within which to model decisions by
multiple entities where the outcome for each depends on the choices of all. Game theory is increasingly invoked by engineering and computer science as a framework for
multiagent systems [11,12,15,17,18].

A finite, noncooperative, single-stage, strategic-form game consists of (i) a set of
autonomous decision makers, or *players*, denoted $\mathcal{X}_n = \{X_1, \ldots, X_n\}$ where $n \geq 2$,
(ii) a finite action set $\mathcal{A}_i$ for each X_i, and (iii) a *utility* $u_{X_i}: \mathcal{A} \to \mathbb{R}$ for each X_i,
$i = 1, \ldots, n$, where $\mathcal{A} = \mathcal{A}_1 \times \cdots \times \mathcal{A}_n$ is the product action space. For any *action
profile* $\mathbf{a} = (a_1, \ldots, a_n) \in \mathcal{A}$, the utility $u_{X_i}(\mathbf{a})$, defines the benefit to X_i as a consequence of the instantiation of $\mathbf{a}$. These utilities are *categorical* in the sense that $u_{X_i}(\mathbf{a})$
unconditionally defines the benefit to X_i of the group instantiating the action profile $\mathbf{a}$.

X_i also must possess a logical structure that defines how it should play the game. The
most widely used logical structure is the doctrine of *individual rationality*:
each X_i should act in a way that maximizes is own utility. Under the assumption that
each player subscribes to this notion and assumes that all others do so as well, they each
will solve their corresponding constrained optimization problem, resulting in a Nash
equilibrium.

These mathematical and logical structures may provide an appropriate vehicle with
which to model behavior in an environment of competition and market driven expectations since, in that environment, the dominant notion of rational behavior clearly is
self-interest. It is less clear, however, that self-interest is the dominant (and certainly not

J. Filipe and A. Fred (Eds.): ICAART 2011, CCIS 271, pp. 334–349, 2013.

the only) relevant rationality notion in mixed-motive environments, such as those that contain opportunities for cooperation, compromise, and unselfishness. Arrow clearly delimits the context in which individual rationality applies: "[R]ationality in application is not merely a property of the individual. Its useful and powerful implications derive from the conjunction of individual rationality and other basic concepts of neoclassical theory — equilibrium, competition, and completeness of markets ... When these assumptions fail, the very concept of rationality becomes threatened, because perceptions of others and, in particular, their rationality become part of one's own rationality" [2, p. 203].

Despite Arrow's caution, the mathematical and logical structures of game theory are routinely applied to mixed-motive situations, often producing results that are at variance with observed behavior. Behavioral economics (e.g., see [3]) seeks to mitigate this problem by inserting parameters to model fairness, loss aversion, and other such issues into the utilities to provide more psychological realism. Once included, however, the game is still solved according to conventional individual rationality and categorical utilities.

What is missing with conventional game theory is a notion of group benefit. Cooperative multiagent systems are designed such that the individuals work together to accomplish some task, but, unfortunately, group rationality does not derive from individual rationality. As observed by Luce and Raiffa, "the notion of group rationality is neither a postulate of the model nor does it appear to follow as a logical consequence of individual rationality ... general game theory seems to be in part a sociological theory which does not include any sociological assumptions ... it may be too much to ask that any sociology be derived from the single assumption of individual rationality" [10, p. 193, 196]. Consequently, game theory has proceeded by making assumptions about individual preferences only and then using those preferences to deduce information about the choices (but not the values) of a group.

It might be expected that cooperative game theory possesses some notion of group rationality. This version of game theory permits a subset of players to enter into a coalition such that each receives a payoff that is greater than it would receive if it acted alone. However, cooperative game theory employs categorical utilities and its solutions concepts are based squarely upon the assumption of individual rationality. Each player enters into a coalition solely on the basis of benefit to itself and, even though each may be better off for having joined, a notion of group rationality is not an issue when forming the coalition.

Reliance on categorical utilities and individual rationality limits the application of conventional game theory for the design and synthesis of multiagent systems that are intended to be cooperative. The contributions of this paper are (i) to present a new utility structure that overcomes the limitations of categorical utilities as a model of complex social relationships, (ii) to offer a more general concept of rational behavior that simultaneously accounts for both group an individual welfare, and (iii) to address and control the computational complexity of the resulting model.

2   Preference Models

2.1   Neoclassical Preference Models

The most prevalent assumption employed by game theory when considering preference orderings is also the most simple: a preference ordering over alternatives is defined for each individual agent. Arrow put it succinctly: "It is assumed that each individual in the community has a definite ordering of all conceivable social states, in terms of their desirability to him ... It is simply assumed that the individual orders all social states by whatever standards he deems relevant" [1, p. 17]. According to this view, each agent's preference ordering is completely defined and immutable before the game begins — it is categorical. Thus, from the conventional point of view, the starting point of a game is the definition of categorical utilities for each player. Furthermore, as Friedman argues, it is not necessary to consider the process by which the agents arrive at their preference orderings. "The economist has little to say about the formation of wants; this is the province of the psychologist. The economist's task is to trace the consequences of any given set of wants" [7, p. 13].

If we take the Arrow/Friedman division of labor as the starting point when defining a game, we must assume that the individual is able to reconcile all internal conflicts to the point that a unique categorical preference ordering can be defined that corresponds to its own self interest and which is not susceptible to change as a result of social interaction. This is a tall order, but nothing less will do if we are restricted to categorical preference orderings.

2.2   Social Influence Preference Models

When complex social relationships exist for which categorical preferences are not adequate or appropriate, a natural way for a player to take them into account is by the notion of *influence*. There are many ways to account for social influence, but the approach presented in this paper is to apply a set of principles to define a systematic and logically defensible mathematical model that leads to the definition and implementation of a multiagent decision methodology that accounts for influence relationships when they exist and treats conventional game theory as a special case when such relationships are absent.

Principle 1 (Conditioning). *Agents' preferences may be influenced by the preferences of other agents.*

X_j influences X_i if X_i's preferences are affected by X_j's preferences. Without knowledge of X_j's preferences, X_i is in a state of suspense with respect to its own preferences. Essentially, X_j's preferences propagate through the group to affect X_i's preferences, thereby generating a social bond between the two agents. Once such a bond exists, it is possible to define a notion of joint preference for the two agents viewed simultaneously, and it is possible to extract individual preference orderings from this joint preference ordering since, once X_j's preferences are revealed, X_i need no longer remain in suspense. It is thus be possible for both group and individual preferences to co-exist.

Principle 1 represents an important shift in perspective from conventional game theory. With the conventional approach, the utility of an individual defines its self-interest with respect to the instantiation of actions taken by all players. By contrast, we view the utility of an individual as the consequent of a hypothetical proposition whose antecedent is the assumption that those who influence it have identified their most preferred outcomes.

Definition 1. A conjecture *for X_j is an action profile, denoted* a_j, *that is hypothesized as X_j's most preferred outcome. Let* $\mathcal{X}_m = \{X_{j_1}, \ldots, X_{j_m}\}$ *be a subgroup of* $\mathcal{X}_n$. *A joint conjecture for* $\mathcal{X}_m$, *denoted* $\alpha_m = \{a_{j_1}, \cdots, a_{j_m}\}$, *is a collection of action profiles in* $\mathcal{A}^m = \mathcal{A} \times \cdots \times \mathcal{A}$ *(m times), where* a_{j_l} *is a conjecture for X_{j_l}, $l = 1, \ldots, m$.*

Now suppose X_i is influenced by a subgroup $\mathcal{X}_m$. Given a joint conjecture α_m, the consequent of the hypothetical proposition is a conditional preference ordering for X_i.

Definition 2. *Let* $\mathcal{X}_m = \{X_{j_1}, \ldots, X_{j_m}\}$ *be a subgroup of* $\mathcal{X}_n$ *that influences X_i and let* $\alpha_m = \{a_{j_1}, \cdots, a_{j_m}\}$ *be a joint conjecture for* $\mathcal{X}_m$. *A conditional utility $u_{X_i|\mathcal{X}_m}(\cdot|\alpha_m)$ is a real-valued function defined over $\mathcal{A}$ that specifies the preference ordering for X_i given the joint conjecture* α_m. *That is, $u_{X_i|\mathcal{X}_m}(a|\alpha_m) > u_{X_i|\mathcal{X}_m}(a'|\alpha_m)$ means that X_i prefers a to a', given that X_{j_l} conjectures a_{j_l}, $l = 1, \ldots, m$.*

Each X_i must define a conditional utility for the joint conjectures of the subgroup that influences it. This requirement increases the complexity of a problem statement over the conventional requirement of specifying only one categorical utility for each X_i. However, as we shall explore in Section 5, there often will be ways to simplify the specification that keeps the complexity under control. Nevertheless, the inclusion of social influence will generally result in increased complexity.

2.3 Group Preference

Once we extend beyond self-interest via social bonds induced by conditional preferences, it becomes possible to consider a more general notion of group preference. It may happen that the social bonds are so strong that unanimity will result, but that situation will not generally obtain. If agents disagree regarding what outcome is best, then some degree of conflict, or discord, will exist within the group. Thus, when designing a system whose members must coordinate, a critical issue is the *concordance*, or the degree of harmony, among its members. Accordingly, a meaningful notion of group preference is for its members to function concordantly.

Definition 3. *Let* $\mathcal{X}_k = \{X_{i_1}, \ldots, X_{i_k}\}$ *be a subgroup of* $\mathcal{X}_n$. *A concordant utility $U_{\mathcal{X}_k}$ is a real-valued function defined over $\mathcal{A}^k$ such that, for each joint conjecture $\alpha_k = (a_{i_1}, \ldots, a_{i_k}) \in \mathcal{A}^k$, $U_{\mathcal{X}_k}(\alpha_k)$ defines the concordance of* α_k. *When $k = 1$, the concordant utility becomes a conventional categorical utility for X_k, that is, $U_{X_k} \equiv u_{X_k}$.*

When $k > 1$, the concordant utility is a generalization of individual utility which, rather than providing a preference ordering for a single agent over $\mathcal{A}$, provides a concordant ordering for a k-member subgroup over the product space $\mathcal{A}^k$ of joint conjectures. When $a_{i_1} = \cdots = a_{i_k}$, the concordant utility measures the degree of harmony if all members conjecture the same action profile. When the conjectures are different, $U_{\boldsymbol{\mathcal{X}}_k}(\mathbf{a}_{i_1}, \ldots, \mathbf{a}_{i_k})$ measures the degree of concordance that exists among the members of the subgroup if each X_{i_l} were to view $\mathbf{a}_{i_l}$ as its most-preferred outcome. The expression $U_{\boldsymbol{\mathcal{X}}_k}(\boldsymbol{\alpha}_k) > U_{\boldsymbol{\mathcal{X}}_k}(\boldsymbol{\alpha}'_k)$ means that $\boldsymbol{\alpha}'_k$ causes a more severe conflict for the group than does $\boldsymbol{\alpha}_k$.

Definition 4. *Let $\boldsymbol{\mathcal{X}}_k = \{X_{i_1}, \ldots, X_{i_k}\}$ and $\boldsymbol{\mathcal{X}}_m = \{X_{j_1}, \ldots, X_{j_m}\}$ be disjoint subgroups of $\boldsymbol{\mathcal{X}}_n$. For each $\boldsymbol{\alpha}_m \in \boldsymbol{\mathcal{A}}^m$, a conditional concordant utility given $\boldsymbol{\alpha}_m$ is a real-valued function $U_{\boldsymbol{\mathcal{X}}_k|\boldsymbol{\mathcal{X}}_m}(\cdot|\boldsymbol{\alpha}_m)$ defined over $\boldsymbol{\mathcal{A}}^k$ that defines a concordant utility for $\boldsymbol{\mathcal{X}}_k$ given that $\boldsymbol{\mathcal{X}}_m$ jointly conjectures $\boldsymbol{\alpha}_m$. When $k = 1$, the concordant utility becomes a conditional utility for X_k, that is, $U_{X_k|\boldsymbol{\mathcal{X}}_m} \equiv u_{X_k|\boldsymbol{\mathcal{X}}_m}$ (see Definition 2).*

Example 1. Suppose the group $\{X_1, X_2, X_3\}$ is to purchase an automobile. X_1 is to choose the model, either a convertible (C) or a sedan (S), X_2 is to choose the manufacturer, either domestic (D) or foreign (F), and X_3 is to choose the color, either red (R) or green (G). The action spaces are $\mathcal{A}_1 = \{C, S\}$, $\mathcal{A}_2 = \{D, F\}$, $\mathcal{A}_3 = \{R, G\}$.

$U_{X_1 X_2}[(C, F, R), (S, D, G)] \geq U_{X_1 X_2}[(S, D, G), (C, F, R)]$ means that concordance is higher (i.e., it is less conflictive) for the subgroup $\{X_1, X_2\}$ if X_1 were to most-prefer a foreign made red convertible and, simultaneously, X_2 were to most-prefer a domestic-made green sedan, than if X_1 were to most-prefer a domestic green sedan and, simultaneously, X_2 were to most-prefer a foreign-made red convertible. Thus, even though the two stakeholders do not have the same preferences in either case, the severity of the differences in opinion is less for the $(C, F, R), (S, D, G)$ combination than for the $(S, D, G), (C, F, R)$ combination.

$U_{X_1|X_2}(C, F, R|C, F, R) \geq U_{X_1|X_2}(S, D, G|C, F, R)$ means that X_1 prefers a foreign made red convertible to a domestic-made green sedan, given the hypothesis that X_2 most-prefers a foreign-made red convertible. Notice that, since the consequent involves only one stakeholder, the conditional joint conjecture ordering becomes a conditional ordering, and we may more properly rewrite this expression as $u_{X_1|X_2}(C, F, R|C, F, R) \geq u_{X_1|X_2}(S, D, G)|C, F, R)$.

$u_{X_1|X_2 X_3}[S, D, G|(C, F, R), (S, D, G)] \geq u_{X_1|X_2 X_3}[C, F, R|(C, F, R), (S, D, G)]$ means that X_1 prefers a domestic-made green sedan to a foreign-made red convertible, given the hypothesis that X_2 most-prefers a foreign-made red convertible and that X_3 most-prefers a domestic-made green sedan.

$U_{X_2 X_3|X_1}[(C, F, R), (S, F, R)|S, D, G] \geq U_{X_2 X_3|X_1}[(S, D, G), (C, D, G)|S, D, G]$ means that the subgroup $\{X_2, X_3\}$ is less conflicted, given that X_1 most-prefers (S, D, G), for X_2 and X_3 to prefer (C, F, R) and (S, F, R), respectively, than respectively to most-prefer (S, D, G) and (C, D, G).

Computing the concordant utility of a group $\boldsymbol{\mathcal{X}}_n = \{X_1, \ldots, X_n\}$ is a key component of our approach, since that function captures all social relationships that exist in the group. Using this function, we can define notions of rational behavior both for the group as a whole and for each of its members. Rather than define such a function from first principles, however, we propose to synthesize it from more elementary relationships. We begin by imposing the following principle.

Principle 2 (Endogeny). *If a concordant ordering for a group of agents exists, it must be determined by the social interactions among the subsets of the group.*

This principle precludes the exogenous imposition of aggregation structures. For example, a common conventional aggregation procedure is to form the weighted sum of individual utilities. Such a structure, however, is appropriate only under conditions of preferential independence (e.g., see [6,9]). When preferential dependencies exist, we seek an aggregation structure that naturally emerges from within the group.

Given the existence of a concordant utility $U_{\boldsymbol{\mathcal{X}}_m}$ and a conditional concordant utility $U_{\boldsymbol{\mathcal{X}}_k|\boldsymbol{\mathcal{X}}_m}$, our goal is to compute the concordant utility of the union of the two subgroups; i.e., to form $U_{\boldsymbol{\mathcal{X}}_m\boldsymbol{\mathcal{X}}_k}$, the concordant utility for $\boldsymbol{\mathcal{X}}_m \cup \boldsymbol{\mathcal{X}}_k$.

Definition 5. *Let $\boldsymbol{\mathcal{X}}_k = \{X_{i_1}, \dots, X_{i_k}\}$ and $\boldsymbol{\mathcal{X}}_m = \{X_{j_1}, \dots, X_{j_m}\}$ be disjoint subgroups of $\boldsymbol{\mathcal{X}}_n$ such that $U_{\boldsymbol{\mathcal{X}}_m}$ and $U_{\boldsymbol{\mathcal{X}}_k|\boldsymbol{\mathcal{X}}_m}$ are defined. These utilities are endogenously aggregated if there exists a function F such that*

$$U_{\boldsymbol{\mathcal{X}}_m\boldsymbol{\mathcal{X}}_k}(\boldsymbol{\alpha}_m, \boldsymbol{\alpha}_k) = F[U_{\boldsymbol{\mathcal{X}}_m}(\boldsymbol{\alpha}_m), U_{\boldsymbol{\mathcal{X}}_k|\boldsymbol{\mathcal{X}}_m}(\boldsymbol{\alpha}_k|\boldsymbol{\alpha}_m)], \tag{1}$$

When social relationships exist among the members of a group, there may not be a unique way to represent them mathematically. Consider the following example.

Example 2. Consider the two-agent system $\{X_1, X_2\}$ and let us suppose that X_1 possesses a categorical utility u_{X_1} over $\boldsymbol{\mathcal{A}}$, but that X_2 possesses a conditional utility of the form $u_{X_2|X_1}$; that is, X_2 conditions its preferences on the preferences of X_1. Our desire is to define a function F such that

$$U_{X_1X_2}(\mathbf{a}_1, \mathbf{a}_2) = F[u_{X_1}(\mathbf{a}_1), u_{X_2|X_1}(\mathbf{a}_2|\mathbf{a}_1)]. \tag{2}$$

Now let us suppose that there is a well-defined social relationship between X_1 and X_2 such that, when defining their preferences, they both take into consideration that, ultimately, they will be operating in a group environment, and not in isolation. Under these conditions, it is possible to re-frame the scenario by X_2 defining a categorical utility u_{X_2} and X_1 defining a conditional utility $u_{X_1|X_2}$. Under this framing, the aggregation problem requires

$$U_{X_2X_1}(\mathbf{a}_2, \mathbf{a}_1) = F[u_{X_2}(\mathbf{a}_2), u_{X_1|X_2}(\mathbf{a}_1|\mathbf{a}_2)]. \tag{3}$$

Principle 3 (Consistency). *If a multiagent decision problem can be framed in more than one way using exactly the same information, all such framings should yield the same aggregated concordant ordering.*

Definition 6. *Let $\boldsymbol{\mathcal{X}}_k$ and $\boldsymbol{\mathcal{X}}_m$ be disjoint subgroups of $\boldsymbol{\mathcal{X}}_n$ and suppose there exist two framings of the preferences and relationships between the two subgroups of the forms $\{U_{\boldsymbol{\mathcal{X}}_k}, U_{\boldsymbol{\mathcal{X}}_m|\boldsymbol{\mathcal{X}}_k}\}$ and $\{U_{\boldsymbol{\mathcal{X}}_m}, U_{\boldsymbol{\mathcal{X}}_k|\boldsymbol{\mathcal{X}}_m}\}$. The endogenous aggregation is consistent if*

$$F[U_{\boldsymbol{\mathcal{X}}_k}(\boldsymbol{\alpha}_k), U_{\boldsymbol{\mathcal{X}}_m|\boldsymbol{\mathcal{X}}_k}(\boldsymbol{\alpha}_m|\boldsymbol{\alpha}_k)] = F[U_{\boldsymbol{\mathcal{X}}_m}(\boldsymbol{\alpha}_m), U_{\boldsymbol{\mathcal{X}}_k|\boldsymbol{\mathcal{X}}_m}(\boldsymbol{\alpha}_k|\boldsymbol{\alpha}_m)]. \tag{4}$$

3   Aggregation of Preferences

Our aim is to define an aggregation mechanism by which a notion of group preference ordering can emerge from the aggregation of conditional individual preference orderings. An essential characteristic of any such mechanism is that it must possess the following property.

Principle 4 (Monotonicity). *If a subgroup prefers one alternative to another and the complementary subgroup is indifferent with respect to the two alternatives, then the group as a whole must not prefer the latter alternative to the former one.*

Principle 4 invokes the common sense concept that, in the absence of opposition, the group must not arbitrarily override the wishes of individuals. Thus, if X_1 prefers a to a' and X_2 is indifferent between the two profiles, the group $\{X_1, X_2\}$ should not prefer a' to a. In terms of utilities, this condition means that F must be nondecreasing in both arguments.

When modeling influence relationships, it is critical that we delimit generality to ensure computational tractability. We thus propose the following principle.

Principle 5 (Acyclicity). *No cycles occur in the influence relationships among the agents.*

Given two disjoint subgroups $\mathcal{X}_k$ and $\mathcal{X}_m$ of $\mathcal{X}_n$, acyclicity means that it cannot happen that, simultaneously, $\mathcal{X}_m$ directly influences $\mathcal{X}_k$ and $\mathcal{X}_k$ directly influences $\mathcal{X}_m$. The fact that cycles are not permitted does reduce the generality of the model. Nevertheless, restricting to one-way influence relationships is a significant generalization of the neoclassical approach, which assumes that all utilities are categorical and, hence, are trivially acyclic.

3.1   The Aggregation Theorem

It remains to define a function F that complies with the above-mentioned principles. Since positive affine transformations preserve the mathematical integrity of von Neumann-Morgenstern utilities, we may assume, without loss of generality, that all utilities are non-negative and normalized to sum to unity; that is,

$$U_{\mathcal{X}_k}(\alpha_k) \geq 0 \, \forall \alpha_k, \quad U_{\mathcal{X}_m | \mathcal{X}_k}(\alpha_m | \alpha_k) \geq 0 \, \forall \alpha_m, \alpha_k, \tag{5}$$

$$\sum_{\alpha_k} U_{\mathcal{X}_k}(\alpha_k) = 1, \quad \sum_{\alpha_m} U_{\mathcal{X}_m | \mathcal{X}_k}(\alpha_m | \alpha_k) = 1 \, \forall \alpha_k . \tag{6}$$

Theorem 1 (The Aggregation Theorem.). *Let $\mathcal{X}_n = \{X_1, \ldots, X_n\}$ be an n-member multiagent system and let $\mathcal{B}_m$ denote the set of all m-element subgroups of $\mathcal{X}_n$. That is, $\mathcal{X}_m \in \mathcal{B}_m$ if $\mathcal{X}_m = \{X_{i_1}, \ldots, X_{i_m}\}$ with $1 \leq i_1 < \cdots < i_m \leq n$. Let $\{U_{\mathcal{X}_m} : \mathcal{X}_m \in \mathcal{B}_m, m = 1, \ldots, n\}$ be a family of normalized non-negative concordant utilities and let*

$$\{U_{\mathcal{X}_m | \mathcal{X}_k} : \mathcal{X}_m \cap \mathcal{X}_k = \varnothing, \mathcal{X}_m \in \mathcal{B}_m, \mathcal{X}_k \in \mathcal{B}_k, m + k \leq n\} \tag{7}$$

be a family of normalized non-negative conditional concordant utilities associated with all pairs of disjoint subgroups of $\mathcal{X}_n$. These utilities are endogenously aggregated if and only if, for every pair of disjoint subgroups $\mathcal{X}_m$ and $\mathcal{X}_k$,

$$U_{\mathcal{X}_m \mathcal{X}_k}(\alpha_m, \alpha_k) = F[U_{\mathcal{X}_k}(\alpha_k), U_{\mathcal{X}_m \mathcal{X}_k}(\alpha_m | \alpha_k)] \tag{8}$$

$$= U_{\mathcal{X}_m \mathcal{X}_k}(\alpha_m | \alpha_k) U_{\mathcal{X}_k}(\alpha_k). \tag{9}$$

This theorem was originally introduced by [5] as an alternative development of the mathematical syntax of probability theory. The proof below follows [8].

Proof of the Aggregation Theorem. Let $\mathcal{X}_i$, $\mathcal{X}_j$, and $\mathcal{X}_k$ be arbitrary pairwise disjoint subgroups of $\mathcal{X}_n$, and let $U_{\mathcal{X}_i \mathcal{X}_j \mathcal{X}_k}$, $U_{\mathcal{X}_i | \mathcal{X}_j \mathcal{X}_k}$, $U_{\mathcal{X}_i \mathcal{X}_j | \mathcal{X}_k}$, $U_{\mathcal{X}_i \mathcal{X}_j}$, $U_{\mathcal{X}_i | \mathcal{X}_j}$, and $U_{\mathcal{X}_i}$ be endogenously aggregated concordant utilities. That is,

$$U_{\mathcal{X}_i \mathcal{X}_j \mathcal{X}_k}(\alpha_i, \alpha_j, \alpha_k) = F[U_{\mathcal{X}_j \mathcal{X}_k}(\alpha_j, \alpha_k), U_{\mathcal{X}_i | \mathcal{X}_j \mathcal{X}_k}(\alpha_i | \alpha_j, \alpha_k)] \tag{10}$$

$$= F[U_{\mathcal{X}_k}(\alpha_k), U_{\mathcal{X}_i \mathcal{X}_j | \mathcal{X}_k}(\alpha_i, \alpha_j | \alpha_k)]. \tag{11}$$

But

$$U_{\mathcal{X}_j \mathcal{X}_k}(\alpha_j, \alpha_k) = F[U_{\mathcal{X}_k}(\alpha_k), U_{\mathcal{X}_j | \mathcal{X}_k}(\alpha_j | \alpha_k)] \tag{12}$$

and

$$U_{\mathcal{X}_i \mathcal{X}_j | \mathcal{X}_k}(\alpha_i, \alpha_j | \alpha_k) = F[U_{\mathcal{X}_j | \mathcal{X}_k}(\alpha_j | \alpha_k), U_{\mathcal{X}_i | \mathcal{X}_j \mathcal{X}_k}(\alpha_i | \alpha_j, \alpha_k)]. \tag{13}$$

Substituting (12) into (10) and (13) into (11) yields

$$F\left[F[U_{\mathcal{X}_k}(\alpha_k), U_{\mathcal{X}_j | \mathcal{X}_k}(\alpha_j | \alpha_k)], U_{\mathcal{X}_i | \mathcal{X}_j \mathcal{X}_k}(\alpha_i | \alpha_j, \alpha_k)\right] =$$
$$F\left[U_{\mathcal{X}_k}(\alpha_k), F[U_{\mathcal{X}_j | \mathcal{X}_k}(\alpha_j | \alpha_k), U_{\mathcal{X}_i | \mathcal{X}_j \mathcal{X}_k}(\alpha_i | \alpha_j, \alpha_k)]\right]. \tag{14}$$

In terms of general arguments, this equation becomes

$$F[F(x, y), z] = F[x, F(y, z)], \tag{15}$$

called the *associativity equation*. By direct substitution it is easy to see that (15) is satisfied if

$$f[F(x, y)] = f(x)f(y) \tag{16}$$

for any function f. It has been shown by [5] that if F is differentiable in both arguments, then (16) is the general solution to (15). Taking f as the identity function, $F(x, y) = xy$, and

$$U_{\mathcal{X}_i \mathcal{X}_j}(\alpha_i, \alpha_j) = F[U_{\mathcal{X}_i}(\alpha_i), U_{\mathcal{X}_j | \mathcal{X}_i}(\alpha_j | \alpha_i)]$$

$$= U_{\mathcal{X}_i}(\alpha_i) U_{\mathcal{X}_j | \mathcal{X}_i}(\alpha_j | \alpha_i). \tag{17}$$

To prove the converse, we note that F given by (9) is nondecreasing in both arguments since $U_{\mathcal{X}_k}$ and $U_{\mathcal{X}_m | \mathcal{X}_k}$ are nonnegative. Also, since the subgroups $\mathcal{X}_m$ and $\mathcal{X}_k$ are arbitrary, (9) holds if we reverse the roles of m and k. Thus, consistency is satisfied and the aggregation is endogenous. □

The aggregation theorem establishes that, upon compliance with the aforementioned principles, utility aggregation conforms to the same mathematical syntax as does probability. Consequently, the various epistemological properties of probability theory can be accorded analogous interpretations in the praxeological context. Key concepts in this regard are marginalization, independence, and the chain rule.

Marginalization. Let $\mathcal{X}_m = \{X_{j_1}, \ldots, X_{j_m}\}$ and $\mathcal{X}_k = \{X_{i_1}, \ldots, X_{i_k}\}$ be disjoint subgroups of $\mathcal{X}_n$. Then the marginal concordant utility of $\mathcal{X}_m$ is obtained by summing over $\mathcal{A}^k$, yielding

$$U_{\mathcal{X}_m}(\alpha_m) = \sum_{\alpha_k} U_{\mathcal{X}_m \mathcal{X}_k}(\alpha_m, \alpha_k). \tag{18}$$

Independence. Let $\mathcal{X}_m$ and $\mathcal{X}_k$ be disjoint subgroups of $\mathcal{X}_n$. These subgroups are *praxeologically independent* if neither subgroup influences the other; that is,

$$U_{\mathcal{X}_m \mathcal{X}_k}(\alpha_m, \alpha_k) = U_{\mathcal{X}_m}(\alpha_m) U_{\mathcal{X}_k}(\alpha_k). \tag{19}$$

The Chain Rule. Let $\mathcal{X}_m$, $\mathcal{X}_k$, and $\mathcal{X}_l$ be pairwise disjoint subgroups of $\mathcal{X}_n$. Then

$$U_{\mathcal{X}_m \mathcal{X}_k \mathcal{X}_l}(\alpha_m, \alpha_k, \alpha_l) = U_{\mathcal{X}_m | \mathcal{X}_k \mathcal{X}_l}(\alpha_m | \alpha_k, \alpha_l) U_{\mathcal{X}_k | \mathcal{X}_l}(\alpha_k | \alpha_l) U_{\mathcal{X}_l}(\alpha_l). \tag{20}$$

The chain rule is the mechanism by which individual conditional utilities can be aggregated to form the concordant utility. To see, let us first recall that the acyclicity principle ensures that at least one agent possesses a categorical utility. Without loss of generality, let us assume this condition holds for X_1. Successively applying the chain rule, we obtain

$$U_{X_1 \cdots X_n}(a_1, \ldots, a_n) = u_{X_n}(a_n | a_{n-1}, \ldots, a_1) u_{X_{n-1}}(a_{n-1} | a_{n-2}, \ldots, a_1) \cdots u_{X_1}(a_1). \tag{21}$$

3.2   Utility Networks

Since the influence flows are acyclic, we may represent the multiagent system as a *directed acyclic graph* (DAG). Furthermore, since the utilities that comply with the aggregation theorem possess the mathematical syntax of probability mass functions, the edges of the DAG are conditional utilities. We shall term such a graph a *utility network*, and note that it possesses all of the properties of a Bayesian network, albeit with different semantics (e.g., see [13,4]).

Definition 7. *The* parent set *for X_i, denoted* pa(X_i), *is the subgroup of agents whose preferences directly influence X_i. The* child set *of X_i, denoted* ch(X_i), *is the subgroup that is directly influenced by X_I.*

Without loss of generality, we may assume that the vertices of the network are enumerated such that all children of any given node have a higher-numbered index, otherwise the indexing is arbitrary. We may then rewrite (21) as

$$U_{X_1 \cdots X_n}(a_1, \ldots, a_n) = \prod_{i=1}^{n} u_{X_i | \text{pa}(x_i)}[a_i | \text{pa}(a_i)]. \tag{22}$$

where $\mathrm{pa}\,(\mathbf{a}_i) = \{\mathbf{a}_{i_1}, \ldots, \mathbf{a}_{i_{p_i}}\}$ is the joint conjecture corresponding to $\mathrm{pa}\,(X_i) = \{X_{i_1}, \ldots, X_{i_{p_i}}\}$. If $p_i = 0$, then $u_{X_i \mid \mathrm{pa}\,(x_i)} = u_{X_i}$, a categorical utility.

To illustrate, consider the network illustrated in Figure 1. X_1 is a root vertex, and possesses a categorical utility u_{X_1}, $\mathrm{pa}\,(X_2) = \{X_1\}$, and $\mathrm{pa}\,(X_3) = \{X_1, X_2\}$. The concordant utility is

$$U_{X_1 X_2 X_3}(\mathbf{a}_1, \mathbf{a}_2, \mathbf{a}_3) = u_{X_1}(\mathbf{a}_1) u_{X_2 \mid X_1}(\mathbf{a}_2 \mid \mathbf{a}_1) u_{X_3 \mid X_1 X_2}(\mathbf{a}_3 \mid \mathbf{a}_1, \mathbf{a}_2). \qquad (23)$$

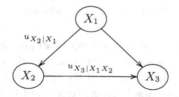

Fig. 1. The DAG for a three-agent system

If the utilities of all agents are categorical, then no social influence exists, the corresponding DAG has no edges, and, hence, no social bonds are generated. The aggregation formula defined by (22) becomes analogous to the creation of the joint distribution of independent random variables as the product of the marginal distributions, and aggregation sheds no additional light on group behavior.

4   Conditional Games

A *conditional game* is a triple $\{\boldsymbol{\mathcal{X}}_n, \boldsymbol{\mathcal{A}}, U_{\boldsymbol{\mathcal{X}}_n}\}$ where $\boldsymbol{\mathcal{X}}_n = \{X_1, \ldots, X_n\}$ is a group of n agents with product action space $\boldsymbol{\mathcal{A}} = \mathcal{A}_1 \times \cdots \times \mathcal{A}_n$ and $U_{\boldsymbol{\mathcal{X}}_n} = U_{X_1 \cdots X_n}$ is a concordant utility. Equivalently, by application of (22), a conditional game can be defined in terms of the conditional utilities $u_{X_i \mid \mathrm{pa}\,(X_i)}$, $i = 1, \ldots, n$. If all utilities are categorical, a conditional game becomes a conventional game.

With a conditional game, the possibility exists for an expanded notion of rational behavior. To proceed, we observe that, since each agent can control only its own actions, what is of interest is the utility for the group if all agents *make conjectures over, and only over, their own action spaces.*

Definition 8. *Consider the concordant utility* $U_{X_1 \cdots X_n}(\mathbf{a}_1, \ldots, \mathbf{a}_n)$. *Let* a_{ij} *denote the jth element of* $\mathbf{a}_i$; *that is,* $\mathbf{a}_i = (a_{i1}, \ldots, a_{in})$ *is X_i's conjecture. Next, form the action profile* $(a_{11}, \ldots, a_{nn})$ *by taking the ith element of each X_i's conjecture, $i = 1, \ldots, n$. Now let us sum the concordant utility over all elements of each* $\mathbf{a}_i$ *except the ii-th elements to form the* group welfare function *for* $\{X_1, \ldots, X_n\}$, *yielding*

$$w_{X_1 \cdots X_n}(a_{11}, \ldots, a_{nn}) = \sum_{\sim a_{11}} \cdots \sum_{\sim a_{nn}} U_{X_1 \cdots X_n}(\mathbf{a}_1, \ldots, \mathbf{a}_n), \qquad (24)$$

where $\sum_{\sim a_{ii}}$ *means the sum is taken over all a_{ij} except a_{ii}. The* individual welfare function *of X_i is the i-th marginal of* $w_{X_1 \cdots X_n}$, *that is,*

$$w_{X_i}(a_{ii}) = \sum_{\sim a_{ii}} w_{X_1 \cdots X_n}(a_{11}, \ldots, a_{nn}). \qquad (25)$$

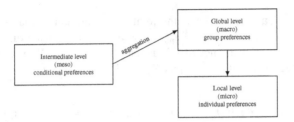

Fig. 2. Flow of social influence

The group welfare function provides a complete *ex post* description of the relationships between the members of a multiagent system as characterized by their *ex ante* conditional utilities. Unless the members of the system are praxeologically independent, the *ex post* utility is not simply an aggregation of individual utilities, as is the case with classical social choice theory. Rather, it constitutes a meso to macro to micro propagation of preferences: from the intermediate, or meso, level, derived from local influences between the agents in the form of conditional preferences, up to the global, or macro, level and down to the individual, or micro, level, as illustrated in Figure 2.

We define the *maximum group welfare* solution as

$$(a_1^*, \ldots, a_n^*) = \arg\max_{\mathbf{a} \in \mathcal{A}} w_{X_1 \cdots X_n}(a_1, \ldots, a_n). \tag{26}$$

Also, the *maximum individual welfare* solution is

$$a_i^\dagger = \arg\max_{a_i \in \mathcal{A}_i} w_{X_i}(a_i). \tag{27}$$

If $a_i^\dagger = a_i^*$ for all $i \in \{1, \ldots, n\}$, the action profile is a *consensus* choice. In general, however, a consensus will not obtain, and negotiation may be required to reach a compromise solution.

The existence of group and individual welfare functions provides a rational basis for meaningful negotiations; namely, that any compromise solution must at least provide each agent with its security level; that is, the maximum guaranteed benefit it could receive regardless of the decisions that others might make. The security level for X_i is

$$s_{X_i} = \max_{a_i} \min_{\sim a_i} \sum_{\sim \mathbf{a}_i} U_{X_1 \cdots X_n}(\mathbf{a}_1, \ldots, \mathbf{a}_n). \tag{28}$$

In addition to individual benefit, we must also consider benefit to the group. Although a security level, per se, for the group cannot be defined in terms of a guaranteed benefit (after all, the group, as a single entity, does not actually make a choice), a possible rationale is that the benefit to the group it should never be less than the smallest guaranteed benefit to the individuals. This approach is consistent with the principles of justice espoused by [14], who argues, essentially, that a society as a whole cannot be better off than its least advantaged member. Accordingly, let us define a security level for the group as $s_{X_1 \cdots X_n} = \min_i\{s_{X_i}\}/n$, where we divide by the number of agents since the utility for the group involves n players.

Now define the *group negotiation set*

$$\mathcal{N}_{x_1 \cdots x_n} = \{\mathbf{a} \in \mathcal{A} : w_{x_1 \cdots x_n}(\mathbf{a}) \geq s_{x_1 \cdots x_n}\}, \tag{29}$$

the *individual negotiation sets*

$$\mathcal{N}_{x_i} = \{a_i \in \mathcal{A}_i : w_{x_i}(a_i) \geq s_{x_i}\}, \ i = 1, \ldots, n, \tag{30}$$

and the *negotiation rectangle*

$$\mathcal{R}_{x_1 \cdots x_n} = \mathcal{N}_{x_1} \times \cdots \times \mathcal{N}_{x_n}. \tag{31}$$

Finally, define the *compromise set*

$$\mathcal{C}_{x_1 \cdots x_n} = \mathcal{N}_{x_1 \cdots x_n} \cap \mathcal{R}_{x_1 \cdots x_n}. \tag{32}$$

If $\mathcal{C}_{x_1 \cdots x_n} = \varnothing$, then no rational compromise is possible at the stated security levels. One way to overcome this impasse is to decrement the security level of the group iteratively by a small amount, thereby gradually enlarging $\mathcal{N}_{x_1 \cdots x_n}$ until $\mathcal{C}_{x_1 \cdots x_n} \neq \varnothing$. If $\mathcal{C}_{x_1 \cdots x_n} = \varnothing$ after the maximum reduction in group security has been reached, then no rational compromise is possible, and the system may be considered dysfunctional. Another way to negotiate is for individual members to decrement their security levels iteratively, thereby enlarging the negotiation rectangle.

Once $\mathcal{C}_{x_1 \cdots x_n} \neq \varnothing$, any element of this set provides each member, as well as the group, with at least its security level. One possible tie-breaker is

$$\mathbf{a}_c = \arg \max_{\mathbf{a} \in \mathcal{C}_{x_1 \cdots x_n}} w_{x_1 \cdots x_n}(\mathbf{a}), \tag{33}$$

which provides the maximum benefit to the group such that each of its members achieves at least its security level.

5   Partial Sociation

Our development thus far has assumed the full generality of conditioning; namely, that (i) a conditional utility depends on the entire conjecture profiles of all of the parents, and (ii) an agent's conditional utility is a function of all elements of the action profile. If maximum complexity is required to define social relationships properly, the full power of conditional game theory may be necessary. It is often the case, however, that the influence relationships are sparse, in that an agent does not condition its preferences on the entire conjecture profiles of its parents. It can also be the case that an agent's utility does not depend upon the entire action profile. To account for such situations, we introduce the notion of *sociation*.

Suppose X_i has $p_i > 0$ parents, denoted $\mathrm{pa}(X_i) = \{X_{i_1}, \ldots, X_{i_{p_i}}\}$, with conditional utility $u_{X_i | \mathrm{pa}(X_i)}(\mathrm{pa}(\mathbf{a}_i))$, where $\mathrm{pa}(\mathbf{a}_i) = \{\mathbf{a}_{i_1}, \ldots, \mathbf{a}_{i_{p_i}}\}$ is the joint conjecture for $\mathrm{pa}(X_i)$.

Definition 9. *A conjecture subprofile, for* X_{i_k}, *denoted* $\hat{\mathbf{a}}_{i_k}$, *is the subprofile comprising the elements of of* $\mathbf{a}_{i_k}$ *that influence* X_i. *We then have*

$$u_{X_i| \, \text{pa} \, (X_i)}[\mathbf{a}_i| \, \text{pa} \, (\mathbf{a}_i)] = u_{X_i| \, \text{pa} \, (X_i)}[\mathbf{a}_i| \, \text{pa} \, (\hat{\mathbf{a}}_i)] \,, \tag{34}$$

where $\text{pa} \, (\hat{\mathbf{a}}_i) = (\hat{\mathbf{a}}_{i_1}, \dots, \hat{\mathbf{a}}_{i_{p_i}})$. $\{X_1, \dots, X_n\}$ *is completely conjecture sociated if* $\hat{\mathbf{a}}_{i_k} = \mathbf{a}_{i_k}$ *for* $k = 1, \dots, p_i$ *and* $i = 1, \dots, n$. *It is* completely conjecture dissociated *if* $\hat{\mathbf{a}}_{i_k} = a_{i_k}$ *for* $k = 1, \dots, p_i$ *and* $i = 1, \dots, n$, *in which case,* $\text{pa} \, (\hat{\mathbf{a}}_i) = (a_{i_1}, \dots, a_{i_{p_i}})$. *Otherwise, the group is* partially conjecture sociated.

Definition 10. *A utility subprofile, denoted* $\tilde{\mathbf{a}}_i$, *comprises the subprofile of* $\mathbf{a}_i$ *that affects* X_i's *utility. We then have*

$$u_{X_i| \, \text{pa} \, (X_i)}[\mathbf{a}_i| \, \text{pa} \, (\mathbf{a}_i)] = \tilde{u}_{X_i| \, \text{pa} \, (X_i)}[\tilde{\mathbf{a}}_i| \, \text{pa} \, (\hat{\mathbf{a}}_i)] \,, \tag{35}$$

where $\tilde{u}$ *denotes* u *with the dissociated arguments removed.* $\{X_1, \dots, X_n\}$ *is completely utility sociated if* $\tilde{\mathbf{a}}_j = \mathbf{a}_j$ *for* $i = 1, \dots, n$. *It is* completely utility dissociated *if* $\tilde{\mathbf{a}}_i = a_i$ *for* $i = 1, \dots, n$, *in which case*

$$u_{X_i| \, \text{pa} \, (X_i)}[\mathbf{a}_i| \, \text{pa} \, (\mathbf{a}_i)] = \tilde{u}_{X_i| \, \text{pa} \, (X_i)}[(a_i| \, \text{pa} \, (\hat{\mathbf{a}}_i)] \,. \tag{36}$$

Otherwise, the group is partially utility sociated.

Definition 11. *A group* $\{X_1, \dots, X_n\}$ *is* completely dissociated *if it is both completely conjecture dissociated and completely utility dissociated, in which case,* $\text{pa} \, (\mathbf{a}_i) = \text{pa} \, (a_i) = (a_{i_1}, \dots, a_{i_{p_i}})$, *the profile of conjecture actions of the members of* $\text{pa} \, (X_i)$.

For a partially sociated system, the concordant utility assumes the form

$$U_{X_1 \cdots X_n}(\mathbf{a}_1, \dots, \mathbf{a}_n) = \tilde{U}_{X_1 \cdots X_n}(\tilde{\mathbf{a}}_1, \dots, \tilde{\mathbf{a}}_n) \tag{37}$$

$$= \prod_{i=1}^{n} \tilde{u}_{X_i| \, \text{pa} \, (X_i)}[\tilde{\mathbf{a}}_i| \, \text{pa} \, (\hat{\mathbf{a}}_i)] \,, \tag{38}$$

where $\tilde{U}$ is U with the dissociated arguments removed. For a completely dissociated group, the concordant utility coincides with the group welfare function and assumes the form

$$w_{X_1 \cdots X_n}(a_1, \dots, a_n) = \prod_{i=1}^{n} \tilde{u}_{X_i| \, \text{pa} \, (X_i)}[a_i| \, \text{pa} \, (a_i)] \,. \tag{39}$$

Example 3. Let us now reconsider the automobile buying example introduced in Example 1. We shall assume that the influence flows are as depicted in Figure 1, with the corresponding concordant utility of the form expressed by (23), yielding

$$U_{X_1 X_2 X_3}[(a_{11}, a_{12}, a_{13}), (a_{21}, a_{22}, a_{23}), (a_{31}, a_{32}, a_{33})] =$$
$$u_{X_1}(a_{11}, a_{12}, a_{13}) u_{X_2|X_1}(a_{21}, a_{22}, a_{23}|a_{11}, a_{12}, a_{13})$$
$$u_{X_3|X_1 X_2}[a_{31}, a_{32}, a_{33}|(a_{11}, a_{12}, a_{13}), (a_{21}, a_{22}, a_{23})] \,. \tag{40}$$

If all agents have opinions about the model, manufacturer, and color attributes, then each agent must specify utility valuations for each of the eight possible outcomes. Thus, X_1 would specify eight utility values. X_2, however, would need to define eight conditional utilities, each with eight valuations, and thus would make 64 utility specifications. Finally, X_3 would define 64 conditional utilities, each with eight valuations, and thus would make 512 utility specifications — a formidable task.

Now let us suppose that X_1 is concerned only about the model and manufacturer, but has no opinion about the color. Also, we assume that X_2 is concerned only about the manufacturer given X_1's conjecture about the model and manufacturer. Finally, let us assume that X_3 is concerned only about the color given X_1's conjecture about the model and X_2's conjecture about the manufacturer.

As a result of these simplifications, we see that X_1 is partially utility sociated, thus $\tilde{a}_1 = (a_{11}, a_{12})$. We also see that X_2 is completely utility dissociated and partially conjecture sociated, hence $\tilde{a}_2 = a_{22}$ and pa $(a_{22}) = (a_{11}, a_{12})$. Finally, X_3 is also completely utility dissociated and partially conjecture sociated, thus $\tilde{a}_3 = a_{33}$ and pa $(a_{33}) = (a_{11}, a_{22})$. The concordant utility simplifies to

$$\tilde{U}_{X_1 X_2 X_3}(a_{11}, a_{12}, a_{22}, a_{33}) = \tilde{u}_{X_1}(a_{11}, a_{12})$$
$$\tilde{u}_{X_2 | X_1}(a_{22} | a_{11}, a_{12}) \tilde{u}_{X_3 | X_1 X_2}(a_{33} | a_{11}, a_{22}). \quad (41)$$

Under these simplifications, we see that X_1 need only make four specifications when defining $\tilde{u}_{X_1}$, X_2 need make two specifications for each of X_1's four specifications, resulting in eight specifications when defining $\tilde{u}_{X_2 | X_1}$, and X_3 need make two specifications for each of the four joint specifications of X_1 and X_2, resulting in eight specifications when defining $\tilde{u}_{X_3 | X_1 X_2}$, yielding a grand total of 20 utility specifications — a considerable reduction in complexity from the 584 specifications required under the condition of complete sociation (and even less than the 24 specifications needed to define categorical utilities).

Tables 1(a), (b), and (c) respectively tabulate X_1's categorical utility, X_2's conditional utilities given X_1's conjectures, and X_3's conditional utilities given the conjectures for X_1 and X_2.

Table 1. (a) The categorical utility $\tilde{u}_{X_1}(a_{11}, a_{12})$, (b) the conditional utility $\tilde{u}_{X_2 | X_1}(a_{22} | a_{11}, a_{12})$, and (c) he conditional utility $\tilde{u}_{X_3 | X_1 X_2}(a_{33} | a_{11}, a_{22})$

(a)			(b)					(c)				
	a_{12}			(a_{11}, a_{12})					(a_{11}, a_{22})			
a_{11}	D	F	a_{22}	(C, D)	(C, F)	(S, D)	(S, F)	a_{33}	(C, D)	(C, F)	(S, D)	(S, F)
C	0.1	0.4	D	0.3	0.5	0.6	0.4	R	0.1	0.3	0.5	0.8
S	0.3	0.2	F	0.7	0.5	0.4	0.6	G	0.9	0.7	0.5	0.2

The individual and group welfare functions are illustrated in Tables 2 (a) and (b).

The group negotiation set is $\mathcal{N}_{X_1 X_2 X_3} = \{(C, D, G), (C, F, G), (S, F, R)\}$ and the negotiation rectangle is $\mathcal{R}_{X_1 X_2 X_3} = \{(C, F, G), (S, F, G)\}$, yielding the compromise set $\mathcal{C}_{X_1 X_2 X_3} = \{(C, F, G)\}$: a green foreign-made convertible.

Table 2. (a) The individual welfare function, w_{X_i}, $i = 1, 2, 3$, and (b) the group welfare function $w_{X_1 X_2 X_3}(a_{11}, a_{22}, a_{33})$

(a)	
$w_{X_1}(C) = 05$	$w_{X_1}(S) = 0.5$
$w_{X_2}(D) = 0.49$	$w_{X_2}(F) = 0.51$
$w_{X_3}(R) = 0426$	$w_{X_3}(G) = 0.574$

(b)

a_{11}	(a_{22}, a_{33})			
	(D, R)	(D, G)	(F, R)	(F, G)
C	0.023	0.207	0.081	0.189
S	0.130	0.130	0.192	0.048

6   Conclusions

As acknowledged by many decision theorists [2,10,16], neoclassical game theory is an appropriate model for competitive and market-driven scenarios, but it offers limited capacity for the design and synthesis of multiagent systems that are intended to cooperate, compromise, and negotiate.

This paper (i) presents a principle-based extension to neoclassical game theory that replaces categorical utilities with conditional utilities that encode the social influence relationships that exist among the agents; (ii) develops notions of rational multiagent decision making to define rational behavior simultaneously for groups and for individuals; and (iii) addresses computational complexity by maximally exploiting influence sparseness among the agents. Conditional game theory provides a powerful framework within which to design and synthesize cooperative multiagent systems.

References

1. Arrow, K.J.: Social Choice and Individual Values, 2nd edn. John Wiley, New York (1951,1963)
2. Arrow, K.J.: Rationality of self and others in an economic system. In: Hogarth, R.M., Reder, M.W. (eds.) Rational Choice. University of Chicago Press, Chicago (1986)
3. Camerer, C.: Behavioral Game Theory: Experiments in Strategic Interaction. Princeton Univ. Press, Princeton (2003)
4. Cowell, R.G., Dawid, A.P., Lauritzen, S.L., Spiegelhalter, D.J.: Probabilistic Networks and Expert Systems. Springer, New York (1999)
5. Cox, R.T.: Probability, frequency, and reasonable expectation. American Journal of Physics 14, 1–13 (1946)
6. Debreu, G.: Theory of Value. Yale University Press, New Haven (1959)
7. Friedman, M.: Price theory. Aldine Press, Chicago (1961)
8. Jaynes, E.T.: Probability Theory: The Logic of Science. Cambridge University Press, Cambridge (2003)
9. Keeney, R.L., Raiffa, H.: Decisions with Multiple Objectives. Cambridge University Press, Cambridge (1993); First published by John Wiley & Sons (1976)
10. Luce, R.D., Raiffa, H.: Games and Decisions. John Wiley, New York (1957)
11. Nisan, N., Roughgarden, T., Tardos, E., Vazirani, V.V. (eds.): Algorithmic Game Theory. Cambridge Univ. Press, Cambridge (2007)
12. Parsons, S., Gmytrasiewicz, P., Wooldridge, M. (eds.): Game Theory and Decision Theory in Agent-Based Systems. Kluwer, Boston (2002)

13. Pearl, J.: Probabilistic Reasoning in Intelligent Systems. Morgan Kaufmann, San Mateo (1988)
14. Rawls, J.B.: A Theory of Justice. Harvard University Press, Cambridge (1971)
15. Shoham, Y., Leyton-Brown, K.: Multiagent Systems. Cambridge University Press, Cambridge (2009)
16. Shubik, M.: Game Theory in the Social Sciences. MIT Press, Cambridge (1982)
17. Vlassis, N. (ed.): A Concise Introduction to Multiagent Systems and Distributed Artificial Intelligence. Morgan & Claypool Publishers, San Rafael (2007)
18. Weiss, G. (ed.): Multiagent Systems. MIT Press, Cambridge (1999)

On the Design of Agent-Based Artificial Stock Markets

Olivier Brandouy[1], Philippe Mathieu[2], and Iryna Veryzhenko[3]

[1] Sorbonne Graduate Business School, Paris, France
[2] LIFL, UMR CNRS-USTL 8022, Lille, France
[3] LEM, UMR CNRS 8174, Lille, France
olivier.brandouy@univ-paris1.fr,
philippe.mathieu@lifl.fr,
iryna.veryzhenko@univlille1.fr

Abstract. The purpose of this paper is to define software engineering abstractions that provide a generic framework for stock market simulations. We demonstrate a series of key points and principles that has governed the development of an Agent-Based financial market application programming interface (API). The simulator architecture is presented. During artificial market construction we have faced the whole variety of agent-based modelling issues : local interaction, distributed knowledge and resources, heterogeneous environments, agents autonomy, artificial intelligence, speech acts, discrete or continuous scheduling and simulation. Our study demonstrates that the choices made for agent-based modelling in this context deeply impact the resulting market dynamics and proposes a series of advances regarding the main limits the existing platforms actually meet.

Keywords. Multi-agent systems, Artificial market, Market microstructure, Agents behaviour.

1 Introduction

Multi-agent modelling is nowadays actively applied to financial markets simulation. This is partly due to its ability at reflecting a wide range of complexities arising in these markets, and to its flexibility for exploring the impact of automation in trading. Thus, agent-based computational simulations [1] may contribute to several scientific debates in Finance. One example at the crossroads of multi-agent modelling and machine learning is *Bayesian learning* : this technique can be used by agents to incorporate all available information into the decision making process [2]. It can also be employed to track a moving parameter [3] such as the fundamental value of a given stock. On the other hand, financial markets offer an important field of application for agent-based modelling and machine learning, since agent objectives and interactions are clearly defined. For this reason, financial market environments can help to answer some modelling issues related to agent engineering, or test robustness of existing behavioural models.

At the present, there is a large number of agent-based frameworks, with varying functionalities and architectures, addressing different problems. Generally speaking, there are two major approaches to agent-based financial market simulations. The first one is leads to the realization of a specific market structure, with specific agents' behaviours,

J. Filipe and A. Fred (Eds.): ICAART 2011, CCIS 271, pp. 350–364, 2013.

trading instruments and rules. The second one is geared at designing a general environment with flexible settings and functionalities that can accept heterogeneous agents populations.

We first list a limited set of these models belonging to these philosophies and that point out some design questions. – Altreva Adaptive Modeler is a tool for creating agent-based market simulations for a specific problem: stock price forecasting [4]. The author, among other questions, describes a problem of a memory limitation during the framework processing.

– JLM market simulator is another tool that investors can use to create a market model using their own inputs. The authors (Jacobs Levy Equity Management, Inc.) conclude that it is not an easy task to build a complex asynchronous simulation with reasonably realistic properties. The first highlighted source of potential problems is the diversity of agents' behaviours (for instance, there are only mean-variance investors in JLM); a second one is the specification of user's trading strategies, that does not require the user's programming skills [5].

– Ascape is a general agent-based framework, developed at Brookings Institute in the 90's [6], that is actively used in financial market modelling. Its developers discuss design possibilities to express the same basic modelling ideas in one way and have them tested in many different environments and configurations.

Most of present artificial market platforms suffer from a lack of flexibility and must be viewed as softwares rather than APIs, because they are mainly oriented for solving a given problem and, most of the time, cannot easily be used to explore a wide range of financial issues. This is due to some structural choices that are made by the developers during the coding phase.

In this paper we present the ArTificial Open Market API (here-after ATOM, see http://atom.univ-lille1.fr) and focus our attention on some important issues of agent-based stock market design we have faced during its development. Among others, we make a specific point on the ability of this new, generic architecture to overcome some issues mentioned previously. After a general exposure of our main choices in terms of software engineering, we propose our solution for several problems : i) how should we manage information to reprocess real market order-flows (necessity of a unique order identity?), ii) how should the scheduling system be organized, iii) how can we integrate a human agent in the simulation process, with his own strategy, ("human in the loop") ...? In addition, we tackle an important design question in the representation and structure of the trading agents, from zero-intelligence agents to sophisticated intelligence ones, like technical traders. We emphasize the importance of calibration and validation aspects of agent-based market and run several series of validation tests in order to show the ability of ATOM at replicating real market price motions.

2   The Artificial Trading Open Market API

ATOM is an environment for Agent-Based simulations of stock markets. At the present moment, it is realized based on the architecture close to the Euronext-NYSE Stock Exchange one. Agent-Based artificial stock markets aim at matching orders sent by virtual traders to fix quotation prices. Price formation is ruled by a negotiation system between

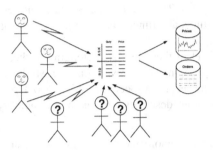

Fig. 1. Double auction market with human and artificial traders

sellers and buyers based on an asynchronous, double auction mechanism structured in an order book (see Figure 1). ATOM is developed as large scale experimental platform with heterogeneous agents populations. It is a concrete implementation of MAS abstract design issues: *agent autonomy* and its *behaviour* (strategy), each agent asses her position and makes decisions individually; price history and orders sets are *emergent phenomenon* of market activities; allowance of *distributed simulations* with many computers interacting through a network as well as local-host, extremely fast simulations; possibility to design experiments mixing *human beings and artificial traders*.

2.1 Distinctive Features

Following distinctive aspects of ATOM can be highlighted:

1. ATOM can use various kind of sophisticated agents with their own behaviours and intelligence (see section 4), including portfolio optimizers as this platform provides the multi-assets (multi-orderbook) organisation. In addition, ATOM contains all variety of the Euronext-Nyse orders: limit, market, stop-limit,iceberg, etc.
2. Thousands of agents can evolve simultaneously, creating a truly heterogeneous population. Once designed, agents evolve by themselves, learning and adapting to their (financial) environment. 100,000 agents have been employed simultaneously for technical analysis evaluation research [7].
3. ATOM can combine human-beings and artificial traders on a single market using its network capabilities. This feature support a wide variety of configurations: from "experimental finance" classrooms with students, to competing strategies run independently and distantly by several banks or research labs. The scheduler can be set so to allow human agents to freeze the market during their decision process or not (see above, section 3.4).
4. ATOM serves as a "replay-engine", meaning that it is possible to re-execute whole trading log file with following information: identification of order book, that corresponds to stock identity, agents identification, the platform time stamp fixed at the very moment orders arrive to a given order-book, prices resulting from the. orders. ATOM takes less than 5 seconds to replay an entire day of trading, that contains 400,000 activities. This tool is extremely important for policy-oriented experiments focusing on the technical features of the market microstructure (tick size, price fixing protocol) and its influence on the price dynamic.

5. Agents can be viewed as simple nutshells in certain cases: they only execute actions predefined by third party, meaning that their behaviour is defined (controlled) by experimenter. These agents are called "Hollow Agents". For example, a human trader can act through such agents. By definition, "Hollow Agents" do not have any artificial intelligence and can be assimilated to human-machine wrappers.

6. Today, there are only few artificial stock markets presenting multi-assets order book ([8], [5], [9]). ATOM is one of them, that allows us to simulate more advanced agents behaviour with risk-return management and wealth optimisation utility functions (see section 4).

7. We overcome a weak point of most existing simulators, that include only limit and market order types (for instance [8], [10]), ATOM contains all spectrum of orders present in Euronext Stock Exchange.

3 Artificial Market Design Issues

Artificial Stock Markets (here-after ASM) are environments that allow to express all classical notions used in multi-agent systems. ASM, like any other MAS, are suited for the study of various emergents phenomena. Using the so-called "vowels" approach [11], the definition of AEIO[1] is straightforward : for example, the environment in which agents evolve is the market microstructure, agents behaviours mimic those of real traders with their own and interaction occurs through orders and prices. Nevertheless, if one wants to build an efficient platform, several issues can be identified and must be precisely and strictly regulated.

3.1 Entities Organization

During an artificial stock market modelling many MAS design principles are employed, for example, modularity and encapsulation, that suggest dividing the system into different sub-organizations. There are different attempts to organize artificial stock markets. For example, *Agents – Rules* and *Securities* in [12] or *Autonomous engine* (platform core) – *Simulation User Interface* and *Agents* separate binary modules in [13].

The choice of ATOM organisation is resulting from intention to introduce fairly tractable markets, to be close to NYSE-EURONEXT stock exchange organisation and to ground modelling on MAS design main concepts. ATOM architecture can be viewed as a system with interacting components: i) *Agents*, and their behaviours, ii) *Market* is defined in terms of microstructure and iii) *Bank* reflects intermediaries and monetary financial institutions iv) the *Artificial Economic World* provides economical indicators. Depending upon the researcher's targets, the *Artificial Economic World* can be plugged or not in the simulations.

We link each system entity with the sets of *Responsibilities* in order to cover all functionality and complexity of real world market. Thus, *Market* is responsible for the generating of market scenarios and price path, it presents set of constraints, rules, regulations, leading participants to activities. *Agents* may play roles of buyer or seller with

---

[1] A Agents, E Environment, I Interactions, O Organization, U Users.

different trading objectives. The agents directly initialize transactions. *Bank* component represents all intermediaries, that maintain information exchange between buyers and sellers. At the same time, Bank can be considered as the special type of buyer or seller, that has unlimited wealth, hence take active part in stock trading. We propose to consider Bank as trading and intermediary agent. *Artificial Economic World* provides external information about perspective corporates development, dividends and coupons changes, tax police modification. This information influences agent decisions. Artificial market architecture (system elements and interaction between them) is presented on the Figure 2. Thanks to its high modularity and its ability to mimic real-world

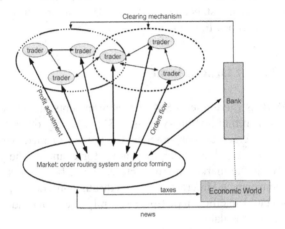

Fig. 2. Market organizations and interactions

environments, it can also serve as a research tool in Portfolio Management, Algorithmic Trading or Risk Management among others.

It is hardly possible to describe the complex algorithmic structures that are necessary for the realization of such multi-agent platforms; therefore we have chosen to introduce three of difficulties one must face while developing an ASM: i) the management of orders' ID, ii) the scheduling system, and iii) the introduction of a human-being in the simulation loop (here-after *"human-in-the-loop"* problem).

3.2   A Unique Identity for Orders

In its simplest form, an order is a triplet of direction (purchase or sale), a quantity and a price. Usually this type of order is called a *"Limit Order"*. In the Euronext-NYSE system, several other orders are used (see *"EURONEXT" Rule Book* at http://www.euronext. com). Once constructed by an agent, the order is sent to the order-book. It is then ranked in the corresponding auction-queue ("Bid" or "Ask" if it is an order to "Buy", respectively to "Sell") where are stacked the other pending orders using a "price-then-time" priority rule. As soon as two pending orders can be matched, they are processed as a "deal", which delivers a new price. Notice that the clearing mechanism implies that cash is transferred from the buyer to the seller and stocks from the seller to the buyer.

For various reasons, financial institutions need to proceed again an historical record of orders (for example, for the optimization of algorithmic trading methods). Such historical records collect the expression of human behaviours in specific circumstances. The first difficulty of order-flow replaying is exact interpretation of the order flow as it is expressed in the real-world. If order set consists of only "*Limit Orders*", it could be perfectly reproducible. Unfortunately, issued orders can be modified or deleted. It implies one must be able to identify clearly reference between an "Update" or a "Delete" and previously issued orders. Thus, a generic platform has to use a unique ID for orders. To our knowledge, this is neither the case for the Genoa Artificial Stock Market (see [14]) nor in the Santa-Fe ASM for example (see [15]). How should an ASM deal with this issue?

In ATOM, each order is sent by an agent. Each agent has a unique name. The Id is then constituted by the agent's name, owner of this order, *plus* a unique number managed by the agent itself. It is the responsibility of the agent to be able to retrieve its own orders. If she does not do that, she will not be able to send a Cancel or an Update order. Two other solutions exist : the first option could use the system time-stamp, however, with classical systems and languages, it is perfectly possible to process several orders during the same tick time; the second option could consist in making the market managing these ID's, but this would imply to use a corresponding table between external and internal Ids. This latter complication is useless.

3.3 Time Handling and Scheduling System

A crucial question for distributed system design is the way one deals with time. There are two aspects for this problem: the modelling choice (sequential/parallel evaluation of agents) and the architecture choice (single stream or multiple streams processes). In ATOM the scheduling system is parametric, thus one can choose between four possible configurations (see figure 3). In each case, one can also decide if the talking possibilities for agents will be balanced among them or not. These possibilities give the designer a real power for answering a wide range of problems and experiments.

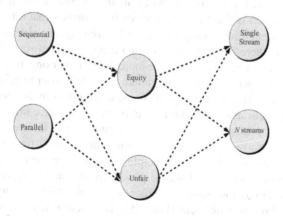

Fig. 3. Modelling and architecture choices in time handling problem

The main difference between the sequential and parallel simulation principle is about the tick time. In MAS philosophy, one considers that time changes along environments changes. In the sequential mechanism, the tick time changes after each order sent by an agent. This means that an agent is always able to see other agents' decisions before making its own decision. In the parallel mechanism, the tick time changes after one decision round processed over all agents. It is a way to simulate parallelism between the agents. This is the principle followed by the Conway's game of Life [16].

However choosing one of these arrangements is just a modelling choice. Any of them can be obtained in a single stream architecture as well as in a multiple stream architecture.

In a single stream architecture, one needs a specific software engineering pattern to code the parallelism. The easiest way is to let the Market collect all the orders before their execution. It is a way to simulate simultaneity in agents decisions. If one needs to ensure fairness among agents, ATOM uses a loop to give the talk to all agents – "equitable round table". An agent is allowed to act only once in each talk round. Of course, if one wants to depart from this fairness, it is sufficient to pick randomly an agent and to offer him the possibility to decide.

Note that an agent can decline the possibility to speak, hence, even if we have a single stream, we can easily simulate different talking frequencies. A possibility to express an intention does not necessarily imply that a new order is issued. Since agents are autonomous, they can evaluate their positions every round, modify trading rules according to new market conditions. Developing an agent that sends twice less orders than the others can be made by programming her behaviour such as she will decline word on odd turns (keep unchanged position), while others accept to talk each time they have the possibility to do so.

The main advantage of the single stream architecture is that the designer can reproduce perfectly all the experiments. He keeps the control on agent's talk. We consider that it is the best way to build and test experiments.

In a multiple stream architecture, parallelism is obvious, but the designer does not have the control about the talking order of agents. This order is defined by the Operating System, and of course, it can produce biases in simulations. Nevertheless, one particularity of this approach is that the time is given in seconds – real time. It is also easy to express the different trading frequencies for different agents, similarly with what is described above. If there is no synchronization mechanism between the streams, the simulation is unfair, an agent can talk twice more than another one. In a fair simulation, one just has to put a synchronization pattern like a Cyclic Barrier to grant this property. This architecture is preferable if one wants to include humans in the loop.

As ATOM is a multi-asset artificial market platform, we have implemented a "one order for one book" rule: during a talk round, agents are just allowed to send at most one single order to a given order book (i.e. one order at most per stock) within the same "round table discussion". This principle helps to keep the equity in agents actions. However, notice that agents have the possibility to send several orders within the same "round table discussion" to several order books: this ability is simply constrained by the "one order for one book" rule. If the ASM is settled such as it runs a multi-stock experiment, an agent can therefore rebalance her portfolio using one order per category of stocks she holds. The proper system scheduler provides this possibility.

It is necessary to stress that ATOM can govern all combinations between sequential and parallel mechanisms, equity or unfairness in agents' actions, one stream or multiple streams processes. The combination of single stream, parallel mechanism, equity in actions is used for most of the experiments concerning financial problems. Parallel mechanism, multiple streams processes are used to allow the human investor to trade in equitable conditions with artificial agents within the platform.

3.4 Human in the Loop

ATOM can include human-beings in the simulation loop. This is an important feature that is seldom offered in multi-agent artificial stock markets, if simply possible with respect to the algorithmic choices made in other platforms. Human agents do not differ from artificial agents in their philosophy: they share the same general characteristics as other agents. The so-called "vowels" approach is respected, even if U (for "Users") is subsumed by A – Agents. A human agent is an interface allowing for human-machine interaction. Through this interface one can create and send orders. Notice that human agents do not have any artificial intelligence: they just embed human intelligence in a formalism that is accepted by the system.

To allow the introduction of human in the loop, ATOM has been designed to deal with communications over the network. Human agents can be run on different machines and the system allows client-server configurations. This approach is particularly fruitful for a pedagogic use of the platform during Finance class for example. In this latter case, several students have their own trading interface on their computers. In other terms, each of them runs a human agent linked to the ATOM server through the network. However, the presence of human agents does not alter the way the scheduler operates.

Two kinds of human agents can co-exist in ATOM: Modal Human Agents (MHA) and Non Modal Human agent (NMH).

- MHA can stop the scheduling system. As long as human-entity does not express her intentions (to issue a new order or to stay unchanged), the simulation is temporary frozen. In a classroom, this aspect is particularly important and leaves time for students to estimate current position and to make decision.
- NMH cannot freeze the simulation, which means that human agents compete in real time with artificial traders. Even if human agents can have a hard time in this situation, it remains realistic in a financial world where algorithmic trading is more and more frequent.

In this section we have presented three major technical points that characterize ATOM and should also concern many ASM. Even if other important technical issues could not be mentioned in this article, we have stressed that the development of artificial stock market platforms put forward a series of complex issues in terms of computer science. In the next section, we introduce some additional elements relative to the artificial intelligence of virtual agents that can be run in our platform. This question is of main concern for computer scientists and for financial researchers alike.

4   Artificial Traders: From Basic Reactive Agents to Highly Sophisticated Entities

Artificial agents, market participants, comply with basic agent-based modelling concepts [17].

- *Autonomy* means that an agent is not passive subject to a global, external flow of control in its actions. An agent has own objectives, abilities to accept information, then to analyze it and based on these results to make decision about further actions.
- *Proactivity* means that the agents act in order to achieve its objectives or goals. In terms of artificial financial market, agents trade (set up the buy and sell orders) to maximize their wealth.

Many ASM can run large populations of homogeneous, respectively heterogeneous artificial traders. This is also the case for ATOM, moreover it allows facilities which are not available in other platforms. Generally speaking, artificial traders are characterized by their a) available set of *actions* (buy, sell) and possibility to switch between these activities (from buyer to seller) b) *decision making rules*, for instance, buyers cannot buy at a price higher than their buyer value and sellers cannot sell for a price below their seller cost c) *scheduling* of action: how often agent is able to send the orders in respond to market request, some agent participate one time per hour, while others trade every minute d) *information* consideration, in the mean which information agent requires from market or external word in order to make decision and what kind of information she shares for others e) possibility to describe *status* in mean of number of assets and available cash or current budget. Agents heterogeneity is driven by different combinations of these properties. For example, the following types of agents can be implemented:

Zero Intelligence Traders (ZIT). This behaviour is merely based on stochastic choices: there are equal possibilities to send ask or bid order, ZIT do not observe and do not ask any information to set up prices and quantities, that are random variable. Concerning scheduling, such traders respond to every market request. This kind of behaviour has been popularized in economics by [18]. Despite their extreme simplicity, these agents are widely used because more sophisticated forms of rationality appear to be useless to explain the emergence of the main financial stylized facts at the intraday level.

Technical Traders. "Chartists" are a specific population of technical traders. These agents try to identify patterns in past prices (using charts or statistical signals) that could be used to predict future prices and henceforth send appropriate orders. One can find an example of such behaviour in [19]. From a software engineering perspective, these agents need to have some feedback from the market and some kind of learning process as well (reinforcement learning for large sets of rules is generally used). At the same time, technical traders ignore the actual nature of the company, currency or commodity. This lead to some complex algorithmic issues. For example, if one considers a population of a few thousand Technical Traders, it is highly desirable to avoid that each agent compute the same indicators, or simply store themselves the whole price series.

Sophisticated Intelligence Traders (SIT). Several kinds of SIT can evolve in ATOM:

i) *Cognitive Agents* generally have a full artificial intelligence, although it can be designed to be rather minimal (usual features to develop such agents are memory, information analysis processes, expectations, strategies and learning capacities). For example, an agent buying at a specific price and sending immediately a "stop order" to short her position if the price drops under $\theta\%$ times the current price, will fall in this category. Agents using strategic order splitting (see for example [20]) or exploiting sophisticated strategies (for instance, [21]) can also be considered as Cognitive Agents.

ii) *Evolutionary Agents* are the ultimate form of SIT; they outperform Cognitive Agents in terms of complexity since they are able to evolve with their environment. These agents can also generate new rules or strategies (this can require genetic algorithms for example).

iii) *Risk Averse Agents* set up portfolio models in which the individual chooses a set of assets in order to maximise some function of wealth. For each investment possibility I from the set of alternatives F, the agent will undertake one I_{opt}, that will maximize resulting wealth $W(I)$ or $E[W(I_{opt})] = max_{I \in F} E[W(I)]$. Utility function provides relative measure of investor's preferences for wealth and the amount of risk they are willing to undertake in order to maximize their wealth. In ATOM, the agents have a choice between different utility functions: Constant Absolute Risk Aversion (CARA), Constant Relative Risk Aversion (CRRA), logarithmic and quadratic.

iv) *Mean-variance Agents* are investors trading over several order books, hence refer to portfolio optimisation aspects. When an investor wants to reoptimize her portfolio, she chooses and "ideal" portfolio from a mean-variance efficient frontier [22], that is based on analysis of internal and external information. The choice of portfolio depends on the trader's risk aversion. These agents send buy and sell orders in order to get closer to "ideal" portfolio. Such population of the agents is heterogeneous due to their initial cash available, reoptimization (trading) frequency and risk aversion.

We introduce environment-based interactions, where market restricts agents behaviour and at the same time it evolves in response to agents activities. Thus, the environment has its own state and rules of changes [23]. Traders submit orders depending on the state of the order book or best quotes (environment). These orders may result in a change in the best prices. The state of the market changes over time. A feedback loop is formed: a trader submits orders which affect the state of the market which affects the decisions of the trader on what order to submit. This aspect relies to *Interaction Movement Computation* (MIC^*) [24], where the environment defines actions sets of autonomous agents to achieve their goals. Agents interact with one another in order to achieve either a common or individual objectives through environment. The agents can also interact through the common variable of the past price history, but they are not directly affected by the actions of others. In order to keep agents equality and to avoid the biases in the internal information access, agents should be informed about book order changes simultaneously. Notification method is realized in ATOM in accordance with Influence Reaction Model for Simulation (IRM4S)[25], [26]: all orders, as influences, are collected in the order book, once, all agents have sent their orders, price is fixed as reaction. This method is used only for synchronous trading mechanism.

5 Validation Tests

As mentioned previously (see section 2), every ASM should succeed in processing perfectly a given order flow collected from a real-world stock market at a specific date. The result is obtained confronting prices delivered by the market at this date and the prices generated by the ASM using the same set of orders. It should also generate relevant "stylized facts" with regard to their real-world counterpart: these stylized facts are statistical characteristics of financial time series that prove to be systematically observed in various contexts (different assets, periods of time, countries).

This section presents how ATOM fulfill this requirements, moreover, performance tests are considered.

5.1 Performance Test

We ran several experiments to demonstrate running time for realistic price series generation and existing order-flow execution.

To demonstrate ATOM price fixing ability, we use a group of heterogeneous agents. The population consists of Zero Intelligence Traders (ZIT) and Technical Simple Moving Average Traders (in the equal proportions), described in the section 4. Number of fixing prices is 10^{52}. Number of agents varies from 10 to 10^5. Results are introduced in the Figure 4(a). It takes about 12 minutes to run 10^5 agents for price fixing.

To test running time of replaying engine, we use real market order flow. The same agents population is used to read all variety of orders (limit, market, stop-limit, iceberg, etc.) and send them to order book. It is up to the market to fix price in a proper way (according to a fixing protocol). Number of orders vary from 100 to 10^5. It takes 2 minutes to replay 10^5 orders (see figure 4(b)).

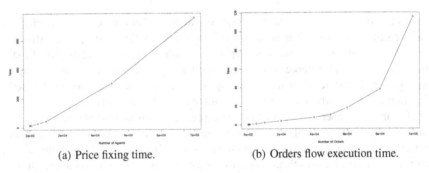

(a) Price fixing time.   (b) Orders flow execution time.

Fig. 4. Results of the performance testing

---

[2] On the Euronext Stock Exchange the number of fixed prices for different stocks varies from 1000 to 5000 per day.

5.2 ATOM Reality-Check

In this section, we report a series of tests conducted to check whether ATOM can generate financial dynamics in line with the ones of the Euronext-NYSE stock-exchange or not. The first series of test is devoted to the ability of ATOM at generating unbiased prices when it deals with a real-world order-flow.

Figure 5(a) and Figure 5(b) reports results of the first reality-check (top Figures report results produced with the ATOM data, bottom Figures being those based on Euronext-NYSE data). We ran ATOM with a Hollow Agent reading the entire set of 83616 orders concerning the French blue-chip France-Telecom (FTE) recorded on June 26th 2008 between H9.02'.14".813''' and H17.24'.59".917'''. As mentioned previously, handling time in simulations is particularly complex and may lead to unsolvable dilemma. We cannot guarantee an exact matching of waiting times but rather a coherent distribution of these values delivered by the simulator engine with regard to the observed waiting times.

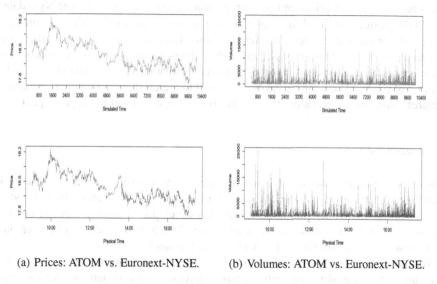

(a) Prices: ATOM vs. Euronext-NYSE.   (b) Volumes: ATOM vs. Euronext-NYSE.

Fig. 5. Results of the "Reality Check" procedure

Notice that ATOM performs rather decently in satisfying the first reality check procedure.

5.3 Stylized Facts

The second subset of tests focuses on ATOM ability to generate realistic artificial prices when populated with artificial agents. We ran a series of simulations to verify if ATOM can generate major stylized facts that are usually reported in the literature (see for example [27]). For the sake of simplicity and space-saving, we only report in a pictorial form of the classical departure from Normality of asset returns at the intraday level (Figures

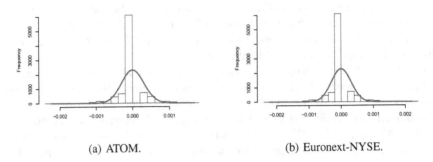

(a) ATOM.   (b) Euronext-NYSE.

Fig. 6. Stylized fact, departure from Normality

6(a) and 6(b)). Notice again that these statistics are reported on the left-hand Figures when based on ATOM prices and on the right hand when based on Euronext-Nyse data.

ATOM produces stylized facts, quantitative as well as qualitative, that is quite difficult task for most of artificial market platforms. Even if some artificial markets are able to reproduce the main stylized facts such as the non Gaussian return distribution or volatility clustering, the corresponding quantitative characteristics (basic statistics) do not fit real ones. ATOM can be easily calibrated to match specific quantitative market features (moments). This calibration facility is described in detail in the paper [28].

6   Conclusions

The recent financial crisis has stressed the need for new research tools that can deal with the high level of complexity of the economic world. Agent based methods propose a powerful alternative to traditional approaches developed in finance. Among others, Artificial Stock Markets offer a completely controlled environment to test new regulations, new exchange structures or new investment strategies.

We showed that building a realistic artificial stock market platform can be efficiently done using the main MAS concepts: agents' behaviours, environment etc. We also discussed a series of software engineering and architecture design issues to implement such systems. We illustrate these points with the ATOM application programming interface (API). The latter can be used to provide a polymorphic platform for a wide range of large scale experiments, including or not artificial agents, sophisticated behaviours, communication over the network...

Along this article, we have tried to show how Finance can benefit from Agent-Based Modelling in tackling complex phenomena emerging in the market. For example, a wide range of heterogeneity in agents' behaviours can be settled, that opens new perspectives in this field where a *representative agent* is traditionally used. Nevertheless, our main point is to show how the power of Agent-Based modelling can be challenged with this field of Economics that necessitates to mobilize the most advanced techniques of the domain.

References

1. Woolridge, M.: Introduction to Multiagent Systems, New York, NY, USA (2002)
2. Mitchell, T.M. (ed.): Machine Learning. WCB/McGraw-Hill, New York (1997)
3. Cesa-Bianchi, N., Lugosi, G.: Worst-case bounds for the logarithmic loss of predictors. Machine Learning 43, 247–264 (2001)
4. Witkam, J.: Altreva adaptive modeler (2003)
5. Jacobs, B.I., Levy, K.N., Markowitz, H.M.: Financial market simulation. The Journal of Portfolio Management 30th Anniversary Issue, 142–151 (2004)
6. Inchiosa, M.E., Parker, M.T.: Overcoming design and development challenges in agent-based modeling using ascape (3), 7304–7308 (2002)
7. Brandouy, O., Mathieu, P.: A conceptual framework for the evaluation of agent-base trading and technical analysis. In: Artificial Markets Modeling, Methods and Applications. Lecture Notes in Economics and Mathematical Systems, vol. 599, pp. 63–79 (2007)
8. Chen, S.H., Yen, C.H., Liao, C.C.: On aie-asm: a software to simulate artificial stock markets with genetic programming. In: Chen, S.-H. (ed.) Evolutionary Computation in Economics and Finance, pp. 107–122. Physica-Verlag (2002)
9. Ponta, L., Raberto, M., Cincotti, S.: A multi-assets artificial stock market with zero-intelligence traders. A Letter Journal Exploring the Frontiers of Physics 93, 28002–p1–28002–p7 (2011)
10. Raberto, M., Cincotti, S., Dose, C., Focardi, S.M.: Price formation in an artificial market: limit order book versus matching of supply and demand. Nonlinear Dynamics and Heterogenous Interacting Agents (2005)
11. Ricordel, P.-M., Demazeau, Y.: Volcano, A Vowels-Oriented Multi-Agent Platform. In: Dunin-Keplicz, B., Nawarecki, E. (eds.) CEEMAS 2001. LNCS (LNAI), vol. 2296, pp. 253–262. Springer, Heidelberg (2002)
12. LeBaron, B.: Evolution and time horizons in an agent based stock market. Macroeconomic Dynamics 5, 225–254 (2001)
13. Muchnik, L., Solomon, S.: Markov nets and the natlab platform; application to continuous double auction. New Economic Windows (2006)
14. Raberto, M., Cincotti, S., Focardi, S., Marchesi, M.: Traders' long-run wealth in an artificial financial market. Computational Economics 22, 255–272 (2003)
15. Le Baron, B.: Building the santa fe artificial stock market. Working Paper, Brandeis University (2002)
16. Garder, M.: Mathematical games: The fantastic combinations of john conway's new solitaire game 'life'. Scientific American (October 1970)
17. Wooldridge, M., Jennings, N.: Intelligent agents: Theory and practice. The Knowledge Engineering Review 10(2), 115–152 (1995)
18. Gode, D.K., Sunder, S.: Allocative efficiency of market with zero-intelligence traders: Market as a partial substitute for individual rationality. Journal of Political Economy 101(1), 119–137 (1993)
19. Arthur, B.: Inductive reasoning and bounded rationality: the el-farol problem. American Economic Review 84, 406–417 (1994)
20. Tkatch, I., Alam, Z.S.: Strategic order splitting in automated markets. SSRN eLibrary (2009)
21. Brandouy, O., Mathieu, P., Veryzhenko, I.: Ex-post optimal strategy for the trading of a single financial asset. SSRN eLibrary (2009)
22. Markowitz, H.: Portfolio selection. The Journal of Finance 7(1), 77–91 (1952)
23. Weyns, D., Omicini, A., Odell, J.: Environment as a first-class abstraction in multiagent systems. JAAMAS 14(1), 5–30 (2007)

24. Gouaïch, A., Michel, F., Guiraud, Y.: MIC*: A Deployment Environment for Autonomous Agents. In: Weyns, D., Van Dyke Parunak, H., Michel, F. (eds.) E4MAS 2004. LNCS (LNAI), vol. 3374, pp. 109–126. Springer, Heidelberg (2005)
25. Michel, F.: The irm4s model: The influence/reaction principle for multi-agent based simulation. In: AAMAS 2007. Sixth International Joint Conference on Autonomous Agents and Multiagent Systems, pp. 908–910 (2007)
26. Ferber, J., Muller, J.P.: Influences and reaction: a model of situated multiagent systems. In: Second International Conference on Multiagent Systems, ICMAS 1996, pp. 72–79 (1996)
27. Cont, R.: Empirical properties of asset returns: stylized facts and statistical issues. Quantitative Finance 1, 223–236 (2001)
28. Veryzhenko, I., Brandouy, O., Mathieu, P.: Agent's minimal intelligence calibration for realistic market dynamics. In: Progress in Artificial Economics Computational and Agent-Based Models. Lecture Notes in Economics and Mathematical Systems, vol. 645, pp. 3–14 (2010)

Defining Virtual Organizations Following a Formal Approach

Sergio Esparcia and Estefanía Argente

Grupo de Tecnología Informática - Inteligencia Artificial
Departamento de Sistemas Informáticos y Computación
Universitat Politècnica de València, Camino de Vera, s/n - 46022, Valencia, Spain
{sesparcia, eargente}@dsic.upv.es
http://users.dsic.upv.es/grupos/ia/ia.html

Abstract. One of the main approaches to define a Virtual Organization, which is a concrete type of Organization Centered Multi-Agent System, is to follow a formal approach. Using a formalization involves introducing mathematical and logical concepts. In this paper, we present a formalization for Virtual Organizations, named Virtual Organization Formalization, which allows defining this kind of systems organized by means of the Organizational Dimensions, extracted after studying the human organization theory, where entities are distributed by the functionality they are providing to the Virtual Organization. Additionally, this work studies different proposals that are aimed to formally define an Organization Centered Multi-Agent System, comparing them with our own proposal.

Keywords: Virtual organizations, Formalization, Multi-agent systems.

1   Introduction

During the last years, different trends to develop Multi-Agent Systems (MAS) have been appeared. One of the most important approaches is the organizational approach. Organizations describe system functionality, structure, environment and dynamics. In Organization Centered MAS (OCMAS), the organization exists as an explicit entity of the system [1], defined by its designers following a top-down approach. In OCMAS, agents are aware of the organization in which they are participating and they are provided with a representation of it. Agents can use this knowledge to reason about it and to establish relationships and interactions to reach their objectives.

OCMAS can be defined and described by means of a formal approach. To formalize them it is necessary to introduce concepts taken from mathematical and logic theories, such as LAO [2], whose syntax to define a system follows the temporal logic language CTL [3]. Other proposals not only provide a formal way to define an OCMAS, but also a language to describe it, such as the proposal by Grossi *et al.* [4], which employs a multimodal propositional logic language to model agent organizations, based on Kripke models. Formal approaches are very useful in order to obtain a clear definition of OCMAS, improving the study and analysis of the different issues regarding them. Additionally, these formalizations are commonly used to check the correctness and integrity of an OCMAS, by means of techniques like model checking [5].

J. Filipe and A. Fred (Eds.): ICAART 2011, CCIS 271, pp. 365–381, 2013.
© Springer-Verlag Berlin Heidelberg 2013

However, formal approaches are not always able to represent all concepts that compose an agent organization. Using the current proposals it is not possible to completely define a paradigm for developing agent systems such as Virtual Organizations (VO) [6], which are sets of individuals and institutions that need to coordinate resources and services. Thus, they are open systems [7] formed by the grouping and collaboration of heterogeneous entities, and allowing model systems at a high level of abstraction. They include the integration of organizational and individual perspectives and also the dynamic adaptation of models to organizational and environmental changes.

OCMAS can be structured by splitting them in different dimensions [8]. Concretely, VOs can be structured by means of the Organizational Dimensions [9], which should be considered when modeling an organization. These dimensions describe all the entities that compose the organization, distributed by the functionality of the entities that they are providing to it. These dimensions are: *structural*, *functional*, *dynamical*, *environment* and *normative*. Current formal proposals only define a subset of the dimensions and concepts presented in the Organizational Dimensions. Thus, it seems necessary to be provided with a formalization that clearly models the Organizational Dimensions, making a clear difference between them.

The objective of this work is to present a formal framework to define a VO, taking the Organizational Dimensions as a basis. Using this representation, we are able to verify the correctness and completeness of the defined VOs, by means of techniques based on logical and mathematical approaches like model checking. Additionally, this formalization will be useful when dealing with self-adaptive and self-organization concepts, since it will be established how the system changes through time. The rest of this work is structured as follows: Section 2 describes the Organizational Dimensions. Section 3 describes formal frameworks related with our work. Section 4 presents the *Virtual Organization Formalization* (VOF), a formal framework to define VOs. Section 5 presents a discussion between our proposal and the analyzed frameworks. Finally, section 6 gives our conclusions and future work on this work.

2 Organizational Dimensions

When modeling an organization, the following dimensions should be taken into account [9]: (i) *structural*, describing the entities that structure the system; (ii) *functional*, which details the functions, goals and services of the organization; (iii) *dynamical*, which considers the interactions between elements, and their effects; (iv) *environment*, describing the elements that surround the system; and (v) *normative*, which defines the mechanisms used by the society to influence the behavior of its members.

The **Structural Dimension** comprises all the elements of the organization that are independent from the agents that are part of it. Thus, it is based on roles, groups and their patterns of interrelationship (inheritance, compatibility, communication, and so on). Additionally, the topology of the system is established.

The **Functional Dimension** specifies the global goals of the organization, its offered functions and services, the goals followed by different components of the organization and the tasks and plans that must be executed to reach these goals.

The **Dynamical Dimension** specifies how the organization evolves through time, detailing the way in which agents enter and leave it, how they adopt certain roles according to their capabilities and abilities, and how they can participate in the units or groups of the organization where they are admitted. This dimension also details the interactions that take place between internal and external entities.

The **Environment Dimension** describes how agents are connected with other types of entities such as artifacts, applications or resources; and how agents can perceive and act on the environment.

Finally, the **Normative Dimension** determines the set of defined actions and rules to manage the behavior of the members of the organization. Norms are widely used to limit the autonomy inside societies and to solve coordination problems, especially when it is not possible to exercise a total social control.

3   Related Work

Based on different logics and formal methods, some proposals to model OCMAS have been defined, each giving its particular vision and adapting its formalization to the specific kind of system that they are looking to build. In this section, a set of relevant proposals on this field has been reviewed: OperA [10], LAO [2], Process-Oriented Modeling Framework (POMF) [11], MOISE^{Inst} [12], MACODO [13], PopOrg [14] and the proposals by Grossi *et al.* [4] and Jonker *et al.* [15]. All these proposals are analyzed following the Organizational Dimensions described in section 2, in order to check whether they are taking into account the entities and concepts from each dimension.

Table 1 compares these proposals, analyzing the organizational elements that they take into account. Next, we will depict in detail the contents of this table, describing each studied proposal.

OperA proposes an Organizational Model to describe organizations that defines the social, normative, interaction and communicative structures of the society. The *Social Structure* of OperA is related to the Structural Dimension, since it contains roles, groups and dependency relations between roles. Also, its Social Structure is related to the Functional Dimension since it takes into account the objectives associated with roles. The *Normative Structure* is obviously related to the Normative Dimension, as both consider norms. The *Interaction Structure* models the activity of the system, which is considered as the dynamics taken from the Dynamical Dimension. Finally, the *Communicative Structure* manages communication between agents, like interactions in the Dynamical Dimension. Nevertheless, OperA does not model the environment.

MOISE^{Inst} is composed of four specifications, distributed in a similar way that the Organizational Dimensions are. The *Structural Specification* (SS) defines the roles that agents will play, including the relations between them, and an additional structural level named group, where roles belong to and interactions are carried out. The SS contains elements from the Structural Dimension, but it does not model the topology of the system. The *Functional Specification* (FS), related to the Functional Dimension, only defines the goals that the system must achieve. The *Contextual Specification* (CS) defines the different contexts that influence the organizational dynamics and the transitions between

Table 1. Comparison between different formal representations

	OperA	MOISE	PopOrg	LAO	POMF	MACODO	Grossi	Jonker
Organizational concepts								
Structural Dimension								
Roles	✓	✓	✓	✓	✓	✓	✓	✓
Groups	✓	✓		✓		✓		
Agents		✓	✓	✓	✓	✓	✓	✓
Relations	✓	✓	✓	✓			✓	✓
Topology	✓			✓				
Functional Dimension								
Capabilities			✓	✓	✓	✓		
Abilities		✓	✓					
Services			✓		✓			
Objectives	✓	✓		✓	✓		✓	
Dynamical Dimension								
Interactions	✓	✓	✓				✓	✓
Dynamics	✓	✓	✓	✓		✓		✓
Environment Dimension								
Environment		✓		✓		✓		✓
Resources					✓			
Normative Dimension								
Norms	✓	✓				✓	✓	
Syntax				CTL	L_{PR}	Z	Org	TTL
Semantics	LCR			CTL*	T_{PR}		Org	

them. This specification defines the environment, taken from the Environment Dimension, and its dynamics, just like the Dynamical Dimension does. The CS does not model the resources populating the environment or the interactions between agents. Finally, the *N ormative Specification* (NS) defines the rights and duties of roles and groups inside the organization, which are known as norms in the Normative Dimension. All agents that have adopted a role from the SS compose the *Organizational Entity* (OE), which is the element of the system that controls the dynamic elements of the organization, including agents and all events that they generate, such as their interactions.

In order to manage the structural dynamics of a MAS, PopOrg is a model based on two basic concepts: the *population* of an organization and its *structure*. The population of a MAS is its set of agents, as well as the behaviors and actions, which represent the capabilities and abilities from the Functional Dimension; and the exchange processes (services from the Functional Dimension) that agents are able to carry out. Therefore, the population of a PopOrg organization mainly takes concepts from the Functional Dimension plus agents from the Structural Dimension. Moreover, the structure of the organization is composed of roles and the links between them, which are elements that belong to the Structural Dimension. To relate the population and the structure, PopOrg has a third element called *implementation* that relates roles with agents, and links with exchange processes. Also, PopOrg stores the different states that the system goes through during its execution. Unfortunately, PopOrg does not model any of the entities from the Environment or Normative Dimensions.

The *logic for agent organizations* (LAO) is an extension of CTL logic. The Functional Dimension is completely represented in LAO, including agents, objectives, groups, topology (establishing links between agents), and roles, which are represented by means of capabilities and abilities, elements taken from this dimension. LAO additionally defines different states of the world where the system is located (related to the Environment Dimension) and its transitions (related to the Dynamical Dimension). LAO is a very complete proposal, since it takes into account a large subset of the elements of the Organizational Dimensions, but it does not formalize the Normative Dimension.

The *Process Oriented Modeling Framework* (POMF) is structured by means of four views. The main one is the *process oriented view*, where tasks, processes and workflows are defined. This view includes the concept of service from the Functional Dimension, being a workflow divided into processes that are split into tasks. It also includes the resources of the Environment Dimension. The *organization oriented view* includes the role entity, which describes the set of capabilities of the organizational processes in a concrete workflow that are then assigned to agent entities, defined in the *agent view*, where groups of agents are not able to be modeled. Therefore, the organization oriented view is related with both Structural and Functional Dimensions and the agent view is related to the Structural Dimension. Finally, the *performance oriented view* describes the organizational goals, such as the Functional Dimension does. However, POMF does not provide a formalization for neither Dynamical nor Normative Dimensions.

The MACODO framework is centered on the dynamics of organizations with self-organization concepts, offering a model with concepts related to the Functional Dimension, such as roles (establishing the concept of position, similar to a job offer), contracts of roles (an agreement between an agent and an organization for a concrete position to control the access to an available role), agents (including their context and local environment, which is related to the Environment Dimension), and organizations (groups of agents defined by a set of open role positions and current role contracts). Relations of hierarchy and communication between roles are not considered. A role is described as a set of capabilities, which is the only entity from the Functional Dimension that MACODO takes into account. Since MACODO is focused on self-organization, dynamics of the system from the Dynamical Dimension, including changes in its context or in its set of agents, are formalized. To control the activities that the organization carries out, MACODO is enhanced with a set of laws, similar to norms from the Normative Dimension. Although MACODO does not model other relevant organizational concepts such as objectives, it deals with elements from all the Organizational Dimensions.

The organizational formalization proposed by Grossi *et al.* pursues to represent the organizational structure. This formal method takes the concepts of *role*, establishing relations between them; and *agent* from the Structural Dimension. The roles of the organization are conceived around three basic notions: objectives, norms and information. Objectives are the only elements related to the Functional Dimension that are presented in this proposal; and the Normative Dimension is taken into account using norms. Regarding information, knowledge about the current state of the organization can be given to agents by other agents, so this is a type of interaction from the Dynamical Dimension. Since this proposal is focused on modeling the organization structure, it does not take into account the Environment Dimension.

Finally, Jonker *et al.* defined a framework for modeling and providing a formal analysis of organizations, based on a generic representation of them by means of a set of roles. Apart from the roles, this proposal also formalizes two concepts from the Structural Dimension: agents and relations between roles. These relations enable the interactions from the Dynamical Dimension, which is completed by taking into account the dynamics of the organization. One of the main advantages of this work is that it is able to explicitly model the environment of the Environment Dimension, although it does not model environmental resources. The main lack of this formalization is that it does not formalize any concept from the Functional and Normative Dimensions, so designers are not able to model concepts such as objectives and norms.

Generally, all the analyzed formalizations present a good approach to define an organization in a formal way. Nevertheless, none of the proposals take into account all entities taken from Organizational Dimensions, but these formalizations propose different ways to structure an OCMAS. Thus, it seems interesting to be provided with an explicit description of the Organizational Dimensions, which are useful for representing organizational elements. Therefore, in section 4 our proposal to model organizations will be presented, which models an organization clearly defining its dimensions. This proposal also integrates some features taken from some proposals presented in this section.

4   Formal Description of a Virtual Organization

In this section the concept of Virtual Organization (VO) will be defined in a formal way, taking into account its organizational dimensions. This formalization, named VOF (*Virtual Organization Formalization*), will be focused on three elements: (i) the Organizational Specification (*OS*), which details the set of elements that specify the organization; (ii) the *Organizational Entity* (*OE*), which represents the instantiation of the elements in *OS*; and (iii) the *Organizational Dynamics* (ϕ), which relates elements from *OS* with elements from *OE*.

Definition 1. *A Virtual Organization is defined, at a given time t, as a tuple $O^t = \langle OS^t, OE^t, \phi^t \rangle$ where:*

- *OS refers to the Organizational Specification. It is defined as $OS = \langle SD, FD, ED, ND \rangle$ where:*
  - *SD is the Structural Dimension.*
  - *FD is the Functional Dimension.*
  - *ED is the Environment Dimension.*
  - *ND is the Normative Dimension.*
- *OE refers to the Organizational Entity, which represents the dynamic elements of the system.*
- *ϕ allows to relate OS with OE, thus defining the Dynamic Dimension, together with the OE.*

The VO will change through time by modifying its states, occurred after a change in the environment, and it will change from one state to another by means of a transition. The following subsections will describe in detail these three elements.

4.1 Organizational Specification

The Organizational Specification details the set of elements that specify the organization, containing organizational units, roles, norms, and the rest of elements that build the dimensions of a Virtual Organization. The Organizational Structure is composed of: (i) the *Structural Dimension*, which contains roles, organizational units and their relationships; (ii) the *Functional Dimension*, describing objectives, functionalities and services of an organization; (iii) the *Environment Dimension*, which describes the artifacts and workspaces from the environment of the organization; and (iv) the *Normative Dimension*, which defines the norms that rule the system.

Structural Dimension. The *Structural Dimension* describes the components of the system and their relations. It allows defining the structural components of an organization, i.e. all the elements that are independent from the entities that are finally executed. In a more specific way, it defines the organizational units and the structural elements, roles and relationships between roles.

Definition 2. *The Structural Dimension (SD) of a Virtual Organization is defined as* $SD = \langle R, OU, Relations \rangle$ *where:*

- *R refers to the roles of the organization.*
- *OU is the set of organizational units.*
- *Relations is a set of relationships, defined as Relations = $\langle SocialRelations, StructRelations, DimRelations \rangle$ where:*
    - **SocialRelations** *refers to the social relationships between roles, which can be formalized as:*

$$SocialRelations = \begin{cases} inf: & R \to R \\ col: & R \to R \\ sup: & R \to R \\ comp: & R \to R \end{cases}$$

    where: **inf** *(information) refers to the information relation, which allows communication between roles;* **col** *(collaboration) allows a role to monitor the activities of other roles;* **sup** *(supervision) defines that an agent playing a specific role can transfer or delegate one or some of his objectives to a subordinate role; and* **comp** *(compatibility) depicts that an agent playing a specific role can also play another compatible role in the organization at the same time.*
    - **StructRelations** *refers to the structural relationships defined by the structure of the organization, which can be formalized as:*

$$StructRelations = \begin{cases} RoleHier: & R \to R \\ Contains: & OU \to 2^{OU} \\ Roles: & OU \to 2^{R} \end{cases}$$

where: **RoleHier** represents the hierarchy between roles of the organization; **Contains** defines the topology of the organization by means of relations between organizational units; and **Roles** defines the roles that are located inside an organizational unit.

- **DimRelations** allows relating this dimension with others, through the element OU, and can be formalized as:

$$DimRelations = \begin{cases} Norms: & OU \to 2^N \\ Services: & OU \to 2^S \\ Goals: & OU \to 2^G \\ Workspaces: & OU \to 2^{WS} \end{cases}$$

where: **Norms** defines the norms, described in the normative dimension, which rule an OU; **Services** relates an OU with the services that it contains; **Goals** describe the objectives that are necessary to be reached inside an OU; and **Workspaces** details the workspaces (see Definition 4) where an OU can be located.

Properties of the Relations. The social relation *inf* is symmetrical, since a role can provide information to a second role, and viceversa; transitive, since agents can build an information chain, and reflexive as an agent can send information to himself. The relations *col* and *sup* are both asymmetrical, since an agent cannot monitor or supervise the agent which is monitoring or supervising him; reflexive, because an agent can collaborate or supervise himself; and transitive, allowing to create a command chain inside the organization.

The compatibility relation (*comp*) has reflexive and transitive properties, because a role is compatible with itself and a role is compatible with the roles that have a compatibility relation with its compatible roles. It is interesting to notice that the *comp* relation is not symmetrical (e.g. $comp(r_1, r_2)$ not always implies $comp(r_2, r_1)$). For example, the relation $comp(Professor, Teacher)$ is correct, because a professor can work as a teacher in every moment, but a teacher might not be capable of playing the role of professor. Finally, relations *RoleHier* and *comp* are related, since an agent playing a specialized role is capable of playing its generalized role. Formally:

$$\forall r_1, r_2 \in R : RoleHier(r_1, r_2) \to comp(r_2, r_1) \tag{1}$$

Let $r_1, r_2 \in R$ be two roles belonging to OS. The information, collaboration and supervision relations define the following relations in an implicit way:

$$sup(r_1, r_2) \to col(r_2, r_1) \tag{2}$$

$$col(r_1, r_2) \to inf(r_1, r_2) \land comp(r_2, r_1) \tag{3}$$

This means that a supervision relation between two agents implies that a supervised agent will collaborate with a supervisor agent to help him to reach his objectives. Also, a collaboration relation between two roles implies that an information link between them exists and the second role of the relation is compatible with the first one.

The relation *Contains* from the *StructRelations* set has the following properties: (i) asymmetrical, since an OU cannot be contained in another OU that contains it; (ii) transitive, because it is considered that an OU contained inside another OU is also contained inside the predecessors of the OU that contains it; and (iii) irreflexive, since an OU cannot contain itself. In a similar way, the *RoleHier* relation has the same properties of the *Contains* relation, because a role cannot have an inheritance relation with itself, the relations between roles are transitive to allow defining a complete role hierarchy and a subordinated role cannot be the supervisor of its supervisor.

Properties of the Entities. Firstly, an organizational unit is contained inside another OU, this implies that the roles from this OU are compatible with those of its predecessor OU. Formally:

$$\forall OU_1, OU_2 \in OU : Contains(OU_1, OU_2) | \forall r_1 \in \tag{4}$$

$$Roles(OU_1) \land \forall r_2 \in Roles(OU_2) \to comp(r_2, r_1)$$

It should be noted that the *Roles* relation is recursive: the roles that an OU offers are not only its own roles, but also those from its predecessor OUs. Formally:

$$\forall o \in OU : \forall r \in Roles(o) \to r \in Roles(o) \lor r \in Roles(o_1) : o \in Contains(o_1) \tag{5}$$

Properties of the OU. The relations between organizational units allow defining three different types of structures of an organization:

- *'hierarchy'*. A hierarchy implies that there is a supervisor role, with supervision relations to all the other members of its same organizational unit (OU). Formally, $\exists r \in Roles(OU) : \forall r_i \neq r \in Roles(OU) \to sup(r, r)$. If a designer wants to make his system tighter, he can also prohibit communications between subordinated roles.
- *'team'*. In this kind of structure, all roles have coordination relations between them. Formally, it is defined as $\forall r_1, r_2 \in Roles(OU) : col(r_1, r_2)$.
- *'plain'*. This structure establishes information relationships between roles. Formally, $\exists r_1, r_2 \in Roles(OU) : inf(r_1, r_2)$.

Functional Dimension. The *Functional Dimension* details the specific functionality of the system, based on services, tasks and objectives, as well as the interactions of the system, activated by means of objectives or service usage. It allows defining the functionality of organizational units, roles and agents of the MAS, including services and objectives that these entities offer or consume.

Definition 3. *The Functional Dimension (FD) from the Organizational Structure of a Virtual Organization is defined as* $FD = \langle G, S, Ta, FuncRel \rangle$ *where:*

- *G represents the goals followed by the organization.*
- *S is the set of services that the system offers or requires.*
- *Ta are the tasks that compose the services.*
- *FuncRel = ⟨GT, Client, Provider, Obtains, Achieves, Task, Invoke, Plan⟩ is the set of relations of this dimension, where:*

- $GT : G \rightarrow 2^G$ is the *Goal Tree* of the organization, describing the dependencies between different goals of the organization.
- *Client* : $S \rightarrow 2^R$ relates a service with the set of roles that use it.
- *Provider* : $S \rightarrow 2^R$ relates a services with the set of roles that offer it.
- *Obtains* : $S \rightarrow 2^G$ describes the set of goals that can be achieved by a service, thus defining the functionality of the system.
- *Achieves* : $Ta \rightarrow 2^G$ defines the set of goals that are reached when a task is executed.
- *Task* : $S \rightarrow 2^{Ta}$ shows how services are split in different tasks.
- *Invoke* : $S \rightarrow 2^S$ describes the dependencies between services, showing which services need to be invoked by other services to complete their functionality, thus allowing the composition of services.
- *Plan* : $G \rightarrow 2^S$ represents the sequence of services that must be followed in order to achieve a goal.

Properties of the Relations. The Goal Tree relation is irreflexive, asymmetrical and transitive, since a goal cannot be related with itself, neither with its predecessor but it can be related with the successors of its successors.

It must be assured that the provider of a service must be a role contained in the same OU as the service. Formally:

$$\forall o \in OU \wedge \forall s \in Services(o) \rightarrow Provider(s) \subseteq Roles(o) \tag{6}$$

This restriction assures that the services of an OU will be provided only inside it, but they can be accessed by agents from other OUs (e.g. using the *Invoke* relation).

As pointed out in this section, the *Invoke* relation allows services to invoke other services to reach their goals. In order to execute this operation, it must be assured that the provider of the invoker service must be a client of the invoked service. Formally:

$$\forall s_1, s_2 \in S, s_2 \in Invokes(s_1) : \forall r_1 \in Provider(s_1) \rightarrow \exists r_2 \in Client(s_2) \wedge r_1 = r_2 \tag{7}$$

A key issue for the system designer is to assure that the services located in an organizational unit must help to reach its goals. Formally, it is described as:

$$\forall o \in OU; \forall s \in Services(o) \wedge \forall g_1 \in Pursues(s) : \exists g_2 \in Goals(o) \rightarrow g_2 \in GT(g_1) \tag{8}$$

It is is possible that a specific goal of a service could not be reached by any of the tasks that compose it, so then this service must invoke another which should include at least a task that achieves this desired goal. Formally, it is expressed as:

$$\forall g \in Obtains(s_1) \rightarrow (\exists t \in Task(s_1) \wedge g \in Achieves(t))$$
$$\vee(\exists s_2 \in S \wedge g \in Obtains(s_2) \wedge s_2 \in Invokes(s_1)) \tag{9}$$

Environment Dimension. The Environment Dimension describes the artifacts, i.e. entities that populate the environment of a MAS. This dimension uses the concept of artifact [16], an element introduced by the Agents & Artifacts (A&A) conceptual framework. These elements are employed by agents in order to reach their goals, since artifacts have no associated goals. Additionally, the A&A framework presents the concept of workspace, used to define the topology of the environment of a MAS.

Definition 4. *The Environment Dimension of a Virtual Organization is defined as $ED = \langle WS, AR, EnvFunc \rangle$ where:*

- *WS is the set of workspaces that build the environment of a MAS, where $ws \in WS$ is defined as $ws = \langle Loc \rangle$ and Loc is referred to the location of the workspace inside the environment.*
- *AR is the set of artifacts, where an artifact $ar \in AR$ is defined as $ar = \langle PR, OP, LO, St \rangle$, where:*
  - *PR are the observable properties of an artifact that agents can check without executing any operation on it.*
  - *OP is the set of operations that agents can execute when interacting with the artifact.*
  - *LO refers to the link operations, which allows the composition and distribution of artifacts.*
  - *St is the internal state of an artifact.*
- *$EnvFunc = \langle Located, Composition \rangle$ is the set of functions that act on the environmental elements, where:*
  - *$Located : AR \to 2^{WS}$ describes the set of workspaces where an artifact is located.*
  - *$Composition : WS \to 2^{WS}$ allows defining intersection and nesting relations between workspaces that build the environment.*

Properties of the Relations. The *Composition* relation is reflexive and symmetrical, since a workspace can intersect with itself. Additionally, it is necessary for an artifact to be contained, at least, in a workspace. Formally:

$$\forall ar \in Ar : \exists ws \in WS \to ws \subseteq Located(ar) \tag{10}$$

Normative Dimension. The Normative Dimension describes normative restrictions on the behavior of the entities of the system, including sanctions and rewards, based on the work by Criado *et al.* [17].

Definition 5. *The Normative Dimension of a Virtual Organization is defined as $ND = \langle N, >_n \rangle$ where:*

- *N is the set of norms of the system.*
- *$>_n$ is an order relationship between norms, defining the priority between them. This relation establishes a total relation order between the norms governing the system, avoiding the priority confusion when a norm is executed.*

Formally, a norm is defined as:

Definition 6. *A norm* $n \in N$ *is defined as* $n = \langle D, CO, AC, EX, SA, RE \rangle$ *where:*

- $D = \{O, F\}$ *is the deontic operator, i.e. obligations (O) and prohibitions (F) that impose restrictions on the behavior of the agents.*
- *CO is a logical formula that represents the action that must be carried out in case of obligations, or has to be avoided in case of prohibitions.*
- *AC, EX are well-formed formulas that determine the conditions of norm activation and expiration, respectively.*
- *SA, RE* $\in S$ *are expressions that describe the actions (sanctions, SA; and rewards, RE) that will be carried out in case of violation or fulfilment of norms, respectively.*

Properties of the Relations. The priority function $>_n$ is asymmetrical and transitive, defining an univocal relation between the norms governing the system.

The topology of the system will also define new order relationships between norms. If an OU called ou_2 is contained in an OU named ou_1, its norms must have higher priority than the norms of ou_1. Formally,

$$\forall ou_1, ou_2 : n_1 \in Norms(ou_1) \wedge n_2 \in Norms(ou_2) \wedge Contains(ou_1, ou_2) \rightarrow n_2 >_n n_1 \tag{11}$$

4.2 Organizational Entity

The Organizational Entity of a Virtual Organization is the set of active elements of the organization. These elements can change through time. They are considered as the dynamic elements of the system.

Definition 7. *The Organizational Entity of a Virtual Organization is defined as* $OE = \langle A, GR, AN, AS \rangle$ *where:*

- *A is the set of agents that populate the VO.*
- *GR is the set of groups that are currently in the system. A group is an instantiation of an organizational unit.*
- *AN* $\subseteq N$ *is the set of active norms of the system, i.e. all those norms whose activation condition is true but its expiration condition has not been reached yet* ($AC \wedge \neg EX$).
- *AS* $\subseteq S$ *is the set of services that the agents of the organization are currently providing.*

The agents (A) populating the system are playing roles, they are located into groups (GR) and provide services (S), as described in the next subsection. An OU defines an organizational pattern for the agents that are inside it, but this does not define the concrete agents that must populate it. On the contrary, a group is a concrete instantiation of an OU, defining a set of agents that populate it. Thus, an OU can be instantiated by different groups.

4.3   Organizational Dynamics

The Organizational Dynamics presents the relations between the elements of the Organizational Structure and the Organizational Entity.

Definition 8. *The Organizational Dynamics of a Virtual Organization is defined as* $\phi = \langle plays, inUnit, provides, perceives, isUnit \rangle$ *where:*

- *plays* : $A \rightarrow 2^R$ *is a function that relates an agent with the set of roles that he is playing inside the organization.*
- *inUnit* : $A \rightarrow 2^{GR}$ *is the function that describes the groups where an agent is located.*
- *provides* : $A \rightarrow 2^S$ *represents the set of services that an agent provides.*
- *perceives* : $A \rightarrow 2^{WS}$ *represents the set of workspaces that an agent is able to perceive.*
- *isUnit* : $GR \rightarrow OU$ *defines the type of organizational unit instantiated by a group.*

Properties of the Relations. The *plays*, *inUnit*, *provides* and *perceives* relations allow agents to play different roles, be located in different groups, provide different services and perceive different workspaces in the organization, respectively. The *isUnit* relationship allows knowing the type of organizational unit that a group is instantiating.

The situation where an agent plays a role inside a unit and a scenario where an agent is inside an organizational unit playing a role can be checked in equations 12 and 13. It must be noted that the *Roles* function is recursive, as explained in subsection 4.1.

$$\forall r \in plays(a) \rightarrow \exists o \in OU \wedge \exists g \in inUnit(a) : isUnit(g) = o \wedge r \in Roles(o) \quad (12)$$

$$\forall g \in inUnit(a) \rightarrow \exists o \in OU \wedge isUnit(g) = o \wedge \exists r \in Roles(o) : r \in plays(a) \quad (13)$$

Equation 12 establishes that an agent can only play the roles provided by the groups where he is located. These roles are the ones provided by the organizational units instantiated by these groups. Equation 13 defines that an agent must play at least one role from each group where he is located.

In addition, using the *provides* relationship from ϕ, it is possible to define the set of active services (*AS*) from *OE*. Formally,

$$AS = \bigcup_{a \in A} s \in provides(a) \quad (14)$$

4.4   Multi-Agent Systems Based on Virtual Organizations

In previous sections, the different dimensions and entities that compose the state of a VO at a given time were defined in a formal way. Nevertheless, a VO changes through time, passing from one state of the organization to another. Thus, it is necessary to define all the possible states of the organization as well as the allowed transitions between these states. For this issue, we based our work in the proposal by da Rocha Costa and Dimuro [14].

To model the states of a VO and their transitions, let VO be the universe of all the possible organizations O. A multi-agent system based on virtual organizations is a structure $MAS = (VO, D)$ where, for every time $t \in T, D^t \subseteq VO \times VO$ defines transitions between different states of the system. In every state of the organization $O \in VO$, in a given time $t \in T$, there is a set of possible next states of the organization, denoted by $D^t(O) \subseteq VO$. Thus, for every $t \in T$, it holds that $O^{t+1} \in D^t(O^t)$, so an organization will only change to another state when it is allowed to reach from the initial state.

Since the organization is composed by three elements (OS, OE and ϕ), before executing a change of state it is necessary to check that these elements are able to change from the initial state to the possible destination state. Formally:

$$((OS^{t+1}, OE^{t+1}, \phi^{t+1}) \in D^t(OS^t, OE^t, \phi^t)) \leftrightarrow$$

$$((OS^{t+1} \in D^t_{OS}(OS^t)) \wedge (OE^{t+1} \in D^t_{OE}(OE^t)) \wedge (\phi^{t+1} \in D^t_\phi(\phi^t))) \qquad (15)$$

However, in order to swap from one state to another, it is not necessary to produce a change in all three elements that compose the VO. A change ranges from a very small variation in one or few of the elements building the organization to a big amount of changes in a large amount of entities from the VO. Formally:

$$(O^t \wedge \bigcirc O^{t+1}) \rightarrow \Box(\neg((OS^t \wedge \bigcirc OS^t) \wedge (OE^t \wedge \bigcirc OE^t) \wedge (\phi^t \wedge \bigcirc \phi^t))) \qquad (16)$$

The above formula (that uses LTL [18]) helps us to formalize how D^t is build.

$$D^t = \bigcup (O^t, O^{t+1}) | O^{t+1} \neq O^t \qquad (17)$$

D^t is composed by the set of all possible transitions of the MAS, where each new transition is generated every time that in OS, OE or ϕ (or in a combination of these elements) an atomic change is produced.

5   Discussion

In section 3 an analysis of the most relevant formalization proposals was presented. The *Virtual Organization Formalization* (VOF) takes inspiration from features taken from some of the analyzed proposals. In this section, we depict a comparison between VOF and these background proposals.

Firstly, the organizational temporal evolution proposed by VOF is mainly based on PopOrg, which models the dynamics of the *population* (similar to our OE) and the *organization* (similar to our OS).

Regarding the structure of an organization, OperA offers relations between roles that are similar to those included in VOF. The supervision relation of VOF is similar to the combination of the power and authorization relations of OperA, expressing that an agent is able to delegate its objectives to a subordinated agent, like the power relation does (the authorization relation expresses the power relation, but as a temporal situation). Also, in OperA, the objective that a subordinated agent can take from a superior agent is determined by the type of existing relation between roles, which establishes their hierarchy. However, VOF defines this hierarchy using the *RoleHier* relation.

In MOISE*^{Inst}*, the structural levels of the organization are split into: (i) *individual level*, built by organizational roles, and presents hierarchy relations between roles (similar to our *RoleHier* relation); (ii) *social level*, which is built from link relationships between roles, classified as *acq* (*acquaintance*), i.e. having a representation of other agents, *com* (similar to our *inf* relation), in which agents are able to communicate between them, and *aut* expressing authority over other agents, thus combining *col* and *sup* relations from VOF; and (iii) *collective level*, which defines *groups* of agents, establishing the *compatibility* between roles and their *cardinalities*. VOF adds the *comp* relation, in order to express whether an agent can take a given role if he is taking another one.

LAO models the topology of the system by means of dependency chains between agents. The topology can be a *hierarchy*, if there is a chain of command, or a *network*, if every agent is responsible for an organizational goal and has a delegation relationship to another agent. VOF models the topology of the system using *Contains* relations between OUs. These relations allow defining three types of organizations: *hierarchy*, similar to the structure defined by LAO; *team*, when all agents collaborate with each other; and *plain*, which assumes information relationships between roles.

Regarding the Functional Dimension, although PopOrg and POMF model concepts that are similar to services (by means of exchange processes or workflows, respectively), they are better described in VOF. PopOrg focuses on the actions developed by the process and the agents that are carrying them out, while POMF is focused on describing the tasks that compose a given workflow. VOF goes beyond, (as it follows a Service Oriented Approach) and formalizes a service by means of the roles that it can provide and consume, the goals that can be achieved with this service, the invoke relationships between services, and the tasks that compose each service (as well as the goals that these tasks help to reach).

The environment used in VOF is based on the Agents & Artifacts conceptual framework, which was included in the SODA metamodel [19].

VOF model norms in a very similar way to the proposal of MOISE*^{Inst}*, although they use different languages to describe them. VOF is able to relate a norm to a set of OUs, using the *Norms* relation from *DimRelations*, limiting its effect only to this set.

The Organizational Entity from VOF can be also compared with other proposals. For example, a specified group is defined in MOISE*^{Inst}* as a 'group specification', while VOF defines it as an Organizational Unit. On the other hand, a group instantiation is named 'group' in both MOISE*^{Inst}* and VOF. In addition, the *OE* from VOF defines the set of norms and services that are currently active in the organization.

Finally, VOF clearly divides the elements that specify the system (i.e. elements that will produce a structural change if they are modified) and the more dynamic elements of the MAS, represented in the *OE*. Our specification gives agents the possibility to belong to a specific group and provide or use a service.

6   Conclusions and Future Work

This work presented a formal specification for Virtual Organizations, named VOF (*Virtual Organization Formalization*), which is composed of: (i) the Organizational

Specification (OS), which details the components that specify the system and divides them by means of the organizational dimensions; (ii) the Organizational Entity (OE), which defines the active elements of the system; and (iii) the Organizational Dynamics (ϕ), which details the relationships between elements from OS and OE.

Additionally, we have analyzed a set of different formalizations, focusing on organizational concepts taken from the Organizational Dimensions. After this analysis, we noticed that the analyzed formalizations do not take into account all concepts from Organizational Dimensions. Therefore, our proposal is aimed to cover all these concepts and to provide a formalization as much complete as possible.

As a future work, this formalization will help us when dealing with concepts related to adaptation in Organization Centered Multi-Agent Systems, being easier for us to identify the entities of the system that would change through time. VOF will be integrated into the reasoning process of the BDI agents from the THOMAS framework [20], in order to develop agents that are able to know whether an organization is working in a correct way, or they need to execute an adaptation process. Moreover, using this formalization we will be able to check the correctness of a defined OCMAS.

Acknowledgements. This work is supported by TIN2009-13839-C03-01 and PROM-ETEO/2008/051 projects of the Spanish government and CONSOLIDER-INGENIO 2010 under grant CSD2007-00022.

References

1. Lemaître, C., Excelente, C.B.: Multi-agent organization approach. In: Proceedings of II Iberoamerican Workshop on DAI and MAS (1998)
2. Dignum, V., Dignum, F.: A logic for agent organizations. In: Proc. FAMAS, pp. 83–100 (2007)
3. Emerson, E.A.: Temporal and modal logic. In: Handbook of Theoretical Computer Science (vol. B): formal models and semantics (1991)
4. Grossi, D., Dignum, F., Dastani, M., Royakkers, L.: Fundations of organizational structures in multiagent systems. In: Proc. AAMAS, pp. 690–697 (2005)
5. Clarke, E.M.: Model Checking. In: Ramesh, S., Sivakumar, G. (eds.) FST TCS 1997. LNCS, vol. 1346, pp. 54–56. Springer, Heidelberg (1997)
6. Foster, I., Kesselman, C., Tuecke, S.: The anatomy of the grid: Enabling scalable virtual organizations. Int. J. High Perform. Comput. Appl. 15(3), 200 (2001)
7. Gonzalez-Palacios, J., Luck, M.: Towards Compliance of Agents in Open Multi-agent Systems. In: Choren, R., Garcia, A., Giese, H., Leung, H.-f., Lucena, C., Romanovsky, A. (eds.) SELMAS. LNCS, vol. 4408, pp. 132–147. Springer, Heidelberg (2007)
8. Hubner, J.F., Sichman, J.S., Boissier, O.: A Model for the Structural, Functional, and Deontic Specification of Organizations in Multiagent Systems. In: Bittencourt, G., Ramalho, G.L. (eds.) SBIA 2002. LNCS (LNAI), vol. 2507, pp. 118–128. Springer, Heidelberg (2002)
9. Criado, N., Argente, E., Julián, V., Botti, V.: Designing Virtual Organizations. In: Demazeau, Y., Pavón, J., Corchado, J.M., Bajo, J. (eds.) PAAMS 2009. AISC, vol. 55, pp. 440–449. Springer, Heidelberg (2009)
10. Dignum, V.: A model for organizational interaction: based on agents, founded in logic. PhD thesis, Utrecht University (2003)
11. Popova, V., Sharpanskykh, A.: Process-Oriented Organization Modeling and Analysis Based on Constraints. Technical report (2006)

12. Gâteau, B., Boissier, O., Khadraoui, D., Dubois, E.: Moise^{Inst}: An organizational model for specifying rights and duties of autonomous agents. In: Proc. EUMAS, pp. 484–485 (2005)
13. Haesevoets, R., Weyns, D., Holvoet, T., Joosen, W.: A formal model for self-adaptive and self-healing organizations (2009)
14. da Rocha Costa, A., Dimuro, G.: Semantic Concepts for a Formal Structural Dynamics of Situated Multiagent Systems. In: Sichman, J.S., Padget, J., Ossowski, S., Noriega, P. (eds.) COIN 2007. LNCS (LNAI), vol. 4870, pp. 139–154. Springer, Heidelberg (2008)
15. Jonker, C.M., Sharpanskykh, A., Treur, J., Yolum, P.I.: A framework for formal modeling and analysis of organizations. Appl. Intell. 27(1), 49–66 (2007)
16. Ricci, A., Viroli, M., Omicini, A.: Give agents their artifacts: the A&A approach for engineering working environments in MAS. In: Proc. AAMAS, p. 150 (2007)
17. Criado, N., Argente, E., Botti, V.: Rational Strategies for Norm Compliance in the n-BDI Proposal. In: De Vos, M., Fornara, N., Pitt, J.V., Vouros, G. (eds.) COIN 2010. LNCS, vol. 6541, pp. 1–20. Springer, Heidelberg (2011)
18. Pnueli, A.: The temporal logic of programs. In: 18th Annual Symposium on Foundations of Computer Science, pp. 46–57. IEEE (1977)
19. Nardini, E., Molesini, A., Omicini, A., Denti, E.: SPEM on test: the SODA case study. In: Proc. of the 2008 ACM Symposium on Applied Computing, pp. 700–706. ACM (2008)
20. Argente, E., Botti, V., Carrascosa, C., Giret, A., Julian, V., Rebollo, M.: An Abstract Architecture for Virtual Organizations: The THOMAS approach. Knowledge and Information Systems, 1–35 (2011)

Reinforcement Learning for Self-organizing Wake-Up Scheduling in Wireless Sensor Networks

Mihail Mihaylov[1], Yann-Aël Le Borgne[1], Karl Tuyls[2], and Ann Nowé[1]

[1] Vrije Universiteit Brussel, Brussels, Belgium
[2] Maastricht University, Maastricht, The Netherlands
{mmihaylo,yleborgn,ann.nowe}@vub.ac.be
k.tuyls@maastrichtuniversity.nl

Abstract. Wake-up scheduling is a challenging problem in wireless sensor networks. It was recently shown that a promising approach for solving this problem is to rely on reinforcement learning (RL). The RL approach is particularly attractive since it allows the sensor nodes to coordinate through local interactions alone, without the need of central mediator or any form of explicit coordination. This article extends previous work by experimentally studying the behavior of RL wake-up scheduling on a set of three different network topologies, namely line, mesh and grid topologies. The experiments are run using OMNET++, a the state-of-the-art network simulator. The obtained results show how simple and computationally bounded sensor nodes are able to coordinate their wake-up cycles in a distributed way in order to improve the global system performance. The main insight of these experiments is to show that sensor nodes learn to synchronize if they have to cooperate for forwarding data, and learn to desynchronize in order to avoid interferences. This synchronization/desynchronization behavior, referred to for short as (de)synchronicity, allows to improve the message throughput even for very low duty cycles.

Keywords: Reinforcement learning, Synchronicity and desynchronicity, Wireless sensor networks, Wake-up scheduling.

1 Introduction

A Wireless Sensor Network is a collection of densely deployed autonomous devices, called *sensor nodes*, which gather data with the help of sensors [4]. The untethered nodes use radio communication to transmit sensor measurements to a terminal node, called the *sink*. The sink is the access point of the observer, who is able to process the distributed measurements and obtain useful information about the monitored environment. Sensor nodes communicate over a wireless medium, by using a multi-hop communication protocol that allows data packets to be forwarded by neighboring nodes to the sink. A typical multi-hop communication protocol is to rely on a shortest path tree with respect to the hop distance [4]. Such a tree is obtained by letting nodes broadcast packets after deployment, in order identify their neighbors. The nodes then determine the neighbor node which is the closest (in terms of hops) to the sink, and use it

J. Filipe and A. Fred (Eds.): ICAART 2011, CCIS 271, pp. 382–396, 2013.
© Springer-Verlag Berlin Heidelberg 2013

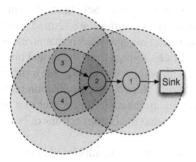

Fig. 1. Sensor nodes connected to a base station by means of a multi-hop routing tree. Grayed circles indicate overlapping communication regions.

as the relaying node for the multi-hop routing. An example of multi-hop shortest path routing structure is given in Fig. 1, together with the radio communication ranges of sensor nodes.

Since communication is the most energy expensive action, it is clear that in order to save energy, a node should turn off its antenna (or go to sleep) [5]. However, when sleeping, the node is not able to send or receive any messages, therefore it increases the latency of the network, i.e., the time it takes for messages to reach the sink. High latency is undesirable in any real-time applications. On the other hand, a node does not need to listen to the channel when no messages are being sent, since it loses energy in vain. As a result, nodes should determine on their own *when* they should be awake within a frame. This behavior is called *wake-up scheduling*. Once a node wakes up, it remains active for a predefined amount of time, called *duty cycle*.

Wake-up scheduling in wireless sensor networks is an active research domain [7,12,8,2]. A good survey on wake-up strategies in WSNs is presented in [14]. The standard approach is S-MAC, a synchronized medium access control (MAC) protocol for WSN [17]. In S-MAC, the duty-cycle is fixed by the user, and all sensor nodes synchronize in such a way that their active periods take place at the same time. This synchronized active period enables neighboring nodes to communicate with one another. The use of routing then allows any pair of nodes to exchange messages. By tuning the duty-cycle, wake-up scheduling therefore allows to adapt the use of sensor resources to the application requirements in terms of latency, data rate and lifetime [14].

Recently, we showed that the wake-up scheduling problem could be efficiently tackled in the framework of multi-agent systems and reinforcement learning. In wireless sensor networks, the sensor nodes can be seen as agents, which have to logically self-organize in groups (or *coalitions*). The actions of agents within a group need to be synchronized (e.g., for data forwarding), while *at the same time* being desynchronized with the actions of agents in other groups (e.g., to avoid radio interferences). We refer to this concept for short as *(de)synchronicity*.

Coordinating the actions of agents (i.e., sensor nodes) can successfully be done using the reinforcement learning framework by rewarding successful interactions (e.g., transmission of a message in a sensor network) and penalizing the ones with a negative outcome (e.g., overhearing or packet collisions) [10]. This behavior drives the nodes

to repeat actions that result in positive feedback more often and to decrease the probability of unsuccessful interactions. *Coalitions* are formed when agents select the same
successful actions. A key feature of our approach is that no explicit notion of coalition is
necessary. Rather, these coalitions emerge from the global objective of the system, and
agents learn by themselves with whom they have to (de)synchronize (e.g. to maximize
throughput in a routing problem). Here desynchronization refers to the situation where
one agent's actions (e.g. waking up the radio transmitter of a wireless node) are shifted
in time, relative to another, such that the (same) actions of both agents do not happen at
the same time.

In this article, we extend our previous results by illustrating the benefits of our self-
adapting RL approach in three wireless sensor networks of different topologies, namely
line, mesh and grid. We show that nodes form coalitions which allow to reduce packet
collisions and end-to-end latency, even for very low duty cycles. This (de)synchronicity
is achieved in a decentralized manner, without any explicit communication, and without any prior knowledge of the environment. Our simulations are implemented using
OMNET++, a state-of-the-art simulator [11].

The paper is organized as follows. Section 2 presents the reinforcement learning approach for solving the wake-up scheduling problem in WSN. Section 3 analyzes and
discusses the performances of the RL approach on three different topologies, namely
line, mesh and grid topologies. We also compare it to the standard S-MAC protocol and
briefly discusses future work. Section 4 concludes this paper.

2   (De)synchronicity with Reinforcement Learning

This section presents our decentralized approach to (de)synchronicity using the reinforcement learning framework. The proposed approach requires very few assumptions
on the underlying networking protocols, which we discuss in Section 2.1. The subsequent sections detail the different components of the reinforcement learning mechanism.

2.1   Motivations and Network Model

Communication in WSNs is achieved by means of networking protocols, and in particular by means of the Medium Access Control (MAC) and the routing protocols [4]. The
MAC protocol is the data communication protocol concerned with sharing the wireless transmission medium among the network nodes. The routing protocol allows to
determine where sensor nodes have to transmit their data so that they eventually reach
the sink. A vast amount of literature exists on these two topics [4], and we sketch in the
following the key requirements for the MAC and routing protocols so that our reinforcement learning mechanism presented in Section 2.2 can be implemented. We emphasize
that these requirements are very loose.

We use a simple MAC protocol, inspired from S-MAC [17], that divides the time
into small discrete units, called frames. We further divide each frame into time slots.
The frame and slot duration are application dependent and in our case they are fixed by
the user prior to network deployment. The sensor nodes then rely on a standard duty cycle mechanism, in which the node is awake for a predetermined number of slots during

each period. The duration of the awake period is fixed by the user, while its position is initialized randomly within the frame for each node. These active slots will be shifted as a result of the learning, which will coordinate nodes' wake-up schedules in order to ensure high data throughput and longer battery life. Each node will learn to be in active mode when its parents and children are awake, so that it forwards messages faster (synchronization), and stay asleep when neighboring nodes on the same hop are communicating, so that it avoids collisions and overhearing (desynchronization).

The routing protocol is not explicitly part of the learning algorithm and therefore any multi-hop routing scheme can be applied without losing the properties of our approach. In the experimental results, presented in Section 3, the routing is achieved using a standard shortest path multi-hop routing mechanism. The forwarding nodes need not be explicitly known, as long as they ensure that their distance to the sink is *lower* than the sender. Communication is done using a Carrier Sense Multiple Access (CSMA) protocol. Successful data reception is acknowledged with an ACK packet. We would like to note that the acknowledgment packet is necessary for the proper and reliable forwarding of messages. Our algorithm does use this packet to indicate a "correct reception" in order to formulate one of its reward signals (see Subsection 3.1). However, this signal is not crucial for the RL algorithm and thus the latter can easily function without acknowledgment packets. Subsection 2.3 will further elaborate on the use of reward signals.

It is noteworthy that the communication partners of a node (and thus the formation of coalitions) are influenced by the communication and routing protocols that are in use and not by our algorithm itself. These protocols only implicitly determine the *direction* of the message flow and not *who* will forward those messages, since nodes should find out the latter by themselves.

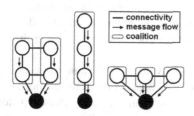

Fig. 2. Examples of routing and coalition formation

Depending on the routing protocol, coalitions (e.g., synchronized groups of nodes) logically emerge across the different hops, such that there is, if possible, only one agent from a certain hop within a coalition. Figure 2 illustrates this concept in three different topologies. It shows as an example how coalitions form as a result of the routing protocol. Intuitively, nodes from one coalition need to synchronize their wake-up schedules. As defined by the routing protocol, messages are not sent between nodes from the same hop, hence these nodes should desynchronize (or belong to separate coalitions) to avoid communication interference. The emergence of coalitions will be experimentally illustrated for different topologies in Section 3.

2.2   Reinforcement Learning Approach: Methodology

Each agent in the WSN uses a reinforcement learning (RL) [15] algorithm to learn an efficient wake-up schedule (i.e. when to remain active within the frame) that will improve throughput and lifetime in a distributed manner. It is clear that learning in multi-agent systems of this type requires careful exploration in order to make the action-values of agents converge. We use a value iteration approach similar to single-state Q-learning [16] with an implicit exploration strategy, as subsection 2.5 will further elaborate on. However, our update scheme differs from that of traditional Q-learning (cf. subsection 2.4). The battery power required to run the algorithm is marginal to the communication costs and thus it is neglected. The main challenge in such a decentralized approach is to define a suitable reward function for the individual agents that will lead to an effective emergent behavior as a group. To tackle this challenge, we proceed with the definition of the basic components of the reinforcement learning algorithm described in this section.

2.3   Actions and Rewards

The actions of each agent are restricted to selecting a time window (or a wake period) within a frame for staying awake. Since the size of these frames remains unchanged and they constantly repeat throughout the network lifetime, our agents use no notion of states, i.e. we say that our learning system is stateless (or single-state). The duration of this wake period is defined by the duty cycle, fixed by the user of the system. In other words, each node selects a slot within the frame when its radio will be switched on for the duration of the duty cycle. Thus, the size of the action space of each agent is determined by the number of slots within a frame. In general, the more actions agents have, the slower the reinforcement learning algorithm will converge [6]. On the other hand, a small action space might lead to suboptimal solutions and will impose an energy burden on the system. Setting the right amount of time slots within a frame requires a study on itself, that we shall not undertake in this paper due to space restrictions (see subsection 3.1 for exact values).

Every node stores a "quality value" (or Q-value) for each slot within its frame. This value for each slot indicates how beneficial it is for the node to stay awake during these slots for every frame, i.e. what is an efficient wake-up pattern, given its duty cycle and considering its communication history. When a communication event occurs at a node (overheard, sent or received a packet) or if no event occurred during the wake period (idle listening), that node updates the quality-value of the slot(s) when this event happened. The motivation behind this scheme is presented in subsection 2.5.

2.4   Updates and Action Selection

The slots of agents are initiated with Q-values drawn from a uniform random distribution between 0 and 1. Whenever events occur during node's active period, that node updates the quality values of the slots, at which the corresponding events occurred, using the following update rule:

$$Q_s^i \leftarrow (1 - \alpha) \cdot \hat{Q}_s^i + \alpha \cdot r_{s,e}^i$$

where $Q_s^i \in [0, 1]$ is the quality of slot s within the frame of agent i. Intuitively, a high Q_s^i value indicates that it is beneficial for agent i to stay awake during slot s. This quality value is updated using the previous Q-value ($\hat{Q}_s^i$) for that slot, the learning rate $\alpha \in [0, 1]$, and the newly obtained reward $r_{s,e}^i \in [0, 1]$ for the event e that (just) occurred in slot s. Thus, nodes will update as many Q-values as there are events during its active period. In other words, agent i will update the value Q_s^i for each slot s where an event e occurred. The latter update scheme differs from that of traditional Q-learning [16], where only the Q-value of the selected action is updated. The motivation behind this update scheme is presented in subsection 2.5. In addition, we set here the future discount parameter γ to 0, since our agents are stateless (or single-state).

Nodes will stay awake for those consecutive time slots that have the highest sum of Q-values. Put differently, each agent selects the action $a_{s'}$ (i.e., wake up at slot s') that maximizes the sum of the Q-values for the D consecutive time slots, where D is the duty cycle, fixed by the user. Formally, agent i will wake up at slot s', where

$$s' = \arg\max_{s \in S} \sum_{j=0}^{D} Q_{s+j}^i$$

For example, if the required duty cycle of the nodes is set to 10% ($D = 10$ for a frame of $S = 100$ slots), each node will stay active for those 10 consecutive slots within its frame that have the highest sum of Q-values. Conversely, for all other slots the agent will remain asleep, since its Q-values indicate that it is less beneficial to stay active during that time. Nodes will update the Q-value of each slot for which an event occurrs within its duty cycle. Thus, when forwarding messages to the sink, over time, nodes acquire sufficient information on "slot quality" to determine the best period within the frame to stay awake. This behavior makes neighboring nodes (de)synchronize their actions, resulting in faster message delivery and thus lower end-to-end latency.

2.5   Exploration

As explained in the above two subsections, active time slots are updated individually, regardless of when the node wakes up. The reason for this choice is threefold. Firstly, this allows each slot to be explored and updated more frequently. For example, slot s will be updated when the node wakes up anywhere between slots $s - 1$ and $s - D + 1$, i.e. in D out of S possible actions. Secondly, updating individual Q-values makes it possible to alter the duty cycle of nodes at run time (as suggest some preliminary results, not displayed in this paper) without invalidating the Q-values of slots. In contrast, if a Q-value was computed for each start slot s, i.e. the reward was accumulated over the wake duration and stored at slot s only, changing the duty cycle at run-time will render the computed Q-values useless, since the reward was accumulated over a different duration. In addition, slot s will be updated only when the agent wakes up at that slot. A separate exploration strategy is therefore required to ensure that this action is explored sufficiently. Thirdly, our exploration scheme will continuously explore and update not only the wake-up slot, but all slots within the awake period. Treating slots individually results in an implicit exploration scheme that requires no additional tuning.

Even though agents employ a greedy policy (selecting the action that gives the highest sum of Q-values), this "smooth" exploration strategy ensures that all slots are explored and updated regularly at the start of the application (since values are initiated randomly), until the sum of Q-values of one group of slots becomes strictly larger than the rest. In that case we say that the policy has converged and thus exploration has stopped. The speed of convergence is influenced by the duty cycle, fixed by the user, and the learning rate, which we empirically chose to be 0.1. A constant learning rate is in fact desirable in a non-stationary environment to ensure that policies will change with respect to the most recently received rewards [15].

3   Results

We proceed with the experimental comparison between our (de)synchronization approach and a fully synchronized state-of-the-art MAC protocol, viz. S-MAC [17]. All components of the compared networks, such as the routing and CSMA communication protocols, remain the same. The S-MAC protocol illustrates network performance under synchronized behavior, where all nodes are active at the same time. In other words, we compare our RL technique to networks with no coordination mechanism, but which employ some means of time synchronization, the small overhead of which will be neglected for the sake of a clearer exposition. This synchronized approach ensures high network throughput, but as we will demonstrate in subsection 3.2, it fails at short duty cycles.

3.1   Experimental Setup

We applied our approach on three networks of different size and topology. In particular, we investigate two extreme cases where nodes are arranged in a 4-node line (Figure 3(a)) and a 6-node single-hop mesh topology (Figure 4(a)). The former one requires nodes to synchronize in order to successfully forward messages to the sink. Intuitively, if any one node is awake while the others are asleep, that node would not be able to forwarded its messages to the sink. Conversely, in the mesh topology it is most beneficial for nodes to fully desynchronize to avoid communication interference with neighboring nodes. Moreover, the sink is able to communicate with only one node at a time. The third topology is a 4 by 4 grid (Figure 5(a)) where sensing agents need to both synchronize with some nodes and at the same time desynchronize with others to maximize throughput and network lifetime. The latter topology clearly illustrates the importance of combining synchronicity and desynchronicity, as neither one of the two behaviors alone achieves the global system objectives. Subsection 3.2 will confirm these claims and will elaborate on the obtained results.

Each of the three networks was run for 3600 seconds in the OMNeT++ simulator [11] and results were averaged over 30 runs. This network runtime was sufficiently long to eliminate any initial transient effects. To illustrate the performance of the network at high data rates, we set the sampling period of nodes to one message every 10 seconds. For each node the start of this period is at a uniformly random time within the first frame of the simulation and thereafter messages in that node are periodically generated at the same slot every frame. Frames have the same length as the sampling period and were divided in $S = 2000$ slots of 5 milliseconds each. The duration of the slot was chosen such that only one DATA packet can be sent and acknowledged within that time. All hardware-specific parameters, such as transmission power, bit rate, etc., were set according to the data sheet of our radio chip — CC2420 [1]. In addition, we chose the protocol-specific parameters, such as packet header length and number of retransmission retries as specified in the IEEE 802.15.4 communication protocol [3].

Since collisions constitute the biggest obstacle in the pursuit of low latency, each node contends for the channel for a small random duration within a fixed contention window of 5 slots. To facilitate the throughput of messages at high data rates, we deviated from the contention policy of S-MAC that uses the entire active time as a contention window. Instead, in our simulations we fixed the maximum contention window of S-MAC to 5 slots for a more fair comparison.

We modeled five different events, namely overhearing ($r = 0$), idle listening ($r = 0$ for each idle slot), successful transmission ($r = 1$ if ACK received), unsuccessful transmission ($r = 0$ if no ACK received) and successful reception ($r = 1$). Maximizing the throughput requires both proper transmission as well as proper reception. Therefore, we treat the two corresponding rewards equally. Furthermore, most radio chips require nearly the same energy for sending, receiving (or overhearing) and (idle) listening [5], making the three rewards equal. We consider these five events to be the most energy expensive or latency crucial in wireless communication. Additional events were also modeled, but they were either statistically insignificant (such as busy channel) or already covered (such as unsuccessful transmissions instead of collisions).

Due to the exponential smoothing nature of the reward update function (cf. subsection 2.4) the Q-values of slots will be shifted towards the latest reward they receive. We would expect that the "goodness" of slots will decrease for negative events (e.g. transmission was not acknowledged), and will increase for successful communication. Therefore, the feedback agents receive is binary, i.e. $r^i_{s,e} \in \{0, 1\}$, since it carries the necessary information. Other reward signals were also evaluated, resulting in similar performance.

3.2 Evaluation

We would like to point out that both S-MAC and our approach are controlled by the same parameter — the duty cycle, which is fixed by the user of the system. Since the active time of nodes in both approaches is the same, the energy consumption of the two protocols is nearly identical. The only difference to S-MAC is that with our approach nodes *learn when* to hold their duty cycle within the frame, as opposed to S-MAC, where all nodes are awake at the beginning of the frame. Therefore, in the following evaluation we vary the duty cycle of the nodes and monitor the average end-to-end latency across the different simulation runs.

Figure 3(b) displays an example of the resulting schedule of the line topology (Figure 3(a)) after the action of each agent converges for 5% duty cycle. The results indicate that all four nodes have successfully learned to stay awake at the same time in order for messages to be properly forwarded to the sink. In other words, we observe that all nodes belong to the same coalition, as suggested in Figure 2. If any one node in the

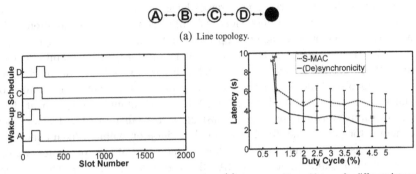

(a) Line topology.

(b) Example of a learned wake-up schedule for duty cycle of 5%.

(c) Average end-to-end latency for different duty cycles.

Fig. 3. Experimental results for the line topology

line topology had remained active during the sleep period of its immediate neighbors, its messages, together with those of its higher hop neighbors would not have been delivered to the sink. Even though neighboring nodes are awake at the same time (or have synchronized), one can see that schedules are slightly shifted in time. The reason for this desynchronicity is to reduce the overhearing of higher hop communication and to increase throughput by compensating for propagation delays — a behavior that nodes have learned by themselves.

Figure 3(c) displays the average end-to-end latency of the learning and the synchronized nodes respectively, where error bars signify one standard deviation across 30 runs. Since the learned wake-up schedules of our approach closely resemble the prescribed behavior of S-MAC, the latency improvement over different duty cycles is marginal. Nevertheless, the end-to-end latency of our learning agents is on average 2 seconds less than under the S-MAC protocol. The reason for this improvement lies in the fact that with S-MAC all nodes wake up at the beginning of the frame, while with our approach agents learn when it is best to wake up. Since each node periodically generates messages at a different time within the 10-seconds frame, the latency of S-MAC is on average 5 seconds. Learning, however, allows flexibility in the wake-up times, such that a node could wake up immediately after generating its messages (and all other nodes will learn to wake nearly at the same time) and therefore reduce the queuing time of messages for at least one node.

As evident in Figure 3(c), both approaches are inefficient at very low duty cycles. The reason for this high latency is the fact that the active period of nodes is too short compared to the propagation delay. Therefore, messages need to be queued for more than one frame on average, which results in traffic congestion.

In contrast to the previous topology, our second set of experiments investigate the performance of the network where all nodes lie on the same hop from the sink. This setup presents agents with the opposite challenge, namely to find an active period where no other node is awake. The latter behavior will eliminate communication interference with neighboring nodes and will ensure proper reception of messages at the sink. Figure 4(b) displays an example of the wake-up schedule of the learning nodes for a duty

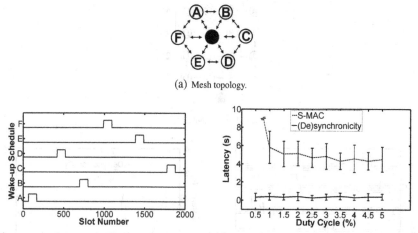

(a) Mesh topology.

(b) An example of a learned wake-up schedule for duty cycle of 5%.

(c) Average end-to-end latency for different duty cycles.

Fig. 4. Experimental results for the mesh topology

cycle of 5% after the actions of agents converge. One can observe that the state of desynchronicity has been successfully achieved where each node is active at a different time within a frame. Put differently, each node has chosen a different wake-up slot and therefore belongs to different coalition. The benefit of this desynchronized pattern is clearly evident in Figure 4(c) where we compare it to the average end-to-end latency of the synchronized system. Error bars represent one standard deviation across 30 runs. Since all nodes lie within one hop of the sink, the performance of the learning agents is not dependent on the duty cycle for this topology. Each node independently learns to hold its active period immediately after it generates a message, as long as no neighbor is awake at the same time. Therefore, the average end-to-end latency is slightly more than the duration of one transmission. Similarly to the line topology, when nodes use the S-MAC protocol, the end-to-end latency of the system is on average half the sampling period, for reasons outlined above. Moreover, for duty cycle of 0.5%, the S-MAC nodes are unable to deliver their messages to the sink, since all nodes try to transmit during the same short awake period and thus all messages collide. This effect is indicated with the discontinued dashed line in Figure 4(c).

Lastly, we investigate a combination of the above two topologies, namely the grid shown in Figure 5(a). Nodes here need to synchronize with those that lie on the same branch of the routing tree to ensure high throughput, while at the same time desynchronize with neighboring routing branches to avoid communication interference. An example of the wake-up schedule of the learning nodes at 5% duty cycle is displayed in Figure 5(b). As expected, the four columns of nodes belong to four different coalitions, where nodes in one coalition are synchronized with each other (being active nearly at the same time) and desynchronized with the other coalitions (sleeping while others are active). This is the state of (de)synchronicity. Nodes in one coalition exhibit comparable behavior to those in a line topology, i.e. they have synchronized with each other (while

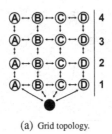

(a) Grid topology.

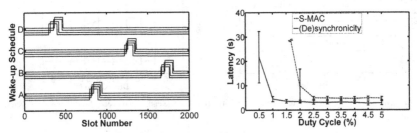

(b) An example of a learned wake-up schedule for duty cycle (c) Average end-to-end latency for different duty cycles.
of 5%.

Fig. 5. Experimental results for the grid topology

still slightly shifted in time). At the same time nodes on the same hop have learned to desynchronize their active times similar to the mesh topology.

The result of applying our learning approach in a grid topology for various duty cycles can be observed in Figure 5(c). It displays the average end-to-end latency of the network when using synchronicity and (de)synchronicity respectively. Here again error bars signify one standard deviation across 30 runs. Due to the high data rate, when using S-MAC nodes are incapable of delivering all packets for duty cycles lower than 2%. This reduced performance at low duty cycles is due to the large number of collisions and re-transmissions necessary when all nodes wake up at the same time. The learning approach on the other hand drives nodes to coordinate their wake-up cycles and shift them in time, such that nodes at neighboring coalitions desynchronize their awake periods. In doing so, nodes effectively avoid collisions and overhearing, leading to lower end-to-end latency. When nodes coordinate their actions, they effectively reduce communication interference with neighboring nodes. This behavior results in lower amount of overheard packets, less collisions and therefore fewer retries to forward a message, as compared to the fully synchronized network. Nevertheless, at very low duty cycles the active time of nodes is too short to forward all messages and therefore, similar to the line topology, the network experiences traffic congestion.

3.3   Discussion

We would like to discuss here the convergence time of the learning agents. The implicit exploration scheme, described in subsection 2.5 makes nodes select different actions in the beginning of the simulation in order to determine their quality. As time progresses, the Q-values of slots are updated sufficiently enough to make the policy of the agents converge. We measured that after 80 iterations (or frames) on average the actions of agents do not change any more and thus the state of (de)synchronicity has been reached. In other words, after 800 seconds each node finds the wake-up schedule that improves message throughput and minimizes communication interference. This duration is sufficiently small compared to the lifetime of the system for a static WSN, which is in the order of several days up to a couple of years depending on the duty cycle and the hardware characteristics [4]. However, it is still unclear under which conditions convergence proofs can be brought. Further research is therefore required to better characterize the convergence criteria.

Despite the improvements that our approach offers over the standard S-MAC protocol, we discuss here two shortcomings that need to be addressed. First of all, the duty cycle set by the user of the system affects all nodes equally. In other words, all nodes are active for the same amount of time. Depending on their position in the network, however, nodes require different duration for their active periods. Nodes close to the sink are subject to heavier traffic load compared to leaf nodes, whose active time need not be as high. The second shortcoming of our technique concerns the coordination of actions among active agents. Clearly, being awake at the same time is not sufficient for two nodes to successfully exchange messages. If two agents on the same routing branch attempt to transmit at the same slot, their messages will collide. Agents therefore need to learn not only the time of their active period within a frame, but also when to transmit and when to listen during that active period.

The above two shortcomings are being addressed in an extension of our algorithm, which we call DESYDE [9]. The three main differences to the proposed approach are outlined below:

1. In DESYDE we let agents learn two quality values for each slot, instead of one. One quality value indicates how beneficial it is for the node to *transmit* during that slot, while the other value indicates how good it is to *listen* for messages. In slots where it is neither good to transmit nor to listen, the node will turn off its antenna and enter *sleep* mode. Thus, each node learns the quality of three actions: *transmit*, *listen* and *sleep*, as opposed to only *wake-up* and *sleep*.

2. The algorithm in DESYDE differs from the one proposed in this paper also in the value of the learning rate α. In DESYDE we set this value to 1, which dramatically alters the learning behavior of nodes. With $\alpha = 1$, nodes remember only the most recently observed feedback signal for each slot and discard old observations. In this way the behavior of nodes resembles a Win-Stay Lose-Shift strategy [13] where in our setting agents at each slot repeat the action that was successful at the same slot in the previous frame and try a different action if it was unsuccessful.

3. The last difference is the action selection method — in DESYDE nodes select at each slot the action with the highest expected reward, rather than staying awake for the slots with the highest sum of Q-values. If none of the two quality values are above 0 for a given slot, the agent selects *sleep* in that slot in the next frame. In this way nodes adapt their duty cycle to the traffic load of the network and may wake up at different slots within a frame, as opposed to holding only one active period.

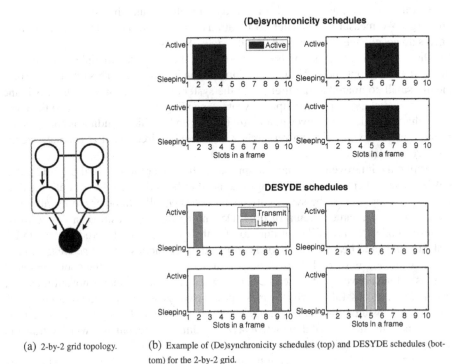

(a) 2-by-2 grid topology.   (b) Example of (De)synchronicity schedules (top) and DESYDE schedules (bottom) for the 2-by-2 grid.

Fig. 6. Comparison between (De)synchronicity and DESYDE schedules on the 2-by-2 grid

To illustrate the effect of the above three differences, consider the 2 by 2 grid in Figure 6(a). An example of the resulting wake-up schedule of the four nodes is illustrated in Figure 6(b) for both (De)synchronicity (top) and DESYDE (bottom). In this example, the frame contains 10 slots, and the four schedules reported (for each approach) are those of the four nodes in the grid, arranged in the same order as in Figure 6(a). Figure 6(b) (top) shows the schedules of nodes using the algorithm presented in this paper. One can see that the left nodes are synchronized for communication at slots $2-4$, while the right ones are active at slots $5-7$. In other words, upper nodes are synchronized with lower ones (being active at the same slots) and left nodes are desynchronized with right.

The schedules of nodes when using DESYDE are presented in Figure 6(b) (bottom). Here upper transmission slots are synchronized with lower reception slots and left active slots are desynchronized with right active slots. More precisely, at slot 2, the upper left node transmits when the lower left node receives, while the right nodes are synchronized for communication at slot 5. The lower left node sends its data to the base station at slot 7 and forwards that of the upper left node at slot 9. The lower right node does the same at slots 4 and 6, respectively. Thus, with both approaches we observe the same coalitions as in our schematic model in Figure 6(a). On the one hand, DESYDE allows nodes to adapt their active time based on the (static) traffic load and therefore prolongs the network lifetime. However, one disadvantage is that DESYDE is not well suited for irregular data traffic, since nodes tightly adapt their schedules to match the expected traffic. For the same reason, DESYDE is also vulnerable to clock drifts. The approach presented in this paper, on the other hand, is more flexible in these respects, since most nodes remain active longer than it is necessary to forward their packets.

4   Conclusions

In this paper we presented a decentralized reinforcement learning (RL) approach for self-organizing wake-up scheduling in wireless sensor networks (WSNs). Our approach improves the throughput of the system even for very low duty cycles, as compared to the standard S-MAC protocol. When using our RL policy, agents independently learn to synchronize their active periods with nodes on the same routing branch, so that message throughput is improved. At the same time, nodes desynchronize with other routing branches in order to reduce communication interference. We demonstrated how initially randomized wake-up schedules successfully converge to the state of (de)synchronicity based only on local interactions and without any form of explicit coordination. As a result, our approach makes it possible that sensor node coordination *emerges* rather than is *agreed* upon.

The proposed approach provides a basis for a number of extensions that we are currently investigating. In particular, the wake-up schedules of individual nodes may be adapted on the basis of their own traffic load, as illustrated by the DESYDE strategy. This adapted version of the protocol allows to further reduce the convergence time and the end-to-end latency of the system.

Acknowledgements. The authors of this paper would like to thank the anonymous reviewers for their useful comments and valuable suggestions. This research is funded by the agency for Innovation by Science and Technology (IWT), project DiCoMAS (IWT60837); and by the Research Foundation - Flanders, Belgium (FWO), project G.0219.09N.

References

1. CC2420: Data sheet,
   http://focus.ti.com/docs/prod/folders/print/cc2420.html
2. Cohen, R., Kapchits, B.: An optimal wake-up scheduling algorithm for minimizing energy consumption while limiting maximum delay in a mesh sensor network. IEEE/ACM Trans. Netw. 17(2), 570–581 (2009)
3. Gutierrez, J., Naeve, M., Callaway, E., Bourgeois, M., Mitter, V., Heile, B.: IEEE 802.15. 4: a developing standard for low-power low-cost wireless personal area networks. IEEE Network 15(5), 12–19 (2002)
4. Ilyas, M., Mahgoub, I.: Handbook of sensor networks: compact wireless and wired sensing systems. CRC (2005)
5. Langendoen, K.: Medium access control in wireless sensor networks. Medium Access Control in Wireless Networks 2, 535–560 (2008)
6. Leng, J.: Reinforcement learning and convergence analysis with applications to agent-based systems. Ph.D. thesis, University of South Australia (2008)

7. Liang, S., Tang, Y., Zhu, Q.: Passive wake-up scheme for wireless sensor networks. In: Proceedings of the 2nd ICICIC, p. 507. IEEE Computer Society, Washington, DC, USA (2007)

8. Liu, Z., Elhanany, I.: Rl-mac: a reinforcement learning based mac protocol for wireless sensor networks. Int. J. Sen. Netw. 1(3/4), 117–124 (2006)

9. Mihaylov, M., Le Borgne, Y.A., Tuyls, K., Nowé, A.: Distributed cooperation in wireless sensor networks. In: Yolum, Tumer, K., Stone, P., Sonenberg (eds.) Proceedings of the 10th International Conference on Autonomous Agents and Multiagent Systems (AAMAS 2011), Taipei, Taiwan (May 2011)

10. Mihaylov, M., Le Borgne, Y.A., Tuyls, K., Nowé, A.: Self-organizing synchronicity and desynchronicity using reinforcement learning. In: Filipe, J., Fred, A. (eds.) Proceedings of the 3rd International Conference on Agents and Artificial Intelligence (ICAART), Rome, Italy, pp. 94–103 (January 2011)

11. OMNET++: Project, http://www.omnetpp.org/ – a C++ simulation library and framework

12. Paruchuri, V., Basavaraju, S., Durresi, A., Kannan, R., Iyengar, S.S.: Random asynchronous wakeup protocol for sensor networks. In: Proceedings of the 1st BROADNETS, pp. 710–717. IEEE Computer Society, Washington, DC, USA (2004)

13. Posch, M.: Win-Stay, Lose-Shift Strategies for Repeated Games–Memory Length, Aspiration Levels and Noise. Journal of Theoretical Biology 198(2), 183–195 (1999)

14. Schurgers, C.: Wakeup Strategies in Wireless Sensor Networks. In: Wireless Sensor Networks and Applications, p. 26. Springer, Heidelberg (2007)

15. Sutton, R.S., Barto, A.G.: Reinforcement Learning: An Introduction. MIT Press (1998)

16. Watkins, C.: Learning from delayed rewards. Ph.D. thesis, University of Cambridge, England (1989)

17. Ye, W., Heidemann, J., Estrin, D.: Medium access control with coordinated adaptive sleeping for wireless sensor networks. IEEE/ACM Trans. Netw. 12(3), 493–506 (2004)

Self-Organizing Logistics Process Control: An Agent-Based Approach

Jan Ole Berndt

Center for Computing and Communication Technologies (TZI)
Universität Bremen, Am Fallturm 1, 28359 Bremen, Germany
joberndt@tzi.de

Abstract. Logistics networks face the contradictory requirements of achieving high operational effectiveness and efficiency while retaining the ability to adapt to a changing environment. Changing customer demands and network participants entering or leaving the system cause these dynamics and hamper the collection of information which is necessary for efficient process control. Decentralized approaches representing logistics entities by autonomous artificial agents help coping with these challenges. Coordination of these agents is a fundamental task which has to be addressed in order to enable successful logistics operations. This paper presents a novel approach to self-organization for multiagent system coordination. The approach avoids a priori assumptions regarding agent characteristics by generating expectations solely based on observable behavior. It is formalized, implemented, and applied to a logistics network scenario. An empirical evaluation shows its ability to approximate optimal supply network configurations in logistics agent coordination.

1 Introduction

Logistics plays a major role in globalized economy. Industrial production and trade require efficient and reliable supply networks. Growing interrelations between these networks and the inherent dynamics of the logistics domain result in a high complexity of global supply processes [9]. The application of conventional centralized planning and control approaches to these processes suffers from that complexity. Therefore, decentralized methods become necessary which employ autonomous actors representing logistics entities and objects [10].

From the artificial intelligence point of view, these autonomous entities can be represented by intelligent software agents to model logistics networks as multiagent systems (MAS). These systems enable simulations, evaluations, and actual implementations of new approaches in autonomous logistics [17].

In order to develop the aforementioned approaches, coordination and cooperation of autonomous entities is a challenging task. In the logistics domain, coordination faces the contradictory requirements of achieving high operational efficiency while retaining the system's ability to adapt to a changing environment. On the one hand, supply networks have to achieve high performance rates concerning asset utilization, cost reduction, and customer satisfaction. On the other hand, they require flexible and robust structures in order to react to unforeseen changes caused by the domain's inherent dynamics.

J. Filipe and A. Fred (Eds.): ICAART 2011, CCIS 271, pp. 397–412, 2013.

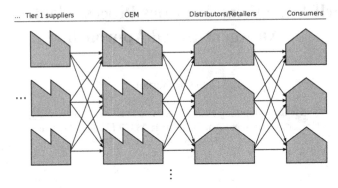

Fig. 1. Schematic diagram of a supply network showing all possible relationships between the participants

This paper presents a novel approach for self-structuring multiagent systems. Section 2 further explores the challenges in logistics network configuration and operation. Section 3 examines agent coordination mechanisms for organizing decentralized behavior in logistics networks. Motivated by these considerations, Section 4 introduces *expectation-based self-organization* as an adaptive structuring paradigm for multiagent systems based on sociological theory. That approach is evaluated in Section 5 in a simulated supply network scenario with regard to coordination effort and logistics performance. Finally, Section 6 recapitulates the achievements of this paper in a concluding summary and gives an outlook on possible future work.

2   Self-Organizing Supply Networks

In order to efficiently solve repeatedly occurring coordination problems in decentralized systems, organizational structures have to be established [8]. Yet, it is unclear which kind of structure is applied best, given a particular coordination problem. Consider, for instance, a logistics network as depicted in Figure 1. In this network, the participants must choose which subset of the possible relationships between each two tiers (shown as arrows in the direction of material flows) actually to establish. This decision has to consider transaction costs (e.g., interaction effort and transportation costs) as well as the responsiveness and reliability of possible business partners in order to enable efficient operations within the network.

A supply network can be represented as a graph consisting of logistics entities as its nodes and their possible business relationships as edges. Establishing an organizational structure refers to the choice of a subgraph restricting the set of edges to a subset of all possible ones. An efficient organizational structure furthermore minimizes the actually instantiated relationships while maximizing the achieved operations outcome with regard to logistics performance measures.

However, due to the dynamics of logistics processes, conventional design time evaluation and optimization of these organizational structures is not sufficient in terms of flexibility and robustness. Increasing demands of the final consumers, for example, require structural modifications in the distribution part of the supply network in order to fulfill those demands: Additional storage capacity has to be allocated and even completely new channels of product distribution must be established. Thus, the structures in that part the supply network must be refined, i.e., additional or alternative options of business relationships must be instantiated.

This is but an example for the dynamics in logistics that is further aggravated by the *openness* of those systems [1]: Not only consumer demand changes as well as unforeseen failures of scheduled operations may happen (leading to the need of dynamic replanning and reallocation of resources), but the logistics market itself may alter. New competitors as well as new customers may enter, causing further changes in demand, prices, and requirements of products and services. These developments evoke the necessity for each participant to constantly adapt his relationships to customers and suppliers in order to secure market shares and to fulfill the customers' demands. Such an adaption, furthermore, affects other business relationships within the network, requiring an extended refinement of supply partnerships therein.

Thus, modeling and operating supply networks with multiagent systems requires the agents' ability to establish organizational configurations that allow for efficient operation, while being flexible enough (i.e., alterable) to cope with the dynamics of logistics processes. Hence, self-organizing MAS become necessary which autonomously arrange their structure in accordance with dynamically changing conditions. In this context, self-organization is therefore considered as the emergent evolvement and modification of organizational structures defining business relationships between supply network partners.

3 Agent Coordination

In order to be able to autonomously coordinate their activities (e.g., to establish and operate logistics networks), artificial agents have to interact with each other. For this purpose, agent communication languages modeling speech acts between the agents are commonly used [4,5]. Based on these speech acts, a range of interaction and negotiation protocols have been developed which coordinate agent behavior. Patterns of interaction reflect relationships between the participants and, thus, express the structure of the multiagent system. Vice versa, structuring a supply network, modeled as a MAS, means defining channels and modes of agent communication.

A wide variety of different structuring paradigms for MAS has been proposed [7]. These structures range from strict hierarchies [12] to market-based methods [2]. The former use centralized decision-making at the top and distributed processing of specific tasks at the bottom; the latter are completely decentralized and rely on negotiations for each single task rather than on any middle or long term relationships. These predefined mechanisms differ in their ability to handle changing conditions as well as in their necessary effort for coordinating the actions of a network's members [16]. Therefore, the expected dynamics of the application domain must be estimated in order to make use of them.

However, choosing a prototypical organization approach for a whole network may not be sufficient. In fact, heterogenous relationships may be required between agents in different parts of the supply network. Moreover, predetermining agent interaction patterns will necessarily lead to a compromise between efficient operation and adaptive behavior: For example, negotiation based interaction paradigms are highly adaptive when it comes to changing behavior of participating agents (as they allow for determining the best result under any given conditions). Nevertheless, they lead to a large overhead of communication and computation effort as every interaction task involves all possible participants among the agents.

In order to confine the interaction effort [18], a MAS can be subdivided into teams of agents with similar properties or joint objectives [21,22]. Team building and joint action among autonomous agents for distributed problem solving includes determination of potentials for cooperative acts, formation of teams, distributed planning, and the actual processing of plans [22]. In the logistics domain, team formation methods have shown benefits in terms of increased resource utilization efficiency while reducing the communication effort of agents performing similar tasks [17,19].

However, clustering agents in teams usually focuses on short term behavior and tasks, rather than on middle and long term structures in agent interaction. Furthermore, team formation processes rely on the exchange of information about agent properties and goals among the potential team members. Hence, they assume any participating agents to behave benevolently, i.e., to be trustworthy. In an open system, however, agents may be confronted with deceitfully behaving participants [13] or others, simply not willing to share information.

Thus, potential interaction partners in open MAS cannot be assumed a priori to exhibit particular behavioral characteristics. In fact, they appear as *black boxes* and therefore must be observed by the other agents or the system designer in order to determine their characteristics during runtime of the system. Based on such observations, a structuring approach for MAS has been proposed, using explicit modeling of *expectations* concerning communication flows [1,14]. This approach, which is inspired by the sociological theory of communication systems [11], establishes a notion of communicative agent behavior that is reflected by the modeled expectations.

Feeding those expectations back into the decision-making process of interacting agents offers a promising foundation for self-structuring MAS, as they reflect other agents' characteristics inferred from their observable behavior. Customer demands, for instance, can be observed from the incoming orders on the supplier's side. The supplier can establish expectations regarding the customers' behavior and subsequently adapt his own behavior with regard to these expectations. Hence, the system as a whole is enabled to adapt to implicit characteristics and external impacts by the agents refining their communication patterns in terms of business relationships, i.e., the system organizes itself.

To summarize, agent coordination refers to communication processes between these agents. Prototypical coordination mechanisms lead to a compromise between operational efficiency and flexibility while dynamic team formation requires additional behavioral assumptions to overcome this problem. However, the systems-theoretical perspective of expectations structuring agent interaction (rather than assumptions and

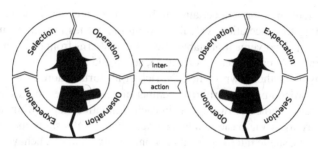

Fig. 2. Agent decision-making in a feedback loop of behavior observation, expectation, action selection, and operation

commitments) provides a promising foundation for self-organization as a paradigm for multiagent coordination.

Nevertheless, in the aforementioned approach [1,14], expectations reflecting and guiding agent behavior are modeled by the system designer as an external observer. However, self-organization requires organizational structures to emerge from the system's operations without external intervention; i.e., the mentioned feedback loop must be closed within the multiagent system. Thus, the next section introduces the notion of *double contingency* which describes the emergence of mutual expectations structuring communication flows between agents appearing as black boxes. In the following, this concept is operationalized in order to demonstrate its ability to enable autonomous coordination of agent communication systems.

4   Expectation-Based Self-Organization

According to the sociologist Niklas Luhmann, *double contingency* denotes both the fundamental problem of social order generation as well as its own solution leading to the emergence of social order [11, pp. 103–136]. Referring to Parsons and Shils [15], he points out: Given two entities *alter* and *ego*, mutually appearing as black boxes to each other, "if alter makes his action dependent on how ego acts, and ego wants to connect his action to alter's" [11, p. 103], they reciprocally block their ability to act at all.

However, the solution to that problem lies in the interdependency of actions, as well. As soon as alter or ego behave in whatever way, action becomes not only possible, but social structures emerge from the self-referential circle of mutually dependent activities. Those structures consist of expectations evolving from, e.g., ego's observation of his own actions as well as of alter's behavior. These expectations, in turn, guide ego's selection of subsequent actions. Hence, a feedback loop of observation, expectation, selection, and operation (action) emerges as depicted in Figure 2.

In the context of multiagent systems, double contingency can be viewed analogically as the problem of determining interaction opportunities. It also denotes its own solution through the emergence of expectations guiding agent communication as the fundamental operation in MAS. As a starting point serves the simulation model by Dittrich et al. [3]: They simulate and analyze Luhmann's concept of double contingency in a scenario of two agents interacting with each other by exchanging messages with varying content. The agents memorize a certain number of these messages and select their response according to expectations calculated from the entries in their memories. That approach shows the evolvement of stable interaction patterns from the agents' behavior under a wide range of parameter conditions [3, sec. 3 and 5].

In an extension of their own model, Dittrich et al. furthermore examine the emergence of social order among an arbitrary number of agents [3, sec. 6]. To this end, they introduce a random choice of two agents in each simulation step, letting them interact in the same way as in the basic dyadic setting. Their results show that, for growing numbers of agents, stable interaction patterns only evolve if alter's behavior reflects the average agent behavior within the system and if the agents are able to observe more pairwise encounters than they are involved in themselves [3].

However, those requirements as well as their abstract model of message contents prevent an application of that approach for self-organization in MAS following particular purposes. Choosing agent pairs for interaction at random contradicts the objective of emerging agent relationships which define interaction channels. In fact, self-organization refers to the systematic choice of interaction partners among the set of all agents in a MAS in its very core. Thus, that selection must be based on expectations regarding interaction outcomes. In applied self-organizing MAS (e.g., for modeling supply networks), the semantics of message contents depending on the respective application domain is a crucial factor for the determination of such outcomes. Hence, it has to be considered when generating agent expectations.

Therefore, in the following, a model of double contingency is developed, based on the basic approach by Dittrich et al. [3], allowing for the application of self-organizing coordination of an arbitrary number of agents (Section 4.1). Moreover, the original model using meaningless messages is enriched with semantics derived from the logistics domain, being compatible with a standard agent interaction protocol (Section 4.2).

4.1 Modeling Double Contingency

In this model, agent operations consist of sending FIPA-ACL compliant messages [5]. Observing them refers to their storage in an agent's memory which is used to calculate expectations for possible further communicative acts. The observing agent subsequently selects its next message to be sent according to these expectations. Thus, an agent's communicative behavior exclusively depends on its memorized observations of other agents' behavior, avoiding any further assumptions of their internal properties and characteristics. Hence, the basic steps enabling the agents to self-organize are as follows.

1. The *observation* of incoming messages sent by other agents.
2. The *selection* of messages to be sent to other agents.

An agent's memory is a vector $MEM = (mem_1, \ldots, mem_n)$ with a fixed length n, where each entry mem_i denotes a tuple of messages $m \in M$ (M being the set of all possible messages). The second message is the observed response to the first one: $mem_i = \langle m_{received,i}, m_{sent,i} \rangle$. An agent possesses two of those memories, MEM_{ego} and MEM_{alter}, storing its own reactions to perceived messages and observed others' reactions to its own messages, respectively. Thus, observation takes place when sending a message

m_{sent} by adding it to MEM_{ego} together with the last received message $m_{received}$ as well as when receiving a message $m_{received}$ by adding it to MEM_{alter} together with the last message m_{sent} the agent sent itself. Each time, a tuple of messages is memorized, if this would lead to a memory size $> n$ the oldest entry is removed from the memory.

This way to model an agent's memory is an important modification of that by Dittrich et al. [3], differing in alter not only being considered one single agent, but the whole community of agents other than ego. This reflects Luhmann's understanding of double contingency as a phenomenon not restricted to an encounter of two individuals, but occurring between systems in a generalized manner [11, pp. 105–106]. Thus, expectations may well be established regarding the behavior of the whole MAS, considering it as a social system. The entries in its memory, therefore, reflect an agent's observations of its interactions with any of its fellow agents.

Moreover, this interpretation of double contingency between an agent and the whole agent community allows not only for the content of a message to be selected according to memorized experience from former agent interactions. In fact, it also enables the agent to determine a message's receivers (i.e., the interaction partners) in the selection process. Hence, the advantages of the dyadic model by Dittrich et al. [3] regarding structural emergence are retained while avoiding the aforementioned drawbacks of its extension for an arbitrary number of agents.

In order to calculate expectations from an agent's memory MEM, the memory access function $lookup : MEM^* \times M \times M \longrightarrow [0,1]$ (with MEM^* denoting the set of all possible agent memories MEM) estimates the probability of one message being observed as the response to another:

$$lookup(MEM, m_{received}, m_{sent}) = \frac{l_{m_{received}, m_{sent}}}{\sum\limits_{m_j \in M} l_{m_{received}, m_j}} \tag{1}$$

where

$$l_{m_{received}, m_{sent}} = \frac{c_M}{|M|} + \sum\limits_{i=1}^{n} \frac{n+1-i}{n} \cdot \begin{cases} 1 & \text{if } \langle m_{received}, m_{sent} \rangle \equiv mem_i \in MEM \\ 0 & \text{else} \end{cases} \tag{2}$$

Here, $\equiv$ is an equivalence relation on the message tuples $\langle M, M \rangle \times \langle M, M \rangle$. Therefore, $\langle m_{received}, m_{sent} \rangle \equiv mem_i$ denotes the pairwise equality of the received and sent messages, compared to those in memory entry mem_i, with regard to their performatives, sets of receivers, and contents. This is the second major modification of the original model, allowing for considering advanced message semantics (in contrast to the very abstract message representation by Dittrich et al. [3]). Especially the content of messages depends on the application domain. Thus, domain dependent equality measures (e.g., the distinction of orders for different product types) are required. The constant c_M is used to avoid message combinations to be regarded completely impossible in case of missing observations [3, sec. 9.4]. With $mem_1 \in MEM$ being the most recent observation, this function uses a linear discount model to reflect the agent gradually forgetting past observations.

Two kinds of expectations are subsequently calculated for selecting an agent's next message. On the one hand, the *expectation certainty* (EC) denotes an agent's assuredness about which reaction to expect from the MAS following its own message. On the other hand, the *anticipated expectation* (AE)[1] reflects an agent's estimation of other agents' expectations towards its own behavior.

The EC is calculated based on a modified version of the standard deviation, estimating an agent's certainty over the possible reactions to its next message m_{sent} [3, sec. 2.1 and 9.5]:

$$EC_{m_{sent}} = \sqrt{\frac{|M|}{|M|-1} \sum_{m_j \in M} \left(\frac{1}{|M|} - lookup(MEM_{alter}, m_{sent}, m_j)\right)^2} \qquad (3)$$

This linear function returns a value of 0 for uniformly distributed probability estimations over the others' possible reactions to an agent's message. Contrastingly, the most inhomogenous distribution of those estimated probabilities leads to a value of 1. Thus, the function reflects the certainty of the agent expecting a particular response to its message. However, note that the *lookup* of each value for the possible reactions of the MAS is used with the sent message as its first argument. This is because MEM_{alter} contains ego's observations of himself from alter's perspective. Thus, as ego's m_{sent} is what alter receives from him, it is treated as the received message in MEM_{alter}.

On the other hand, the AE is calculated directly using the *lookup*-function as the estimated probability of the agent's next message m_{sent} in response to the last received message $m_{received}$ [3, sec. 2.1]:

$$AE_{m_{sent}} = lookup(MEM_{ego}, m_{received}, m_{sent}) \qquad (4)$$

As MEM_{ego} stores all observations of ego's responses to received messages, Equation 4 reflects ego's anticipation of alter's perception of his behavior. Hence, the AE denotes an agent's estimation of what is expected from itself by the community of its fellow agents.

Finally, a weighted sum combines both types of expectations to a selection value V for each possible next message $m_{sent} \in M$. This value represents the potential of a given message to stabilize the interaction flows within the MAS. High selection values reproduce themselves when an agent chooses a corresponding message and thereby feeds it back into the control loop. This leads to an emergence of interaction patterns (repeatedly occurring communication flows between the agents) which represent the social structures in a MAS. However, differing from Luhmann's theory and the model by Dittrich et al. [3], goal-directed agent interaction requires social structures which facilitate the fulfillment of the agents' objectives. Therefore, at this stage, a utility function $utility: M \longrightarrow \mathbb{R}_+$ is additionally introduced. This function enables V not only to reflect communicative stability within the system, but also directs the agent's behavior towards domain dependent performance criteria. Thus, $V_{m_{sent}}$ is given by the following equation.

$$V_{m_{sent}} = (\alpha EC_{m_{sent}} + (1-\alpha)AE_{m_{sent}}) \cdot utility(m_{sent}) + \frac{c_f}{|M|} \qquad (5)$$

---

[1] Dittrich et al. [3] call this *expectation-expectation* (EE), literally translating Luhmann's original German term. Luhmann, however, uses *anticipated expectation* in the English edition of his main work [11].

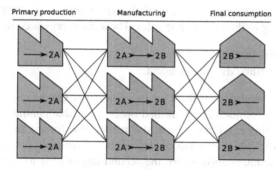

Fig. 3. A simple supply network depicting agent roles and relationships in the logistics domain

The parameter $\alpha \in [0,1]$ weights the balance between EC and AE. The constant c_f avoids marginal differences in the weighted sum to cause overly high effects on the message selection in order to retain an agent's ability to try out alternative messages, i.e., to occasionally explore the possibility space [3, sec. 9.1].

Calculating $V_{m_{sent}}$ for all possible message options $m_{sent} \in M$ enables an agent to select its operations (i.e., the messages to be sent) according to its expectations which are based on observations of its interaction with other agents. As the selection of an operation leads to further observations, the aforementioned feedback loop is closed. However, the method of actually choosing an operation in accordance with the calculated selection values remains to be determined. That method depends on an agent's role in the MAS and is introduced in the next subsection.

4.2   Representing the Logistics Domain

When modeling supply network participants as autonomous agents, these agents may have different capabilities. As shown in Figure 3, they can be classified in *primary producers* that produce raw materials without consuming anything, *final consumers* that only consume products, and *manufacturers* that consume materials and semi-finished parts in order to transform them into new parts and products. Concerning the business relationships between the entities, it is sufficient to distinguish the agents by their roles as producers and/or consumers of certain goods (manufacturers acting both as producers and consumers). Their respective possible relationships as *suppliers* and *customers* are depicted by the edges between the entities in Figure 3 (with the left hand side of an edge being attached to a supplier and its right hand side being connected to the respective customer).

These relationships denote possible occurrences of order/delivery processes, that form the fundamental operations of a logistics system. They are modeled using the FIPA-REQUEST interaction protocol [6]: An order is placed by sending a REQUEST message containing a product type and the requested amount of that good to any subset of the possible suppliers for this product. An answer with the REFUSE or FAILURE performative is considered a failure to deliver while an INFORM leads to the supplier agent removing the specified amount of products from its inventory and the customer adding it to its own one.

For selecting their messages based on their expectations, the agents have different objectives, according to their respective roles. These are represented in:

1. An agent's utility function.
2. The selection method used by an agent.

From a customer's point of view, there are two objectives. On the one hand, a customer strives to maximize the number of fulfilled orders to enable continuous product consumption. On the other hand, this role is also responsible for the amount of messages occurring in the MAS which depends on the number of receivers per message. In order to ensure a low communication effort, the second objective is to minimize the number of order receivers. Thus, for calculating the selection values for each message, the following utility function is employed.

$$utility(m_{sent}) = \frac{1}{|rec(m_{sent})|} \cdot eor(m_{sent}) \tag{6}$$

In this function, $rec(m_{sent})$ denotes the set of receivers of message m_{sent} and $eor(m_{sent})$ is the estimated order fulfillment rate, calculated as follows.

$$eor(m_{sent}) = \sum_{m_j \in M} lookup(MEM_{alter}, m_{sent}, m_j) \cdot \begin{cases} 1 & \text{if } perf(m_j) = \text{INFORM} \\ 0 & \text{else} \end{cases} \tag{7}$$

As $perf(m_j)$ indicates the performative of message m_j, the eor represents the estimated probability of a positive answer to the given order. Hence, this utility function favors those orders that have a small number of receivers while having a high estimated probability to be fulfilled.

Finally, a message m_{sent} is randomly chosen out of the set of all possible messages with a probability based on its selection value. In order to be able to adjust the level of randomness in this selection, the selection value is further modified by an exponent γ, allowing for choosing from a range between completely random selection ($\gamma = 0$) and deterministically selecting the maximum value ($\gamma = \infty$). Therefore, following Dittrich et al. [3, sec. 2.1] again, selection is done using a probability distribution over all possible messages m_{sent}, calculated as follows.

$$p(m_{sent}) = \frac{V_{m_{sent}}^{\gamma}}{\sum_{m_j \in M} V_{m_j}^{\gamma}} \tag{8}$$

From a supplier's point of view, on the other hand, the objectives are easier to represent. A supplier is assumed to be generally interested in fulfilling an order, if possible. If it is not possible to fulfill all orders, a supplier prefers to maximize the system's stability in terms of predictability of further incoming orders and anticipated expectations of the customers. In other words, a supplier favors orders by his regular customers as he can expect them to place further orders in the future and he can anticipate the expectation of their orders being fulfilled. This setting is directly represented in the weighted sum of EC and AE. Thus, the supplier's utility function remains unused ($utility(m_{sent}) = 1$).

For the choice of a message, the selection value $V_{m_{sent}}$ is calculated for each answer $m_{sent} \in M$ with $perf(m_{sent}) =$ INFORM. Starting with the highest selection value, the messages are processed in descending order. As long as the supplier's inventory stock level allows for fulfilling the processed order, an INFORM message is sent. If that is no longer possible, all subsequent orders are refused.

5 Empirical Evaluation

In order to validate the ability of expectation-based self-organization to efficiently structure and operate multiagent systems modeling supply networks, that approach will be compared to the performance of a system with a previously defined communication structure. For this purpose, the approach is implemented and applied to an example scenario using the multiagent-based simulation system PlaSMA [20].

5.1 Experimental Setup

In this evaluation, a network with three tiers and three parallel operating entities is modeled as depicted in Figure 3. Each agent produces and/or consumes an amount of two units of the product types A and/or B (two A being transformed into two B by the agents at the manufacturing tier). Furthermore, every agent has an outbound inventory capacity of four units per product type, restricting the amount of goods that can be produced and stored by a single logistics entity. The agents acting as customers pursue a policy of ordering an amount of four units if the respective inventory stock level reaches six or less.

In the simulation, a message sent by an agent can be received and processed in the next time slice at the earliest. Therefore, sending an order and receiving the response takes two simulation cycles. In that time, four units of the required type of products can be consumed. Thus, the chosen order batch size enables maximal utilization of production and consumption processes while requiring minimal outbound storage capacity on the suppliers' side. However, the threshold of six units for placing an order enables the agents at the manufacturing tier to build up inbound safety stocks, allowing for continued production in case of supply shortfalls and thus compensating disturbances at the early network tiers.

Knowing these mentioned capabilities of the participating agents, it is easy to pre-structure this network by choosing an arbitrary bijection out of the possible relationships between each two tiers. For each order following the mentioned policy, this ensures the number of receivers being one (the possible minimum) and the supplier to be able to fulfill that order as soon as enough raw material has been produced in an initialization phase (as the amount of consumed goods equals that of produced ones). Thus, such an arrangement of relationships necessarily leads to a maximized operation efficiency of the modeled supply network using a minimal number of sent messages. Regarding these objectives, it therefore guarantees optimal results making it especially suitable as a reference for the self-organizing approach.

However, without prior knowledge of other agents' capabilities and relationships, the choice of interaction partners leading to an efficient and reliable network structure is not an obvious one. As the possible configurations of message receivers for each order correspond to the power set of the set of available suppliers (without the empty set), in a network with n tiers and m parallel actors at each tier, the total number of potential relationships is $(m \cdot (2^m - 1))^{n-1}$ (the possible communication paths through the network)[2]. Thus, in the chosen scenario the self-organizing agents can choose between 441 possible interaction patterns leading to different performance rates. Therefore, in this simple scenario, agent coordination is already complex enough to make it suitable for evaluating the emergence of communication structures.

For this purpose, the expectation-based agents are configured as follows. The set of possible orders to be sent by a customer is given by the possible combinations of their receivers, their performatives, and their content. As there is only one type and a fixed amount of units to order per customer, there is only one possible content. The same holds for the performative, as an order is always a REQUEST message. Thus, the set of possible orders is determined by the possible combinations of a message's receivers (the power set of the set of possible suppliers). For the replies, on the other hand, the receiver as well as their contents are preassigned by the incoming orders. Hence, a supplier's only choice is between the message performatives according to the FIPA-REQUEST interaction protocol.

For generating the results presented in the following subsection, the constant values are based on those used by Dittrich et al. [3]: $c_M = 2$ and $c_f = 0.02$. The agent memory size is set to $n = 25$ for both MEM_{ego} and MEM_{alter}, the balance between EC and AE to $\alpha = 0.5$, and the customers' selection value gain to $\gamma = 3$. All agent memories are initially populated with randomly chosen messages in order to reflect the agents not having any specific prior information about promising interaction channels.

In order to validate the approach to expectation-based self-organization, it is compared with an optimal configuration as outlined above. The performance is measured with regard to the number of receivers per order, the final consumers' customer satisfaction rate (i.e., the number of fulfilled orders), and the utilization of the final consumers' product consumption. The first two criteria directly reflect the customers' utility function. They give information about the communication effort needed to operate the network (message receivers) as well as about the reliability of the emerging relationships between the agents (customer satisfaction). Thus, these measures reflect the extend of stability of the evolving network structures. The consumers' utilization, on the other hand, is an additional logistics performance measure that allows for validating the supply network's overall operating efficiency in terms of product throughput rates.

5.2   Results and Discussion

The results depicted in Figures 4–6 show the number of receivers, the customer satisfaction, and the consumer utilization as average values over 200 simulation runs. Each run consists of 1000 production and/or consumption operations. For the calculation of the order fulfillment rate, the last ten messages received are considered for each time slice

---

[2] There are m agents at a tier with $2^m - 1$ possible interaction partners, each. The potential paths through the network are given by the combination of those options over all $n - 1$ links between two tiers.

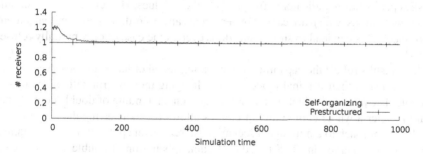

Fig. 4. Number of message receivers (orders of final consumers)

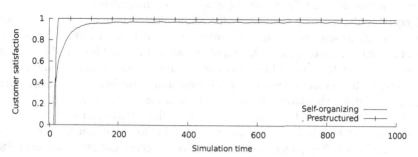

Fig. 5. Customer satisfaction among the final consumers

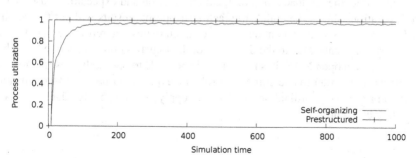

Fig. 6. Consumption rate (utilization) among the final consumers

while the utilization is measured over the last ten attempts to consume the respective amount of products.

For the prestructured reference configuration, Figures 5 and 6 show that there is a short initialization phase until the inventories of the suppliers are filled high enough to be able to fulfill the customers' orders. After that phase, the optimal values are reached for the order fulfillment rate and the customers' utilization while the number of receivers per order is always one by definition (Figure 4).

In the self-organizing network, these levels are not reached completely. However, the values converge near the optimum, showing that the agents autonomously establish one to one interaction relationships (Figure 4) that still lead to a near optimal order fulfillment rate of more than 97% (Figure 5). The process utilization (Figure 6) as a logistics performance indicator corresponds to these values. However, it shows slightly higher fluctuations, which are caused by the agents always ordering the minimal amount of products. This can lead to supply shortfalls even in the case of only partially refused orders[3].

These results reflect the capability of generating social order as it is observed by Dittrich et al. [3] in their original model. Thus, changing their interpretation of a dyadic encounter between individuals to a more general understanding of double contingency regarding alter a whole community of entities allows for transferring the properties of their basic approach to a multiagent scenario. Therefore, an application of expectation-based self-organization in MAS based on Luhmann's notion of double contingency is possible without the requirement for a reduction of interaction to pairwise communication processes or the need for extended agent observation activities.

Concerning the logistics application, the results demonstrate that the expectation-based approach to self-organizing agent interaction is not only capable of efficiently structuring and operating the modeled supply network. In fact, it is even able to establish an optimal configuration of agent communication channels (one to one relationships), leading to similar performance rates compared to the benchmark arrangement in the course of the simulation. As the agents occasionally explore alternative interaction options, delivery failures occur from time to time, leading to slightly less than optimal customer satisfaction and utilization rates due to the minimal order size and inventory capacities. Regarding these measures, safety stocks and increased order sizes may compensate the disturbances to further improve logistics performance.

To summarize, the feedback loop of agent observation and expectation-based selection of operations shows the ability to reach near optimal results without the requirement for a priori assumptions about agent characteristics[4] or repetitive negotiations between several agents. Due to the dynamics of the logistics domain and the black box nature of agents in open MAS, it is not generally possible to optimally prestructure a logistics network. In order to overcome this problem, expectation-based self-organization provides a promising coordination method for supply systems, being adaptive as well as operating efficiently.

6 Conclusions

This paper has identified the requirement for both adaptive and efficient supply networks. As multiagent systems provide a means for decentralized modeling of logistics networks, possible coordination techniques have been investigated in terms of their applicability to address the challenges in supply network organization. In this context,

---

[3] When exploring alternative sets of suppliers, an agent may split its orders over, e.g., two suppliers. If one of the suppliers refuses that order and the other one sends a delivery message, the production process utilization suffers from a supply shortage. The customer satisfaction, however, is less affected by this partially refused order. Therefore, the product consumption varies to a higher extend than the customer satisfaction.

[4] In contrast to that, e.g., determining the benchmark configuration requires knowledge of the agents' production and consumption rates.

expectations regarding observable behavior have been presented as a means for dynamically structuring agent relationships, avoiding the necessity of a priori assumptions regarding agent properties and behavior.

Based on theoretical foundations from sociology [11], a simulation approach to emerging interaction patterns using expectations has been adapted and generalized to be applicable in multiagent systems. That method has been evaluated in a supply network scenario according to coordination efficiency and reliability as well as logistics performance. The results illustrate that self-organized agent coordination based on mutual expectations is able to establish organizational structures which allow for near optimal performance rates regarding the evaluation criteria. Hence, the approach has been shown to enable efficient interaction of autonomous entities to emerge solely based on locally observable agent behavior.

However, there are still questions open for future examination. While the presented approach performs very well in a stable agent community with repeating interaction contents (i.e., a static supply network setup), it remains to be analyzed in a setting with dynamically changing agent memberships and activities. In such a scenario, a self-organizing network can be assumed to actually outperform a predefined structure as the latter is not able to adapt to changing conditions. Furthermore, in that context, an examination of the different parameters' impact on the predictability and speed of convergence (learning rate) and the limits of overall performance of the emerging system structure will give further insights into the capabilities of expectation-based self-organization. This may motivate further refinements of that approach to agent coordination.

Acknowledgements. This research is supported by the German Research Foundation (DFG) within the Collaborative Research Center 637 "Autonomous Cooperating Logistic Processes: A Paradigm Shift and its Limitations" (SFB 637) at Universität Bremen, Germany.

References

1. Brauer, W., Nickles, M., Rovatsos, M., Weiss, G., Lorentzen, K.F.: Expectation-Oriented Analysis and Design. In: Wooldridge, M.J., Weiß, G., Ciancarini, P. (eds.) AOSE 2001. LNCS, vol. 2222, pp. 226–244. Springer, Heidelberg (2002)
2. Collins, J., Youngdahl, B., Jamison, S., Mobasher, B., Gini, M.: A Market Architecture for Multi-Agent Contracting. In: Agents 1998, pp. 285–292. ACM Press, Saint Paul (1998)
3. Dittrich, P., Kron, T., Banzhaf, W.: On the Scalability of Social Order: Modeling the Problem of Double and Multi Contingency Following Luhmann. JASSS 6(1) (2003)
4. Finin, T., Fritzson, R., McKay, D., McEntire, R.: KQML as an agent communication language. In: Adam, N.R., Bhargava, B.K., Yesha, Y. (eds.) CIKM 1994, pp. 456–463. ACM, New York (1994)
5. Foundation for Intelligent Physical Agents: FIPA Agent Communication Language Specifications. Standard (2002)
6. Foundation for Intelligent Physical Agents: FIPA Request Interaction Protocol Specification. Standard (2002), document No. SC00026H
7. Horling, B., Lesser, V.: A Survey of Multi-Agent Organizational Paradigms. Know. Eng. Rev. 19(4), 281–316 (2005)

8. Horling, B., Mailler, R., Lesser, V.: A Case Study of Organizational Effects in a Distributed Sensor Network. In: Sonenberg, L., Sierra, C. (eds.) AAMAS 2004, pp. 1294–1295. IEEE Computer Society, New York (2004)

9. Hülsmann, M., Scholz-Reiter, B., Austerschulte, L., Wycisk, C., de Beer, C.: Autonomous Cooperation – A Way to Cope with Critical Incidents in International Supply Networks (ISN)? An Analysis of Complex Adaptive Logistic Systems (CALS) and their Robustness. In: 24th EGOS Colloquium, Vrije Universiteit Amsterdam (2008)

10. Hülsmann, M., Scholz-Reiter, B., Freitag, M., Wycisk, C., De Beer, C.: Autonomous Cooperation as a Method to cope with Complexity and Dynamics? – A Simulation based Analyses and Measurement Concept Approach. In: Bar-Yam, Y. (ed.) ICCS 2006, Boston (2006)

11. Luhmann, N.: Social Systems. Stanford University Press, Stanford (1995)

12. Montgomery, T.A., Durfee, E.H.: Search Reduction in Hierarchical Distributed Problem Solving. Group Decis. Negot. 2, 301–317 (1993)

13. Nickles, M., Rovatsos, M., Weiss, G.: Expectation-oriented modeling. Eng. Appl. Artif. Intel. 18(8), 891–918 (2005)

14. Nickles, M., Weiss, G.: Multiagent Systems Without Agents – Mirror-Holons for the Compilation and Enactment of Communication Structures. In: Fischer, K., Florian, M., Malsch, T. (eds.) Socionics. LNCS (LNAI), vol. 3413, pp. 263–288. Springer, Heidelberg (2005)

15. Parsons, T., Shils, E. (eds.): Toward a General Theory of Action. Harvard University Press, Cambridge (1951)

16. Schillo, M., Spresny, D.: Organization: The Central Concept for Qualitative and Quantitative Scalability. In: Fischer, K., Florian, M., Malsch, T. (eds.) Socionics. LNCS (LNAI), vol. 3413, pp. 84–103. Springer, Heidelberg (2005)

17. Schuldt, A.: Multiagent Coordination Enabling Autonomous Logistics. Doctoral dissertation, Universität Bremen (2010)

18. Schuldt, A.: Team Formation for Agent Cooperation in Logistics: Protocol Design and Complexity Analysis. In: Filipe, J., Fred, A. (eds.) ICAART 2011, pp. 398–405. SciTePress, Rome (2011)

19. Schuldt, A., Berndt, J.O., Herzog, O.: The Interaction Effort in Autonomous Logistics Processes: Potential and Limitations for Cooperation. In: Hülsmann, M., Scholz-Reiter, B., Windt, K. (eds.) Autonomous Cooperation and Control in Logistics. Springer, Heidelberg (to appear)

20. Schuldt, A., Gehrke, J.D., Werner, S.: Designing a Simulation Middleware for FIPA Multiagent Systems. In: Jain, L., Gini, M., Faltings, B.B., Terano, T., Zhang, C., Cercone, N., Cao, L. (eds.) WI-IAT 2008, pp. 109–113. IEEE Computer Society Press, Sydney (2008)

21. Tambe, M.: Towards Flexible Teamwork. J. Artif. Intell. Res. 7, 83–124 (1997)

22. Wooldridge, M., Jennings, N.R.: The Cooperative Problem-solving Process. J. Logic Comput. 9(4), 563–592 (1999)

Manipulation of Weighted Voting Games
and the Effect of Quota

Ramoni O. Lasisi and Vicki H. Allan

Department of Computer Science, Utah State University
UT 84322-4205, Logan, U.S.A.
ramoni.lasisi@aggiemail.usu.edu, vicki.allan@usu.edu

Abstract. The *Shapley-Shubik, Banzhaf,* and *Deegan-Packel* indices are three prominent power indices for measuring voters' power in weighted voting games. We consider two methods of manipulating weighted voting games, called *annexation* and *merging*. These manipulations allow either an agent, called an *annexer* to take over the voting weights of some other agents, or the coming together of some agents to form a *bloc* of manipulators to have more power over the outcomes of the games. We evaluate the extent of susceptibility to these forms of manipulation and the effect of the quota of a game on these manipulation for the three indices. Experiments on weighted voting games suggest that the three indices are highly susceptible to annexation while they are less susceptible to merging. In both annexation and merging, the Shapley-Shubik index is the most susceptible to manipulation among the indices. Further experiments on the effect of quotas of weighted voting games suggest the existence of an inverse relationship between the susceptibility of the indices to manipulation and the quotas for both annexation and merging. Thus, weighted voting games with large quota values closer to the total weight of agents in the games may be less vulnerable to annexation and merging than those with corresponding smaller quota values.

keywords: Agents, Weighted voting games, Manipulation, Annexation, Merging, Power indices.

1 Introduction

Weighted voting games (WVGs) are classical cooperative games which provide compact representation for coalition formation models in multiagent systems. WVGs are mathematical abstractions of voting systems. In a voting system, voters express their opinions through their votes by electing candidates to represent them or influence the passage of bills. Each member of the set of voters V, has an associated weight $w : V \rightarrow Q^+$. A voter's weight is the number of votes controlled by the voter, and this is the maximum number of votes she is permitted to cast. A subset of agents, called a *coalition*, wins in a WVG, if the sum of the weights of the individual agents in the coalition meets or exceeds a certain threshold called the *quota*. Such coalitions whose weight meets or exceeds the quota are referred to as the *winning coalitions*. It is natural to naively think that the numerical weight of an agent directly determines the corresponding strength of the agent in a WVG. The measure of the strength of an agent is

J. Filipe and A. Fred (Eds.): ICAART 2011, CCIS 271, pp. 413–428, 2013.

414   R.O. Lasisi and V.H. Allan

termed its *power*. Consider, for example, a WVG of three voters, a_1, a_2, and a_3 with respective weights $6, 3$, and 1. When the quota for the game is 10, then a coalition consisting of the three voters is needed to win the game. Thus, each of the voters are of equal importance in achieving the winning coalition. Hence, they each have equal power irrespective of their weight distribution. The power of each agent in a WVG reflects its significance in the elicitation of winning coalitions. A widely accepted method for measuring such power is using *power indices* [9,10,15,16]. Three prominent power indices are the *Shapley-Shubik, Banzhaf*, and *Deegan-Packel* indices [15,16].

This paper discusses WVGs and two methods of manipulating those games, called *annexation* and *merging* [2]. In annexation, a strategic agent, termed an *annexer*, may alter a game by taking over the voting weights of some other agents in order to use the weights in her favor. As a straightforward example of annexation, consider when a shareholder buys up the voting shares of some other shareholders [15]. We refer to agents whose voting shares were bought over as *assimilated voters*. The new game consists of the previous agents in the original game whose weights were not annexed and the *bloc* of agent made up of the annexer and the assimilated voters. The annexer also incurs some *annexation cost* to allow purchasing the votes of the assimilated voters. In this situation, only the annexer benefits from annexation as the power of the bloc in the new game is compared to the power of the annexer in the original game. On the other hand, merging is the voluntary coordinated action of would-be manipulators who come together to form a bloc. The agents in the bloc are also assumed to be assimilated voters since they can no more vote as individual voters in the new game, rather as a bloc. The new game consists of the previous agents in the original game that were not assimilated as well as the bloc formed by the assimilated voters. The power of the bloc in the new game is compared to the sum of the individual powers of all members of the assimilated bloc in the original game. No annexation costs occur as individual voters in the bloc are compensated via increase of power. All the agents in the bloc benefit from the merging in the case of power increase, having agreed on how to distribute the gains of their collusion. In both annexation and merging, strategic agents who agree to assimilation anticipate that the value of their power in the new games to be at least the value of their power in the original games.

We evaluate the susceptibility to manipulation via annexation and merging in WVGs of the following power indices: Shapley-Shubik, Banzhaf, and Deegan-Packel indices. Susceptibility to manipulation is the extent to which strategic agents may gain power with respect to the original games they manipulate. We provide empirical analysis of susceptibility to annexation and merging in WVGs among the three indices. The remainder of the paper proceeds as follows. Section 2 discusses related work. Section 3 provides definitions and notations used in the paper. In Section 4, we provide examples using the three indices to illustrate manipulation via annexation and merging. Section 5 considers unanimity and non unanimity WVGs. Section 6 provides empirical evaluation of susceptibility of the indices to manipulation via annexation and merging for non unanimity WVGs. In Section 7, we empirically evaluate the effects of quota of WVGs on annexation and merging for non unanimity WVGs. We conclude in Section 8.

2   Related Work

Weighted voting games and power indices are widely studied [1,2,5,14,16]. WVGs have many applications, including economics, political science, neuroscience, threshold logic, reliability theory, distributed systems [3], and multiagent systems [4]. Prominent real-life situations where WVGs have found applications include the United Nations Security Council, the Electoral College of the United States and the International Monetary Fund [1,14]. The study of WVGs has also necessitated the need to fairly determine the power of players in a game. This is because the power of a player in a game provides information about the relative importance of that player when compared to other players. To evaluate players' power, prominent power indices such as Shapley-Shubik, Banzhaf, and Deegan-Packel indices are commonly employed [16]. These indices satisfy the axioms that characterize a power index, have gained wide usage in political arena, and are the main power indices found in the literature [10]. These power indices have been defined on the framework of subsets of winning coalitions in the game they seek to evaluate. A wide variation in the results they provide can be observed. Then, comes the question of which of the power indices is the most resistant to manipulation in a WVG. The choice of a power index depends on a number of factors, namely, the a priori properties of the index, the axioms characterizing the index, and the context of decision making process under consideration [10].

The three indices we consider measure the influence of voters differently. There are many situations where their values are the same for similar games. However, there exists an important example of the US federal system in using the Shapley-Shubik and Banzhaf indices where they do not agree [9]. According to Laruelle and Valenciano [11], and Kirsch [8], the decision of which index to use in evaluating a voting situation is largely dependent on the assumptions about the voting behavior of the voters. When the voters are assumed to vote completely independently of each other, the Banzhaf index has been found to be appropriate. On the other hand, Shapley-Shubik index should be employed when all voters are influenced by a common belief affecting their choices. Deegan-Packel index is appealing in that it assigns powers based on size of the winning coalition, thus giving preference to smaller coalitions (which may be easier to form).

Very little work exists on manipulation via annexation and merging in WVGs, and the more detailed analysis of players merging into blocs, until now, has remained unexplored [2]. Machover and Felsenthal [15] proved that if a player annexes other players, then the annexation is always advantageous for the annexer using the Shapley-Shubik index. Annexation can be advantageous or disadvantageous using the Banzhaf index. For the case of merging, in both the Shapley-Shubik and Banzhaf indices, merging can be advantageous or disadvantageous. Aziz and Paterson [2] show that for some classes of WVGs, for both Shapley-Shubik and Banzhaf indices, it is disadvantageous for a coalition to merge, while advantageous for a player to annex. They also prove some NP-hardness results for annexation and merging. They show that for both Shapley-Shubik and Banzhaf indices, finding a beneficial annexation is NP-hard. Also, determining if there exists a beneficial merge is NP-hard for the Shapley-Shubik index. Machover and Felsentha [15], and Aziz and Paterson [2] have shown that it can be advantageous for agents to engage in annexation or merging for Shapley-Shubik and Banzhaf indices in some classes of WVGs. The authors stop short of addressing the question of upper bounds on the extent to which strategic agents may gain with respect to the games they manipulate. In view of this, our work differ from those of these authors. We extend the work of Lasisi and Allan [12], as we study the susceptibility of the three power indices to manipulation via annexation and merging. We empirically consider the extent to which strategic agents may gain by engaging in such manipulation and show how the susceptibility among the indices compares for different WVGs.

3 Definitions and Notations

Weighted Voting Game. Let $I = \{1, \cdots, n\}$ be a set of n agents. Let $\mathbf{w} = \{w_1, \cdots, w_n\}$ be the corresponding positive integer weights of the agents. Let a *coalition* $S \subseteq I$ be a non empty subset of agents. A WVG G with *quota* q involving agents I is represented as $G = [w_1, \cdots, w_n; q]$. Denote by $w(S)$, the weight of a coalition S derived from the summation of the individual weights of agents in S i.e., $w(S) = \sum_{i \in S} w_i$. A coalition S, wins in the game G if $w(S) \geq q$ otherwise it loses. q is constrained as follows $\frac{1}{2}w(I) < q \leq w(I)$.

Simple Voting Game. Each coalition S, has an associated value function $v : S \rightarrow \{0, 1\}$. The value 1 implies a win for S and 0 a loss. In the WVG G above, $v(S) = 1$ if $w(S) \geq q$ and 0 otherwise.

Dummy and Critical Agents. An agent $i \in S$ is *dummy* if its weight in S is not needed for S to be a winning coalition, i.e., $w(S\backslash\{i\}) \geq q$. Otherwise, it is *critical* to S, i.e., $w(S) \geq q$ and $w(S\backslash\{i\}) < q$.

Unanimity Weighted Voting Game. A WVG in which there is a single winning coalition and every agent is critical to the coalition is a *unanimity* WVG.

Shapley-Shubik Power Index. This index quantifies the marginal contribution of an agent to the grand coalition. Each agent in a permutation is given credit for a win if the agents preceding it do not form a winning coalition but, by adding the agent in question, a winning coalition is formed. The index is dependent on the number of permutations for which an agent is critical. Denote by Π the set of all permutations of n agents in a WVG G. Let $\pi \in \Pi$ define a one-to-one mapping onto itself where $\pi(i)$ is the position of the ith agent in the permutation order. Denote by $S_\pi(i)$, the predecessors of agent i in π, i.e., $S_\pi(i) = \{j : \pi(j) < \pi(i)\}$. The Shapley-Shubik index, $\varphi_i(G)$, of agent i in G is

$$\varphi_i(G) = \frac{1}{n!} \sum_{\pi \in \Pi} [v(S_\pi(i) \cup \{i\}) - v(S_\pi(i))] \tag{1}$$

Banzhaf Power Index. Another index that has also gained wide usage in the political arena is the Banzhaf power index. Unlike the Shapley-Shubik index, its computation depends on the number of winning coalitions in which an agent is critical. There can be more than one critical agent in a particular winning coalition. The Banzhaf index, $\beta_i(G)$, of agent i in a game G, is given by

$$\beta_i(G) = \frac{\eta_i(G)}{\sum_{i \in I} \eta_i(G)} \tag{2}$$

where $\eta_i(G)$ is the number of coalitions in which i is critical in G.

Deegan-Packel Power Index. The Deegan-Packel power index is also found in the literature for computing power indices. The computation of this power index for an agent i takes into account both the number of all the minimal winning coalitions (MWCs) in the game as well as the sizes of the MWCs having i as a member [16]. A winning coalition $C \subseteq I$ is a MWC if every proper subset of C is a losing coalition, i.e., $w(C) \geq q$ and $\forall T \subset C, w(T) < q$. The Deegan-Packel power index, $\gamma_i(G)$, of an agent i in a game G, is given by

$$\gamma_i(G) = \frac{1}{|MWC|} \sum_{S \in MWC_i} \frac{1}{|S|} \tag{3}$$

where MWC_i are the sets of all MWCs in G that include i.

Susceptibility of Power Index to Manipulation. Consider a coalition $S \subset I$, let $\&S$ define a bloc of assimilated voters formed by agents in S. Let Φ be a power index. Denote by $\Phi_i(G)$, the power of an agent i in a WVG G. Let G' be the resulting game when a WVG G is manipulated via annexation or merging.

    <u>Annexation</u>: Let an agent i alter G by annexing a coalition S. We say that Φ is susceptible to manipulation via annexation if there exists a G', such that $\Phi_{\&(S \cup \{i\})}(G') > \Phi_i(G)$; the annexation is termed *advantageous*. If $\Phi_{\&(S \cup \{i\})}(G') < \Phi_i(G)$, then it is *disadvantageous*.

    <u>Merging</u>: Let a coalition S alter G by merging into a bloc $\&S$. We say that Φ is susceptible to manipulation via merging if there exists a G', such that $\Phi_{\&S}(G') > \sum_{i \in S} \Phi_i(G)$; the merging is termed *advantageous*. If $\Phi_{\&S}(G') < \sum_{i \in S} \Phi_i(G)$, then it is *disadvantageous*.

Factor of Increment (Decrement). The factor of increment (resp. decrement) of the original power from a manipulation is $\frac{\Phi_i(G')}{\Phi_i(G)}$. The value represents an increment (or gain) if it is greater than 1 and decrement (or loss) if it is less than 1. The factor of increment provides an indication of the extent of susceptibility of power indices to manipulation. A higher factor of increment indicates that the index is more susceptible to manipulation in that game.

Domination of Manipulability. Let Θ be another power index other than Φ. Let agent i alter a game G by annexing agents S. Suppose the power of the agent in a new game G' are $\Phi_{\&(S\cup\{i\})}(G')$ and $\Theta_{\&(S\cup\{i\})}(G')$ as determined by Φ and Θ respectively. We say that the manipulability of one index say $M_{\Phi_i}(G', G)$, dominates the manipulability of another index $M_{\Theta_i}(G', G)$ for a particular game G, if the factor by which i gain in Φ is greater than the factor by which it gain in Θ, i.e., $M_{\Phi_i}(G', G) > M_{\Theta_i}(G', G)$ which implies that $\frac{\Phi_{\&(S\cup\{i\})}(G')}{\Phi_i(G)} > \frac{\Theta_{\&(S\cup\{i\})}(G')}{\Theta_i(G)}$. Hence, Φ is more susceptible to manipulation via annexation in G than Θ. The domination of manipulability can be similarly defined for manipulation via merging.

4 Annexations and Merging

This section provides examples illustrating manipulation via annexation and merging in WVGs. The power of the strategic agents i.e., the annexer or the bloc of manipulators, and the factor of increment (decrement) are also summarized in a table for each example using the three power indices.

4.1 Manipulation via Annexation

Example 1. Annexation Advantageous.

Let $G = [5, 8, 3, 3, \mathbf{4}, 2, 4; 18]$ be a WVG. The assimilated agents are shown in bold, with agent 1 being the annexer. In the original game, the Deegan-Packel index of the annexer is $\gamma_1(G) = 0.1722$. In the new game, $G' = [\mathbf{9}, 8, 3, 3, 2, 4; 18]$, its Deegan-Packel index is $\gamma_1(G') = 0.2604$, a factor of increase of 1.51.

Table 1. The annexer power in the game $G = [5, 8, 3, 3, \mathbf{4}, 2, 4; 18]$, the altered game $G' = [\mathbf{9}, 8, 3, 3, 2, 4; 18]$, and the factor of increment for the three indices

Power Index	G	G'	Factor
Shapley-Shubik	0.1714	0.3500	2.04
Banzhaf	0.1712	0.3400	1.99
Deegan-Packel	0.1722	0.2604	1.51

Example 2. Annexation Disadvantageous.

Let $G = [\mathbf{8}, 9, 9, 5, 7, \mathbf{3}, 9; 29]$ be a WVG. The assimilated agents are shown in bold, with agent 1 being the annexer. In the original game, the Deegan-Packel index of the annexer is $\gamma_1(G) = 0.1711$. In the new game, $G' = [\mathbf{11}, 9, 9, 5, 7, 9; 29]$, its Deegan-Packel index is $\gamma_1(G') = 0.1591$, a factor of decrease of 0.93.

Table 2. The annexer power in the game $G = [8, 9, 9, 5, 7, 3, 9; 29]$, the altered game $G' = [11, 9, 9, 5, 7, 9; 29]$, and the factor of increment (decrement) for the three indices

Power Index	G	G'	Factor
Shapley-Shubik	0.1786	0.2167	1.21
Banzhaf	0.1774	0.2167	1.22
Deegan-Packel	0.1711	0.1591	0.93

4.2 Manipulation via Merging

Example 3. Merging Advantageous.

Let $G = [4, 2, 1, 1, 8, 7, 4; 17]$ be a WVG. The assimilated agents are shown in bold. In the original game, the Deegan-Packel indices of these agents are, $\gamma_2(G) = 0.0926$, $\gamma_6(G) = 0.1889$, and $\gamma_7(G) = 0.1704$. Their cummulative power is 0.4519. In the new game, $G' = [13, 4, 1, 1, 8; 17]$, the Deegan-Packel index of the bloc is $\gamma_1(G') = 0.5000$, a factor of increase of 1.11.

Table 3. The cummulative power of the assimilated agents in the original game $G = [4, 2, 1, 1, 8, 7, 4; 17]$, the power of the bloc in the altered game $G' = [13, 4, 1, 1, 8; 17]$, and the factor of increment for the three indices

Power Index	G	G'	Factor
Shapley-Shubik	0.4881	0.6667	1.37
Banzhaf	0.4851	0.6000	1.24
Deegan-Packel	0.4519	0.5000	1.11

Example 4. Merging Disadvantageous.

Let $G = [5, 8, 3, 4, 9, 1, 5; 30]$ be a WVG. The assimilated agents are shown in bold. In the original game, the Deegan-Packel indices of these agents are, $\gamma_2(G) = 0.1833$, $\gamma_5(G) = 0.1333$, and $\gamma_7(G) = 0.1417$. Their cummulative power is 0.5083. In the new game, $G' = [22, 5, 3, 4, 1; 30]$, the Deegan-Packel index of the bloc is $\gamma_1(G') = 0.3056$, a factor of decrease of 0.60.

Table 4. The cumulative power of the strategic agents in the original game $G = [5, 8, 3, 4, 9, 1, 5; 30]$, the power of the bloc in the altered game $G' = [22, 5, 3, 4, 1; 30]$, and the factor of decrement for the three indices

Power Index	G	G'	Factor
Shapley-Shubik	0.6762	0.4667	0.69
Banzhaf	0.5789	0.3684	0.64
Deegan-Packel	0.5083	0.3056	0.60

5   Weighted Voting Games

5.1   Unanimity Weighted Voting Games

Recall that a WVG in which there is a single winning coalition and every agent is critical to the coalition is a unanimity WVG. Manipulation via annexation and merging in unanimity WVGs is less interesting compared to the non unanimity WVGs that provides more complex and realistic scenarios that are not well-understood. Aziz and Paterson [2] show that for unanimity WVGs, for both the Shapley-Shubik and Banzhaf indices: *it is disadvantageous for a coalition to merge and advantageous for a player to annex other players*. These results extend to the Deegan-Packel index too [13]. Contrary to Aziz and Paterson [2] however, for the case of annexation, Lasisi and Allan [13] contend that it is not true in its entirety that *it is advantageous for an annexer to annex other players in unanimity* WVGs. They argue that apart from the fact that annexation always increases the power of other agents that are not annexed by the same factor of increment as the annexer achieved, the annexer also incurs annexation costs that reduce the benefit the annexer thought it gained. Finally, Lasisi and Allan [13] bound the extent to which a strategic agent may gain in annexation. They show that for any unanimity WVG of n agents, the upper bound on the extent to which a strategic agent may gain while annexing other agents is at most n times the power of the agent in the original game using any of the three indices.

5.2   Non Unanimity Weighted Voting Games

For the sake of simplicity, we assume that only one of the agents is engaging in annexation at a time. However, we are not oblivious of the fact that other agents also have similar motivations to engage in annexation in anticipation of power increase. For the case of manipulation via merging, we assume that the assimilated agents in the bloc can easily distribute the gains from their collusion among themselves in a fair and stable way. Thus, paving way for manipulation.

Consider a WVG G of I agents with quota q. If any agent $i \in I$ has weight $w_i \geq q$, then the agent will always win without forming coalitions with other agents. The more interesting games we consider are those for which $w_i < q$, and such that q satisfies the inequality $q \leq w(I) - m$, where m is chosen randomly to be the weight of exactly one of the agents in the game. When the grand coalition (i.e., a coalition of all the agents) emerges, it will always contain some agents that are not critical in the coalition. It is easy to see that all the winning coalitions in this type of games are non unanimity. In order to evaluate the behaviors of the power indices for non unanimity WVGs, we conduct experiments to evaluate the effects of manipulation when a strategic agent annexes other agents in the games or when manipulators merge to form blocs using each of the three indices.

6   Experiments

This section provides detail descriptions of the simulation environment used for the conduct of experiments and analysis of the experimental results used for the evaluation of the effects of annexation and merging in non unanimity WVGs.

6.1 Simulation Environment

We perform experiments to evaluate the effects of manipulation via annexation and merging by agents using each of the three power indices. To facilitate comparison, we have 15 agents in each of the original WVGs. The weights of agents in these games are chosen so that all weights are integers not larger than ten. These weights are reflective of realistic voting procedures as the weights of agents in real votings are not too large [4]. When creating a new game, all agents are randomly assigned weights, and the quota of the game is also generated to satisfy the inequality of non unanimity WVGs of Subsection 5.2. For the case of manipulation via annexation, we randomly generate WVGs and assume that only the first agent in the game is engaging in the manipulation, i.e., the annexer. Then, we determine the power derived by each of the three power indices (i.e., Shapley-Shubik, Banzhaf, and Deegan-Packel power index) for this agent in the game. After this, we consider annexation of at least one agent in the game by the annexer, while the weights of other agents not annexed remain the same in the altered games. For a particular game, the annexer may annex $1 \leq i \leq 10$ other agents; we refer to i as the *bloc size*. The bloc size and the members of the bloc are randomly[1] generated for each game. The weight of the annexer in the new game is the sum of the weights of the agents it annexed plus the annexer's initial weight in the original game. We compute the new power index of the annexer in the altered games next. Now, we determine the factor of increment by which the annexer gains or loses in the manipulation for the corresponding bloc sizes i, in the range $1 \leq i \leq 10$.

We use the same procedure as described above for the case of manipulation via merging with the following modifications. Since merging requires coordinated action of the manipulators, we randomly select strategic agents among the agents in the WVGs to form the blocs of manipulators. The bloc size $2 \leq i \leq 10$, for merging is also randomly generated for each game. The weight of a bloc in a new game is the sum of the weights of the assimilated agents in the bloc. The bloc participates in the new game as though a single agent. We compute the new power index of the bloc in the altered games next. We determine the factor of increment by which the bloc gains or loses in the manipulation for the corresponding bloc sizes. Unlike in annexation, the power of the bloc is compared with the sum of the original powers of the individual agents in the bloc. For our study, we generate $2,000$ original WVGs for various bloc sizes and allow manipulation by the annexer or the bloc of manipulators. For each game, we compute the factor of increment by which the annexer or the bloc gains or loses. Finally, we compute the average value of these factors of increment over all the games for each bloc size. We use $2,000$ WVGs in order to capture a variety of games that are representative of the non unanimity WVGs and to minimize the standard deviation from the true factors when we compute the average values. The average value of the factors of increment provides the extent of susceptibility to manipulation by each of the three indices. We estimate the domination of manipulability among the three indices by comparing their average factors of increment simultaneously in similar games.

---

[1] We note that randomly generating members of the blocs fails to consider the benefits of a more strategic approach to manipulation. We plan to address this in future work.

6.2   Simulation Results

Experiments confirm the existence of advantageous annexation and merging for the non unanimity WVGs when agents engage in manipulation using the three indices. However, the extent to which agents gain varies with both annexation and merging, and among the indices. Consider manipulation via annexation first. We provide a comparison of susceptibility to manipulation among the three power indices by comparing the population of factors of increment attained by strategic agents in different games for each of the indices. A summary of susceptibility to manipulation via annexation among the three indices for 2,000 WVGs is shown in Figure 1. The x-axis indicates the bloc sizes while the y-axis is the average factor of increment achieved by agents in the 2,000 WVGs for corresponding bloc sizes.

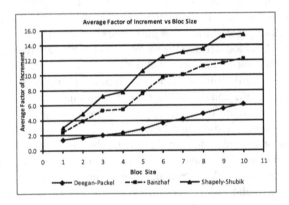

Fig. 1. Susceptibility to manipulation via annexation among Shapley-Shubik, Banzhaf, and Deegan-Packel indices for non unanimity WVGs

The effect of manipulation via annexation is pronounced for the three power indices, as all the indices are highly susceptible to manipulation. However, the higher susceptibility of the Shapley-Shubik and Banzhaf indices than the Deegan-Packel index can be observed from Figure 1. While the average factor of increment for manipulation rapidly grows with the bloc sizes for the Shapley-Shubik and Banzhaf indices, that of the Deegan-Packel index grows more slowly. By the average factor of increment, the Shapley-Shubik index manipulability dominates that of Banzhaf index, which in turn dominates that of Deegan-Packel index. Also, there is a positive correlation between the average factor of increment and the bloc sizes for the three indices. The average factor of increment increases with the bloc sizes. This analysis suggests that the Shapley-Shubik and Banzhaf power indices are more susceptible to manipulation via annexation than the Deegan-Packel power index. Since all the three power indices are susceptible to manipulation via annexation, this provides some motivations for strategic agents to generally engage in such manipulation for non unanimity WVGs when they are being evaluated using any of the three power indices, and in particular, when the Shapley-Shubik power index is employed.

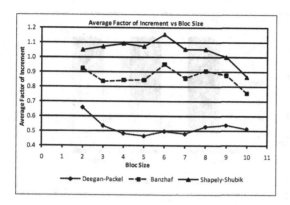

Fig. 2. Susceptibility to manipulation via merging among Shapley-Shubik, Banzhaf, and Deegan-Packel indices for non unanimity WVGs

Figure 2 provides similar results for the non unanimity WVGs for manipulation via merging. We again compare susceptibility to manipulation among the three power indices. Unlike manipulation via annexation, only the Shapley-Shubik index appears to be susceptible to manipulation for this type of game. Also, there appears not to be any correlation between the average factor of increment achieved by the bloc of manipulators and the bloc size for the three power indices. Thus, it is unclear to the would-be manipulators what bloc size would be advantageous or disadvantageous to the bloc, and to what extent. It is easy to see from the trends of the three power indices in Figure 2, that, using the average factor of increment over the games we consider, the Shapley-Shubik index manipulability dominates that of the Banzhaf index, which in turn dominates that of the Deegan-Packel index. Another positive result that is observable from Figure 2 is that the highest average factor of increment for the three power indices is less than a factor of 1.2 as compared to a factor of 15, found for the Shapley-Shubik index, 12 for the Banzhaf index, and 6 for the Deegan-Packel index under manipulation via annexation (see Figure 1).

In Figure 3, examination of the 2, 000 non unanimity WVGs we consider reveals that many of the games are advantageous for Shapley-Shubik index, few for the Banzhaf index, and virtually none for the Deegan-Packel index. The figure shows the percentage of advantageous and disadvantageous games for manipulation via merging among the three indices. Even for the cases where the games are advantageous for the three indices, the factor of increment achieved by the blocs of manipulators are not very high, and in all cases are less than a factor of 2. The experimental evidence suggests that the the Shapley-Shubik index is more susceptible to manipulation via merging than the Banzhaf and Deegan-Packel power indices for non unanimity WVGs, even though the factor of increment is not high. Now, since only the Shapley-Shubik index is more susceptible to manipulations via merging, and also, since the factor by which the bloc of manipulators gains is very low, we suspect that this may provide less motivation for strategic agents to generally engage in manipulation via merging for the non unanimity WVGs when they are being evaluated using any of the three power indices, and in particular, when the Deegan-Packel index is employed.

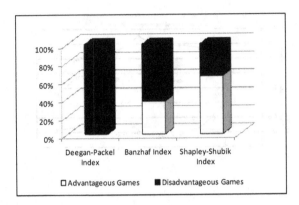

Fig. 3. Percentage of advantageous and disadvantageous games for manipulation via merging among the three indices for 2, 000 non unanimity WVGs

7   Effect of Quota on Annexation and Merging

The choice of the quota of a WVG is crucial in determining the distribution of power of agents in a WVG [10]. Thus, this choice of the quota naturally extends to the determination of the effects of manipulation in WVGs. We conduct another set of experiments and provide empirical evaluation of the effects of the quota of a game on annexation and merging for non unanimity WVGs. Recall that the non unaninimty WVGs we consider are those for which the weight of each agent $w_i < q$, where q is the quota of the game; and such that q satisfies the inequality $q \leq w(I) - m$, where m is randomly chosen to be the weight of exactly one of the agents, and $w(I)$ is the total weight of all agents in the game (see Subsection 5.2). Again, since the more interesting games are those for which the quota is not too small [5], and in order to ensure that disjoint winning coalitions does not emerge, q is constrained as $\frac{1}{2}w(I) < q \leq w(I) - m$.

The simulation environment for the set of experiments in this section is similar to our previous environments with the following modifications. When creating a new game (termed, *original game*), agents are randomly assigned weights and the quota q of the game is set as $q = \frac{1}{2}w(I) + 1$, where I is the set of agents in the game. As before, for the altered WVGs, we allow the manipulation of the original game via annexation of at least one agent by the annexer or merging of at least two agents by the assimilated agents. The bloc size $1 \leq i \leq 10$ and the members of the blocs in both cases are randomly generated. Unlike in previous experiments where the quota of the altered WVGs remain fixed while the weights of the blocs in annexation and merging vary, we fix the weights and members of blocs for the altered WVGs and vary the quota as follows $q+1, q+2, \cdots, w(I) - m$, with q being the quota of the original game. Thus, the number of altered games for different original games varies since q depends on the total weight of agents in the original games. We compute the power index of the blocs in the altered games, compare with the original game, and determine the factor of increment or decrement as appropriate for annexation or merging.

In order to facilitate the ease of characterization of the effects of the quota of WVGs on annexation and merging in our experiments, we partition the altered WVGs into *four equal* groups using the quotas. We refer to the groups as *Range1, Range2, Range3,* and *Range4*. First, we consider the effect of the quota on annexation. We generate 400 original WVGs, which in turn generate approximately 3,800 altered WVGs in each of the four ranges. As expected for annexation (based on results of previous sections), all the altered WVGs are advantageous for the three indices with variations in the factors of increment achieved by annexers. So, we evaluate the factors of increment for the ranges.

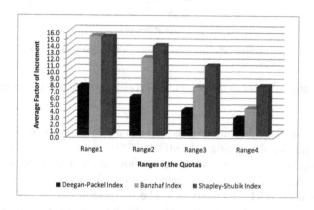

Fig. 4. Effect of quota on annexation for WVGs using average factor of increment

Figure 4 shows the average factor of increment for manipulation via annexation among the three power indices in the altered WVGs for the four ranges. The x-axis indicates the ranges while the y-axis is the average factor of increment. The figure shows that the average factor of increment is high for the altered WVGs in Range1. Our explanation for this is as follow. The quotas of all the altered WVGs in Range1 have the least values compared to other ranges. So, it is easy for annexers to quickly acquire the quota requirement of the games. More importantly, we observe many cases where the weight of the blocs of assimilation due to annexation are greater than the quota for this range. In such cases, an annexer claims the maximum power of 1 that is available in the game. However, as the quota increases into other ranges, it becomes increasingly difficult for an annexer to achieve high values since the annexer needs to form coalitions with other agents to gain some power. Summarily, we observe a fairly uniform degradation of the factors of increment from Range1 to Range4, and an inverse relationship between the quotas of the games and the factors of increment using the three power indices; at least for the non unanimity WVGs that we consider.

Now, we consider the effect of the quota on merging. Figure 5 provides similar results as above for manipulation via merging. Unlike annexation, where the factors of increment is very high for Range1 i.e., greater than 15 for both Banzhaf and Shapley-Shubik indices, the corresponding values for Range1 here is less than a factor of 1.5 for both Banzhaf and Shapley-Shubik indices. Similar explanations for degradation of average factors of increment as the quotas increase into other ranges observed for annexation above hold here too.

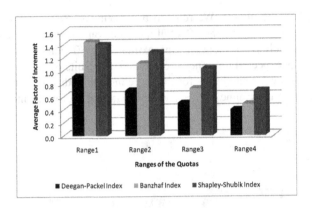

Fig. 5. Effect of quota on merging for WVGs using average factor of increment

We do not have here such situation as found in annexation where all the games are advantageuos. So, we also evaluate the effects of the quota in merging by considering the proportion of advantageous or disadvantageous games in the four ranges. Figure 6 shows the percentage of advantageous games for manipulation via merging among the indices in the altered WVGs for the ranges. The x-axis indicates the ranges while the y-axis is the percentage of advantageous games. All the altered WVGs are advantageous for Banzhaf index, Shapley-Shubik index, and about 30% of the games for Deegan-Packel index in Range1. This we observe may be partly due to the lower quota values for the games in this range. As the quotas of the altered WVGs increase into Range2, all the games remain advantageous for Shapley-Shubik index while those of Banzhaf and Deegan-Packel indices decrease to about 64% and 11% respectively. Further decrease is noticed for Range3. Finally, in Range4 where the quotas of the altered WVGs are the highest, virtually non of the games are advantageous for both Deegan-Packel and Banzhaf indices with about 13% of the games still advantageous for Shapley-Shubik index. As before, it appears that the proportion of the advantageous games decreases with increasing quotas but no clear pattern can be observed except that the proportion

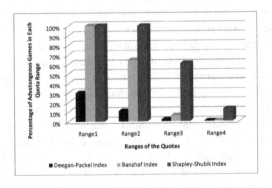

Fig. 6. Effect of quota on merging for WVGs using proportion of advantageous games

of advantageous games for Shapley-Shubik index is the highest in all cases. This result is consistent with our results of susceptibility of manipulation via merging of Subsection 6.2 (see Figure 2).

8 Conclusions

We consider two methods of manipulating weighted voting games, called *annexation* and *merging* while focusing on the susceptibility of three of the power indices used in evaluating agents' power in such games. The following prominent power indices are used to evaluate agents' power: Shapley-Shubik, Banzhaf, and Deegan-Packel indices. We consider the extent to which strategic agents may gain by engaging in such form of manipulation and show how the susceptibility among the three indices compares for non unanimity weighted voting games. Experiments on weighted voting games suggest that the games are less vulnerable to manipulation via merging, while they are extremely vulnerable to manipulation via annexation for the three power indices. Also, while the average factor of increment of power due to manipulation grows with bloc sizes for manipulation via annexation, there appears to be no correlation between the average factor of increment and the bloc size for manipulation via merging. Again, the Shapley-Shubik index manipulability (i.e., the extent of susceptibility to manipulation) dominates that of the Banzhaf index, which in turn dominates that of the Deegan-Packel index for both manipulation via annexation and merging. Hence, the Shapley-Shubik index is more susceptible to manipulation via annexation and merging than the Banzhaf and the Deegan-Packel indices, with Deegan-Packel index being the least susceptible among the three power indices.

Further experiments on the effect of quotas suggest the existence of an inverse relationship between the susceptibility of the three power indices to manipulation and the quotas of weighted voting games for both manipulation via annexation and merging. Thus, weighted voting games with large quota values closer to the total weight of agents in the games may be less vulnerable to manipulation via annexation and merging than those with corresponding smaller quota values.

Acknowledgements. This work is supported by NSF research grant #0812039 entitled "Coalition Formation with Agent Leadership".

References

1. Alonso-Meijide, J.M., Bowles, C.: Generating Functions for Coalitional Power Indices: An Application to the IMF. Annals of Operations Research 137, 21–44 (2005)
2. Aziz, H., Paterson, M.: False-name Manipulations in Weighted Voting Games: splitting, merging and annexation. In: 8th International Conference on Autonomous Agents and Multiagent Systems, Budapest, Hungary, pp. 409–416 (2009)
3. Aziz, H., Paterson, M., Leech, D.: Combinatorial and Computational Aspects of Multiple Weighted Voting Games. The Warwick Economics Research Paper Series (TWERPS) 823, University of Warwick, Department of Economics (2007)

4. Bachrach, Y., Elkind, E.: Divide and Conquer: False-name Manipulations in Weighted Voting Games. In: 7th International Conference on Autonomous Agents and Multiagent Systems (AAMAS 2008), Estoril, Portugal, pp. 975–982 (2008)
5. Bachrach, Y., Markakis, E., Procaccia, A.D., Rosenschein, J.S., Saberi, A.: Approximating Power Indices - Theoretical and Empirical Analysis. Autonomous Agents and Multiagent Systems 20(2), 105–122 (2010)
6. Deng, X., Papadimitriou, C.H.: On the Complexity of Cooperative Solution Concepts. Mathematics of Operations Research 19(2), 257–266 (1994)
7. Garey, M.R., Johnson, D.S.: Computers and Intractability: A Guide to the Theory of NP-Completeness. W.H. Freeman, San Fransisco (1979)
8. Kirsch, W.: On Penrose's Squareoot Law and Beyond. Homo Oeconomicus 24(3,4), 357–380 (2007)
9. Kirsch, W., Langner, J.: Power Indices and Minimal Winning Coalitions. Social Choice and Welfare 34(1), 33–46 (2010)
10. Laruelle, A.: On the Choice of a Power Index. Instituto Valenciano de Investigaciones Economicas 2103, 99–10 (1999)
11. Laruelle, A., Valenciano, F.: Assessing Success and Decisiveness in Voting Situations. Social Choice and Welfare 24(1), 171–197 (2005)
12. Lasisi, R.O., Allan, V.H.: False Name Manipulations in Weighted Voting Games: Susceptibility of Power Indices. In 13th International Workshop on Trust in Agents Societies, Toronto, Canada, pp. 139–150 (2010)
13. Lasisi, R.O., Allan, V.H.: Annexations and Merging in Weighted Voting Games: The Extent of Susceptibility of Power Indices. In: 3rd International Conference of ICAART, Rome, Italy, pp. 124 –133 (2011)
14. Leech, D.: Voting Power in the Governance of the International Monetary Fund. Annals of Operations Research 109(1), 375–397 (2002)
15. Machover, M., Felsenthal, D.S.: Annexation and Alliances: When Are Blocs Advantageous a Priori. Social Choice and Welfare 19(2), 295–312 (2002)
16. Matsui, T., Matsui, Y.: A Survey of Algorithms for Calculating Power Indices of Weighted Majority Games. Journal of the Operations Research Society of Japan 43(1) (2000)

Distributed Consequence Finding:
Partition-Based and Cooperative Approaches

Katsumi Inoue[1], Gauvain Bourgne[1], and Takayuki Okamoto[2]

[1] National Institute of Informatics
2-1-2 Hitotsubashi, Chiyoda-ku, Tokyo 101-8430, Japan
[2] Graduate School of Science and Technology, Kobe University
1-1 Rokkodai-cho, Nada-ku, Kobe 658-8501, Japan
{ki,bourgne}@nii.ac.jp

Abstract. When knowledge is physically distributed, information and knowledge of individual agents may not be collected to one agent because they should not be known to others for security and privacy reasons. We thus assume the situation that individual agents cooperate with each other to find useful information from a distributed system to which they belong, without supposing any master or mediate agent who collects all necessary information from the agents. Then we propose two complete algorithms for distributed consequence finding. The first one extends a technique of theorem proving in partition-based knowledge bases. The second one is a more cooperative method than the first one. We compare these two methods and other related approaches in the literature.

1 Introduction

There is a growing interest in building large knowledge bases. Dealing with a huge amount of knowledge, two problems can be encountered in real domains. The first case is that knowledge is originally centralized so that one can access the whole knowledge but the size of the knowledge base is too huge to be handled. The second case is that knowledge is *distributed* in several sources so that it is hard or impossible to immediately access the whole or part of knowledge. The former case is studied in the line of research on *parallel* or *partition-based* reasoning. For example, partition-based theorem proving by Amir and McIlraith [3] divide a knowledge base into several parts each of which is easier to be handled so that the scalability of a reasoning system is improved.

On the other hand, in the second case we suppose *multi-agent systems* or *peer-to-peer systems* [2], in which an agent does not want to expose all its information to other agents for security and privacy reasons. Sometimes, it is inherently impossible to tell what other agents want to know and to ask what can be obtained from others. In such a case, each agent must give up gathering all necessary information from other agents, and moreover, no master or mediate agent can be assumed to exist to collect all information from agents. That is, we need to solve the problem with knowledge distributed as it is. In this research, we mainly deal with such distributed knowledge bases, but hope that those algorithms considered for distributed reasoning can be applied to the first case to gain efficiency.

J. Filipe and A. Fred (Eds.): ICAART 2011, CCIS 271, pp. 429–444, 2013.

In this work, we consider the problem of *distributed consequence finding*. Consequence finding [18,12,20] is a problem to discover an interesting theorem derivable from an axiom set, and is a promising method for problem solving in AI such as query answering [17], abduction [12,16,22], induction [23,13], diagnosis, planning, recognition and understanding. There are some *complete* procedures for consequence finding in first-order clausal theories [12,8] and efficient systems have also been developed [21,22]. Our concern here is to design a complete method in the distributed setting, that is, to obtain every consequence that would be derived from the whole knowledge base if it were gathered together. In this paper, we propose two new methods for distributed consequence finding.

The first method here is a generalization of partition-based theorem proving by [3] to consequence finding. The whole axiom set is partitioned into multiple sets called *partitions*, each of which can be associated with one agent. In this method, a pair of partitions must be connected with their *communication language*. The connections between partitions constitute a graph, but cycles must be removed so that the graph are transferred to a tree. Consequence finding is firstly performed in the leaves of the connection tree, and its consequences are sent to the parent if they belong to the communication language. This process is repeated until the root. To get a complete procedure in this method, it is important to decide the communication languages between two partitions, so we propose the method to determine them. It should be stressed that, although partition-based theorem proving by [3] also uses a consequence finding procedure in each individual reasoning task of an agent, the aim of [3] is not consequence finding from the knowledge base but is used for theorem proving tasks.

The second proposed method is a more cooperative one. In this method, we do not presuppose graph structures of agents, but any agent has a chance to communicate with other agents, hence the framework is more dynamic than the first method. Firstly, a new clause is added to an agent A_1, either as a top clause of the given problem or as a newly sent message from other agents, then triggers consequence finding from that clause with the axioms of A_1. Then, for each such newly derived clause C, if there is a clause D in the axiom set of another agent A_2 such that C and D can be resolved, then C is sent to A_2 and is added there. This process is repeated until no more new clause can be resolved with any clause of any other agent. We will compare these two methods and centralized approaches, and discuss the merits and demerits of both methods. We will also discuss relations with other previously proposed approaches to consequence finding in distributed settings [14,1,2].

The rest of this paper is organized as follows. Section 2 reviews the background of consequence finding and SOL resolution. Section 3 proposes partition-based consequence finding. Section 4 proposes a more cooperative algorithm for consequence finding and Section 5 compares the two proposed approaches. Section 6 discusses related work, and Section 7 gives a summary and future work.

2   Consequence Finding

In this section, we review consequence finding from an axiom set and a complete procedure for it. The task of consequence finding is related with many AI reasoning problems, and is indispensable in partition-based theorem proving in Section 3.1 too.

A *clause* is a disjunction of literals. Let C and D be two clauses. C *subsumes* D if there is a substitution θ such that $C\theta \subseteq D$. C *properly subsumes* D if C subsumes D but D does not subsume C. A *clausal theory* is a set of clauses, which is often identified with the *conjunctive normal form* (CNF) formula composed by taking the conjunction of all clauses in it. Let Σ be a clausal theory. $\mu\Sigma$ denotes the set of clauses in Σ not properly subsumed by any clause in Σ. A *consequence* of Σ is a clause entailed by Σ. We denote by $Th(\Sigma)$ the set of all consequences of Σ.

The *consequence finding* problem was first addressed by Lee [18] in the context of the resolution principle, which has the property that the consequences of Σ that are derived by the resolution principle includes $\mu Th(\Sigma)$. To find "interesting" theorems for a given problem, the notion of consequence finding has been extended to the problem to find *characteristic clauses* [12]. Each characteristic clause is constructed over a sub-vocabulary of the representation language called a "production field". Formally, a *production field* $\mathcal{P}$ is a pair, $\langle \mathbf{L}, Cond \rangle$, where $\mathbf{L}$ is a set of literals closed under instantiation, and $Cond$ is a certain condition to be satisfied, e.g., the maximum length of clauses, the maximum depth of terms, etc. When $Cond$ is not specified, $\mathcal{P} = \langle \mathbf{L}, \emptyset \rangle$ is simply denoted as $\langle \mathbf{L} \rangle$. A production field $\mathcal{P}$ is *stable* if, for any two clauses C and D such that C subsumes D, D belongs to $\mathcal{P}$ only if C belongs to $\mathcal{P}$.

A clause C *belongs to* $\mathcal{P} = \langle \mathbf{L}, Cond \rangle$ if every literal in C belongs to $\mathbf{L}$ and C satisfies $Cond$. For a set Σ of clauses, the set of logical consequence of Σ belonging to $\mathcal{P}$ is denoted as $Th_{\mathcal{P}}(\Sigma)$. Then, the *characteristic clauses* of Σ with respect to $\mathcal{P}$ are defined as: $Carc(\Sigma, \mathcal{P}) = \mu Th_{\mathcal{P}}(\Sigma)$. We here exclude any tautology $\neg L \vee L$ ($\equiv$ *True*) in $Carc(\Sigma, \mathcal{P})$ even when both L and $\neg L$ belong to $\mathcal{P}$. When $\mathcal{P}$ is a stable production field, it holds that the empty clause $\square$ is the unique clause in $Carc(\Sigma, \mathcal{P})$ if and only if Σ is unsatisfiable. This means that theorem proving is a special case of consequence finding. The use of characteristic clauses enables us to characterize various reasoning problems of interest to AI, such as nonmonotonic reasoning, diagnosis, and knowledge compilation as well as abduction and induction. In the propositional case [20], each characteristic clause of Σ is a *prime implicate* of Σ.

When a new clause C is added to a clausal theory Σ, further consequences are derived due to this new information. Such a new and "interesting" clause is called a "new" characteristic clause. Formally, the *new characteristic clauses* of C with respect to Σ and $\mathcal{P}$ are: $Newcarc(\Sigma, C, \mathcal{P}) = \mu [Th_{\mathcal{P}}(\Sigma \wedge C) - Th(\Sigma)]$.

When a new formula is not a single clause but a clausal theory or a CNF formula $F = C_1 \wedge \cdots \wedge C_m$, where each C_i is a clause, $Newcarc(\Sigma, F, \mathcal{P})$ can be computed as:

$$Newcarc(\Sigma, F, \mathcal{P}) = \mu \left[\bigwedge_{i=1}^{m} Newcarc(\Sigma_i, C_i, \mathcal{P}) \right], \tag{1}$$

where $\Sigma_1 = \Sigma$, and $\Sigma_{i+1} = \Sigma_i \wedge C_i$, for $i = 1, \ldots, m - 1$. This incremental computation can be applied to get the characteristic clauses of Σ with respect to $\mathcal{P}$ as follows.

$$Carc(\Sigma, \mathcal{P}) = Newcarc(True, \Sigma, \mathcal{P}). \tag{2}$$

Several procedures have been developed to compute (new) characteristic clauses. *SOL resolution* [12] is an extension of the Model Elimination (ME) calculus to which *Skip* operation is introduced along with *Resolve* and *Ancestry* operations. With Skip operation, SOL resolution focuses on deriving only those consequences belonging to the production field $\mathcal{P}$. *SFK resolution* [8] is based on a variant of ordered resolution, which is enhanced with Skip operation for finding characteristic clauses. SOL resolution is complete for finding $Newcarc(\Sigma, C, \mathcal{P})$ by treating an input clause C as the *top clause* and derives those consequences relevant to C directly. SOLAR (SOL for Advanced Reasoning) [21,22] is a sophisticated deductive reasoning system based on SOL resolution [12] and the connection tableaux, which avoids producing non-minimal consequences as well as redundant computation using state-of-the-art pruning techniques. Consequence enumeration is a strong point of SOLAR as an abductive procedure because it enables us to compare many different hypotheses [16].

3   Partition-Based Consequence Finding

This section proposes partition-based consequence finding. We start from a review of the basic terminology and the message passing algorithm between partitioned knowledge bases in [3], whose basic idea is from Craig's Interpolation Theorem [6,24].

3.1   Partitions and Message Passing

We suppose the whole axiom set $\mathcal{A} = \bigcup_{i \leq n} \mathcal{A}_i$, in which each axiom set $\mathcal{A}_i$ $(i \leq n)$ is called a *partition*. We denote as $S(\mathcal{A}_i)$ the set of (non-logical) symbols appearing in $\mathcal{A}_i$. A *graph induced from the partitions* $\bigcup_{i \leq n} \mathcal{A}_i$ is a graph $G = (V, E, l)$ such that (i) the set V of nodes are the same as the partitions, that is, $i \in V$ iff the partition $\mathcal{A}_i$ exists; (ii) the set E of edges are constructed as $E = \{(i, j) \mid S(\mathcal{A}_i) \cap S(\mathcal{A}_j) \neq \emptyset\}$, that is, the edge $(i, j) \in E$ iff there is a common symbol between $\mathcal{A}_i$ and $\mathcal{A}_j$; and (iii) the mapping l determines the label $l(i, j)$ of each edge (i, j) called the *communication language* between the partitions $\mathcal{A}_i$ and $\mathcal{A}_j$. In partition-based theorem proving by [3], $l(i, j)$ is initially set to the *common language* of $\mathcal{A}_i$ and $\mathcal{A}_j$, which is $C(i, j) = S(\mathcal{A}_i) \cap S(\mathcal{A}_j)$. The communication language $l(i, j)$ is then updated by adding symbols from some other partitions when cycles are broken (Algorithm 2). In Section 3.3, $l(i, j)$ is further extended by including the language for consequence finding.

Given the partitions $\bigcup_{i \leq n} \mathcal{A}_i$ and its induced graph $G = (V, E, l)$, we now consider the query Q to be proved in the partition $\mathcal{A}_k$ $(k \leq n)$. Given a set S of non-logical symbols, the set of formulas constructed from the symbols in S is denoted as $\mathcal{L}(S)$.

Definition 1. For two nodes $i, k \in V$, the length of a shortest path between i and k is denoted as $dist(i, k)$. Given k, we define $i \prec_k j$ if $dist(i, k) < dist(j, k)$. When k is clear from the context, we simply denote $i \prec j$ instead of $i \prec_k j$. For a node $i \in V$, a node $j \in V$ such that $(i, j) \in E$ and $j \prec_k i$ is called a *parent* of i (with respect to $\prec_k$). In the ordering $\prec_k$, the node k is called the *root* (with respect to $\prec_k$), and a node i that is not a parent of any node is called a *leaf* (with respect to $\prec_k$).

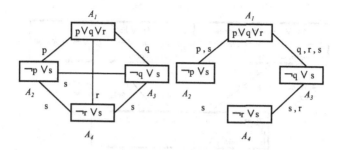

Fig. 1. Translation of a cyclic graph to a tree [3]

Algorithm 1 (Message Passing) [3]

1. *Determine $\prec_k$ according to Definition 1.*
2. *Perform consequence finding in each $\mathcal{A}_i$ in parallel. If $\mathcal{A}_k \models Q$, then return YES.*
3. *For every $i, j \in V$ such that j is the parent of i, if there is a consequence φ of the partition $\mathcal{A}_i$ such that $\varphi \in \mathcal{L}(l(i, j))$, then add φ to the axiom set $\mathcal{A}_j$.*
4. *Repeat Steps 2 to 3 until no more new consequence is found.*

Algorithm 1 works well for theorem proving at $\mathcal{A}_k$ when the induced graph is a tree. However, if there is a cycle, we need to break it to transform the graph to a tree.

Algorithm 2 (Cycle Cut) [3]

1. Find a shortest cycle $v_1, \ldots, v_c(= v_1)$ $(v_i \in V)$ in G. If there is no cycle, return G.
2. Select a such that $a < c$ and $\Sigma_{j<c, j \neq a} \mid l(v_j, v_{j+1}) \cup l(v_a, v_{a+1}) \mid$ is smallest.
3. For every $j < c$, $j \neq a$, let $l(v_j, v_{j+1}) := l(v_j, v_{j+1}) \cup l(v_a, v_{a+1})$.
4. Put $E := E \setminus \{(v_a, v_{a+1})\}$ and $l(v_a, v_{a+1}) := \emptyset$, then go to Step 1.

When there are multiple shortest cycles, common edges should be removed. But if there is no common edge, edges are removed so that the sum of the sizes of communication languages becomes the smallest. It is important to decide the order to remove edges although any ordering results in a translation to a tree. Cycle Cut Algorithm 2 is designed to minimize the total size of the communication languages.

Figure 1 shows an example of cycle cut. The left figure is translated to the right figure. Firstly, the shortest cycle (1,3), (3,4), (4,1) is considered, and then the edge (4,1) is deleted. The communication language of (4,1) is then added to those of (1,3) and (3,4). Next, from the cycle (1,3), (3,2), (2,1), the edge (3,2) is removed, and s is added to $l(1, 3)$ and $l(2, 1)$. Then, the cycle (1,3), (3,4), (4,2), (2,1) is taken, and the edge (4,2) is removed from it, but s is already in $l(3, 4)$ and $l(4, 2)$. Now Algorithm 1 is applied; $\neg p \vee s$ is sent from $\mathcal{A}_2$ to $\mathcal{A}_1$, deducing $q \vee r \vee s$ (as the resolvent of $\neg p \vee s$ and $p \vee q \vee r$), which is then sent from $\mathcal{A}_1$ to $\mathcal{A}_3$, deducing $r \vee s$ (as the resolvent of $q \vee r \vee s$ and $\neg q \vee s$), which is then sent from $\mathcal{A}_3$ to $\mathcal{A}_4$. Finally, the conclusion s is obtained at $\mathcal{A}_4$.

Theorem 1. [3] *Suppose an axiom set and its partitions $\mathcal{A} = \bigcup_{i \leq n} \mathcal{A}_i$ and a formula $Q \in \mathcal{L}(\mathcal{A}_k)$ $(k \leq n)$. If the consequence finding procedure in each partition is sound and complete, applying Algorithm 2 and then Algorithm 1 returns YES iff $\mathcal{A} \models Q$.*

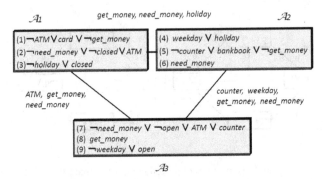

Fig. 2. Partitions for "Getting Money"

Partition-based theorem proving of [3] cannot be directly applied to consequence finding problems for $Q \notin \mathcal{L}(\mathcal{A}_k)$ although [3, Section 2.3] briefly mentions how to apply their MP algorithm to such a query constructed from languages in different partitions (a more detailed discussion will be given later in Section 3.3). Hence, we will extend the partition-based reasoning framework to a complete method for distributed consequence finding.

3.2 Example

We now show an example to see that the partition-based theorem proving method cannot be directly applied to consequence finding. The problem is to find means to withdraw money from one's bank account. The intended solution is that one must have either a cash card or a bankbook, which is represented as $card \lor bankbook$. The knowledge base of this problem consists of the following clauses.

- $\neg holiday \lor closed$ (The bank is closed on holidays.)
- $\neg weekday \lor open$ (The bank is open on weekdays.)
- $holiday \lor weekday$ (Any day is either a holiday or a week day.)
- $\neg need\_money \lor \neg open \lor ATM \lor counter$ (If one needs money and the bank is open, then (s)he goes to an ATM or a counter of the bank.)
- $\neg need\_money \lor \neg closed \lor ATM$ (If one needs money and the bank is closed, then (s)he goes to an ATM.)
- $\neg ATM \lor card \lor \neg get\_money$ (One cannot get money if (s)he does not have a cash card at an ATM.)
- $\neg counter \lor bankbook \lor \neg get\_money$ (One cannot get money if (s)he does not have a bankbook at a counter.)
- Input facts: $need\_money$ (One needs money.)
- Input facts: get_money (One gets money.)

Here we assume that the partitions are constructed as in Fig. 2, in which clauses are distributed in a scattered way. Algorithm 2 removes the edge $(1,3)$ and then adds ATM to the labels of other edges. However, the clause $card \lor bankbook$ cannot be deduced by Message Passing Algorithm 1: since $l(1,2)$ and $l(1,3)$ do not contain $card$, the clause (1) cannot be resolved with any clause in other partitions. In fact, it is necessary to resolve all clauses (1) to (7).

3.3   Partition-Based Consequence Finding

We here propose a new method to construct the communication language so that Message Passing Algorithm can be made complete for consequence finding.

Suppose the whole axiom set and its partitions $\mathcal{A} = \bigcup_{i \leq n} \mathcal{A}_i$. Recall that the common language of $\mathcal{A}_i$ and $\mathcal{A}_j$ is $C(i,j) = S(\mathcal{A}_i) \cap S(\mathcal{A}_j)$ $(i, j \leq n$ $i \neq j)$. Here, we construct the communication language $l(i,j)$ between $\mathcal{A}_i$ and $\mathcal{A}_j$ for consequence finding by extending $C(i,j)$. Let $\mathcal{P} = \langle \mathbf{L} \rangle$ be the given production field. By adding the literals appearing in $\mathbf{L}$ to the common language, each communication language in the case of trees is defined as

$$l(i,j) = C(i,j) \cup S(\mathbf{L}). \tag{3}$$

When there are cycles in the graph G, the final communication language is set after all cycles are cut using Algorithm 2. For example, suppose that an edge (s, r) is removed from a cycle. Then, the communication language of an edge (i, j) $(\neq (s, r))$ in the cycle is defined in the same way as before:

$$l(i,j) = C(i,j) \cup l(s,r) \cup S(\mathbf{L}). \tag{4}$$

Two remarks are noted here. Firstly, in (3) and (4), the polarity of each literal in $\mathbf{L}$ from the production field $\mathcal{P}$ are lost within the symbols $S(\mathbf{L})$. Although this does not harm soundness and completeness of distributed consequence finding, there is some redundancy in communication. Then, unlike the case of common language $C(i,j)$, we can keep the polarity of each literal in $\mathbf{L}$ in any $l(i,j)$ so that unnecessary clauses possessing literals that do not belong to $\mathcal{P}$ are not communicated between partitions. Second, when $C(i,j) = \emptyset$ the edge (i,j) does not exist in the graph G. In this case, $l(i,j)$ need not be updated as $S(\mathbf{L})$ using (3) and actually the edge is kept unnecessary. In fact, if we could add the literals from the production field to those non-existent edges in G, then the resulting graph G' would become strongly connected. By applying Cycle Cut Algorithm 2 to G', the minimal way is to cut those added edges again. However, other edges already have the literals $S(\mathbf{L})$ so their communication languages do not change. Hence, we do not have to reconsider non-existent edges in G.

Algorithm 3 (Partition-based Consequence Finding)

1. If there is a cycle in the induced graph G, select some $k \leq n$ and apply Cycle Cut Algorithm 2 to G and transform it to a tree.
2. Determine the communication language between all pairs of partitions. For each leaf partition $\mathcal{A}_i$, do the following.
3. If $\mathcal{A}_i$ is the root partition, let $\mathcal{P}_i$ be the original production field $\mathcal{P} = \langle \mathbf{L} \rangle$. Otherwise, let j be the parent of i, and define the production field of $\mathcal{A}_i$ as $\mathcal{P}_i = \langle l(i,j)^{\pm} \rangle$, where $l(i,j)^{\pm}$ is the set of literals constructed from $l(i,j)$. Perform consequence finding in $\mathcal{A}_i$ with the production field $\mathcal{P}_i$, and let $Cn_i := Carc(\mathcal{A}_i, \mathcal{P}_i)$. Output each characteristic clause $C \in Cn_i$ if C belongs to the original production field $\mathcal{P} = \langle \mathbf{L} \rangle$.
4. For each clause $C \in Cn_i$, check if $C \in \mathcal{L}(l(i,j))$. If so, send C to $\mathcal{A}_j$ and let $\mathcal{A}_j := \mu(\mathcal{A}_j \cup \{C\})$. Put $i := j$.
5. As long as there is a clause to be sent to the parent partition, repeat Steps 3 to 5.

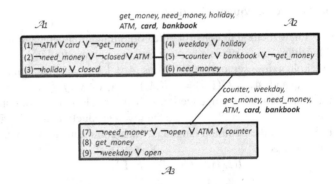

Fig. 3. Updating Communication Languages

Step 1 in Algorithm 3 can be run in parallel for each partition that is most distant from the root partition. Step 4 can be computed in parallel for each characteristic clause.

For the example in Section 3.2, applying Cycle Cut and the decision method of the communication language results in Fig. 3. By this way, the clause (1) consists of the symbols in $l(1, 2)$, then can be resolved with other clauses in A_2. Applying Algorithm 3, the intended consequence $card \lor bankbook$ can be obtained.

Termination of Algorithm 3 is guaranteed under some finiteness conditions. For this, (1) if there is a finite number of cycles in the induced graphs, the maximum depth of a tree is finite after applying Algorithm 2, and (2) if there are no recursive theories in each partition, consequence finding in the partition produces a finite number of characteristic clauses. The second condition is satisfied if ground consequences are only produced and there are no function symbols in the language.

The correctness of a distributed consequence finding algorithm A is defined as follows. Suppose the whole knowledge base A and a production field P. A is *sound* if any clause derived by A is a logical consequence of A and belongs to P. A is *complete* if it holds for any partitioning of A that: for any clause C belonging to $Th_P(A)$, there is a clause D derived by A such that D subsumes C.

Theorem 2 (Soundness and Completeness of Partition-based Consequence Finding). *Suppose an axiom set and its partitions $A = \bigcup_{i \leq n} A_i$, its induced graph $G = (V, E, l)$, and a stable production field $P = \langle \mathbf{L} \rangle$. We assume that every partition A_i has a sound and complete algorithm for consequence finding. Then, Algorithm 3 is sound and complete for distributed consequence finding.*

Proof. *Any clause derived by Algorithm 3 refers to a subset of A and belongs to P. Then, soundness follows from the monotonicity of first-order logic. Completeness can be proved by induction on the length of any clause $C \in Th_P(A)$. When $|C| = 1$, let A_k be a partition of A such that $C \in \mathcal{L}(A_k)$. Then, by Theorem 1, a clause D subsuming C can be derived by Algorithm 3, which works in the same way as Algorithm 1. Suppose that completeness holds for $|C| \leq m$, and we prove the case of $|C| = m + 1$. Let $C = C' \lor L$, where $|C'| = m$ and L is a literal. Let A' be $A \cup \{\neg L\}$. Since C' belongs to P and $C \in Th_P(A)$, $C' \in Th_P(A')$ holds. Then, assume a partition $A' = \bigcup_{i \leq n} A'_i$*

where $\mathcal{A}'_j = \mathcal{A}_j$ for $j \neq k$ and $\mathcal{A}'_k = \mathcal{A}_k \cup \{\neg L\}$ for some $k \leq n$. By induction hypothesis, a clause D' subsuming C' can be derived from $\mathcal{A}'$ at $\mathcal{A}'_k$ by Algorithm 3. In fact, if D' is derived at some $\mathcal{A}'_j$ ($j \neq k$), then it can be sent to $\mathcal{A}'_k$ because D' belongs to $\mathcal{P}$ and hence $D' \in \mathcal{L}(l(j,k))$. We now construct a distributed proof of a clause D subsuming C from $\mathcal{A}$ by adding L to C' appearing in the distributed proof of D' from $\mathcal{A}'$. This is possible because $L \in \mathcal{L}(l(i,j))$ for any $i, j \in V$ by the constructions (3) and (4). Hence, D can be derived at $\mathcal{A}_k$ by Algorithm 3.

Algorithm 3 can be seen as a simple extension of partition-based theorem proving by Amir and McIlraith [3] since the communication languages are extended to include the literals from the production field. However, this small change is essential for consequence finding. For theorem proving, Amir and McIlraith [3, Section 2.3] have mentioned how to deal with a query Q that comprise symbols drawn from multiple partitions. For this, a new partition $\mathcal{A}_Q$ is added with the language $S(\mathcal{A}_Q) = S(Q)$ and $\mathcal{A}_Q$ consists of the clausal form of $\neg Q$. Following addition of this new partition, Cycle Cut must be run on the new graph, and then refutation is performed at $\mathcal{A}_Q$. This method, however, cannot be elegantly applied for consequence finding in general since we do not know the exact theorems or even the possible candidates of theorems to be found in consequence finding. Of course, we can consider the production field $\mathcal{P} = \langle \mathbf{L} \rangle$ for restricted consequence finding. But even with a small $\mathcal{P}$, say $\mathcal{L} = \{a, b, c\}$, to find all consequences with theorem proving we need to query for a, b, c, then possibly $a \vee b$, $a \vee c$ and $b \vee c$, and eventually $a \vee b \vee c$ (though querying the last clause $a \vee b \vee c$ and checking all possible proofs would also work but have high complexity too). Alternatively, considering the new partition $\mathcal{A}_\mathcal{P}$ with the language $S(\mathbf{L})$ makes the graph more tightly connected and cyclic. Applying Cycle Cut would then the resulting communication language of an existing edge to include $S(\mathbf{L})$, which has a similar effect as the equation (3).

Another important change from the MP algorithm by Amir and McIlraith [3] is to use the production field $\mathcal{P}_i = \langle l(i,j)^\pm \rangle$ in Step 3 of Algorithm 3 for consequence finding. This restricts the computations that needs to be done and thus improves efficiency. The use of production fields also enables us to emulate *default reasoning* by adding each default literal in a production field to be skipped [14,15]. Hence, our algorithm can be extended to *partition-based default reasoning*.

4   Cooperative Consequence Finding

Partition-based distributed consequence finding is particularly useful when we have a large knowledge base that should be divided to easily handle each piece of knowledge. However, the algorithm can also be applied to naturally distributed knowledge-based systems in which each theory of an agent grows individually so that multiple agents may have the same knowledge and information simultaneously. Although such possessed knowledge is considered to be redundant in partition-based theories, there is no problem in *decentralized*, *multi-agent* and *peer-to-peer* systems. In such naturally distributed systems, one problem would be to break cycles in the induced graph because no agent should not know an optimal way to minimize the cost of cutting cycles (although we could devise a decentralized version of Cycle Cut algorithm).

In this section, we thus consider an alternative approach to distributed consequence finding that is suitable for such *autonomous* agent systems. The new method is more *cooperative* than the previous one in the sense that agents are always seeking other agents who can accept new consequences for further inference. In this method, we do not presuppose network structures of agents, but any agent can have a chance to communicate with other agents. As the language and knowledge of each agent evolves through interactions, this framework is more *dynamic* than the first method. Since the method is not partition-based, we do not call each distributed component as a partition, but call it an *agent* in this section.

Algorithm 4 (Cooperative Consequence Finding)

1. Suppose a set I of *input clauses* is given. This consists of a query or a goal clause in the case of query answering or abduction as well as any clause input to the whole system $\mathcal{A}$. Let $\mathcal{A}_i$ be a newly created agent whose axiom set is I. Let $\mathcal{P}_\mathcal{A} = \langle Lit_\mathcal{A} \rangle$ be the (stable) production field, where $Lit_\mathcal{A}$ is the set of all literals in the language of $\mathcal{A}$. Perform consequence finding in $\mathcal{A}_i$, and let $N := Carc(\mathcal{A}_i, \mathcal{P}_\mathcal{A})$.
2. For each clause $C \in N$, decide an agent $\mathcal{A}_j$ to which C is sent from $\mathcal{A}_i$. Put $i := j$.
3. In $\mathcal{A}_i$, consequence finding is performed by SOL resolution with the top clause C. Let $N := Newcarc(\mathcal{A}_i, C, \mathcal{P}_\mathcal{A})$. Put $\mathcal{A}_i := \mu(\mathcal{A}_i \cup N)$.
4. Repeat Steps 2 to 3 until no more new characteristic clause is derived.

Algorithm 4 repeats the process of (a) and (b): (a) new consequences obtained in an agent are sent to others, and (b) then they trigger consequence finding in those agents. An advantage of this method is that we only need to compute *new characteristic clauses* $Newcarc(\mathcal{A}_i, C, \mathcal{P}_\mathcal{A})$ in Step 3. In fact, computation of new characteristic clauses is easier than computation of the whole characteristic clauses by SOL resolution. The whole characteristic clauses are still obtained by accumulating the new ones with subsumption checking by simulating (1) and (2) in Step 3. Note that computation of $Carc(\mathcal{A}_i, \mathcal{P}_\mathcal{A})$ in Step 1 is not necessary when I is a single clause or contains no complimentary literals.

In Step 2 of Algorithm 4, we assume that any agent can decide to which agent each clause $C \in N$ should be sent. One such implementation is to associate with each agent $\mathcal{A}_i$ the set of predicates with their polarities appearing in the axiom set. This set must be updated each time a new characteristic clause is computed in the agent. Then, it becomes easier to find a literal that is complementary to a literal in C in other agents. One can also use the current communication language $l(i,j)$ between two agents: if $C \cap l(i,j) \neq \emptyset$ holds, then C can be sent from $\mathcal{A}_i$ to $\mathcal{A}_j$. Note here that we do not need to break cycles, but $l(i,j)$ needs to be updated whenever the axioms are updated. In another way, a blackboard architecture like [5] can be considered as a place to store new characteristic clauses deduced by agents. An agent should check whether (s)he has a clause which can be resolved with a new characteristic clause.

Note that implementation of Steps 2 to 4 can be parallelized provided that synchronization is properly done. The first message passing for unit clauses is illustrated in Fig. 4 for the example of Section 3.2.

Termination of Algorithm 4 is similar to the case of Algorithm 3. For the correctness of Algorithm 4, the following theorem holds.

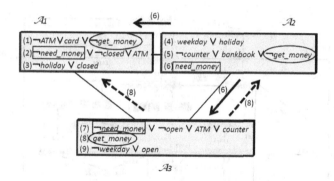

Fig. 4. First Message Passing in Cooperative Consequence Finding

Theorem 3 (Soundness and Completeness of Cooperative Consequence Finding).
Suppose an axiom set and its partitions $\mathcal{A} = \bigcup_{i \leq n} \mathcal{A}_i$. We assume that every agent $\mathcal{A}_i$ has a sound and complete algorithm for consequence finding. Then, Algorithm 4 is sound and complete for distributed consequence finding of $Newcarc(\mathcal{A}, I, \mathcal{P}_A)$.

Proof. Soundness is proved in the same way as Theorem 2. For completeness, $Newcar$ $c(\mathcal{A}, I, \mathcal{A}_P)$ can be decomposed into multiple clause-by-clause $Newcarc$ operations by (1). Since we use the production field $\mathcal{P}_A$ in which all literals appearing in $\mathcal{A}$ can be skipped, all Skip operations in any SOL deduction from the whole $\mathcal{A}$ are also applied by Algorithm 4. On the other hand, all Resolve operations of SOL deductions can be simulated by sending resolving clauses to other agents. Ancestor resolution in SOL deductions can also be done by sending back to previous agents. As a result, any SOL deduction can be simulated in a distributed setting by Algorithm 4.

5   Comparing Two Approaches

We here compare the two proposed methods for distributed consequence finding. We first note that the two methods are not designed to compute the same consequences as long as Theorems 2 and 3 are concerned. Given an axiom set $\mathcal{A}$, partition-based consequence finding computes $Carc(\mathcal{A}, \mathcal{P})$ belonging to a given production field $\mathcal{P}$ in Theorem 2, while cooperative consequence finding computes $Newcarc(\mathcal{A}, I, \mathcal{P}_A)$ for a given set of inputs I in Theorem 3. We could extend both methods to deal with any case by considering the same conditions for them. However, the current conditions are natural in both methods. The partition-based approach is based on Interpolation Theorem [6], which refers to the set of consequences of an axiom set of one partition, yet a language restriction can be used effectively. On the other hand, the cooperative approach is more dynamic and reflective so that ramification from the new input is propagated to other agents, but the language restriction is not easily set since every agent could be related to any other. Nevertheless, an obvious merit of the cooperative method is that we do not need to break cycles and assume no agent who does this.

We compare these two methods with the centralized approach with an accessibility problem in a metabolic networks (the citric acid cycle), depicted in Fig. 5. Given a

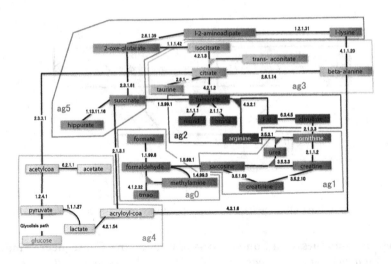

Fig. 5. Decomposed metabolic pathway problem

source metabolite (here glucose, in red), some target metabolites (here, tmao, acetate and transaconitate, in blue) and some possible additional metabolites (here arginine and orthinine, in white), the problem is to determine whether the target metabolites can be produced by the source metabolite (with eventual addition of the specified additional metabolites). The production field is then [tmao, acetate, transaconitate, -arginine, -orthinine], and the rules about the networks are divided among 6 agents as shown on Fig. 5. The problem is logically formalized with each reaction gives one activation rule [-reactant, reaction], one production rule [-reaction, product], plus some composition or decomposition rules if respectively the reactant or the product is a compound (indicated by triangles on our depiction). Each agent's theory contains all the rules that concerns of one of its metabolites (so communication languages initially contains only reaction symbols). A last clause indicates that glucose is initially present ([glucose]. This clause can be used as a top-clause for computing $Newcarc$, otherwise it is considered as a part of the initial theory. In this example, both $Carc$ and $Newcarc$ contains 4 consequences: [acetate], [transconitate], [-arginine, tmao], [-ornithine, tmao]. We then compare our two approaches (partition-base for computing $Carc$, and cooperative one for $Newcarc$) with the centralized case (computing $Carc$).

Table 1 shows the number of resolution steps in each method. As partition-based approach is affected by the choice of the root, and cooperative one by the order in which agents are considered for sending of a clause[1], we give here 6 sets of results, considering in turn each agents as possible root (or first element in the agent ordering).

Comparing two distributed methods with the centralized one, the total number of resolution steps becomes fewer in both methods. This is because (i) the partition-based

---

[1] To limit variability, we consider that when sending a clause, an agent always check potentially relevant neighbour in the same order.

Table 1. Comparison of three methods for "Getting Money"

Approach	# resolution steps					
Centralized (*Carc*)	872 228					
Root	ag0	ag1	ag2	ag3	ag4	ag5
Partition-based	30 344	25 664	**16 346**	26 126	188 143	20 286
Order	0-1-2-3-4-5	1-0-2-3-4-5	2-1-0-3-4-5	3-2-1-0-4-5	4-3-2-1-0-5	5-1-0-4-2-3
Cooperative	12 381	**6 290**	10 851	10 851	10 908	10 908

method restricts clauses sent to its parent to those constructed with the communication language between those partitions and (ii) the cooperative method performs consequence finding in each agent only with top clauses sent from other agents.

6   Comparison with Other Distributed Consequence Finding

Consequence finding has been investigated in a distributed setting [14,3,1]. Inoue and Iwanuma [14] consider a multi-agent framework which performs *speculative computation* under incomplete communication environments. This is a master-slave style multi-agent system, in which a master agent asks queries to slave agents in problem solving and proceeds computation with default answers when answers from slave agents are delayed. Speculative computation is implemented with SOL resolution with the conditional answer method to update agents' beliefs according to situation changes. On the other hand, distributed consequence finding in this paper does not assume any master agent to control the whole system. Amir and McIlraith [3] propose distributed theorem proving to improve the efficiency of theorem proving for structured theories. Their message-passing algorithm reasons over these theories using consequence-finding, and our first (partition-based) approach in this paper also uses it. As already stated in Section 3.3, the main difference between [3] and our partition-based approach is that the goal of the former is theorem proving while our goal is consequence finding. Another difference is that [3] considers how to partition a problem to minimize the intersection of the languages, while we suppose the situation that such optimal partitioning cannot be applied because of inherent distribution of knowledge and impossibility to collect all information to one place. This last observation directed us to the second, cooperative approach to distributed consequence finding, which is quite different from the first one.

The *peer-to-peer* (P2P) consequence finding system proposed by Adjiman *et al.* [1,2] is perhaps closest to our work. Their method is related to our both first (partition-based) and second (cooperative) approaches to consequence finding. [1] composes an *acquaintance graph* from the peers using information of shared symbols, which is similar to a graph induced from the partitions in our first approach. The difference is that [1] does not break cycles in a graph while we do. Also, [1] performs case splitting in goal-oriented reasoning of a peer P_1 by sending to other peer P_2 only those subgoals contained in the shared symbols between P_1 and P_2, then the new consequences of P_2 are returned to P_1, which is then composed in P_1 by replacing the subgoal. Combining the results derived from the subgoals often would result in a huge combination of clauses when the length of the goal is long, yet [2] analyzes the scalability of large P2P systems. On the other hand, we send the clause itself without splitting and no recollection is made. Our second approach can be regarded as a dynamic version of the first approach, in which messages are sent whenever new clauses are derived, and there is no presupposed network structures of agents. Such dynamic aspects are not seen in the P2P setting. Another difference is that [1] can only deal with propositional knowledge bases, while SOL resolution and SOLAR in our paper can be used for consequence finding in first-order clausal theories.

Although not in the context of consequence finding, abduction has also been considered in a distributed setting. Since abduction in clausal theories can be implemented with consequence finding, such work is somehow related to distributed consequence finding. Greco [10] considers how to build *joint explanations* from multiple agents in a P2P setting like [1], but incorporates preference handling to have an agreement between agents. By extending a blackboard architecture of [5], Ma *et al.* [19] address distribution of abductive logic programming agents by allowing agents to enter and exit proofs done by other agents. Those works do not use consequence finding, and communication between agents are fully guaranteed. More recently, Bourgne *et al.* [4] propose the *learner-critique* approach in which the role of each agent dynamically changes between a generator and a tester of hypotheses when each agent never knows which symbols are shared with other agents. In our methods, all agents work uniformly as a reflective inference system that derives consequences upon input of new formulas, although shared symbols are assumed to be known to both agents. Fisher [9] shows that certain forms of negotiation can be characterized by distributed theorem proving in which agents act as theorem-proving components. Analogously, distributed consequence finding might contribute to extended types of negotiation between agents.

In this work, we have focused on distributed reasoning systems in which a clause set is partitioned and the common symbols between partitions are associated with links. In contrast, there is another formalization of distribution in which variables or symbols are partitioned and clauses containing symbols from different partitions are associated with links between those partitions. The former formalization is called *clause-set partitioned distribution*, while the latter is called *variable-set partitioned distribution*. Most works on *distributed constraint satisfaction problems* (DCSP) are based on the latter formalization [25,11]. It is known that these two formalizations can be converted into each other in the propositional case (cf., [7]), yet the effect of the latter case is unknown for consequence finding and the former case can be seen more often in the real situations.

7   Conclusions

In this paper, we have proposed the two complete approaches for distributed consequence finding. The first one extends the method of partition-based theorem proving in a suitable way, and the second one is a more cooperative method for inherently distributed systems. This paper rather focuses on completeness of inference systems, and both approaches have merits and demerits. Partition-based approaches can utilize communication languages to realize restricted consequence finding between the partitions, while the cooperative approach does not need Cycle Cut algorithm. On the negative side, it is important to determine an appropriate ordering in the partition-based method, while the number of messages sent between agents tends to become larger in the cooperative approach.

We could consider a third approach by inheriting the merits of both approaches, such that each agent is autonomous and cooperates each other like the cooperative approach, yet each consequence finder incorporates production fields and communication languages between agents to enhance efficiency. Consideration of such a new approach is left as an important future work. Another future task includes more experiments with large distributed knowledge bases by refining details of two algorithms and by changing topological properties of agent links.

Acknowledgements. This research is supported in part by the 2008-2011 JSPS Grant-in-Aid for Scientific Research (A) No. 20240016.

References

1. Adjiman, P., Chatalic, P., Goasdoué, F., Rousset, M.-C., Simon, L.: Scalability study of peer-to-peer consequence finding. In: Proc. IJCAI 2005, pp. 351–356 (2005)
2. Adjiman, P., Chatalic, P., Goasdoué, F., Rousset, M.-C., Simon, L.: Distributed reasoning in a peer-to-peer setting: Application to the semantic web. J. Artif. Intell. Res. 25, 269–314 (2006)
3. Amir, A., McIlraith, S.: Partition-based logical reasoning for first-order and propositional theories. Artif. Intell. 162, 49–88 (2005)
4. Bourgne, G., Maudet, N., Inoue, K.: Abduction of distributed theories through local interactions. In: Proc. ECAI 2010 (2010)
5. Ciampolini, A., Lamma, E., Mello, P., Toni, F., Torroni, P.: Cooperation and competition in ALIAS: A logic framework for agents that negotiate. Ann. Math. Artif. Intell. 37(1-2), 65–91 (2003)
6. Craig, W.: Linear reasoning: A new form of the Herbrand-Gentzen theorem. J. Symbolic Logic 22, 250–268 (1957)
7. Dechter, R., Pearl, J.: Tree clustering for constraint networks. Artif. Intell. 38, 353–366 (1989)
8. del Val, A.: A new method for consequence finding and compilation in restricted languages. In: Proc. AAAI 1999, pp. 259–264 (1999)
9. Fisher, M.: Characterizing simple negotiation as distributed agent-based theorem-proving—a preliminary report. In: Proc. 4th Int'l Conf. on Multi-Agent Systems, pp. 127–134 (2000)
10. Greco, G.: Solving abduction by computing joint explanations. Ann. Math. Artif. Intell. 50(1-2), 143–194 (2007)
11. Hirayama, K., Yokoo, M.: The distributed breakout algorithms. Artif. Intell. 161, 89–115 (2005)
12. Inoue, K.: Linear resolution for consequence finding. Artif. Intell. 56, 301–353 (1992)
13. Inoue, K.: Induction as consequence finding. Machine Learning 55, 109–135 (2004)
14. Inoue, K., Iwanuma, K.: Speculative computation through consequence-finding in multi-agent environments. Ann. Math. Artif. Intell. 42(1-3), 255–291 (2004)
15. Inoue, K., Iwanuma, K., Nabeshima, H.: Consequence finding and computing answers with defaults. J. Intell. Inform. Systems 26, 41–58 (2006)
16. Inoue, K., Sato, T., Ishihata, M., Kameya, Y., Nabeshima, H.: Evaluating abductive hypotheses using an EM algorithm on BDDs. In: Proc. IJCAI 2009, pp. 810–815 (2009)

17. Iwanuma, K., Inoue, K.: Minimal Answer Computation and SOL. In: Flesca, S., Greco, S., Leone, N., Ianni, G. (eds.) JELIA 2002. LNCS (LNAI), vol. 2424, pp. 245–257. Springer, Heidelberg (2002)

18. Lee, C.T.: A completeness theorem and computer program for finding theorems derivable from given axioms, Ph.D. thesis, Department of Electrical Engineering and Computer Science, University of California, Berkeley, CA (1967)

19. Ma, J., Russo, A., Broda, K., Clark, K.: DARE: A system for distributed abductive reasoning. Autonomous Agents and Multi-Agent Systems 16(3), 271–297 (2008)

20. Marquis, P.: Consequence finding algorithms. In: Handbook for Defeasible Reasoning and Uncertain Management Systems, vol. 5, pp. 41–145. Kluwer (2000)

21. Nabeshima, H., Iwanuma, K., Inoue, K.: SOLAR: A Consequence Finding System for Advanced Reasoning. In: Cialdea Mayer, M., Pirri, F. (eds.) TABLEAUX 2003. LNCS, vol. 2796, pp. 257–263. Springer, Heidelberg (2003)

22. Nabeshima, H., Iwanuma, K., Inoue, K., Ray, O.: SOLAR: An automated deduction system for consequence finding. AI Communications 23(2-3), 183–203 (2010)

23. Nienhuys-Cheng, S.-H., de Wolf, R.: Foundations of Inductive Logic Programming. LNCS, vol. 1228. Springer, Heidelberg (1997)

24. Slagle, J.R.: Interpolation theorems for resolution in lower predicate calculus. J. ACM 17(3), 535–542 (1970)

25. Yokoo, M., Durfee, E.H., Ishida, T., Kuwabara, K.: The distributed constraint satisfaction problem: Formalization and algorithms. IEEE Trans. Know. & Data Eng. 10(5), 673–685 (1998)

Author Index

Åkesson, Knut   161
Allan, Vicki H.   413
Amato, Giuseppe   224
Argente, Estefanía   365

Badeig, Fabien   192
Balbo, Flavien   192
Bassiliades, Nick   240
Belmonte, M.V.   287
Berndt, Jan Ole   397
Bogon, Tjorben   72, 255
Boone, Edward L.   101
Borrajo, Daniel   146
Bourgne, Gauvain   429
Brandouy, Olivier   350

Canessa, Enrique   271
Chaigneau, Sergio   271
Chryssanthacopoulos, James P.   86

da Silva, Viviane Torres   176
de Lucena, Carlos J.P.   176
Díaz, M.   287
dos Santos Neto, Baldoino F.   176

Esparcia, Sergio   365

Falchi, Fabrizio   224
Faltings, Boi   3
Fei, Zhennan   161
Fernández, Susana   146
Fischer, Klaus   11

Grossi, Valerio   208

Hindriks, Koen V.   115

Imoussaten, A.   56
Inoue, Katsumi   429

Jonker, Catholijn M.   115

Klęsk, Przemysław   31
Kochenderfer, Mykel J.   86
Kużelewska, Urszula   131

Lasisi, Ramoni O.   413
Lattner, Andreas D.   72, 255
Le Borgne, Yann-Aël   382
Lennartson, Bengt   161

Mathieu, Philippe   350
Mihaylov, Mihail   382
Miremadi, Sajed   161
Montmain, J.   56

Nowé, Ann   382

Okamoto, Takayuki   429

Pelachaud, Catherine   302
Pérez-Pinillos, Daniel   146
Pini, Maria Silvia   319
Poursanidis, Georgios   72
Prepin, Ken   302

Quezada, Ariel   271

Reyna, A.   287
Ricanek, Karl   101
Rico, A.   56
Rico, F.   56
Rossi, Francesca   319

Saunier, Julien   192
Simmons, Susan J.   101
Stirling, Wynn   334

Timm, Ingo J.   72, 255
Turini, Franco   208
Tuyls, Karl   382

Venable, Kristen Brent   319
Veryzhenko, Iryna   350
Visser, Wietske   115
Vrakas, Dimitris   240

Walsh, Toby   319
Warwas, Stefan   11

Ziaka, Eva   240
Zinnikus, Ingo   11

Author Index